Electrical Inspection Manual *with Checklists*

Noel Williams and Jeffrey S. Sargent

nec 2014

JONES & BARTLETT
LEARNING

World Headquarters
Jones & Bartlett Learning
5 Wall Street
Burlington, MA 01803
978-443-5000
info@jblearning.com
www.jblearning.com

National Fire Protection Association
1 Batterymarch Park
Quincy, MA 02169-7471
www.NFPA.org

International Association of Electrical Inspectors
901 Waterfall Way, Suite 602
Richardson, TX 75080-7702
www.IAEI.org

Jones & Bartlett Learning books and products are available through most bookstores and online booksellers. To contact Jones & Bartlett Learning directly, call 800-832-0034, fax 978-443-8000, or visit our website, www.jblearning.com.

Production Credits
Chief Executive Officer: Ty Field
President: James Homer
Chief Product Officer: Eduardo Moura
Executive Publisher: Kimberly Brophy
Executive Acquisitions Editor: William Larkin
Associate Editor: Olivia MacDonald
Associate Director of Production: Jenny L. Corriveau

Associate Marketing Manager: Jessica Carmichael
VP, Manufacturing and Inventory Control: Therese Connell
Composition: Cenveo Publisher Services
Cover Design: Kristin E. Parker
Director of Photo Research and Permissions: Amy Wrynn
Printing and Binding: Edwards Brothers Malloy
Cover Printing: Edwards Brothers Malloy

ISBN-13: 978-1-284-04183-5
ISBN-10: 1-284-04183-2
ISSN: 1938-4009

6048

Printed in the United States of America
18 17 16 15 14   10 9 8 7 6 5 4 3 2 1

Contents

ABOUT THE AUTHORS

Noel Williams has held licenses as a master electrician in Utah, Colorado, and Wyoming for over 30 years. He is certified as an Electrical Inspector by the International Code Council (ICC) and International Association of Electrical Inspectors (IAEI) and has supervised and managed electrical construction projects for over 25 years. For over 10 years, he was chairman of the Electricians' Licensing Board for the state of Utah and is a former chairman of the *National Electrical Code®* (*NEC®*) Advisory Committee for the Utah Uniform Building Code Commission. Mr. Williams is a senior associate member of IAEI and has served as an officer of the Utah chapter of IAEI. Currently an electrical codes specialist, he also taught electrical apprentices at Salt Lake Community College for 6 years and has developed and taught electrical courses for electricians and inspectors for over 30 years. He authored or coauthored three other National Fire Protection Association (NFPA) publications—*NEC Q & A, Limited Energy Systems,* and *1999 NEC Changes* as well as serving as a seminar instructor and subject matter expert for development of numerous electrical training programs.

Jeffrey S. Sargent is a regional electrical code specialist with NFPA's Regional Operations Division. In that role he is responsible for promoting and supporting the adoption and use of the *National Electrical Code®*. He has been affiliated with the electrical industry for over 25 years as an electrician, inspector, instructor, consultant, author, and electrical special-ist. He was managing editor of the 2011 *NEC Handbook* and was co-editor of the 2012 *NFPA 70E Handbook*. He is a technical contributor to necplus™, NFPA's online source for information on the *NEC, NFPA Journal, the official magazine of NFPA,* and a coauthor of *NEC Q&A*. A a long-time member of the International Association of Electrical Inspectors (IAEI), Mr. Sargent is a licensed master electrician.

Foreword

If electrical inspections are part of your job or if you oversee an electrical inspection function, the *National Electrical Code®* (*NEC®*) and the *Electrical Inspection Manual with Checklists, 2014 Edition* are valuable resources. Electrical inspectors with many years of experience, knowledge, and training as well as those new to the field will find the *Electrical Inspection Manual with Checklists, 2014 Edition* beneficial, as it covers the gamut of topics related to the *NEC*. This manual provides inspectors with a systematic approach to ensuring electrical installations are safe from fire and shock hazards. It also provides uniform interpretation and application of the *NEC*. When safety standards are applied in a consistent and uniform manner, consumers, installers, designers, and inspectors benefit greatly.

Both electrical code regulations and electrical inspectors are extremely important in protecting the public against electrical hazards. Code regulations and electrical inspectors go hand in hand but are not always in sync. The *Code* is not always easy to interpret, and without a clear understanding and knowledge of the *Code*, electrical inspectors cannot perform their job effectively.

In my early years working as an inspector, there were numerous occasions when I completed an inspection but left the job site wondering if I had missed something that did not comply with the applicable code. This edition of the *Electrical Inspection Manual* would have been an excellent reference to put my mind at ease. The information and checklists contained in this manual would have been helpful in identifying potential code violations.

Inspectors will find this manual easy to use. Each chapter includes a section that is intended to help inspectors plan in advance to inspect for a particular type of installation. Information is included on installations ranging from a typical single-family residence to complex installations such as those found in hospitals and hazardous locations. At the end of each chapter is a specific checklist that will help ensure the applicable rules are not overlooked.

This is an excellent publication, and the International Association of Electrical Inspectors (IAEI) highly endorses this manual and commends the National Fire Protection Association (NFPA) for its participation.

David E. Clements
CEO and Executive Director
International Association of Electrical Inspectors

How to Use This Text

This manual is intended to be an aid to organizing and conducting electrical inspections in commonly encountered types of installations and occupancies. It also is intended to assist designers, insurance inspectors, architects, installers, project managers, safety officers, authorities having jurisdiction, or, in short, anyone who conducts, receives, is responsible for, or may wish to perform self-inspections of electrical installations.

The book is not intended as a tutorial on NFPA 70, *National Electrical Code*® (*NEC*®), nor should it be used in place of the *NEC*. Rather, it is intended to be used with the *NEC* as a tool for field inspections. The benefit of this manual is enhanced when users have a basic working knowledge of the organization and scope of the *NEC*.

The book is based on the 2014 edition of the *NEC*. In most cases, it will also be helpful in making inspections under previous editions of the *NEC*, but the details of some rules and the precise code references sometimes vary between code editions.

HOW THE TEXT IS ORGANIZED

With the exception of Chapter 1: Introduction, each chapter follows the same general outline.

- *Overview* offers a brief summary of the installations and of various issues covered by the chapter.
- *Specific Factors* begins the discussion of conducting actual inspections by outlining factors specific to the type of installation and inspection covered in that chapter. This part explains how the type of installation discussed in the chapter differs from other types of installations and inspections. The explanation may include particular hazards related to the type of installation and issues that need special consideration by a designer, installer, or inspector.
- *Key Questions* presents crucial considerations when preparing for or conducting an inspection. Most of the questions are accompanied by an explanation of why the question is important, how the answer to the question is used in the inspection process, and how or where the answer may be found. The questions usually help in the planning of or preparation for an inspection, but they may also be important questions that should be answered during the course of an inspection. The *Key Questions* are listed again at the end of each chapter.
- *Planning from Start to Finish* is intended to help the inspector plan an inspection by considering the normal or possible steps in construction, the timing of inspections, the relationship of one type of inspection to others, and the issues that are likely to be covered at each inspection. This part also helps to identify a logical and suitable starting point for each type of inspection.
- *Working through the Checklist(s)* is the most comprehensive section of each chapter. In this section, each numbered checklist item is stated exactly as it appears in the checklist itself and is then explained in detail. The explanation (or the checklist item itself) contains a short summary or statement of the applicable rule and, for many items, a discussion of the concept behind, or reason for, a particular rule.

Where applicable, this section also covers the relationship between each item and other items in the same or other checklists.

- *Key Terms* are in bold and underlined throughout each chapter. While most definitions are taken directly from Article 100 of the *NEC,* some are quoted from other articles, or they are derived from the scope or other language in a specific article. Supplementary discussion, given in italics for some definitions, explains the significance of the term to the chapter. Following these introductory sections, each chapter is devoted to actual inspections of a particular type of installation and to the checklists and their use.
- *Key Questions* repeats the key questions in each chapter.
- *Checklists* are intended to be duplicated and used in the field. You can find the checklists at: www.go.jblearning.com/Inspection14. They consist of tables of numbered inspection activities, with a single sentence describing each item as an issue or aspect of an installation to be checked, verified, reviewed, determined, or otherwise examined for *Code* compliance. For each checklist item, other columns in the table provide references to the applicable *NEC* article and section and spaces for comments and for "checking off" each inspection activity after it has been completed. It should be noted that the checklists contained in this book are not intended to be all-inclusive, covering all special equipment and occupancies.

This manual is intended to assist the inspector by helping to identify many of the important rules of the *Code* and by organizing the checklists by occupancy type. Part of the planning process is picking the applicable chapters and checklists. The determination of occupancy type identifies the applicable chapters in this manual and the applicable articles of the *NEC.* Chapters 2 through 5 in this book apply to all occupancies, although not all of the checklist items in these chapters will apply to all occupancies. Chapters 6 through 12 are more specialized and apply only to specific types of occupancies or installations, as outlined within each chapter.

Preface

First published in 1999, the *Electrical Inspection Manual* has become an established member of the growing family of National Fire Protection Association (NFPA) guides, handbooks, pocket guides, and other publications and products designed to enhance users' understanding of the *National Electrical Code®* (*NEC®*) requirements and provide practical guidance in applying the provisions of the *NEC*.

As the developer and publisher of the *NEC* since 1911, the NFPA is committed to safeguarding persons and property from potential hazards arising from unsafe electrical installations. It is the role of the electrical inspector to ensure that safe installations in compliance with *Code* requirements are in place. The *Electrical Inspection Manual* is designed to help inspectors perform this vital safety function. The manual is also intended to serve as an inspection planner and organizational tool for contractors, project managers, or anyone else who conducts, receives, or is responsible for inspections or who may wish to perform self-inspections of electrical installations. Because the manual is organized by stage and type of installation, it may also be used as a practical guide for installers and designers in applying the *NEC* to common electrical installations.

Building on the very positive response to past editions, the 2014 edition of the *Electrical Inspection Manual* has been revised throughout to reflect changes in the 2014 *NEC*. It is the authors' hope that the 2014 edition of the *Electrical Inspection Manual* will continue to fill the need for a comprehensive source of information on the electrical inspection process.

CHAPTER 1

Introduction

OVERVIEW

This chapter covers the general details of planning an inspection. It includes one checklist that addresses the electrical safety of the inspector rather than describing the specific details of the inspection activities or the code requirements for a specific installation. The checklist is equally applicable to any other person who will be exposed to an electrical hazard when adjusting, examining, maintaining, or servicing energized electrical equipment.

WHO IS THE ELECTRICAL INSPECTOR?

According to the *National Electrical Code®* (*NEC®*), one primary responsibility of the authority having jurisdiction (often referred to as the AHJ) is to interpret the *Code.* The *NEC* defines and explains the term as follows:

> **Authority Having Jurisdiction (AHJ).** An organization, office, or individual responsible for enforcing the requirements of a code or standard, or for approving equipment, materials, an installation, or a procedure.

> *Informational Note:* The phrase "authority having jurisdiction," or its acronym AHJ, is used in National Fire Protection Association (NFPA) documents in a broad manner, since jurisdictions and approval agencies vary, as do their responsibilities. Where public safety is primary, the AHJ may be a federal, state, local, or other regional department or individual such as a fire chief; fire marshal; chief of a fire prevention bureau, labor department, or health department; building official; electrical inspector; or others having statutory authority. For insurance purposes, an insurance inspection department, rating bureau, or other insurance company representative may be the AHJ. In many circumstances, the property owner or his or her designated agent assumes the role of the AHJ; at government installations, the commanding officer or departmental official may be the AHJ.

In this definition and the Informational Note, the *NEC* does not say precisely whom the AHJ might be. In fact, since the *NEC* is intended to be adopted and enforced as law, whom the AHJ is depends on who adopts and enforces the *NEC*. In most cases, it is a statewide government agency or a local jurisdiction, such as a municipality, that adopts the *NEC*. The AHJ is then commissioned and charged with enforcing the *Code* on behalf of the adopting jurisdiction. As noted, the AHJ is often an employee of the state or municipality. However, many jurisdictions contract with private companies who perform inspections. In effect, the AHJ is whomever is selected by the adopting jurisdiction to make inspections and interpret and enforce the *Code*.

The AHJ can often be an organization, such as a building or safety department. Furthermore, one or more persons or groups may share the duties of the AHJ. For example, the fire marshal may have jurisdiction over certain items, such as fire alarms or fire pumps, while other responsibilities are shared or divided among building officials, electrical inspectors, or municipal engineers or planners. On an industrial or public works project, the project engineer may assume the duties of the AHJ. On military installations, the authority is held by the commanding officer, who usually delegates that authority and responsibility to an engineering staff or to some other group, such as the Army Corps of Engineers. Electric utility companies sometimes act as the AHJ where no local or state-wide agency promulgates or enforces the *Code*. Occasionally, inspections are made by insurance underwriters, in which case the insurance company may act as the AHJ or as one of the AHJs. Federal agencies such as the Mine Safety and Health Administration (MSHA) and the Occupational Safety and Health Administration (OSHA) also use the *NEC* or incorporate parts of the *NEC* into their own published regulations. In these cases, the MSHA or OSHA inspector is the AHJ.

Some installations fall under the regulatory authority of more than one jurisdiction. Hospitals are good examples of installations with multiple AHJs. Frequently, hospitals are inspected by local, state, and regional groups, each being responsible for certain aspects of the facility.

In this book, the terms *authority having jurisdiction, AHJ,* and *inspector* are used interchangeably, though they may not always be one person or the same person, even on a single project.

ELECTRICAL INSPECTIONS AND ELECTRICAL SAFETY

The primary function of the *NEC* and those who enforce its provisions is to ensure the practical safeguarding of persons and property from hazards arising from the use of electricity. In short, the electrical inspector protects the public from unsafe electrical installations. The electrical inspector ensures a safe electrical installation by making sure it conforms to the requirements in the *NEC*. Since the *NEC* has many requirements, the electrical inspector has to understand the rules and be able to apply the rules to the actual installation. This book is intended to assist an inspector or other user in applying the *NEC* to actual installations.

FIGURE 1-1 illustrates the typical approach to electrical safety used in the United States. It comprises four major components:

1. Installation codes (the *NEC* with any requirements unique to the enforcing jurisdiction)
2. Product standards and performance evaluations
3. Installations of electrical equipment performed by qualified electrical contractors and installers
4. Inspection of the electrical installations performed by qualified inspection authorities

To be qualified, inspection authorities should be trained in understanding the relationship between product standards and installation codes and how to apply each to their inspection duties.

Each of these four components has a distinct and vital role in supporting electrical safety. In addition to installation codes, product standards, and qualified inspection personnel, an important factor in ensuring safe electrical installations is the qualifications of those persons who perform the installation. The *Code* is intended to be applied only by qualified persons (see the definition for "Qualified Person" in *NEC* Article 100). Proper training of electricians or other installers is vital to the safety of the installation and those involved in the installation or in the use of the completed installation. This system of four components is time tested, and its success

FIGURE 1-1 Relationship of electrical installation codes, product standards, qualified contractors and installers, and inspection/enforcement to safe electrical installations.

can be measured by the high level of safety that U.S. citizens have come to expect in their homes, workplaces, businesses, and recreational venues.

In addition to its uniform acceptance throughout most of the United States, the *NEC* has been adopted as the national standard for electrical installations in several other countries. A comparison of the requirements contained in the *NEC* with those contained in other international standards for electrical installations, such as International Electrotechnical Commission (IEC) standard 60364-1, confirms that the *NEC* addresses the same fundamental issues in establishing a safe electrical system that are addressed in the other electrical installation standards. The NEC contains fundamental safety principles that protect against:

- Electrical shock
- Thermal effects
- Overcurrent
- Fault currents
- Overvoltage

When viewing an installation of electrical equipment, the inspector must ask, "Is the installation safe?" and "Does this installation comply with the *Code*?" If the answers to these questions are "yes," then the installation should pass inspection. However, in some instances, the electrician or electrical contractor might not have installed the equipment the same way the inspector would have or the same way the inspector has seen it installed elsewhere. This situation sometimes results in a difference of opinion regarding compliance between the installer and inspector. The inspector should overlook such differences and assess the installation with *Code* compliance in mind. The key to a successful and correct electrical inspection lies in applying the rules of the *Code,* not the inspector's personal preference of installation. If the installation meets the *Code* requirements (including any local amendments) and is safe, then the installation should pass inspection.

An inspector must judge both the equipment and the installations for safety. According to Section 110.2 of the *Code*, all equipment and materials used in an electrical installation must be approved. *Approved* is defined in Article 100 as "acceptable to the AHJ." So how does an inspector determine what is acceptable? Experience and education are the primary solutions. A number of resources are also available to help inspectors with this determination. The *NEC* provides helpful guidance in Sections 90.7 and 110.3(A). Third-party product certification organizations, professional organizations, and the NFPA are also major sources of help.

The third-party product certification organizations work with manufacturers and standards organizations such as the NFPA to help develop criteria for the testing of equipment and materials. The resulting product standards are used to design and test equipment to ensure that it can be relied on to function safely. Such equipment is then *listed* and *labeled* so that the inspector as well as users and designers can readily identify equipment that meets the requirements of the product standards, the *NEC*, and other applicable codes and standards (refer to the General Requirements Inspections chapter of this manual or Article 100 of the *NEC* for definitions of "listed" and "labeled.")

Professional organizations such as the International Association of Electrical Inspectors (IAEI) are a very important resource for inspectors. The IAEI has regional chapters that provide a forum for inspectors to meet and discuss common issues, not only with other inspectors, but also with contractors, electricians, engineers, representatives of utilities, manufacturers, and suppliers. Becoming involved in a professional association is invaluable for the exposure to different points of view, the combined experience of the participants, and the educational opportunities that are offered.

The NFPA is another important resource, as it is the sponsor of over 300 codes, standards, recommended practices, and guides related to fire, electrical, and building safety, including, since 1911, the *NEC*. The NFPA's education, training, publications, on-line information, and technical staff are valuable resources available to AHJs to assist them in understanding and correctly applying the requirements of the *Code.*

Safety for the Electrical Inspector

The primary purpose of an electrical inspection is ensuring the safety of a complete installation. That is, an installation must be safe for the typical user under normal conditions. "Normal conditions" in a complete installation are conditions in which there are no exposed live parts that could be a shock hazard.

The *NEC* uses three primary methods to protect persons from electrical hazards: insulation, guarding, and isolation (see the definition of "Exposed (as applied to live parts)" in the General Requirements and Inspections chapter of this manual or Article 100 of the *NEC*). For an installation to be safe, all live parts must be insulated, such as current-carrying conductors in a cable; guarded, such as live parts in an enclosure such as a panelboard; or isolated from contact, such as overhead lines or equipment located in vaults. Often, more than one method is used and other techniques such as grounding of enclosures are also incorporated to ensure a safe installation. Therefore, under normal conditions, all electrical parts should also be insulated and/or contained in enclosures or isolated so that exposure to shock, arc flash, or arc blast injury is minimized.

When covers on electrical enclosures, such as switchgear, switchboards, panelboards, control panels, or motor control centers, are opened or removed for inspection, some of the normal safety features of a complete installation are defeated, and the risk of an electrical injury increases. The only safe alternative is de-energizing the equipment before it is opened and before live parts are exposed for inspection. Otherwise, employees, including electrical inspectors, must be trained and qualified to work on or near energized parts, and they must be properly equipped with personal protective equipment (PPE) and other appropriate tools and equipment so that they will be reasonably protected from shock, arc flash, and arc blast injury. Electrical inspection personnel are subject to the same shock, burn, and blast hazards as electricians when they perform their jobs in the vicinity of open, energized electrical equipment.

In order to select appropriate PPE for a given inspection task, the hazards must be identified and an analysis of the risk must be completed. The necessary information derived from this analysis should include the voltage level for selection of shock protection PPE and insulated tools and test equipment. It should also include an estimate of the incident energy that would be imposed on the face and chest areas in the event of an arcing event, or enough information to assign a Hazard Risk Category from NFPA 70E®, so that appropriate PPE for arc flash protection can be selected. Obtaining the necessary details for selection of PPE is relatively easy for shock protection, since the primary issue is voltage. However, the determination of incident energy levels or Hazard Risk Categories is a significantly more detailed study and requires much more information. This information may not be readily available to most electrical inspectors, and the calculation requires significant engineering expertise. Therefore, whenever possible, inspectors should insist on making inspections only of equipment that has not been energized or that has been de-energized and rendered electrically safe. NFPA 70E, *Standard for Electrical Safety in the Workplace*, provides detailed requirements and methods of providing for the safety of workers exposed to possible hazards from electricity. Generally, task-specific training in both the intended task and the safety-related aspects of the task is required for employees who will work "on or near exposed live parts" (OSHA language) or "while exposed to electrical hazards," as described in NFPA 70E.

This training is part of what is required of a qualified person as defined in the *NEC* and NFPA 70E. In order to understand how NFPA 70E addresses this work and the training required, we will first review some concepts and definitions found in NFPA 70E.

In general, live parts should be put into an electrically safe working condition before an employee works within the Limited Approach Boundary or works where interaction with the electrical equipment increases the risk of an exposure to arc flash hazards. This is intended to minimize exposure to electrical hazards.

An electrically safe working condition is defined in NFPA 70E as "a state in which the conductor or circuit part to be worked on or near has been disconnected from energized parts, locked/tagged in accordance with established standards, tested to ensure absence of voltage, and grounded if determined necessary." The general process of establishing an electrically safe working condition is often called *lockout/tagout* or just LO/TO by electrical workers. The precise procedure that should be used is the one established by a specific employer for a specific site. Therefore, in many cases, an inspector would have to insist that the employees at the site establish the safe working condition, since they will be the ones most likely to be trained in the specific procedures for any given site.

To understand when a person is considered to be "working on or near live parts," the concept of *approach boundaries* must be understood. NFPA 70E uses a very simple concept with regard to shock hazards: The closer a person gets to an exposed part, the more likely that inadvertent (or planned) contact with the live part will occur. Therefore, NFPA 70E establishes three boundaries for approach to such parts, from farthest to nearest:

- Limited Approach Boundary
- Restricted Approach Boundary
- Prohibited Approach Boundary

The dimensions of these boundaries from live parts are given in Table 130.4(C)(a) for AC voltages and Table 130.4(C)(b) in NFPA 70E. The dimensions depend primarily on the voltage. For example, an exposed fixed live part that is 50 to 750 volts (measured between conductors) has a Limited Approach Boundary of 3 ft 6 in. (1.07 m). This boundary may be crossed only by qualified persons or persons continuously escorted by a qualified person. In the 2004 edition of NFPA 70E, this boundary also specified "working near live parts" and that any tools or equipment used within this boundary must be insulated. In the 2009 edition, the term "working near" was deleted, and such work is simply defined as "within the Limited Approach Boundary."

An intermediate boundary that is strictly reserved for qualified persons is defined as the Restricted Approach Boundary. Prior to crossing this boundary, the qualified person must have in place all planned methods of shock protection, such as voltage-rated insulated or insulating tools and gloves or other techniques that insulate either the person from the live parts or the live parts from the person. For 480-volt systems, this boundary is 1 ft 0 in. (304.8 mm).

The Prohibited Approach Boundary is defined as a distance within which a person is considered to have made contact with the live part. Work within this boundary, even if only a tool or probe crosses the boundary, is considered to be "working on live parts." For 480-volt systems, this boundary is 0 ft 1 in. (25.4 mm).

Some inspections could be made from outside these approach boundaries, making the inspection task much safer even where exposed parts remain energized. However, other hazards, such as arc flash and arc blast, may still present significant risks well beyond the approach boundaries. The analysis of these hazards involves many variables but results in two essential estimates: the location of the Arc Flash Boundary and the incident energy exposure of the person at the location where the task takes place.

Incident energy is defined in NFPA 70E as "the amount of energy impressed on a surface, a certain distance from the source, generated during an electrical arc event. One of the units used to measure incident energy is calories per centimeter squared (cal/cm^2)." At some distance, the incident energy will be sufficient to cause burns. At closer distances, the burn is likely to be much worse since the incident energy increases as the distance is decreased.

The Arc Flash Boundary is defined in NFPA 70E as "When an arc-flash hazard exists; an approach limit at a distance from a prospective arc source within which a person could receive a second degree burn if an electrical arc flash were to occur." This boundary is based on some amount of incident energy exposure that is considered enough to heat the skin to a temperature that will result in the onset of a second-degree burn. Obviously, a somewhat greater distance may still result in less severe first-degree burns. The line or boundary is based on the worst burn that is "curable," a second-degree burn. (Third-degree burns are not considered curable because the skin is destroyed.) Persons observing a task that results in an arc flash can be susceptible to burn injury even if they are located farther away than the presumed working distance. In fact, many more people are injured and hospitalized due to arc flash burns than are injured due to electrical shock; some of them are only observers or standby personnel.

In effect, the Arc Flash Boundary establishes a distance within which arc flash protection (arc-rated) PPE is required. The amount or rating of the PPE must be at least equal to the incident energy at the working distance. Locating yourself farther away from the source is safer.

NFPA 70E specifies that the arc flash boundary "is the distance at which the incident energy is 1.2 cal/cm^2." This boundary may be determined by incident energy analysis, defined as "A component of an arc flash hazard analysis used to predict the incident energy of an arc flash for a specified set of conditions." When Hazard/Risk Categories are used to select PPE, the "potential arc flash boundary" is provided for a given set of conditions, and if those conditions are not met, an incident energy analysis is required. To determine a Hazard/Risk Category, certain information is required. A Hazard/Risk Category is based on the type of equipment and the task to be performed on the equipment, as well as assumed maximum available short-circuit current (available bolted fault current) and an assumed maximum clearing time of the upstream overcurrent device. This information may not be available to an electrical inspector, and without it there is no legitimate way to determine where the boundary actually is other than by incident energy analysis. Thus, as stated previously, electrical inspections can be truly safe only when performed on equipment that has never been energized or that has been de-energized and put into an electrically safe working condition.

FIGURE 1-2 shows an idealized example of these four types of boundaries. Limited, Restricted, and Prohibited Approach Boundaries are fixed boundaries based on voltage, but the Arc Flash Boundary depends on other factors such as available fault current, clearing time, and the shape of the enclosure, if applicable.

Electrical inspectors who must do some examination of energized electrical equipment should wear only nonmelting natural fiber materials such as cotton, wool, silk, or rayon, and they should supplement this clothing with arc-rated clothing, whose selection is based on the incident energy exposure related to the specific location, task, and equipment. Such arc-rated clothing may appear no different from ordinary work clothing or may consist of a flash suit that, along with gloves and boots, covers the entire body. **FIGURE 1-3** shows an example of a flash suit with a switching hood.

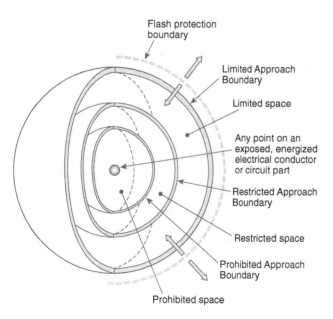

Flash protection boundary

Limited Approach Boundary

Limited space

Any point on an exposed, energized electrical conductor or circuit part

Restricted Approach Boundary

Restricted space

Prohibited Approach Boundary

Prohibited space

FIGURE 1-2   Summary illustration of limits of approach. Note: The dotted line for the flash boundary in the diagram could be either outside or inside the approach boundaries for shock.

WORKING THROUGH THE CHECKLIST

The checklist in this chapter is intended to be used to verify that an inspection involving exposed energized conductors or circuit parts is necessary and can be done safely. It is intended for use in planning for an inspection to verify that an electrical inspector is properly trained and equipped for the hazards in such inspections. The checklist is based on NFPA 70E, *Standard for Electrical Safety in the Workplace, 2012 Edition*. NFPA 70E contains detailed requirements and methods of providing for the safety of workers exposed to possible hazards from electricity.

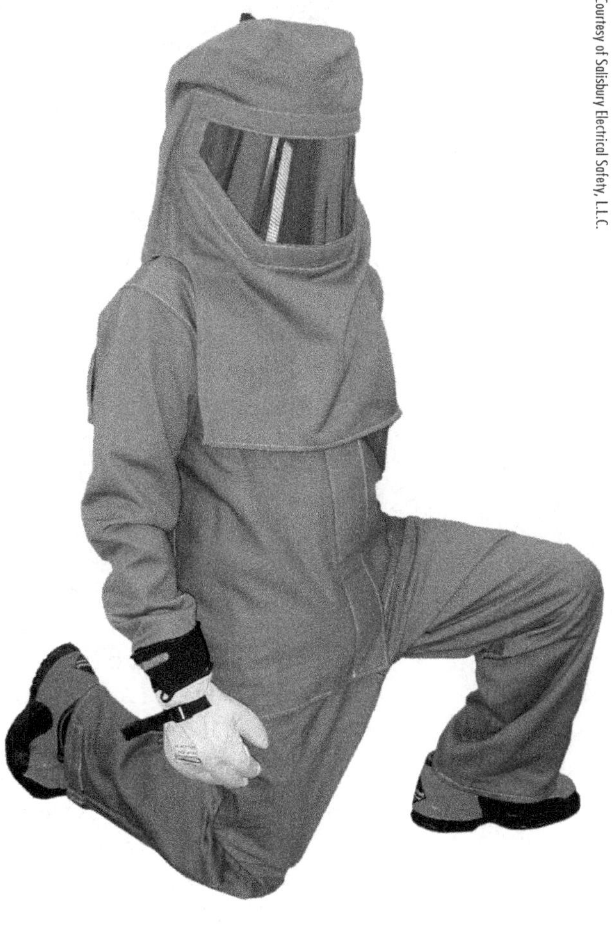

Checklist 1-1: General Safety Checklist for Electrical Inspections

1. Does the inspection task involve exposed energized conductors or circuit parts?

If there will be no exposure to electric shock, arc flash burns, or arc blast, there will be no requirement for special clothing or other PPE for electrical hazards. If the answer to this question is "no," the rest of the checklist may not be applicable. However, this question should be asked before performing any type of inspection because some equipment may be energized, even by temporary power supplies, while other equipment that was specifically considered was de-energized. Also, a dangerous arc

FIGURE 1-3 Arc-rated flash suit with switching hood.

Courtesy of Salisbury Electrical Safety, LLC.

flash hazard may be created by interaction with electrical equipment even if no live parts are exposed. Enclosures alone cannot be considered to be reliably adequate protection from an arc flash that might occur inside the enclosure. Remember that equipment that is de-energized is not necessarily electrically safe unless all steps in the applicable lockout/tagout procedure have been completed. For the purpose of this question, a plug-in tester that is used to check wiring integrity or polarity at a receptacle is not considered to involve exposure to energized parts.

2. Can the risk of exposure to electrical hazards be justified?

OSHA and NFPA 70E require that electrical equipment be put into an electrically safe working condition before work is done that might expose a worker to electrical hazards. Exposing a worker (inspector) to electrical hazards without justification is prohibited. Some tasks on energized equipment can be justified. OSHA and NFPA 70E provide two essential ways to justify the work.

First, energized work is justified if putting the equipment in an electrically safe working condition involves increased or additional hazards. Typically, this applies to equipment that causes a greater hazard or perhaps a definite hazard if the equipment is turned off. Examples include some life-support equipment or some ventilation systems.

Second, energized work is justified if putting the equipment in a safe and de-energized state prevents the work that needs to be done. That is, the work is infeasible in a de-energized condition. This may be due to the fact that the task, such as troubleshooting or start-up, requires that the equipment be running in order to get the needed information or to prove the function of the equipment. This justification is also applied where making one piece of equipment electrically safe would require the shutdown of a continuous industrial process. "Continuous" in this sense does not refer to the hours of operation; it refers to the nature of the process.

If exposure to electrical hazards cannot be justified in one of these two ways, the equipment must be made electrically safe before the work is done. Generally, the justification must be in the form of an Energized Electrical Work Permit. However, those tasks that are most likely to be justified for inspectors (e.g., testing and voltage measuring) do not require written permits but do require appropriate PPE and qualified people.

3. What is the voltage of the equipment that requires inspection?

Shock protection requirements are based entirely on voltage. Voltage must be determined to do any shock hazard analysis. This information can usually be obtained without taking direct measurements. (Shock protection measures would be required to measure the voltage.) Most shock-protection PPE is rated based on voltage that can most easily and safely be determined from equipment nameplates.

4. Where are the approach boundaries for shock protection?

Once the voltage is known, the dimensions for the Limited, Restricted, and Prohibited Approach Boundaries can be found in a table in NFPA 70E. These boundaries are the basis for requirements for qualified persons and shock-protection PPE. Typical dimensions for systems between 250 volts and 600 volts would be 3 ft 6 in. (1.07 m) for the Limited Approach Boundary, 1 ft 0 in. (304.8 mm) for the Restricted Approach Boundary, and 0 ft 1 in. (25.4 mm) for the Prohibited Approach Boundary.

5. Will the inspection involve crossing any of the approach boundaries?

Only qualified persons with their chosen shock protection technique (usually insulated gloves) in place may cross the Restricted Approach Boundary. Generally, an inspector should be able to do a visual inspection without crossing the Restricted Approach Boundary. If this boundary is never crossed, shock-protection PPE is not required. The inspector is required to be electrically qualified for the task or escorted by a qualified person when inside the Limited Approach Boundary. Arc flash protection PPE is a completely separate issue from PPE for shock protection.

6. Has an incident energy analysis been performed for the equipment?

An engineering analysis must be done if we want to estimate the incident energy exposure in a given situation. If such an analysis has been done, the estimated incident energy at the face and chest areas of the electrical worker is used to select a minimum rating for arc-rated PPE. If the analysis has not yet been completed, NFPA 70E offers an alternative way to select PPE. This alternative, called Hazard/Risk Categories, is not "incident energy analysis" and does not produce a value of incident energy exposure, and some calculated values must be known. Thus, some level of analysis or calculation has to be done for either alternative.

7. Are the available short-circuit current and clearing times known?

If an incident energy analysis has not been done for a task, NFPA 70E offers an alternative method of selecting arc-rated PPE. The alternative "table" method uses Hazard/Risk Categories (HRCs) to select arc-rated and some other PPE and assumes some maximum available fault current and some maximum clearing time for the overcurrent device. This approach requires much less calculation than an incident energy analysis, but some calculations or assumptions must be made. The table method of selecting PPE is limited to short-circuit currents and overcurrent protective device clearing times that are within the defined limits specified in the notes corresponding to a specific equipment and task in the tables. The tables offer a pragmatic general approach based on experience, while a detailed incident energy analysis provides a result that is equipment and facility specific.

8. Where is the Arc Flash Boundary?

The Arc Flash Boundary defines a distance from a prospective arc source, within which arc-rated PPE is required. The Arc Flash Boundary is defined as a distance that could result in a second-degree (or "curable") burn if an arc were to occur. Anywhere the burn could be second-degree or worse, arc-rated PPE is required. If an incident energy analysis has been done, this boundary should be available as part of that study. When the tables are used to determine an HRC, the Arc Flash Boundary distance is given in the table. However, this boundary is based on a given maximum fault current and clearing time. Thus, it can be used with the same information required for the use of the tables.

9. Will any part of the body be within the Arc Flash Boundary?

This question and the resulting answer are the basis of a requirement for arc-rated PPE. If a part of the body will be exposed to the possibility of an arc flash that could result in second-degree burns or worse, that part of the body must be protected. If the Arc Flash Boundary is a very short distance, arc flash protection may not be required for a visual observation, but the question continues, where will all parts of the body be while this inspection task is taking place?

10. How will PPE for arc flash protection be selected?

PPE for arc flash protection can be selected based either on the calculated incident energy or on HRCs, using the table method. If the incident energy is known, PPE is selected to have an arc rating not less than the calculated incident energy. NFPA 70E provides guidelines for selecting PPE based on incident energy. Otherwise, if the task is one of the very common tasks covered by the tables, then when requirements in the notes (fault current and clearing time) are met, the table will produce an HRC. The HRC can be used to select clothing and other PPE and to establish a minimum arc rating for the PPE.

11. Is the appropriate arc-rated PPE available?

Generally, the employer is responsible for providing most PPE. Ratings should be checked for compliance with the appropriate standard. Standards are listed in NFPA 70E. Some PPE—mostly PPE for shock protection—will require periodic retesting. PPE should be inspected for damage or contamination before each use.

12. Is the inspector qualified for this specific task and risk?

The best option for any inspection is one that does not involve exposed energized parts. Any other condition that is not electrically safe is obviously a greater risk. Employees who are at greater risk for exposure to electrical shock or arc flash events must be trained and protected. Training to become a qualified person must include safety training in recognizing and avoiding electrical hazards. According to the definition in Article 100 of the *NEC* and in NFPA 70E, the qualified person would also have "skills and knowledge related to the construction and operation of the electrical equipment and installations." Determining who is qualified and providing training is the employer's job. The inspector's task is usually one of observation or examination, but whatever the involvement of the inspector, he or she must make some determination of his or her own qualifications for any given activity and then be able to recognize and avoid hazards, and use the appropriate PPE.

PLAN REVIEW AND INSPECTION PLANNING

Each chapter in this book includes a section that is intended to help an inspector plan inspections of a particular type of installation covered by a specific checklist. Of course, more than one checklist is likely to apply to a given installation, so any one inspection, or the use of any one checklist, may well overlap with other inspections and checklists. Determining the correct approach requires some overall pre-inspection planning.

FIGURE 1-4   Steps in the electrical inspection process.

FIGURE 1-4 illustrates the electrical inspection process as it flows from the submission of original plans to the final inspection and approval. Not every project involves all of the steps shown. On the other hand, some specialized projects involve more steps than are illustrated here.

During pre-inspection planning and prior to beginning any work, an electrical permit/application is issued (**FIGURE 1-5**). The issuance of permits and the collection of fees are administrative functions of the organization responsible for electrical inspections within that jurisdiction.

During pre-inspection planning, an electrical inspector should ask, "What type of building or facility will I be inspecting?" The answer to this question will begin the preliminary steps that aid the inspector in planning an organized and complete inspection. Depending on occupancy or type of facility, the electrical inspector will expect to see certain things in the electrical installation. If the facility is a single-family residence, the inspector will expect to see types of electrical equipment and wiring methods that are common to these occupancies; if the facility is a hospital, the inspector will see a much different type of electrical installation. For example, for the typical single-family residence, an electrical inspector would not need to research the requirements of an essential electrical system (as explained in the Special Occupancies chapter). Nor would the inspector research the requirements for small-appliance branch circuits in a hospital kitchen.

The occupancy and type of building or facility will provide some insight into some of the electrical loads that the inspector will see. For example, an office building will typically have a larger receptacle load than a restaurant. The restaurant will typically have more cooking loads. However, both of these facilities will probably have heating, ventilation, and air-conditioning (HVAC) loads. The loads in a building or facility will help to determine the basic design and size of the electrical system. Therefore, the type of occupancy is an important consideration during the planning phase of the inspection.

The next general question an electrical inspector should ask is, "How big is this building or facility?" The answer to this question will give the inspector an idea of the quantity of electrical load in the building or facility. The size and type of load will also give a clue to the size of the service entrance and the quantity of feeders and branch circuits in the facility. Since much of this information is available from the construction plans, the electrical inspector should request a copy of the plans in advance of the actual site visit.

If the electrical construction job requires the services of a professional engineer, typically drawings, plans, and specifications are created. Prior to the inspection, there are many types of information about the electrical installation that are available from the plans and specifications generated by the engineer. In some jurisdictions, an electrical plan review is a required step in the inspection process. If a plan review is required by the jurisdiction, the engineer, electrical contractor, or general contractor makes the plans available. However, many jurisdictions do not require an electrical plan review. In this case, the electrical inspector should ask to see the information prior to conducting the inspection. Many checklist items in this manual refer to the plans or other information that should be provided to the inspector by the engineer, designer, or installer. If the full set of drawings is available, it is worthwhile to request them and examine them for important details to be viewed during the site visit. Otherwise, the plans can be reviewed on the job site or prior to a specific inspection.

ELECTRICAL PERMIT

Permit # _____
Date: _____
Bldg. Permit # _____

Plans Submitted _____

Address: _____   Contractor: _____
Owner: _____   Address: _____
Fee Amount: _____   License # _____ Phone # _____
Paid: Check   Cash   Other_____   Amount $_____   Collected By _____   Date _____

BUILDING DATA

Use: Residential   Business   Industrial   Other _____   No. of Stories _____

No. of Residential Units _____

Type of Installation: New   Alteration   Repair   Other _____

Wiring Method: NM   AC   MC   Busduct   Conduit _____   Other _____
                                                  Type

Services
Voltage/Phase	Amperage	Conductors
____	____	____
____	____	____
____	____	____
____	____	____

Temporary Services
Voltage/Phase	Amperage	Conductors
____	____	____
____	____	____
____	____	____

No. of Meters
Amperage	Phase
____	____
____	____
____	____

Size of Grounding Electrode Conductor _____

Switchboard/Panel Boards
Voltage/Phase	Amperage	Conductors
____	____	____
____	____	____
____	____	____

Electric Heat
Baseboard   Electric Boiler   Other _____
No.	Wattage
____	____
____	____

Lighting Fixtures
No.
A. Fluorescent ____
B. Incandescent ____
C. Other ____
____
____
Receptacles ____
Switches ____
Ranges ____
Dryers ____
Dishwashers ____
Disposals ____
Water Heaters ____
Signs ____
Other ____
____
Swimming Pools ____
(Installed in Conformance with NEC 680)

Oil or Gas Furnace
No.
____
____

Fire Detection:
System   Individual Detectors # _____
____
____
____
____
____
Emergency/Exit Lights # _____

Air Cond./HVAC
Type/Unit	H.P.	Voltage/Phase
____	____	____
____	____	____
____	____	____

Generators Transformers
No.	Size
____	____
____	____

Motors
H.P.	Voltage/Phase	Conductor
____	____	____
____	____	____
____	____	____

Applicant certifies that all information given is correct and that all pertinent electrical ordinances will be complied with in performing the work for which this permit is issued.

Work must begin within six (6) months of permit issuance or the permit shall become invalid.

Description of Work: _____
_____
_____
_____
_____

_____   _____   _____
Signature of Applicant or   Date   Signature of Building Official
Authorized Representative

Courtesy of City of Portsmouth, NH, Bureau of Inspection

FIGURE 1-5   Example of an electrical permit.

Electrical plans vary widely in scope, presentation, and degree of detail. Industrial plans are often more detailed than plans for commercial projects, and commercial plans are usually more detailed than residential ones. Some projects include control and connection drawings; others show only the power distribution. Some projects are not even completely designed until after the basic structure is complete. For example, many office buildings and retail spaces are built without knowing exactly who the tenants will be. This type of construction is often called "shell and core," because only the outside and interior common areas are designed prior to leasing the spaces and completing the "tenant finishes." In fact, the shell and core construction is often treated as a separate project, and permits for tenant finishes are issued separately, perhaps even to different contractors. Many residential projects have nothing more than a load calculation for the service, and the electrical installation details are worked out in the field to meet code requirements and the desires of the homeowner.

A common electrical plan for an electrical construction job is the *one-line,* or *single-line,* drawing (**FIGURE 1-6**). The single-line drawing is defined in NFPA 70E as "A diagram that shows, by means of single lines and graphic symbols, the course of an electric circuit or system of circuits and the component devices or parts used in the circuit or system." A single-line diagram shows how power is distributed from the source, typically the service, to the utilization equipment. Equipment such as switchgear, switchboards, panel boards, transformers, substations, motor control centers, motors, emergency equipment, transfer switches, and HVAC equipment are represented. Service, feeder, and some branch-circuit raceways and cables are also shown.

The single-line drawing usually indicates the raceway or cable type and trade size, the number of wires, the wire sizes, and any other special information. It may also indicate the voltage level, bus capacities, interrupting current, fuse or circuit breaker ratings, system grounding, metering, relaying, and any other information to help identify the

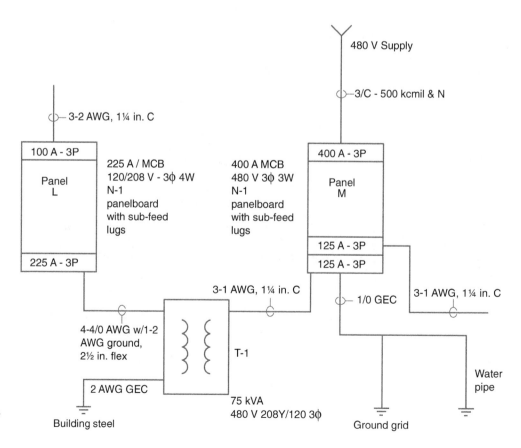

FIGURE 1-6 Typical single-line drawing.

electrical system. A good single-line drawing will show the services, feeders, and major loads and equipment to the branch-circuit panelboard level. Generally, the single-line drawing does not show information beyond a branch-circuit panelboard. For example, the single-line drawing should show the lighting and receptacle panelboards but would not show the lighting or receptacle branch circuits.

The single-line drawing is usually laid out from top to bottom, with the service or other source of power at the top of the page and the loads near the bottom of the page. Sometimes a single-line drawing is laid out from left to right, with the service at the left and loads at the right. Depending on the size of the electrical system, several single-line drawings can be used to depict the electrical system. The electrical inspector may want to make some pre-inspection notes on this drawing to highlight any equipment, raceways, or cables that might require more visual investigation during the site visit.

A *lighting plan* shows the physical placement and types of luminaires used in a building or structure (**FIGURE 1-7**). The lighting plan is drawn to scale and shows fixtures, lighting receptacles, and fixture circuitry. Symbols are often used to designate the different types of luminaires and receptacles. Luminaire symbols are often shown in the legend for the lighting plan. Branch-circuit wiring and fixture-interconnecting raceways or cables also are shown. Phase wires and neutrals in the connecting raceway or cable are shown in a shorthand fashion. The shorthand designation is sometimes included in the legend, but it is not always consistent from one engineer or designer to another. Commonly, short lines drawn diagonally to the circuit run indicate how many ungrounded conductors or "phase" wires are contained in a raceway. Long lines drawn diagonally to the circuit tell how many neutral wires are contained in the raceway. Some designers use other line symbols to indicate grounding conductors or isolated grounding conductors.

Lighting panelboards and panelboard schedules are sometimes included on the lighting plan. A lighting panelboard schedule shows the number of, locations of, and power consumed by the lights on each branch circuit. The panelboard circuit breaker sizes and the connected phase are also shown.

A *power plan* is similar to a lighting plan except that it shows circuits and receptacles for loads other than lighting, such as general-use receptacles, circuit arrangement and size, and the location of special equipment (**FIGURE 1-8**). Both power and lighting drawings are helpful to the inspector in determining the load in the facility. Power and lighting plans may be combined into one drawing when the necessary details can be shown on a single sheet. However, on larger projects, power and lighting plans and panel schedules may require many sheets.

Structural drawings show the building's or structure's physical construction components. At first, they might seem unimportant to the electrical inspector. However, these drawings may provide the answers to some of the questions needed for an inspection. For example, structural drawings are helpful in determining whether the building steel qualifies as a grounding electrode, whether the structure is fire-rated construction, or what the physical attributes of a grid tile ceiling are. The structural requirements for any transformer vaults should be detailed on the structural drawings. Ground grids, grounding rings, grounding electrodes, and underground metal structures and components may be shown on structural drawings. Sometimes a structural or an electrical engineer will include a *grounding drawing* that shows the details for grounding steel columns, connecting concrete-encased electrodes (reinforcing bars, rods, and mesh), or installing grounding electrode conductors.

Floor plans are useful to the inspector to identify the final use of the spaces within the building or structure. The floor plans may trigger questions in the electrical inspector's mind about potential requirements for particular spaces. Floor plans may also answer electrical questions about the structure. For example, floor plans may reveal the location

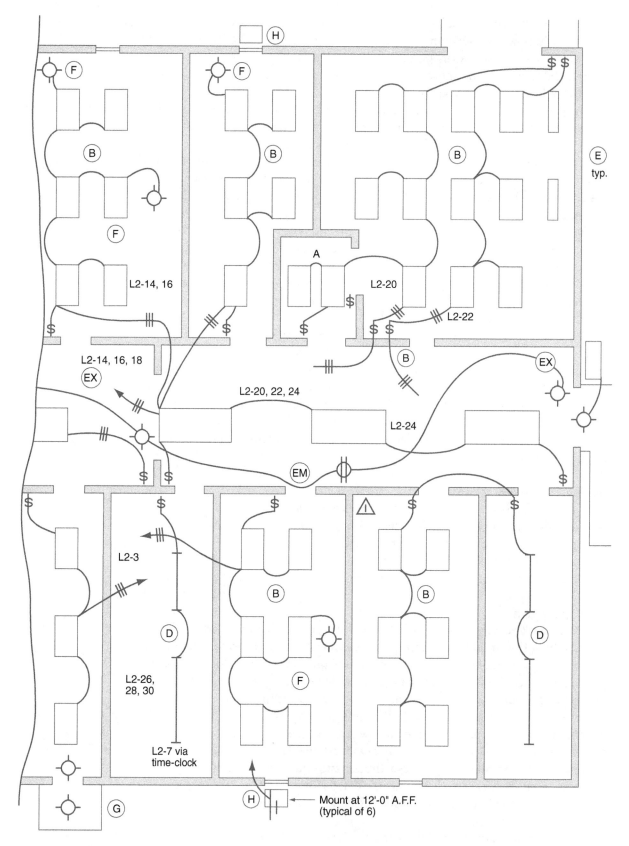

FIGURE 1-7 Typical lighting plan.

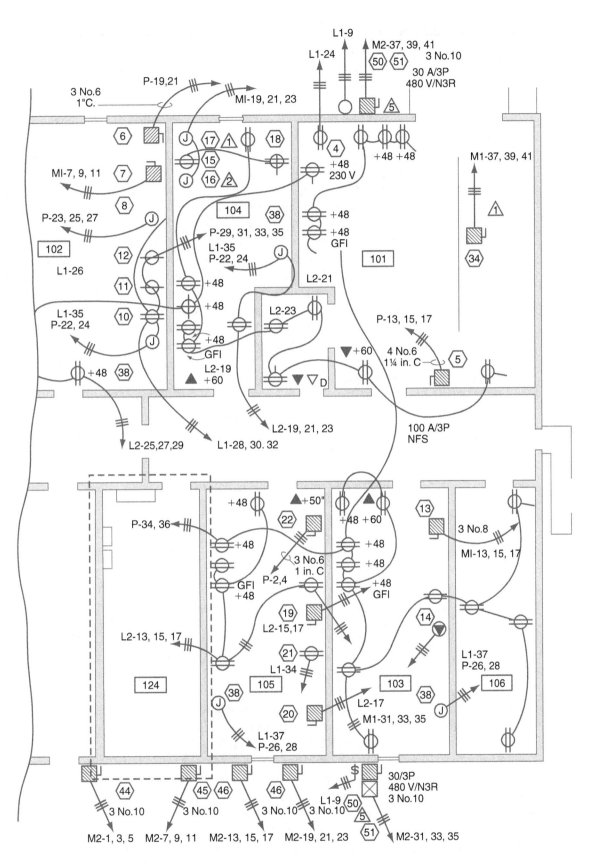

FIGURE 1-8   Typical power plan.

of the patient-care areas of a hospital or areas that might be used as kitchens. The plans should also show hazardous (classified) locations, if any. The location of equipment rooms and elevators and many other architectural details that are related to electrical installations are also shown on floor plans.

The *site plan* is useful in getting a perspective of how the building or structure will sit on the property. It will also give the relative elevations of the building site and may show nearby utility lines, right-of-ways, parking, and other uses of the property. Service transformers, locations of service equipment, communications entrance, outside lighting, and signs are all common items illustrated on a site plan. Separate site plans may be prepared for grading and electrical purposes.

Another drawing that is sometimes available on a construction job is the *equipment placement plan.* This drawing is typically drawn to scale and shows the facility and the physical location of the equipment in the facility. Sometimes, this drawing shows the floor plan of the building under construction. It is usually drawn from the perspective of looking down on the facility from above, called the *plan view.* However, wall perspectives, or *elevation views,* of equipment can be shown on this type of drawing. Conduit and cable runs may also be shown. Raceway and conductor information may be identified directly on the drawing or may be noted and described on a separate section or on a separate sheet, or this information may be coded in a schedule created by the engineer. Since this drawing is done to scale, size and length information is available. The electrical inspector may also want to use this drawing to make notes in the pre-inspection planning phase to identify areas where more visual detail at the site is required.

Electrical specifications are typically text documents created by the engineer that identify the type of equipment and explain how the equipment is to be installed in the facility. These documents can be prescriptive or performance-based and are furnished to provide bid guidance to the electrical contractor and to highlight the installation details not included in the electrical drawings. A specification may prescribe or call for a particular manufacturer's piece of equipment, including model number or style number. The specification may describe individual aspects of equipment performance, ratings, operation, and function.

Most electrical specifications include a scope of work, product lists, installation methods, identification methods, bill of materials, and references, and identify installation responsibilities. Specifications are used to complete the contract documents. Specifications are not always readily available to an inspector, but if they are available, they can provide such information as how the panelboard legends are to be done and how conductors of different voltage systems are to be identified. However, because the specifications are contract documents and not *Code* requirements, they are often subject to contract revisions. Therefore, most inspectors do not verify conformance with specifications as long as the installation meets *Code* requirements; enforcement of contract provisions is generally outside the scope of an inspector's responsibility or jurisdiction.

Compliance with specifications may be enforced if the specifications are written to meet specific requirements of the AHJ, but most AHJs usually do not want to get involved in private contract issues. On the other hand, if the specification calls for the use of a ground ring type of grounding electrode, the inspector is responsible for ensuring that it is incorporated into the grounding electrode system per the requirements of the *NEC.*

Other *miscellaneous drawings and details* may be useful to the electrical inspector. One miscellaneous detail may include the wall where the service entrance is located. Similarly, wall-mounted outdoor lighting may be shown on a miscellaneous drawing or detail.

Control diagrams show how the system is electrically controlled by relays, pushbuttons, limit switches, holding coils, level switches, pressure switches, and/or temperature switches. For example, a control diagram will show how a ground-fault protective relay installed on a 480Y/277-volt, 2000-ampere service will operate the main bolted pressure contact switch. A control diagram will also show a pushbutton and ladder logic scheme for the motors of a particular process. This diagram will show how the system operates but not the physical properties of each element. Control diagrams are often supplied by manufacturers of equipment and may not be included in plans submitted for plan review or for other uses by the inspector.

A *connection diagram* is a control circuit wiring diagram. It shows the general physical location of relays, pushbuttons, switches, and equipment in cabinets or enclosures. The terminals and terminal numbers of the devices are identified on the diagram for field wiring and connection. The electrician uses the connection diagram to properly interconnect the wires and control cables of the electrical equipment. Such drawings are not usually particularly helpful to inspectors, but they may be useful in specific cases.

All of the plans and other documents mentioned in this chapter provide valuable assistance in the planning and completion of electrical inspections. The use of this available information, along with the checklists contained in this manual and the information presented in the *NEC,* helps to ensure that an inspection will be as thorough and accurate as possible.

CHECKLIST

		Checklist 1-1: General Safety Checklist for Electrical Inspections	
✔	Item	Basic Hazard Analysis	Comments
	1.	Does the inspection task involve exposed energized conductors or circuit parts?	
	2.	Can the risk of exposure to electrical hazards be justified?	
	3.	What is the voltage of the equipment that requires inspection?	
	4.	Where are the approach boundaries for shock protection?	
	5.	Will the inspection involve crossing any of the approach boundaries?	
	6.	Has an incident energy analysis been performed for the equipment?	
	7.	Are the available short-circuit current and clearing times known?	
	8.	Where is the Arc Flash Boundary?	
	9.	Will any part of the body be within the Arc Flash Boundary?	
	10.	How will PPE for arc flash protection be selected?	
	11.	Is the appropriate arc-rated PPE available?	
	12.	Is the inspector qualified for this specific task and risk?	

Source: Data from NFPA 70E, *Standard for Electrical Safety in the Workplace*, 2012.

CHECKLISTS

CHAPTER 2

General Requirements Inspections

OVERVIEW

This chapter covers inspections to determine compliance with the general requirements of the *National Electrical Code®* (*NEC®*). The checklists in this chapter cover Articles 90 and 110, in which the most general and broadly applied rules of the *NEC* are located. Article 90 explains the organization, scope, and intent of the *NEC*. The duties of the authority having jurisdiction (AHJ), the approval process, and the role of third-party testing and certification are also covered in Article 90. Article 110 expands on the requirement for approval of **equipment** and materials and furnishes specific elements for consideration in examining equipment for approval. Article 110 also provides fundamental rules that apply to all wiring installations. These rules include requirements for following manufacturers' installation instructions, providing safe working space, furnishing equipment with adequate interrupting and short-circuit current ratings, ensuring the integrity of electrical enclosures, providing secure support for and mounting of equipment, and making splices, connections, and terminations that are safe and reliable.

SPECIFIC FACTORS

Article 110 provides some of the most broadly applied requirements in the *NEC*. The requirements discussed in this chapter apply to virtually all installations. For example, every installation requires a power source and a system to deliver power to a building, structure, or area. The equipment used to distribute power likely requires working space. Such equipment will require **approval** (acceptance by the AHJ), and approval is typically judged on compliance with applicable *NEC* installation requirements and any specific instructions provided by the equipment manufacturer. This chapter covers these basic requirements.

General requirements are, as the term suggests, applicable for all inspections. Most other *Code* requirements are more specific, in that they apply only to certain types of equipment, methods, or systems. The requirements covered by this chapter usually apply in addition to any specific requirements of the *Code*. Unlike the rules of many other articles, many rules of Article 110 are never modified by other rules of the *Code*. For example, there is no modification of the following statement found in Section 110.2: "The conductors or equipment required or permitted by this *Code* shall be acceptable only if approved."

KEY QUESTIONS

1. Is the installation covered by the *NEC*?

Chances are, this question will be resolved before any permit is issued or any inspection is made. Nevertheless, it is a key question, because if the *NEC* does not apply, then neither will the checklists in this manual.

2. What should the interrupting ratings of equipment be?

The precise answer to this question is not always needed as long as the actual calculated value of available short circuit is less than standard equipment ratings—for example, less than 10 kiloamperes interrupting capacity (kAIC) for 120-volt and 240-volt equipment. For large services or for short lengths of service conductors, the answer is often found on the plans. The answer should always be provided by the system designer when requested, and except for individual dwelling units, the available short-circuit current will be required to be marked on service equipment at final inspection. Electric utilities can provide information on the available fault current at the service point, at their transformer, or at another interface between the utility and premises wiring system.

3. At what voltage(s) does the system operate?

This information is critical to determining proper working space for equipment such as panelboards, switchboards, motor control centers, controllers, and many equipment disconnecting means. This answer can be obtained either from the equipment itself or from the plans.

4. What are the electrical loads in the building or structure?

An electrical system is designed to supply electrical energy to the loads within the building or structure. Understanding the types of electrical loads, such as lighting, heating, motor, and appliance loads, and where they are located in a facility will help inspectors to logically understand the electrical system and make it easier to inspect.

5. What are the environmental conditions for the completed installation?

A wet, corrosive, or dusty environment will require wiring methods and enclosure types that are suitable for those conditions. Indoor and outdoor installations also have different requirements for equipment that require dedicated space where the equipment is installed.

PLANNING FROM START TO FINISH

Many of the rules covered by this chapter apply to all types of electrical equipment. Consequently, some items in the checklist will apply on every inspection. For example, everything used in an electrical installation is subject to approval. Similarly, listing and labeling instructions apply to everything that is **listed** and **labeled**, from underground conduits and grounding connections, to luminaires and wiring devices. Although specific dimensioned working clearances do not apply to all equipment, they do apply to a wide variety of equipment. Thus, while the working clearances for panelboards can be reviewed as soon as the panelboard cabinets ("tubs" or "cans") are installed (that is, during the rough inspection), other equipment, such as some disconnects and controllers, will not be in place until the final inspection. Therefore, an electrical inspector must keep many of the general requirements in mind during all inspections.

On the other hand, some items can be checked at specific times. For example, the **interrupting ratings** of equipment can usually be checked at the same time as the inspection of the service and the overcurrent devices and subsequent levels of distribution equipment.

Issues such as interrupting ratings and working clearances can be quite thoroughly investigated in a reasonable amount of time. However, a thorough verification that all instructions included in listing and labeling have been followed would require the inspector to spend as much time on the job as the electrician. In such cases, an inspector must resort to sampling. Workmanship can also be a good indicator of quality. Because the phrase "a neat and workmanlike manner" is subjective, few inspectors will try to enforce workmanship issues in any but the most obvious cases of poor installation practice. Sloppy installations can call for a more careful inspection of such items as terminations, equipment mounting, and general compliance with listing instructions. However, an installer may concentrate exclusively on making a conduit rack or a panelboard look good while ignoring other details that are vital to safety, so an inspector cannot make a reliable judgment based solely on how the installation looks.

WORKING THROUGH THE CHECKLIST

Checklist 2-1: General Requirements for Electrical Inspections

1. Identify installations or parts of installations that are covered by the *NEC*.
This is usually not the subject of much debate for electrical inspectors. In fact, it may be easier to ask, "Is the installation not covered?" The *NEC* probably does not apply if the installation is in a boat or a mine, on part of a railway, under the exclusive control of a communications utility, or under the exclusive control of an electric utility. Other than these exclusions, the *NEC* likely does apply to the installation.

2. Verify that installations have been made in accordance with the instructions included in listing and labeling of materials and equipment.
This is a very broad requirement, and an inspection that could verify adherence to every aspect of listing and labeling would consume too much time for most inspectors. A reasonable method will probably utilize some sampling. One approach is to ask to see the instructions for selected fixtures or equipment. If the installer does not know where the instructions are or knows nothing about them, chances are they have not been followed. Familiarity with requirements does not guarantee compliance, but it is a good indicator. Torquing of terminals is a good example. Sometimes the torque values are on the device, such as on a circuit breaker, or they may be in the instructions included with a panelboard or other type of distribution equipment. If the installer is not familiar with these instructions, it is unlikely the instructions have been followed. Informative Annex I, Recommended Tightening Torque Tables, provides general torquing requirements; however, it is always best to follow any specific instructions marked on the equipment. Asking about any unusual luminaires, control equipment, or appliances is a good way to spot-check for compliance with installation and use instructions. According to the Informational Note in Section 110.3(A)(1), "Special conditions of use or other limitations and other pertinent information may be marked on the equipment, included in the product instructions, or included in the appropriate listing and labeling information."

3. Identify installations and equipment requiring special approval or investigation.
Most equipment is available as listed and labeled equipment. If any equipment is custom made for a specific application, it should be examined for suitability. Some jurisdictions do not allow the installation of such things as custom control panels that are not listed. However, equipment assembled from listed components may be acceptable, based on an inspection of the equipment or application of code requirements such as those provided in Article 409 for Industrial Control Panels. Electrical installations will almost always include some items that are not available as listed products, such as support hardware and many lamps for luminaires. This decision to approve a product that is not listed is the responsibility of the local AHJ. Articles 90 and 110 contain requirements covering the examination of such equipment for safety.

4. Verify that interrupting ratings and short-circuit ratings are adequate for the conditions of the installation.
Most engineered designs include the required fault-current ratings for specific equipment. Section 110.24 requires the maximum available fault current to be marked on service equipment except in dwelling units. This information is required for the specific purpose of verifying appropriate interrupting and short-circuit current ratings of installed

FIGURE 2-1   A circuit breaker with interrupting ratings.

equipment. Caution: The information contained in this marking is not intended to be a substitute for the marking required by NFPA 70E® covering the selection of personal protective equipment (PPE). Comparing the ratings of the installed equipment with the ratings specified on the plans is usually sufficient to verify proper values. **FIGURE 2-1** shows some of the markings located on a circuit breaker. In this example, the circuit breaker is capable of interrupting 10,000 amperes (10 kAIC) at its rated voltage. A panelboard into which this circuit breaker might be installed also has a short-circuit current rating. The rating is dependent on the interrupting ratings of the overcurrent devices that are installed. For example, a panelboard may have a rating of 22 kAIC if a circuit breaker with 22-kAIC ratings is installed, but that same panelboard would have a rating of only 10 kAIC if a circuit breaker like the one shown in Figure 2-1 were installed. Other markings include information on wire sizes, temperature ratings, and torquing values for terminals. If a fault-current calculation has not been provided at the time of inspection, one should be requested from the designer. Equipment interrupting ratings must be verified.

Where such ratings are not provided on plans, a fault-current analysis may be necessary to determine suitability. This is especially important on installations where panelboards and service equipment are very close to the utility transformer, transformers are paralleled, and/or conductors are very large. General guidelines for likely fault-current levels may be obtained from overcurrent device specifications and catalogs, but calculations of fault currents based on actual conditions are the only accurate method of determining suitable interrupting ratings. An example of the information needed to determine the accurate interrupting rating of electrical equipment is provided in **TABLE 2-1**. For electrical service equipment, the electric utility supplying the service can be consulted regarding available fault currents. Table 2-1 is based on an infinite available fault current at the primary side of the transformer, as are many short-circuit calculations. Note that the information given in Table 2-1 may not be the only information needed, especially where large motor loads are supplied. Motors may contribute to the amount of fault current, because when they are suddenly de-energized but still turning, they act for a short time as a contributing source of current in the circuit. In addition, overcurrent protective devices installed in "series-rated systems" must be carefully checked to ensure that the device is suitable for use in this type of overcurrent protection system. In any overcurrent protection scheme, an improperly rated or incompatible device creates a weak link in the chain.

The assumption that the utility fault current is infinite is very useful for determining whether an interrupting rating is adequate, because even if the actual available primary fault current from the utility changes, the device ratings will be adequate if based on infinite primary current. However, Table 2-1 is based on some assumed transformer impedance values, as shown in the second column. If the actual transformer impedance is less than the values given, the fault current will be higher, and interrupting ratings may be inadequate if such a change is made. Also, these values represent maximums and are not an accurate indicator of the actual fault current that may flow, especially when further limited by the internal impedance in an arcing fault. Therefore, the values given in Table 2-1 may not be useful or directly applicable to the evaluation of arc flash hazards for the purpose of selecting PPE.

TABLE 2-1	Typical Short-Circuit (SC) Current Values*

Transformer kVA Rating	Transformer Impedance (% Z)	Line–Line Voltage	Phase (1 or 3)	SC Current at Secondary Terminals	Secondary Conductor Size	Secondary Conductor Length (ft)	Number of Conductors (Parallels)	Conduit Type (Steel, PVC)	SC Current at End of Conductors
50	2	240	1	10417	3/0 Al	50	1	PVC	7055
50	4	240	1	5208	3/0 Al	50	1	PVC	4206
75	2	208	3	10409	3/0 Cu	10	1	PVC	6415
75	4	208	3	5205	3/0 Cu	10	1	PVC	3969
75	4	208	3	5205	3/0 Cu	50	1	PVC	4504
75	2	480	3	4511	1/0 Cu	50	1	Steel	4134
75	4	480	3	2255	1/0 Cu	50	1	Steel	2157
75	4	480	3	2255	1/0 Cu	50	1	PVC	2161
300	2	208	3	41637	250 Cu	25	4	PVC	37291
300	4	208	3	20819	250 Cu	25	4	PVC	19673
300	4	208	3	20819	350 Al	25	4	PVC	19559
750	2	480	3	45106	500 Cu	20	3	PVC	43345
750	4	480	3	22553	500 Cu	20	3	PVC	22104
750	4	480	3	22553	500 Cu	10	3	PVC	22326

*Based on infinite available primary current.

5. Verify that unused openings have been effectively closed.

This is something that can be checked throughout a job simply by looking for openings that have not been sealed. A general rule to keep in mind is that all enclosures for equipment and conductors should be complete and the only openings that may remain open are those intended for mounting or operation of the equipment. Ventilation openings are examples of those permitted as part of the design of listed equipment. Electrical enclosures are intended to keep foreign objects (including fingers) outside and to keep normal, as well as some abnormal, arcing inside; unintended openings will obviously compromise an enclosure. Unused knockouts should be closed, and covers are required to be installed on all types of raceways, enclosures, and boxes. Fillers should also be installed on unused breaker spaces in panelboards. If the factory panel cover is missing, a suitable cover must be installed. Leaving boxes and other enclosures with open knockouts is the most common violation of this rule.

6. Check for broken or damaged parts and contamination by foreign materials.

Failure to mask electrical equipment prior to painting or plastering frequently results in damage to the equipment. Some judgment should be exercised, however. For example, a little stray paint in a device box or on some conductors is not necessarily serious, but a panelboard whose interior has been coated with paint or drywall compound may create serious problems by interfering with electrical connections in the panelboard. Not all broken or damaged parts can be detected with a simple visual inspection, but indications of serious damage are often visible and readily observable. Obvious signs of damage should be investigated more carefully. Total or partial replacement of equipment may be necessary in extreme cases.

7. Check for secure mounting and adequate ventilation space for equipment.

Electrical equipment must be attached firmly to the surface on which it is mounted. Fasteners designed for the purpose should be used. Screws are usually suitable for mounting equipment on wood or metal supports. Equipment mounted to concrete or masonry should be secured with expansion anchors. Toggle bolts or similar fasteners may be adequate for some hollow masonry. Because they will eventually dry out and rot, wooden plugs in masonry are not suitable. Equipment with ventilation openings must be installed so that the openings are not blocked. Some equipment, such as some transformers, are marked with required spacings that must be maintained in order to provide adequate ventilation.

8. Check for proper use and ratings of splices and terminations.

Splices and terminations must be made with devices that are suitable for the conductor materials and the conditions. Terminal blocks, split bolts, compression sleeves and terminals, pressure lugs, and other connection devices must be **identified** for aluminum where used with aluminum conductors. While many such devices are listed and identified for both aluminum and copper, some are suitable only for aluminum or only for copper and are marked accordingly. This marking may be on the actual connector or on the packaging containing the connector. Compliance with such markings is necessary. Generally, terminals for conductors are suitable for only one conductor per terminal unless they are marked otherwise. Splices used for direct burial must be marked for direct burial. Connectors for fine-stranded conductors must be marked for the number of strands. Such conductors are often found in flexible cords and cables and conductors commonly used with photovoltaic systems. "Fine stranded" refers to conductors with stranding finer than Class B or Class C stranding, as shown in Table 10 in Chapter 9 of the *NEC*.

9. Check temperature ratings of terminations.

Most terminations are rated for either 60°C or 75°C (140°F or 167°F). Many pressure terminals are rated for 90°C (140°F) but are applied in or on equipment that have only a 60°C or 75°C rating. In such cases, the lower rating must be observed. Conductors connected to terminals must not supply loads in such a way that the terminal temperature is exceeded. This means that the current in a conductor must be limited to the ampacity of conductors with insulation types corresponding to the temperature rating of the terminal. A common mistake is applying the 90°C conductor ampacities to circuit wiring that is connected to terminations with a 75°C rating. The ampacity selection for the load on the conductor is based on the lowest-rated component. The equipment with 60°C-rated terminals is the basis for using the 8 American Wire Gauge (AWG) type of thermoplastic high heat-resistant nylon-coated (THHN) as a 40-ampere conductor (**FIGURE 2-2**). The 90°C rating is permitted as a starting point for "derating" the conductor due to heating as a result of elevated ambient (correction factors) or number of conductors in a raceway or cable (adjustment factors), or both. The ampacities for conductors that terminate in electrical equipment are based on product testing.

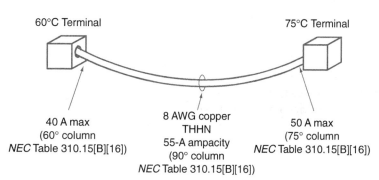

60°C Terminal

75°C Terminal

40 A max
(60° column)
NEC Table 310.15[B][16])

8 AWG copper
THHN
55-A ampacity
(90° column)
NEC Table 310.15[B][16])

50 A max
(75° column)
NEC Table 310.15[B][16])

FIGURE 2-2   Example of selecting ampacity based on lowest-rated component.

In some installations, such as cable trays, the ampacities for conductors in the tray may be based on values other than those shown in Table 310.15(B)(16) of the *NEC*, but Table 310.15(B)(16) is always the basis for selecting ampacities for compatibility with terminal temperature ratings.

10. Check for arc flash protection warning signs.

In addition to shock hazards, people working on **energized** equipment are also **exposed** to hazards from the flash that is produced by arcing faults. Arc flash may cause severe or fatal burns. Where the energy levels are sufficient to cause such burns, NFPA 70E, *Standard for Electrical Safety in the Workplace®, 2012 Edition*, requires PPE for employees working on energized parts and provides assistance in the selection of appropriate PPE. The actual PPE required in any specific case depends on the available energy. The *NEC* requires generic warning signs for switchboards, panelboards, industrial control panels, motor control centers (MCCs), and similar types of equipment that are likely to be examined, adjusted, serviced, or maintained while energized. These types of equipment also require defined and specifically dimensioned working space. The warning signs are not required in individual dwelling units but are required at the common service equipment for a multifamily occupancy.

The *NEC* does not specify the wording on the signs, but the sign must include a warning of the flash hazards. A qualified person can then determine what calculations or other steps may be necessary to select appropriate PPE. American National Standards Institute (ANSI) Z535.4-2007, *Product Safety Signs and Labels*, provides design guidelines for the signs. These general guidelines describe the type of language that must be included on warning labels (**FIGURE 2-3**). Additional details about the hazard, such as the calculated incident energy, the approach and flash protection boundaries, or the specific PPE required, are often included on such labels. However, the additional details are not required by the *NEC*. Additional details are required by NFPA 70E.

Arc Flash and Shock Hazard
Appropriate PPE Required

FIGURE 2-3   Example of a flash hazard warning label.
(Source: *National Electrical Code® Handbook*, 2011, Exhibit 110.8.)

11. Check for markings on enclosures and verify appropriate ratings for the environment.
Table 110.28 in the *NEC* provides enclosure-type designations that provide a degree of protection against specific environmental conditions. These enclosures are for protection against external influences and, according to 110.28, do not provide protection from effects "such as condensation, icing, corrosion, or contamination that may occur within the enclosure or enter via the conduit or unsealed openings." Some enclosures have multiple type ratings. **FIGURE 2-4** shows a listed enclosure identified as Types 4, 4X, 12, and 13. The inspector may use the type marking along with Table 110.28 in the *NEC* to determine whether a specific enclosure type is appropriate for the external influences. Most outdoor types may also be used indoors, but they may provide protection from influences that do not exist indoors.

FIGURE 2-4   Example of an enclosure with multiple type markings.

12. Verify adequate working clearances, dedicated spaces, and headroom around equipment.
The general rule with regard to working space states: "Sufficient access and working space shall be provided and maintained about *all electric equipment* to permit ready and safe operation and maintenance of such equipment." Specific dimensioned working spaces are required around equipment that requires "examination, adjustment, servicing, or maintenance while *energized*" [*NEC* 110.26(A)]. Panelboards, switchboards, MCCs, and many individual motor controllers are good examples of such equipment. Many, but not all, disconnects may need to be opened while energized. Some individual judgment is required. The specific required spaces are determined by the conditions on the other side of the space (behind the worker) and the voltage to ground at which the equipment operates. **Voltage to ground**, even for ungrounded systems, is defined in this chapter and in Article 100 of the *NEC*. The conditions for general working space depth

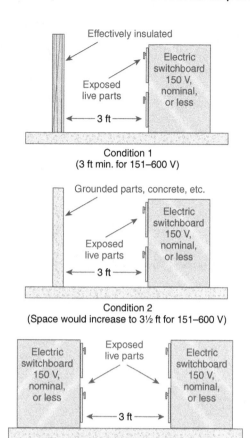

Condition 1
(3 ft min. for 151–600 V)

Condition 2
(Space would increase to 3½ ft for 151–600 V)

Condition 3
(Space would increase to 4 ft for 151–600 V)

FIGURE 2-5 General rules for depth of working space about electrical equipment. Note: For condition 3, where there are enclosures on both sides of the working space, the clearance for only one working space (the largest required clearance) is required.

are shown in **FIGURE 2-5**, where the distances are measured from the **live parts** if the live parts are exposed or from the enclosure front if the live parts are enclosed. If any assemblies, such as switchboards or MCCs, are accessible from the back and expose live parts when they are opened, the working clearance dimensions illustrated would also be required at the rear of the equipment.

FIGURE 2-6 and **FIGURE 2-7** illustrate modifications of the general rules shown in Figure 2-5. In Figure 2-6, where rear or side access is required to work on nonelectric parts only, the minimum working space that may be provided in the rear of equipment is reduced to 30 in. (762 mm). Figure 2-7 illustrates the working clearance allowed where there is a written procedure that prohibits facing doors of equipment from being open at the same time and where only authorized and qualified persons will service the installation. This working clearance "exception" is permitted only for dead-front switchboards, panelboards, or MCCs being replaced in an existing building.

The required width of working space cannot be less than 30 in. (762 mm), regardless of the width of the electrical equipment. However, as shown in **FIGURE 2-8**, the 30-in. (762-mm) wide front working space is not required to be exclusive to or directly centered on the electrical equipment if it can be ensured that the space is sufficient for safe operation and maintenance. The minimum height of the working space is 6½ ft (2.0 m) or the height of the equipment, whichever is greater.

For electrical equipment that does not require service or maintenance while it is energized, sufficient access and working space shall be provided and maintained. This space permits ready and safe operation and maintenance of such equipment (*NEC* 110.26). For example, a duct heater or a variable air volume (VAV) box in a ceiling space may be disconnected prior to servicing, and a snap switch used as a disconnect should not be serviced while energized. Because no specific dimensions are provided for such cases, each case must be judged individually. A 24-in. (610-mm) square opening provided by a removable ceiling panel or a small access door is entirely adequate in cases in which the equipment will not be worked on while energized and may be judged to be adequate in some

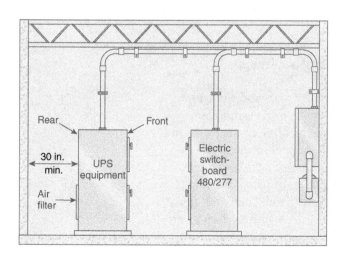

FIGURE 2-6 Example of working space required for nonelectric components of electrical equipment.

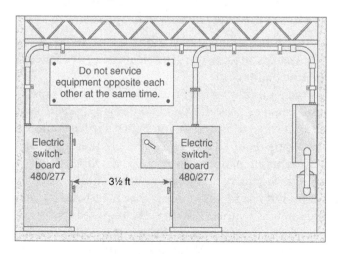

FIGURE 2-7 Example of working space permitted for electrical equipment replacement.

other cases as well. However, if a duct heater has associated controls that are likely to be examined or serviced while energized, Section 424.66(A) and (B) of the *Code* provide working space requirements (including specific provisions for equipment located above a suspended ceiling) unique to this type of heating equipment.

Dedicated space is allocated to the electrical installation, not just the equipment that requires the dedicated space. Unlike working space, dedicated space applies only to specific types of equipment, and the requirement is not based on the likelihood that the equipment will be serviced while energized. The types of equipment that require dedicated space are switchboards, switchgear, panelboards, and MCCs. The dimensions of the dedicated space are based solely on the dimensions of the equipment, not on any voltage or working conditions.

FIGURE 2-9 shows the dedicated spaces above and below a panelboard and the equipment located outside those dedicated spaces. Other electrical equipment such as raceways, boxes, time clocks, and switches could be installed in the dedicated spaces above and below the panelboard, but such other equipment could not occupy the working space of the panelboard. However, in the vertical dimension of the working space, electrical equipment could vary in depth up to 6 in. (152 mm) (**FIGURE 2-10**). This 6-in. (152-mm) limit applies to a

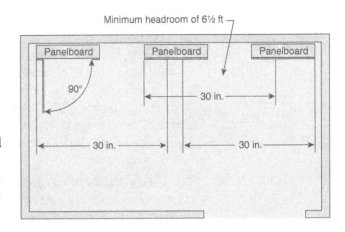

FIGURE 2-8 Example of 30-in. (762-mm) wide working space application with more than one piece of electrical equipment.

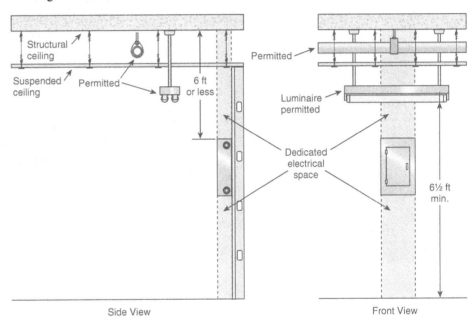

FIGURE 2-9 Dedicated spaces above and below a panelboard.

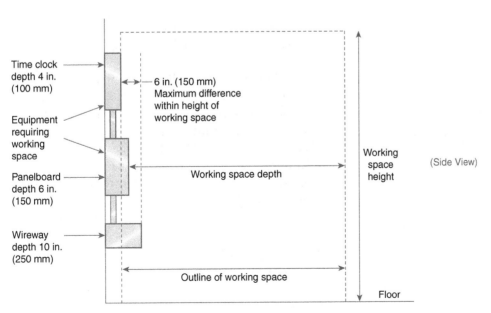

FIGURE 2-10 Equipment of unequal depth permitted in height of working space.

meter socket but not to the meter itself. The meter may project more than 6 in. (152 mm) into the working space.

Nonelectrical equipment (not part of the electrical installation) is prohibited from being installed in the dedicated space. Nonelectrical building construction elements, such as a suspended ceiling support grid and ceiling tiles installed above electrical equipment, are not prohibited by this section. Fire sprinkler protection is permitted for the dedicated electrical equipment space; however, the sprinkler piping is not permitted in the 6-ft (1.8-m) high (or to the structural ceiling) dedicated space measured from the top of the electrical equipment.

13. Verify that working space and dedicated space are not used for storage.
Most violations of this rule occur after inspections are completed and buildings are occupied. Still, any signs of improper use of working space or dedicated space should be noted. This warning will alert users to the safety and fire hazards involved in blocking access to overcurrent devices and control equipment or allowing combustible materials to accumulate in the vicinity of such equipment. The correct initial selection of the location for electrical equipment will go a long way toward continued compliance with this requirement. Some jurisdictions or engineering specifications require that equipment be marked with the minimum clear working space, but the *NEC* does not require such markings.

14. Check adequacy of entrance to and egress from working space in general, and verify that spaces containing large equipment have at least two entrances/exits or the equivalent.
An entrance/egress of "sufficient area" must be provided for all required working spaces. The *Code* does not provide a specific minimum dimension that will provide sufficient area. Equipment installed above a suspended ceiling may be accessible through the removable ceiling panels, which may well be smaller than would be required for other types of equipment or for equipment in other areas. A graphical explanation of access and entrance requirements to a working space is shown in **FIGURE 2-11**, where at least one entrance or egress is required for accessing the working space around electrical equipment.

Note that the installation shown at the bottom of Figure 2-11 would not be acceptable if the electrical equipment were a switchboard or panelboard rated 1200 amperes or more and over 6 ft (1.8 m) wide. For such equipment, called "large equipment," the minimum size of the entry area is clearly stated in Article 110; one entrance not less than 24 in. (610 mm) wide and 6½ ft (2.0 m) high is required at each end (**FIGURE 2-12**). Where the entrances to spaces for large equipment are equipped with doors, and the doors are within 25 ft (7.6 m) of the nearest edge of the working space, the doors must open in the direction of exit

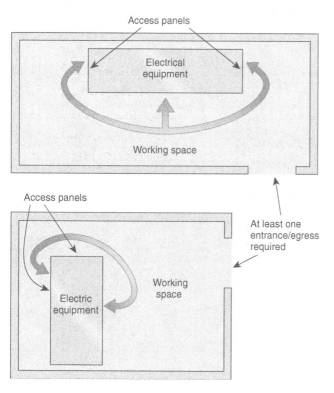

FIGURE 2-11   Entrance to and egress from working space.

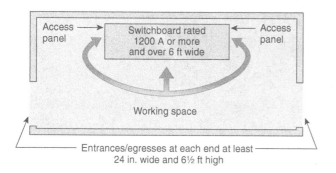

FIGURE 2-12   Entrance to and egress from working space for equipment rated 1200 amperes or more and over 6 ft (1.8 m) wide.

travel and, as noted in the 2014 edition of the *NEC*, must be equipped with listed panic hardware. This requirement applies to large equipment as defined but also to similar equipment that is rated 800 amperes or more but less than 6 ft (1.8 m) wide. Section 110.26 does not require that there be a door or doors. Many entrance and egress pathways to or from working spaces do not have doors, as the equipment may be located in a larger open area. The same personnel door requirement has been added to *Over 600 Volt Installations*, found in *NEC* Section 110.33(A)(1).

There are two alternative approaches to the general entrance and egress rules for large equipment. The first provides that if the location permits a continuous and unobstructed way of exit travel, then only one means of entrance/egress is required (**FIGURE 2-13**). The second provides that if the working space is doubled, then only one entrance/egress is required (**FIGURE 2-14**). **FIGURE 2-15** illustrates a hazardous and unacceptable switchboard (large equipment) arrangement where a person could be trapped behind arcing electrical equipment. This arrangement might be acceptable for equipment that is not considered "large equipment."

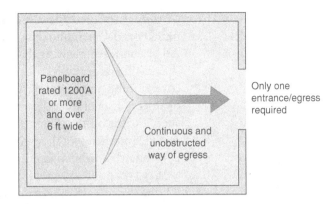

FIGURE 2-13   The equipment location permits a continuous and unobstructed way of egress travel.

15. Verify that working spaces have adequate illumination.

The *Code* does not provide for any specific lighting level in foot-candles; nor does the *Code* require a dedicated light source. If the level of illumination is adequate for inspection purposes, it is probably adequate for servicing equipment. However, where work on energized equipment will take place and PPE must be worn, the required ultraviolet protective face shields and eye protection may reduce light transmittance and a higher lighting level may be necessary. The higher levels required for such energized work could be provided by portable task lighting sources.

Local control of the light source is required and is important in order to ensure that someone servicing equipment is not left in darkness because of some remote- or automatic-control system, such as motion sensors. Occupancy sensors with local manual overrides are usually acceptable.

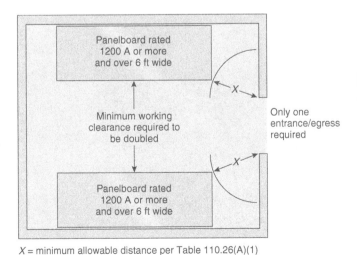

X = minimum allowable distance per Table 110.26(A)(1)

FIGURE 2-14   One entrance/egress is permitted because the working space is doubled.

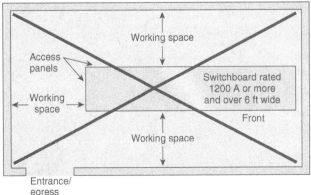

FIGURE 2-15   Arrangement of equipment is unacceptable.

16. Check for identification of disconnecting means and circuit directories for panelboards, switchboards, switchgear, and similar equipment.

All disconnecting means should be marked to indicate their purpose, unless the purpose is evident. For example, the purpose of a disconnect mounted next to or on a single air-conditioning unit is obvious. But if a group of disconnects is mounted next to a group of air conditioners, the disconnects need to be marked to indicate which unit each one supplies. The overcurrent devices installed in most panelboards are also disconnecting means, and each switch or circuit breaker must be clearly marked to indicate its purpose. Panelboard schedules should be filled out in sufficient detail to indicate the nature and location of the loads served. Each circuit in a panelboard schedule is required to be "legibly identified as to its clear, evident, and specific purpose or use." In addition, the *NEC* (in 408.4) requires circuit directories to have "an approved degree of detail that allows each circuit to be distinguished from all others," to be clearly identified without depending "on transient conditions of occupancy," and to also identify "spare positions that contain unused overcurrent devices or switches." These requirements are quite easy to meet in a dwelling unit as long as descriptions (such as "Joe's bedroom") are not used. However, panelboards in commercial occupancies are sometimes provided with markings that do little to indicate the specific loads. A panelboard with all the circuits marked "receptacles" or "lights" does not meet the intent of this rule. Circuit marking that uses a person's name (such as "Joe's office") to identify the location does not comply. Markings that include area descriptions or room numbers are much more useful and safer.

In addition to marking each disconnecting means to indicate its use, each panelboard or switchboard supplied by a feeder must be marked with the origination of the source of power to that panelboard or switchboard. This requirement does not apply to one- and two-family dwellings.

KEY TERMS

Article 100 of the *NEC* defines the key terms pertaining to general requirements as follows:

Approved Acceptable to the authority having jurisdiction (AHJ). [*Approval is a primary responsibility of an electrical inspector. Investigations by a third party and the listing and labeling that result are a great aid to inspectors in this responsibility (see* Labeled *and* Listed).]

Energized Electrically connected to, or is, a source of voltage.

Equipment A general term, including fittings, devices, appliances, luminaires, apparatus, machinery, and the like used as a part of, or in connection with, an electrical installation. [*This very broad term includes virtually everything that is part of an electrical installation. Equipment may have a much narrower meaning as used by installers, but where the word appears in the* NEC, *it must be used as defined in Article 100.*]

Exposed (as applied to live parts) Capable of being inadvertently touched or approached nearer than a safe distance by a person.

> *Informational Note:* This term applies to parts that are not suitably guarded, isolated, or insulated. [*Live parts inside panelboards and similar equipment are not normally considered to be exposed. However, they are exposed during servicing of the panelboard. Therefore, such live parts that are normally enclosed are considered to be exposed for the purposes of determining working clearances. The enclosure of the equipment is used as the basis for establishing clearance distances.*]

Identified (as applied to equipment) Recognizable as suitable for the specific purpose, function, use, environment, application, and so forth, where described in a particular *Code* requirement. (See *Equipment.*)

> *Informational Note:* Some examples of ways to determine suitability of equipment for a specific purpose, environment, or application include investigations by a qualified testing laboratory (listing and labeling), an inspection agency, or other organization concerned with product evaluation.

Interrupting Rating The highest current at rated voltage that a device is identified to interrupt under standard test conditions.

> *Informational Note:* Equipment that is intended to interrupt current at other than fault levels may have its interrupting rating implied in other ratings, such as horsepower or locked rotor current. [*Interrupting ratings are usually measured in amperes and expressed in AIC or kAIC. This rating designates a fuse or circuit breaker's capability to interrupt current safely. A common rating for circuit breakers is 10,000 AIC or 10 kAIC, both of which mean that the breaker can safely interrupt a fault current of 10,000 amperes under specific test conditions. Panelboard, switchboard, and switchgear short-circuit current ratings are related. The strong magnetic fields associated with large fault currents produce significant mechanical stresses that tend to cause conductors or busbars to move around, sometimes violently. Therefore, buses in panelboards, switchboards, and switchgears must be braced against the mechanical forces that occur under short-circuit conditions, and the equipment short-circuit current rating must not be less than the short-circuit current that is available at its line terminals.*]

Labeled Equipment or materials to which has been attached a label, symbol, or other identifying mark of an organization that is acceptable to the AHJ and concerned with product evaluation, that maintains periodic inspection of production of labeled equipment or materials, and by whose labeling the manufacturer indicates compliance with appropriate standards or performance in a specified manner.

Listed Equipment, materials, or services included in a list published by an organization that is acceptable to the AHJ and concerned with evaluation of products or services that maintains periodic inspection of production of listed equipment or materials or periodic evaluation of services, and whose listing states that either the equipment, material, or services meets appropriate designated standards or has been tested and found suitable for a specified purpose.

> *Informational Note:* The means for identifying listed equipment may vary for each organization concerned with product evaluation, some of which do not recognize equipment as listed unless it is also labeled. Use of the system employed by the listing organization allows the AHJ to identify a listed product.

Live Parts Energized conductive components. [*In previous editions of the NEC (prior to 2002), this definition was "Electric conductors, buses, terminals, or components that are uninsulated or exposed, and a shock hazard exists." The new definition recognizes that* live parts *are still live, even when insulated or enclosed.*]

Voltage to Ground For grounded circuits, the voltage between the given conductor and that point or conductor of the circuit that is grounded; for ungrounded circuits, the greatest voltage between the given conductor and any other conductor of the circuit.

KEY QUESTIONS

1. Is the installation covered by the *NEC*?

2. What should the interrupting ratings of equipment be?

3. At what voltage(s) does the system operate?

4. What are the electrical loads in the building or structure?

5. What are the environmental conditions for the completed installation?

CHECKLIST

	Item	Inspection Activity	*NEC* Reference	Comments
Checklist 2-1: General Requirements for Electrical Inspections				
✔				
	1.	Identify installations or parts of installations that are covered by the *NEC*.	90.2(A) and (B)	
	2.	Verify that installations have been made in accordance with the instructions included in listing and labeling of materials and equipment.	90.7, 110.3(B)	
	3.	Identify installations and equipment requiring special approval or investigation.	90.4, 90.7, 110.2, 110.3	
	4.	Verify that interrupting ratings and short-circuit current ratings are adequate for the conditions of the installation.	110.9, 110.10, 110.24	
	5.	Verify that unused openings have been effectively closed.	110.12(A), 408.7	
	6.	Check for broken or damaged parts and contamination by foreign materials.	110.12(C)	
	7.	Check for secure mounting and adequate ventilation space for equipment.	110.13	
	8.	Check for proper use and ratings of splices and terminations.	110.14(A) and (B)	
	9.	Check temperature ratings of terminations.	110.14(C)	
	10.	Check for arc flash protection warning signs.	110.16	
	11.	Check for markings on enclosures and verify appropriate ratings for the environment.	110.28	
	12.	Verify adequate working clearances, dedicated spaces, and headroom around equipment.	110.26(A) and (E)	
	13.	Verify that working space and dedicated space are not used for storage.	110.26(B)	
	14.	Check adequacy of entrance to and egress from working space in general and verify that spaces containing large equipment have at least two entrances/exits or the equivalent.	110.26(C), 110.26(C)(2)	
	15.	Verify that working spaces have adequate illumination.	110.26(D)	
	16.	Check for identification of disconnecting means and circuit directories for panelboards, switchboards, switchgear and similar equipment.	110.22, 408.4	

Source: Data from NFPA 70E, *Electrical Safety in the Workplace*, 2009.

CHAPTER 3

Wiring Methods and Devices

OVERVIEW

Wiring methods include the entire range of materials and techniques used to install electrical circuits from a power source to electrically powered equipment. To provide electricity for power and lighting, a circuit between the power source and the load must be completed. Wiring methods provide the physical means of installing the conductors necessary to complete such circuits. The environmental constraints of the location are a major determining factor in selecting appropriate wiring methods. Beyond the basic function of getting conductors from one place to another, the job of a wiring method is to protect the conductors and their insulation from physical abuse and other damage. Some **locations** have harsh environments that can damage unprotected electrical circuits. These harsh environments may include the presence of moisture or water, elevated or reduced ambient temperatures, dust, chemicals, flammable gases, sunlight, etc. For example, **rainproof enclosures** are constructed, protected, or treated so as to prevent rain from interfering with the successful operation of the apparatus under specified test conditions. Some circuits are installed in the open air, in concrete, or buried in the soil, or they may pass through locations with particular construction ratings. *Code* rules governing the use and installation of wiring methods are intended to ensure that the wiring methods used in a particular location function reliably and safely and do not compromise other building systems.

Wiring devices provide an interface between the electrical system and the user. **Switches** are used to control **utilization equipment** (equipment such as luminaires and appliances that use electricity), and **receptacles** provide users with convenient methods of connecting to the wiring system. *Code* rules covering wiring **devices** are intended to make the user interface safe.

SPECIFIC FACTORS

Wiring methods and their use are a part of all installations and must be considered in all inspections. The key factor to consider for wiring methods is the location where the circuit will be installed.

- How is this location different from typical common ambient conditions?
- What is special about the location?

Wiring methods must be selected to be suitable for environmental conditions and anticipated use. Many of the wiring methods in Chapter 3 of the *National Electrical Code®* (*NEC®*) are referred to both by name and by acronym. In some cases, the acronyms are more commonly used than the actual names. Occasionally, especially where a list of wiring methods is given that may be suitable for a particular purpose, the acronyms may more clearly identify the items and differences in the list than the names. In this book, both names and acronyms are used, and acronyms are sometimes preferred when identifying a list of wiring methods. Both the names and the acronyms are used in the articles that cover the specific wiring methods. **TABLE 3-1** lists a number of cable and **raceway** wiring methods, the article in the *NEC* that covers that method, and the wiring method acronym.

| TABLE 3-1 | Acronyms for Wiring Methods and Relevant *NEC* Article |

Article	Cable Name	Type Acronym
320	Armored cable	AC
322	Flat cable assemblies	FC
324	Flat conductor cable	FCC
326	Integrated gas spacer cable	IGS
328	Medium voltage cable	MV
330	Metal-clad cable	MC
332	Mineral-insulated, metal-sheathed cable	MI
334	Nonmetallic-sheathed cable	NM, NMC, and NMS
336	Power and control tray cable	TC
338	Service-entrance cable	SE and USE
340	Underground feeder and branch-circuit cable	UF
342	Intermediate metal conduit	IMC
344	Rigid metal conduit	RMC
348	Flexible metal conduit	FMC
350	Liquidtight flexible metal conduit	LFMC
352	Rigid polyvinyl chloride conduit	PVC
353	High-density polyethylene conduit	HDPE
354	Nonmetallic underground conduit with conductors	NUCC
355	Reinforced thermosetting resin conduit	RTRC
356	Liquidtight flexible nonmetallic conduit	LFNC
358	Electrical metallic tubing	EMT
360	Flexible metallic tubing	FMT
362	Electrical nonmetallic tubing	ENT

KEY QUESTIONS

1. What wiring methods are being used?

Often the wiring methods are specified in plans or other contract or specification documents. If not, the methods may be determined by visual inspection. Wiring methods suitable for the conditions or locations must be used.

2. What is the occupancy type?

Some occupancies have specific requirements and restrictions for wiring methods. For example, metallic raceways or cables are required in many parts of health care facilities, places of assembly, and theaters. If a nonmetallic wiring method is used in a place of assembly that is required to be of fire-rated construction per the applicable building code, it must either be encased in concrete or, for several specific occupancies, be installed within a wall or ceiling assembly that has a 15-minute finish rating. Conversely, there are few restrictions on wiring methods that may be used on the interior of a single-family dwelling.

3. Are there any fire-rated walls, floors, ceilings, or roofs, and if so, where are they located?

Penetrations through fire-rated barriers must be firestopped. Firestopping is much easier with some wiring methods than with others, rendering some wiring methods impractical (but not necessarily impossible) to use in fire-rated construction. At any rate, penetrations of fire-rated assemblies must always be considered. Locations of fire-rated walls, floors, ceilings, or roofs should be shown on architectural plans.

4. Are there any suspended (grid) ceilings, and if so, are suspended ceilings being used for environmental air?

Suspended grid ceilings provide access to wiring methods and materials that might otherwise be considered inaccessible. For example, wiring in the space above a suspended drywall ceiling is **concealed** unless access is specifically provided, whereas wiring above grid ceilings usually meets the definition of **exposed (as applied to wiring methods)**. If the space above a grid ceiling is used for air handling, the wiring methods must generally be metallic or be specifically **plenum** rated.

"Plenum ceilings" are usually found only in fire-rated construction because the structure forms part of the air-handling enclosure, and standards that regulate the installation of environmental air systems generally do not permit such enclosures to be of combustible construction. Architectural plans will show the locations of grid or other suspended ceilings. "Reflected ceiling" plans, when they are included, are also useful for this purpose. Mechanical plans will indicate the use of ceiling spaces as plenums (the presence or lack of a return air duct system is a good indicator as to whether the ceiling space is used as a return-air plenum). For the purposes of electrical installations, the *NEC* refers to these spaces as "other spaces used for environmental air (plenums)."

5. Is any electrical equipment installed in wet or damp areas or other areas requiring special protection?

Some wiring methods, boxes, and other **equipment** are not suitable for use in wet locations. Others may be used in wet locations only with special finishes or enclosures. Many wiring methods are not suitable for direct burial or encasement in concrete slabs. Wiring methods also must be suitable for corrosive conditions, where such conditions exist. Often the extent of wet locations can be determined only by inspection. Any location underground or outdoors and exposed to weather is a wet location by definition. Protected outdoor areas such as covered porches are usually considered to be damp locations.

6. Are multiwire branch circuits being used?

Multiwire branch circuits are permitted to supply **general-purpose branch circuits** and specific appliance circuits (such as an electric range or dryer) (**FIGURE 3-1**). These circuits are frequently referred to as a *common-neutral* or *shared-neutral circuit.* Multiwire branch

FIGURE 3-1 Wiring configuration and plan detail of multiwire branch circuit.

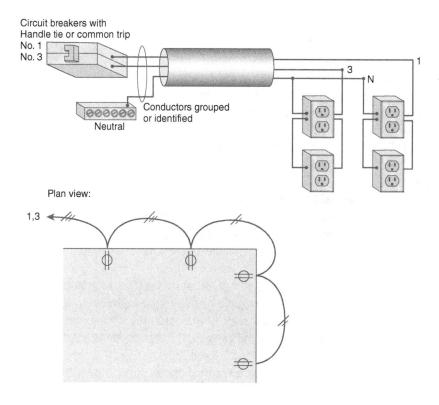

circuits require the use of common-trip circuit breakers or overcurrent devices with **identified** handle ties in order to provide a common disconnecting means for all ungrounded conductors of the circuit. A multiwire branch circuit must receive special attention to ensure the continuity of the grounded circuit conductor. Multiwire branch circuits are often indicated on plans, but inspection of circuits entering panelboards provides a ready answer to this question. Section 210.4(B) requires simultaneous disconnection of all ungrounded conductors of multiwire circuits and grouping of all ungrounded conductors and the neutral conductor of multiwire branch circuits. An alternative to grouping is identification of the conductors to show which grounded (neutral) conductors are associated with which ungrounded conductors.

7. Is any part of the wiring system being installed underground?

Where wiring is installed underground, attention must be given to the suitability of the wiring method and other equipment as well as to burial depths and backfill. Burial depths are not always easy to determine, especially before the final grade has been established. (**TABLE 3-2** notes minimum cover [burial depth] requirements.) Site plans or grading plans may have to be consulted to locate outdoor wiring and to ensure compliance with burial depth rules when underground installations are being inspected. All wiring methods, conductors, splices and terminations, equipment, devices, and covers must be suitable for use in wet locations if installed in wet locations underground or elsewhere.

PLANNING FROM START TO FINISH

The items covered in this chapter will be inspected throughout the course of a project. An examination of some underground wiring and burial depths may be part of the first inspections made. Most issues related to wiring methods and boxes, such as suitability for the use and adequacy of supports, will be inspected at the "rough wiring" stage after a substantial amount of electrical work has been completed but before anything is concealed. Usually, **cabinets** for panelboards or control panels will also be inspected at the rough stage, but occasionally the cabinets may not be placed until a later inspection, perhaps at the final inspection. Where suspended ceilings are used, an inspection is usually made before the ceiling tile is installed. Wiring devices are not usually available for inspection until the final, or "finish," inspection. The actual sequencing of inspections in this chapter will be determined by the progress of the individual project, and the checklist items may be covered under many separate inspections, depending upon the size and complexity of the project.

WORKING THROUGH THE CHECKLISTS

There are four checklists in this chapter:

1. General wiring methods
2. Boxes and conduit bodies
3. Cabinets and cutout boxes
4. Switches and receptacles

Checklist 3-1: General Wiring Methods

1. Identify the wiring methods in use and verify their suitability for the occupancy and conditions.

Uses permitted and not permitted for each wiring method are listed in the applicable article of Chapter 3 in the *NEC*. Certain occupancies in Chapter 5 of the *NEC* also restrict the wiring methods that may be used. In some cases, a wiring method may be suitable for different locations but may require different types of **fittings**. For example, electrical metallic tubing (EMT) may be embedded in concrete or installed outdoors, but must have fittings that are concrete-tight or **raintight**, respectively.

TABLE 3-2   Minimum Cover Requirements, 0 to 1000 Volts, Nominal, Burial in Millimeters (Inches)

Location of Wiring Method or Circuit	Type of Wiring Method or Circuit									
	Column 1 Direct Burial Cables or Conductors		Column 2 Rigid Metal Conduit or Intermediate Metal Conduit		Column 3 Nonmetallic Raceways Listed for Direct Burial Without Concrete Encasement or Other Approved Raceways		Column 4 Residential Branch Circuits Rated 120 Volts or Less with GFCI Protection and Maximum Overcurrent Protection of 20 Amperes		Column 5 Circuits for Control of Irrigation and Landscape Lighting Limited to Not More Than 30 Volts and Installed with Type UF or in Other Identified Cable or Raceway	
	mm	in.	mm	in.	mm	in.	mm	in.	mm	in.
All locations not specified below	600	24	150	6	450	18	300	12	150	6
In trench below 50-mm (2-in.) thick concrete or equivalent	450	18	150	6	300	12	150	6	150	6
Under a building	0 (in raceway or Type MC or Type MI cable identified for direct burial)		0	0	0	0	0 (in raceway or Type MC or Type MI cable identified for direct burial)		0 (in raceway or Type MC or Type MI cable identified for direct burial)	
Under minimum of 102-mm (4-in.) thick concrete exterior slab with no vehicular traffic and the slab extending not less than 152 mm (6 in.) beyond the underground installation	450	18	100	4	100	4	150 (direct burial) 100 (in raceway)	6 4	150 (in raceway or Type MC or Type MI cable identified for direct burial)	6
Under streets, highways, roads, alleys, driveways, and parking lots	600	24	600	24	600	24	600	24	600	24
One- and two-family dwelling driveways and outdoor parking areas, and used only for dwelling-related purposes	450	18	450	18	450	18	300	12	450	18
In or under airport runways, including adjacent areas where trespassing is prohibited	450	18	450	18	450	18	450	18	450	18

Notes:
1. Cover is defined as the shortest distance in millimeters (inches) measured between a point on the top surface of any direct-buried conductor, cable, conduit, or other raceway and the top surface of finished grade, concrete, or similar cover.
2. Raceways approved for burial only where concrete encased shall require concrete envelope not less than 50 mm (2 in.) thick.
3. Lesser depths shall be permitted where cables and conductors rise for terminations or splices or where access is otherwise required.
4. Where one of the wiring method types listed in columns 1–3 is used for one of the circuit types in columns 4 and 5, the shallower depth of burial shall be permitted.
5. Where solid rock prevents compliance with the cover depths specified in this table, the wiring shall be installed in metal or nonmetallic raceway permitted for direct burial. The raceways shall be covered by a minimum of 50 mm (2 in.) of concrete extending down to rock.

Source: Reprinted with permission from NFPA 70, *National Electrical Code®*, Table 300.5, copyright © 2013, National Fire Protection Association, Quincy, MA, 02169. This reprinted material is not the complete and official position of the NFPA or the referenced subject, which is represented only by the Standard in its entirety.

2. Verify that all conductors of a circuit are grouped together.

Conductors of a circuit, including equipment grounding conductors, are kept together to allow the magnetic fields of the conductors to interact. This interaction reduces the overall impedance of an alternating-current (ac) circuit. Low impedance is especially important to the proper operation of overcurrent devices. The interaction of the magnetic fields also effectively eliminates the heating effects of induction on non-current-carrying parts. Otherwise, induced currents could cause excessive temperatures in the non-current-carrying parts that may damage conductor insulation or create other hazards. Where conductors are run to switches, the "feed" wires and "return" wires from the switches along with any "traveler" wires between three-way or four-way switches must be kept together.

The *NEC* requires that a grounded conductor must also be run to switch locations even though there may be nothing connected to it at the switch location. This will accommodate changes to various types of automatic switching devices, some of which need a grounded conductor for operation. However, the 2014 *NEC* provides a list of seven conditions under which a grounded conductor may not be required at switch locations:

1. Switches with integral enclosures
2. Switches that control receptacle loads
3. Lighting loads controlled from multiple locations where at least one switch location is visible from the entire floor area of the room
4. Switches that control lighting loads in rooms that are not habitable (such as bathrooms)
5. Lighting loads controlled by automatic means where the wiring method to the switch allows for the addition of a grounded conductor in the future
6. A switch supplied by a raceway that has space for an additional conductor or where the location of the switch can be accessed without removing finish materials
7. A switch that is located in a framing cavity that is open at the top or bottom or is finished only on one side.

FIGURE 3-2 illustrates a lighting branch circuit in which the conductors of each part of the circuit have been routed together within the same cable or raceway. Notice that if the boxes in Figure 3-2 are connected by an adequately sized raceway, a grounded conductor is not required in the switch box, as shown. If this were a typical cable wiring method in a finished wall, the grounded conductor would be required in the switch box unless one of the alternatives noted above applies. **FIGURE 3-3A** illustrates one method of correctly wiring two three-way switches with conductors together. **FIGURE 3-3B** is an example of conductors not together; this arrangement is not permitted.

Both Figures 3-3A and 3-3B show only current-carrying conductors. Again, the switch boxes would require a grounded conductor in most cases if a cable wiring method is used and the framing is finished on both sides. The grounded conductor is not current-carrying until some device using the grounded conductor is connected at a switch location. Even then, the current is likely to be very small.

In addition to the requirements of 300.3, Section 210.4(D) also requires grouping or identification of conductors, but 210.4 is concerned only with multiwire branch circuits. All of the ungrounded conductors must be grouped with the grounded conductor of the same multiwire branch circuit or otherwise identified. Cable assemblies may serve this purpose where all conductors can be seen

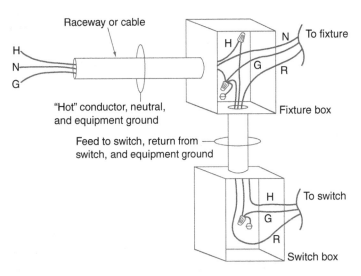

FIGURE 3-2 Conductors of a circuit routed together within the same cable or raceway.

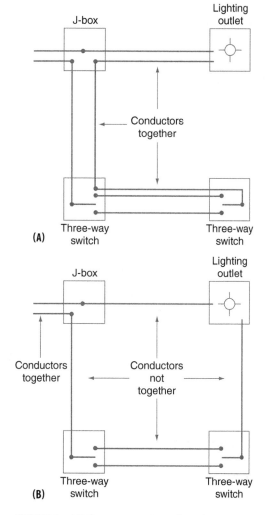

FIGURE 3-3   (A) Correct grouping of conductors.
(B) Incorrect grouping of conductors.

coming from the same cable, but in other cases where the grouping is not obvious, the conductors must be grouped with wire ties or similar means at least one point at the point of origin (where the branch-circuit overcurrent device is located). For example, in Figure 3-1, the three conductors shown connected to the circuit breakers and neutral terminal must be grouped as indicated so that all the associated conductors of the multiwire circuit can be readily identified. If these and only these conductors leave the panelboard in a raceway or cable as shown, no additional grouping or identification is required. One alternative allowed in the 2014 *NEC* is to identify a grounded conductor with a tag that indicates the ungrounded conductor(s) it is associated with. Using this approach, the grounded conductors could be tagged with the circuit numbers of their associated ungrounded conductors.

3. Check insulation values where conductors of different systems share common enclosures.

When conductors of more than one system share a common enclosure or raceway, all of the conductors must be insulated for the highest voltage of any circuit. This minimizes the risk of inadvertent cross-connections. In some cases, circuits of different systems are prohibited from occupying a common enclosure. For example, conductors of systems over 1000 volts are generally required to be physically separated from conductors of systems of 1000 volts or less. The most common branch circuit conductors are insulated at 600 volts. These conductors may occupy the same enclosure in a luminaire for discharge lighting of over 600 volts and up to 1000 volts, and only the secondary wiring is required to be insulated for the secondary voltage. Limited-energy circuits, such as Class 2 and 3 **remote-control circuits** and communications circuits, are also required to be physically separated from conductors used for power or lighting. Conductors for doorbells, room thermostats, and telephones are examples of such limited-energy circuits. Some special circuit types, such as intrinsically safe circuits, are required to be separated even from other intrinsically safe circuits. Photovoltaic (PV) source circuits are required to be separated from PV output circuits.

4. Check wiring methods for spacing from edges of framing and for protection from nails and screws.

Screws and nails, especially those used for drywall installation, are significant threats to the integrity of many concealed wiring methods. Therefore, where nails and screws are likely to penetrate, such as through or alongside metal or wood framing members, spacing or protection must be provided. Some wiring methods, such as EMT, intermediate metal conduit (IMC), rigid metal conduit (RMC), and rigid nonmetallic conduit (PVC or RTRC), do not need such protection and are therefore excluded from this requirement.

Cables and some raceways installed parallel to framing members require a setback of 1¼ in. (32 mm) from the edge(s) of framing members where wall finish will be applied (**FIGURE 3-4**). For nonmetallic-sheathed (NM) cable, a steel plate is used to protect cable less than 1¼ in. (32 mm) from the edge of a wood stud (**FIGURE 3-5**). Although the *Code* specifies the thickness of these steel plates, plates of other materials and lesser thickness are permitted provided they have been tested for equivalency and are listed by a product evaluation organization.

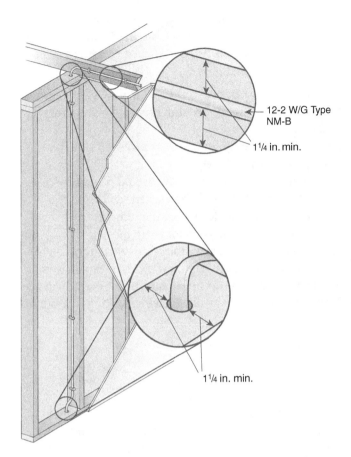

12-2 W/G Type
NM-B

1¼ in. min.

1¼ in. min.

FIGURE 3-4   Minimum clearances for wiring methods installed
through and parallel to framing members.

Courtesy of RACO

FIGURE 3-5   Steel plate used to protect NM cable.

The 2008 edition of the *NEC* introduced a new requirement for maintaining clearances from the underside of
metal-corrugated sheet roofing. The concern in this case is
penetration by screws used to hold down roofing materials. A clearance of at least 1½ in. (38 mm) from the nearest
surface of roof decking is required for all power cables and
raceways except RMC and IMC. In the past, the corruga-
tions have been used as a space for conduit (**FIGURE 3-6A**).
FIGURE 3-6B shows the actual damage resulting from a roof
replacement where the clearance was not maintained and
screws penetrated the raceways. Boxes, like the one in
Figure 3-6A, are equally susceptible to screw penetration. In the 2011 edition, the *NEC*
expanded the 1½-in. (38-mm) clearance to also apply to boxes. The *NEC* now states that
cables, raceways, and boxes may not be installed in concealed spaces in metal-corrugated
sheet roof decking.

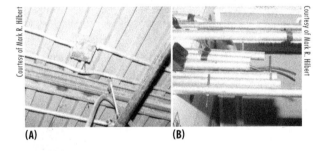

(A)   (B)

Courtesy of Mark R. Hilbert

Courtesy of Mark R. Hilbert

FIGURE 3-6   (A) Wiring installed on the underside of roof
decking. (B) Raceways damaged by screws.

5. Check for insulating bushings or grommets where NM cable is installed through metal studs or where insulated conductors 4 American Wire Gauge (AWG) or larger enter enclosures.

Nonmetallic cable sheaths and conductor insulation must be protected from sharp metal
edges that could otherwise damage the cable. If cables are not protected, the conductors
may be damaged, and enclosures, raceways, or other metal parts, such as metal studs,
could become energized. Insulating bushings or grommets used for this purpose must

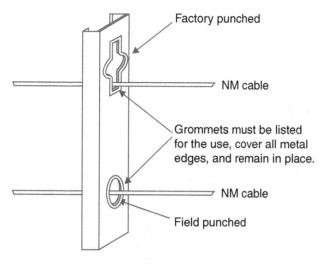

Factory punched

NM cable

Grommets must be listed for the use, cover all metal edges, and remain in place.

NM cable

Field punched

FIGURE 3-7   Listed grommets covering all edges in metal framing.

cover the entire circumference or perimeter of the opening (**FIGURE 3-7**). In addition, bushings used with NM cable in metal studs must be listed and must remain in place after the cable is installed.

6. Check cover, fill, protection, and allowances for ground movement on underground conductors and raceways.

Article 300 provides requirements for cover for various underground wiring methods, conditions, and locations. Backfill must not be of a type that would damage wiring methods. Protection for direct-buried cables and conductors must be provided where they emerge from the ground, and some provision for earth movement must be provided for direct-buried cables. This earth movement refers to things such as freeze/thaw cycles or expansive soils and is not intended to address a catastrophic event such as an earthquake. Direct-buried service conductors generally must have a marking tape buried above them (**FIGURE 3-8**).

7. Verify that electrical raceways and cable trays are used exclusively for electrical conductors.

Electrical raceways and cable trays should not be used to contain or support nonelectrical systems such as pneumatic tubing or water lines. Generally, electrical raceways are not to be used to support other wiring systems either, except for Class 2 control wiring that is directly related to the power wiring inside a raceway. Such wiring may be attached to the associated raceway. Attachment of other systems to wiring methods may both interfere with the dissipation of heat from the wiring method and compromise the integrity of the support method.

8. Check for continuity and completeness in metal raceways and enclosures.

Except where short lengths of raceways are used only for physical protection of other wiring methods, the raceways must be complete between enclosures and fastened to enclosures on both ends. Metal raceways must provide electrical continuity between enclosures and other terminations. This requirement for grounding continuity is reinforced by more specific requirements in Article 250.

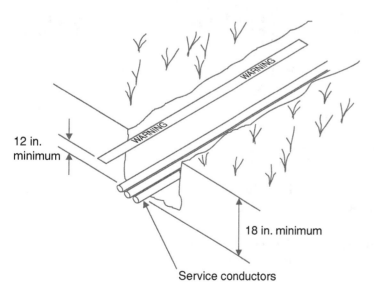

12 in. minimum

WARNING

WARNING

WARNING

18 in. minimum

Service conductors

FIGURE 3-8   Warning ribbon for direct-buried service conductors.

9. Verify that wiring methods are securely fastened in place, supported independently of suspended ceilings, and not used as supports.

As was mentioned in Item 7 of this checklist, except in special cases, raceways may not be used to support other systems or other raceways. Wiring methods are required to be securely fastened in place at the intervals shown in **TABLE 3-3**. Ceiling grid systems may not be used to support wiring raceways or cables. However, separate support wires installed in addition to those required for the ceiling assembly may be used (**FIGURE 3-9**). Such additional wires must be tagged or identified, and since the 2011 update to the *NEC*, this applies to all separate support wires whether

TABLE 3-3 General Securing and Supporting Spacings for Wiring Methods Installed Above Ground*

Wiring Method	Support Intervals, General Rule	Support Spacing from Terminations
Armored cable (AC)	4½ ft (1.4 m)	12 in. (300 mm)
Metal-clad cable (MC)	6 ft (1.8 m)	12 in. (300 mm) for sizes 10 AWG and smaller
Mineral-insulated cable (MI)	6 ft (1.8 m)	6 ft (1.8 m)
Nonmetallic-sheathed cable (NM, NMC, NMS)	4½ ft (1.4 m)	12 in. (300 mm)
Service-entrance cable (SE)		
(a) For services	30 in. (750 mm)	12 in. (300 mm)
(b) For feeders and branch circuits	4½ ft (1.4 m)	12 in. (300 mm)
Underground feeder and branch-circuit cable (UF) where installed above ground	4½ ft (1.4 m)	12 in. (300 mm)
Intermediate metal conduit (IMC)	10 ft (3.0 m) longer intervals permitted for straight runs with threaded couplings—intervals based on size	3 ft (900 mm)
Rigid metal conduit (RMC)	10 ft (3.0 m) longer intervals permitted for straight runs with threaded couplings—intervals based on size	3 ft (900 mm)
Rigid polyvinyl chloride (PVC) conduit	Varies by size	3 ft (900 mm)
Reinforced thermosetting resin conduit	Varies by size	3 ft (900 mm)
Flexible metal conduit (FMC)	4½ ft (1.4 m)	12 in. (300 mm)
Liquidtight flexible metal conduit (LFMC)	4½ ft (1.4 m)	12 in. (300 mm)
Liquidtight flexible nonmetallic conduit—Type B (LFNC-B)	3 ft (900 mm)	12 in. (300 mm)
Electrical metallic tubing	10 ft (3.0 m)	3 ft (900 mm)
Electrical nonmetallic tubing	3 ft (900 mm)	3 ft (900 mm)
Metal wireways	5–10 ft (1.5–3.0 m) horizontally 15 ft (4.5 m) vertically	5 ft (1.5 m)
Nonmetallic wireways	3–10 ft (900 mm to 3.0 m) horizontally 4 ft (1.2 m) vertically	3 ft (900 mm)

*Some wiring methods have unique securing and supporting requirements that apply to specific types of installations.

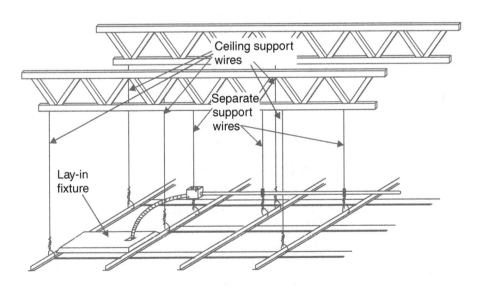

Suspended ceiling

FIGURE 3-9 Separate support wires for wiring methods.

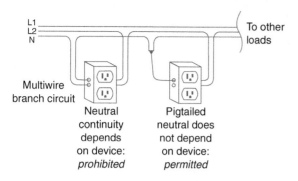

FIGURE 3-10 Examples of correct and incorrect termination of neutral conductor in a multiwire branch circuit.

they are used with fire-rated ceiling assemblies or in non-fire-rated ceilings. Figure 3-9 shows a lay-in-type luminaire connected by a flexible cable or raceway. The cable or raceway must be supported within some maximum distance, usually within 6 ft (1.8 m) of the luminaire (4½ ft [1.4 m] for Type NM cable in dwelling units), but is not required to be supported within the usual distance for terminations shown in Table 3-3. The requirements of 300.11 also apply to communications cables and optical fiber cables, but support spacing for such cables is not specified.

10. Check for continuity of grounded conductors in multiwire branch circuits.

Grounded conductors in multiwire branch circuits can be electrically shared by more than one ungrounded conductor. Continuity of the grounded conductor is especially important in such cases, as a loss of the shared neutral or grounded conductor can cause serious voltage imbalances that may impose excessive voltages on connected equipment, and the open grounded conductor may also present a serious shock hazard. Therefore, grounded (neutral) conductor connections in multiwire branch circuits cannot depend on wiring device terminals such as the feed-through terminals of a duplex receptacle. Instead, grounded conductors must be spliced or "pigtailed" with suitable wire connectors so that continuity is ensured even when devices are removed (**FIGURE 3-10**). Devices may be used for splicing conductors in circuits or portions of circuits where the neutral is not shared by more than one ungrounded (hot) conductor if the device is designed for the purpose. Some wiring devices are clearly marked for this use and may even be marked for more than one wire per terminal.

11. Check for adequate length of free conductors in boxes.

Where conductors are spliced or terminated in a box, at least 6 in. (152 mm) of free conductor must be available in the box for making the splice or termination. The 6 in. (152 mm) is measured from where the conductor emerges from a raceway or cable sheath. However, if the box opening is less than 8 in. (203 mm) in any dimension, the conductor must be long enough for at least 3 in. (76 mm) to extend outside the box opening. **FIGURE 3-11** illustrates these free conductor lengths.

12. Verify that boxes are installed at junction, splice, outlet, switch, and pull points.

Connections between conductors or between conductors and devices or utilization equipment in any wiring method should be **enclosed**. Pulling points in raceway wiring methods should also be enclosed. Exceptions are provided for some limited-energy systems by modification of Article 300 (see Article 725 of the *NEC*, for example) as well as for a number of special situations included in 300.15, such as where direct-buried cables are spliced or where a wiring method is used for physical protection for another wiring method. In some cases, manholes, handholes, integral enclosures, fittings such as conduit bodies, and other enclosures are permitted to be used in lieu of boxes.

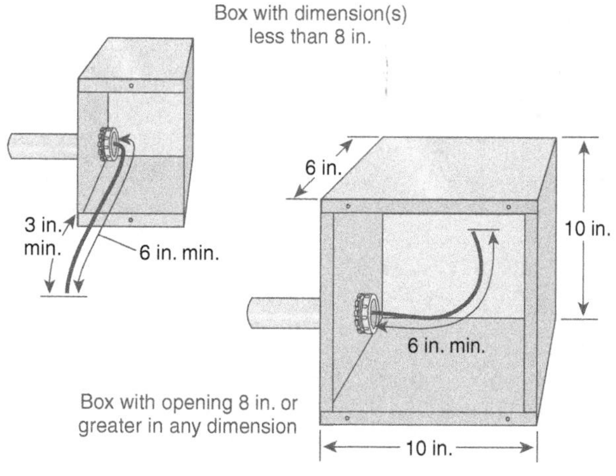

13. Check conductor fill in raceways.

The maximum cross-sectional area that may be occupied by conductors in most raceways is described in Chapter 9 of the *NEC*. The area of a raceway that may be filled by conductors varies according to the number of conductors installed. Greater fill is allowed for short nipples

FIGURE 3-11 Outlet boxes with minimum free conductor lengths.

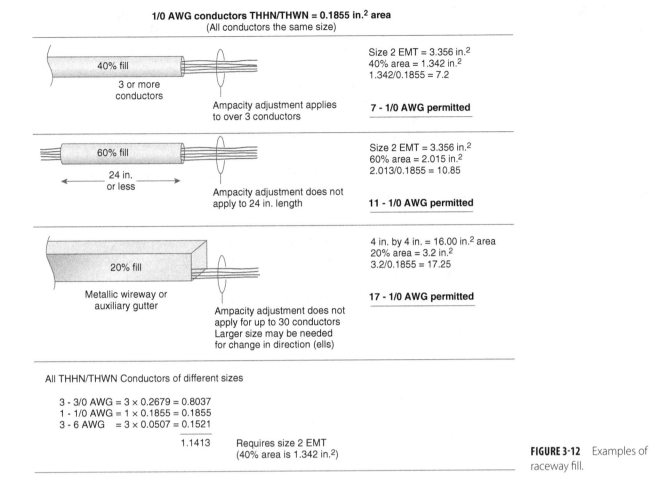

1/0 AWG conductors THHN/THWN = 0.1855 in.² area
(All conductors the same size)

40% fill
3 or more conductors

Ampacity adjustment applies to over 3 conductors

Size 2 EMT = 3.356 in.²
40% area = 1.342 in.²
1.342/0.1855 = 7.2

7 - 1/0 AWG permitted

60% fill
24 in. or less

Ampacity adjustment does not apply to 24 in. length

Size 2 EMT = 3.356 in.²
60% area = 2.015 in.²
2.013/0.1855 = 10.85

11 - 1/0 AWG permitted

20% fill
Metallic wireway or auxiliary gutter

Ampacity adjustment does not apply for up to 30 conductors
Larger size may be needed for change in direction (ells)

4 in. by 4 in. = 16.00 in.² area
20% area = 3.2 in.²
3.2/0.1855 = 17.25

17 - 1/0 AWG permitted

All THHN/THWN Conductors of different sizes

3 - 3/0 AWG = 3 × 0.2679 = 0.8037
1 - 1/0 AWG = 1 × 0.1855 = 0.1855
3 - 6 AWG   = 3 × 0.0507 = 0.1521
                           ‾‾‾‾‾‾
                           1.1413

Requires size 2 EMT
(40% area is 1.342 in.²)

FIGURE 3-12   Examples of raceway fill.

(conduit- or tubing-type raceways not over 24 in. [600 mm] long). The rules are meant to ensure that conductors can be pulled into (or out of) raceways without damage. A certain amount of air space in the raceway also helps the conductors dissipate heat. There are several approaches to calculating or determining raceway fill (**FIGURE 3-12**).

In Figure 3-12, the top three illustrated examples are for conductors that are all the same size. Annex C of the *NEC* could also be used in these cases because the conductors are all the same size. The bottom calculation example shows a raceway fill calculation for conductors of different sizes. Different rules apply depending on the type of raceway used. Although 300.17 states a general requirement and the purpose of the requirement, the actual rules are found in the individual raceway articles as noted in the Informational Note to 300.17. In most cases, the raceway articles refer to Table 1 in Chapter 9 of the *NEC*, but for some raceways, such as wireways, the rule is stated in the raceway article rather than in Chapter 9 in the *NEC*.

14. Verify that raceway systems are complete prior to installation of conductors.
Conductors should not be pulled into raceways until the raceways are complete between termination and pull points. This rule is meant to help protect conductors from damage. Prewired raceways are permitted in certain cases, such as for fixture "whips" made of flexible conduit or tubing, and for some prewired raceway methods, such as nonmetallic underground conduit with conductors, as covered in Article 354 of the *NEC*. This rule does not apply to short sections of raceway used only as physical protection for other wiring methods.

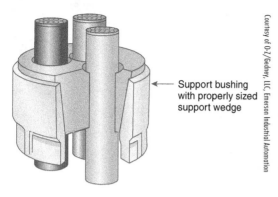

FIGURE 3-13 Example of a bushing used to support conductors in a vertical run.

Courtesy of O-Z/Gedney, LLC, Emerson Industrial Automation

15. Check vertical raceways for adequate conductor supports.
Conductors in long vertical runs must be provided with supports to avoid excessive strain on the conductor and insulation. The heavier the conductors, the more frequently the conductors must be supported. Support may be provided by special clamps, cleats, or fittings (**FIGURE 3-13**) or by deflecting the conductors in pull boxes. Vertical distances and the need for special supports can be reduced by deflecting the raceway to horizontal for a short distance on intervals chosen so that all vertical sections are shorter than the support distance.

In some cases, such as for some fire-rated cables and conductors, the support spacings required by the listing and labeling information may require more frequent vertical supports. As in all other cases where the listing and labeling are more detailed or restrictive than the *NEC*, the listing and labeling instructions must be followed. The support spacings cannot be greater than permitted by 300.19.

16. Verify that fire ratings have been restored at electrical penetrations.
The requirements for fire ratings are covered by building codes. The *NEC* requires only that the spread of fire or smoke is not substantially increased by the installation of wiring methods. Where wiring methods penetrate fire-rated assemblies, either passing through or just into the assembly, the penetrations must be firestopped by using approved methods. In some cases, grout or concrete may be used to restore the fire rating at a penetration made with a steel raceway. In other cases, a specific product or product system must be used. Most of the systems are made of intumescent materials (materials that expand with heat). The product systems may be available in various forms of fittings, including caulk, putty, bags, sheets, or combinations of materials. In any case, the manufacturer's instructions must be carefully followed.

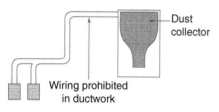

FIGURE 3-14 Ducts for dust, loose stock, or vapor removal.

17. Check installations of wiring in ducts, plenums, and other air-handling spaces for proper methods and materials.
Wiring methods are restricted in air-handling spaces. **FIGURE 3-14** illustrates ducts that carry dust, loose stock, or flammable vapors or that are used with commercial cooking hoods (the restrictions may apply with nonflammable vapors as well). These ducts may not contain any wiring.

FIGURE 3-15 illustrates ducts or plenums specifically fabricated for environmental air. These ducts may contain wiring only as needed for the function of the duct system (for acting on the contained air), and wiring methods are restricted.

FIGURE 3-16 illustrates "other spaces used for environmental air (plenums)," such as joist spaces used for return air in dwellings and areas above suspended ceilings or below raised floors in other occupancies. These spaces may contain wiring that is enclosed in metal raceways or metal-clad cables without nonmetallic coverings or jackets.

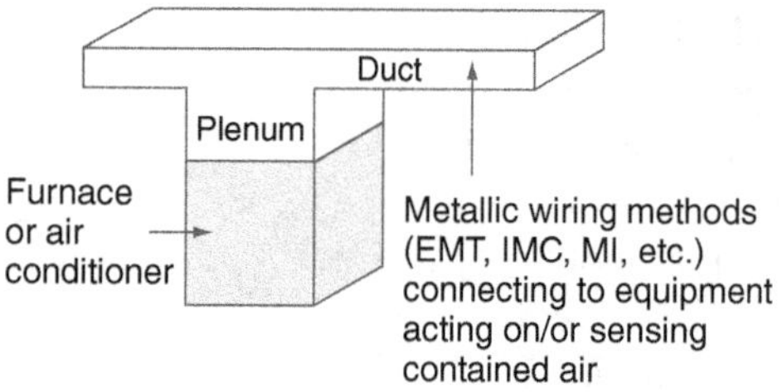

The *Code* specifies wiring methods for the ducts and environmental spaces illustrated. Determining the proper wiring method, and whether the wiring is permitted at all, depends on the function of the duct or space. Multiconductor cables specifically listed for use in an air-handling space may be used. Nonmetallic support means such as cable ties must also be listed as having low smoke and heat release properties. Articles 725, 760, and 800 of the *NEC* include examples of some cables that are "plenum-rated," but that are generally permitted only in "other spaces for environmental air" unless they are in the duct or plenum to connect to equipment that acts directly on the contained air.

FIGURE 3-15 Ducts or plenums for environmental air.

Although certain sections of Articles 725, 760, and 800 may seem to permit wiring to be installed in "ducts and plenums," 725.3(C), 760.3(B), and 800.3(B) and (C) require compliance with 300.22; thus, the so-called plenum cables are still generally restricted to "other spaces for environmental air" plenums and are generally not allowed in plenums specifically fabricated for environmental air. Since the 2011 edition, the *NEC* recognizes that both types of spaces may be called "plenums." Type NM cable is permitted to pass through joist spaces in a dwelling if the spaces are used for return air. The cable can pass through the space only perpendicular to the long dimension of the space.

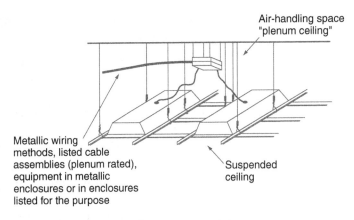

FIGURE 3-16   Other spaces for environmental air.

18. Verify that access to equipment behind removable panels is not compromised by cables, raceways, or equipment.

Wiring methods installed behind removable panels, such as above suspended ceiling grids, must not interfere with access to the space behind the panels. **FIGURE 3-17A** shows a prohibited installation of cables that would interfere with such access. Cables, raceways, and equipment must be separately supported and must not be laid on removable panels (**FIGURE 3-17B**). Some types of equipment, such as light fixtures and smoke detectors, are supported from and attached to ceiling grids and panels. This will make some individual panels or tiles not removable, which should not be considered a violation of the *NEC*.

Some of the articles in *NEC* Chapter 7 and all articles in *NEC* Chapter 8 are generally exempt from the requirements of Article 300, but each of these articles has a similar requirement that cables may not be installed in such a way that access is denied to equipment above removable panels. These articles also require that installations be done in "a neat and workmanlike manner" and provide lists of those rules of Article 300 or Chapter 3 that do apply. Optical fiber cables and communications cables are required to be secured and supported in accordance with 300.11.

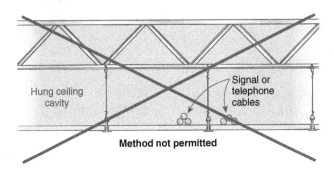

FIGURE 3-17A   Wiring methods incorrectly installed above suspended ceiling grids.

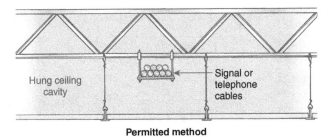

FIGURE 3-17B   Wiring methods correctly installed above suspended ceiling grids.

Checklist 3-2: Boxes and Conduit Bodies

1. Identify wet and damp locations and the suitability of boxes and fittings.

The definitions of <u>wet locations</u> and <u>damp locations</u> are in Article 100 of the *NEC*. Boxes and fittings used in a wet location must be listed for use in wet locations. Boxes and fittings installed in a damp or wet location must be installed so that water cannot accumulate within the enclosure. Some openings may be permitted for drainage. Note that while the interior of a raceway is considered a wet location if the raceway is in a wet location [see Sections 300.5(B) and 300.9], the *NEC* does not say this about enclosures generally unless the enclosure is installed underground. A properly selected enclosure may be installed in a wet location aboveground and still be considered dry (or damp) on the inside, but selection of the box or other enclosure must be compatible with the exterior environment.

2. Check boxes and conduit bodies for adequate space for conductors.

Article 314 of the *NEC* includes the requirements for conductor space in boxes. Primarily, this article requires the installation to "provide free space for all enclosed conductors." The rule lists conductors and various other items that must be counted as conductors.

TABLE 3-4 provides two important pieces of information relative to metal outlet, junction, and device boxes. Based on the standard size and type of box, the table specifies both:

1. The minimum volume for the box
2. The maximum number of conductors (of the same size) that are permitted in the box

It is important to remember that where wiring devices, fixture support hardware, ground wires, or cable clamps are also contained in the box, such items are counted as conductors because they take up space, and the allowable number of actual conductors will be reduced. Box volume may be increased through the use of plaster rings, domed covers, or extension rings if such fittings are marked with their volume.

| TABLE 3-4 | Volume of Standard Metal Boxes |

Box Trade Size			Minimum Volume		Maximum Number of Conductors* (arranged by AWG size)						
mm	in.		cm³	in.³	18	16	14	12	10	8	6
100 × 32	(4 × 1¼)	round/octagonal	205	12.5	8	7	6	5	5	5	2
100 × 38	(4 × 1½)	round/octagonal	254	15.5	10	8	7	6	6	5	3
100 × 54	(4 × 2⅛)	round/octagonal	353	21.5	14	12	10	9	8	7	4
100 × 32	(4 × 1¼)	square	295	18.0	12	10	9	8	7	6	3
100 × 38	(4 × 1½)	square	344	21.0	14	12	10	9	8	7	4
100 × 54	(4 × 2⅛)	square	497	30.3	20	17	15	13	12	10	6
120 × 32	(4¹¹⁄₁₆ × 1¼)	square	418	25.5	17	14	12	11	10	8	5
120 × 38	(4¹¹⁄₁₆ × 1½)	square	484	29.5	19	16	14	13	11	9	5
120 × 54	(4¹¹⁄₁₆ × 2⅛)	square	689	42.0	28	24	21	18	16	14	8
75 × 50 × 38	(3 × 2 × 1½)	device	123	7.5	5	4	3	3	3	2	1
75 × 50 × 50	(3 × 2 × 2)	device	164	10.0	6	5	5	4	4	3	2
75 × 50 × 57	(3 × 2 × 2¼)	device	172	10.5	7	6	5	4	4	3	2
75 × 50 × 65	(3 × 2 × 2½)	device	205	12.5	8	7	6	5	5	4	2
75 × 50 × 70	(3 × 2 × 2¾)	device	230	14.0	9	8	7	6	5	4	2
75 × 50 × 90	(3 × 2 × 3½)	device	295	18.0	12	10	9	8	7	6	3
100 × 54 × 38	(4 × 2⅛ × 1½)	device	169	10.3	6	5	5	4	4	3	2
100 × 54 × 48	(4 × 2⅛ × 1⅞)	device	213	13.0	8	7	6	5	5	4	2
100 × 54 × 54	(4 × 2⅛ × 2⅛)	device	238	14.5	9	8	7	6	5	4	2
95 × 50 × 65	(3¾ × 2 × 2½)	masonry box/gang	230	14.0	9	8	7	6	5	4	2
95 × 50 × 90	(3¾ × 2 × 3½)	masonry box/gang	344	21.0	14	12	10	9	8	7	4
min. 44.5 depth	FS—single cover/gang (1¾)		221	13.5	9	7	6	6	5	4	2
min. 60.3 depth	FD—single cover/gang (2⅜)		295	18.0	12	10	9	8	7	6	3
min. 44.5 depth	FS—single cover/gang (1¾)		295	18.0	12	10	9	8	7	6	3
min. 60.3 depth	FD—single cover/gang (2⅜)		395	24.0	16	13	12	10	9	8	4

*Where no volume allowances are required by 314.16(B)(2) through (B)(5).

Source: *National Electrical Code*®, NFPA, Quincy, MA, 2014, Table 314.16(A).

TABLE 3-5	Volume Allowance Required per Conductor	
	Free Space Within Box for Each Conductor	
Size of Conductor (AWG)	cm³	in.³
18	24.6	1.50
16	28.7	1.75
14	32.8	2.00
12	36.9	2.25
10	41.0	2.50
8	49.2	3.00
6	81.9	5.00

Source: NFPA 70, *National Electrical Code*®, NFPA, Quincy, MA, 2014, Table 314.16(B).

To find the minimum size required in a box or **conduit body**, add the units assigned to the applicable items and multiply the sum by the appropriate volume required per conductor. The volume allowance per conductor in cubic inches (or cubic centimeters) is shown in **TABLE 3-5**. To verify compliance, compare the resulting total volume with the volume of any specific box. These rules apply to boxes containing conductors sized up to 6 AWG.

FIGURE 3-18 shows an example of how to determine the minimum size of a box at a lighting outlet. **FIGURE 3-19** shows an example of how to determine the minimum size of a device box. **FIGURE 3-20** is an example of a calculation to determine the minimum volume for a "ganged" box that contains a mixture of conductor sizes plus two wiring devices (**TABLE 3-6**).

3. Verify that boxes and conduit bodies for conductors 4 AWG and larger are adequately sized.

For conductors sized 4 AWG or larger, the minimum size of boxes and conduit bodies is based on the way the conductors enter and leave the enclosures and the size of the raceways containing the conductors. Cables must be converted to raceway sizes by calculating the appropriate raceway for the conductors contained in the cable. Where conductors are pulled straight through the box or conduit body, the minimum length is greater than it is where angle pulls or U pulls are made. However, the dimensions of a box used for angle pulls or U pulls is required to be increased for additional conduit entries in the same row and wall of the box. Such boxes must also be large enough to provide the minimum required distance between raceways enclosing the same conductors. Sizing boxes that contain spliced conductors is performed the same way as sizing boxes where angle pulls or U pulls are made. **FIGURE 3-21** shows an example of two boxes, one used for angle pulls or splices and the other for straight pulls.

4. Verify that raceways and cables are secured to boxes.
This rule reiterates the requirement that wiring methods must be secured to boxes and other enclosures. Any openings around cable or raceway entries also must be closed. The main difference between this requirement and the similar requirements in Article 300 of the *NEC* is that this

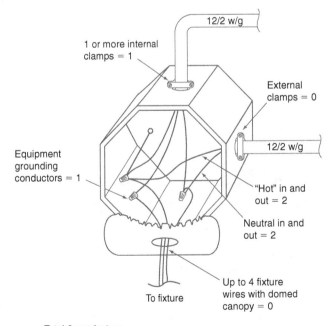

Total 6 conductors
12 AWG = 2.25 cu. in. conductor
6 × 2.25 = 13.5 cu. in. minimum

FIGURE 3-18   Determining the minimum size of a lighting outlet box.

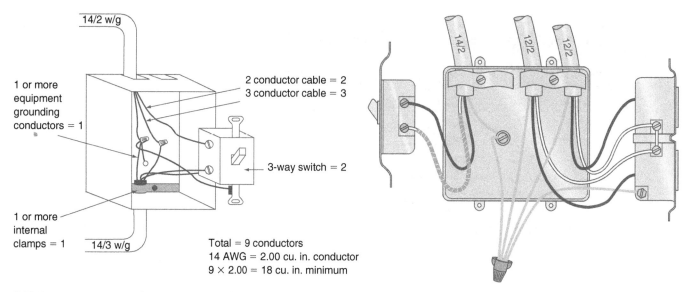

FIGURE 3-19   Determining the minimum size of a device box.

FIGURE 3-20   Example calculation: Two device boxes joined together. (See Table 3-6.)

TABLE 3-6	Tabulation of Conductors and Volume Requirements		
Items Contained Within Box	**Volume Allowance**	**Unit Volume (in³) Based on NEC Table 314.16(B)**	**Total Box Fill (in³)**
Six conductors	2 for 14 AWG conductors	2.00	4.00
	4 for 12 AWG conductors	2.25	9.00
Two clamps	1 (based on 12 AWG conductors)	2.25	2.25
Two devices	2 (based on 14 AWG conductors)	2.00	4.00
	2 (based on 12 AWG conductors)	2.25	4.50
Equipment grounding conductors (all)	1 (based on 12 AWG conductors)	2.25	2.25
Total			26.00 in³

Source: *National Electrical Code®* Handbook, NFPA, Quincy, MA, 2011. Table 314.3, as modified.

requirement is stated from the standpoint of the boxes or conduit bodies and is concerned with the mechanical integrity of an installation and the completeness of an enclosure.

Article 300 is also concerned with electrical continuity. This requirement for boxes provides information on how to deal with some specific cases, such as NM cable used with nonmetallic boxes where there may be no practical method of attaching the cable to the box, so support of the cable near the box is specified.

5. Check for closure of unused openings other than those permitted as part of the design of listed equipment. Unused openings must be closed to complete the box/raceway/cable enclosure for conductors. All boxes must be properly covered. Openings need to be closed to help contain any arcing or sparking that could occur if a fault occurs within a box or conduit body and also to help

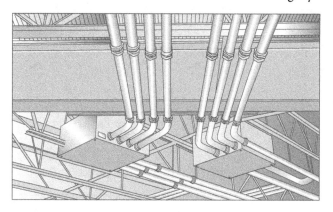

FIGURE 3-21   Boxes used for straight and angle pulls.

prevent contact with the enclosed parts. This requirement applies specifically to openings for conductors and is in addition to the broader requirement for sealing other unused openings that was discussed in the General Requirements Inspections chapter of this manual. All boxes must be accessible.

6. Verify that boxes in walls and ceilings are flush with the finished surface or, if surfaces are noncombustible, within ¼ in. (6 mm) of the finished surface.

Installing boxes precisely flush with finished surfaces is often difficult since the finished surface is not usually there when the box is installed. The *NEC* allows up to ¼ in. (6 mm) of tolerance if the surrounding surface is noncombustible (**FIGURE 3-22**). Because this tolerance effectively creates an extension of the box when a flush-type cover is used, the tolerance is permitted only with noncombustible finishes such as concrete, masonry, or gypsum board. For the same reason, no tolerance is allowed for combustible finishes, and the box must be flush or extend from the surface of the wall (**FIGURE 3-23**). Listed box extenders may be used to effectively bring a recessed box flush with the surface (**FIGURE 3-24**).

7. Check for excessive gaps between edges of boxes and plaster, plasterboard, or drywall surfaces.

This requirement is based on considerations similar to those for Item 6. No gaps of over ⅛ in. (3 mm) are allowed around the edges of boxes and fittings where flush-type covers are used. Flush-type covers (discussed in the next section) are designed to contact the surrounding surface rather than making direct contact with the face of the box. Surface-type covers that contact the facing edges of a box will complete the electrical enclosure independently of the surrounding finish.

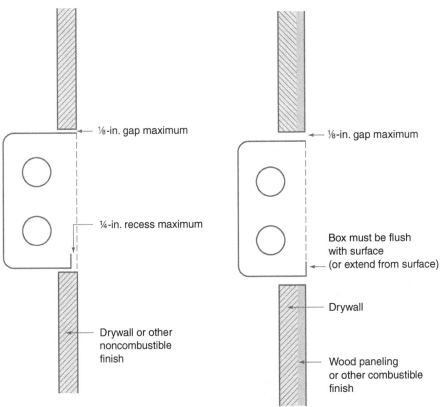

FIGURE 3-22 Box installed with maximum ¼-in. (6-mm) clearance in noncombustible surface.

FIGURE 3-23 Box installed with maximum ⅛-in. (3-mm) gap around edges.

FIGURE 3-24 Listed box extender.

8. Verify that boxes are securely fastened and supported.

The objective of this requirement is to ensure the structural and mechanical integrity of the electrical installation. Generally, boxes are required to be securely and rigidly attached to the surface or structure. In some cases, such as where support is provided by wires or pendants or where boxes or conduit bodies are supported by raceways, the support need not always be rigid but must still be secure. Support methods must not damage boxes or add rough or sharp edges to the interior of boxes and must be consistent with manufacturers' intended methods.

9. Check boxes for adequate depth for the equipment that will be installed within them.

Outlet and device boxes must have enough depth to accommodate any devices or equipment installed and still leave enough depth for conductors that may be between the back of the device or other equipment and the back of the box. Insufficient depth could result in damage to conductors. These rules were expanded and detailed in 2008 and revised again in 2011 and are in addition to the volume requirements of 314.16 discussed in Item 2 in this section. The details are summarized in **TABLE 3-7**.

10. Check for covers or canopies on boxes.

This rule is another requirement that is aimed at ensuring that box enclosures are completed.

11. Check lighting fixture outlet boxes for suitability.

Boxes that are intended for use as luminaire (light fixture) support are required to be marked with the weight to be supported if it is more than 50 lb (23 kg). Boxes that are intended to support fixtures from horizontal surfaces are usually equipped with at least 8–32 screws, in contrast to device boxes, which generally have provisions for 6–32 screws for mounting receptacles and snap switches. However, device boxes may be used to support small luminaires on vertical surfaces or other lightweight utilization equipment without any specific marking. The weight of the equipment is limited to 6 lb (3 kg) if device boxes are used.

12. Check floor boxes and receptacle/cover assemblies for listing.

Boxes that are used for receptacles in floors are required to be listed as *floor boxes*. Show windows and similar locations may be exempted by the authority having jurisdiction, but receptacles and covers must be listed as an assembly for the type of location.

13. Check listing and installation of boxes used for support of ceiling-suspended (paddle) fans.

Ordinary fixture outlet boxes may be used at a paddle fan if the outlet box is supplemented by additional support(s) from the structure. Boxes that are listed specifically for support of

TABLE 3-7   Minimum Depth of Boxes for Outlets, Devices, and Utilization Equipment

Type or Contents of Box	Minimum Depth
Outlet boxes without enclosed devices or utilization equipment	½ in. (12.7 mm)
Large devices or utilization equipment projecting into box over 1⅞ in. (48 mm)	Depth of equipment plus ¼ in. (6 mm)
Utilization equipment or devices supplied by conductors larger than 4 AWG	Box must be identified for the specific function unless dimensions from 312.6 are provided
Utilization equipment or devices supplied by 8, 6, or 4 AWG conductors	2¹/₁₆ in. (52.4 mm)
Utilization equipment or devices supplied by 12 or 10 AWG conductors	1³/₁₆ in. (30.2 mm)
Utilization equipment or devices supplied by 12 or 10 AWG conductors where equipment projects into box over 1 in. (25 mm)	Depth of equipment plus ¼ in. (6 mm)
Utilization equipment or devices supplied by 14 AWG or smaller conductors	¹⁵/₁₆ in. (23.8 mm)

such fans may be used as the sole support. The maximum weight that may be supported by a fan support box is 70 lb (32 kg). Boxes must be marked with the weight they are intended to support if they are used to support weights in excess of 35 lb (16 kg). Thus, a box that is listed for fan support and has no weight marking is limited to 35 lb (16 kg).

A box that is specifically listed for fan support is required at ceiling locations suitable for fan installation if spare, separately switched, ungrounded conductors are provided at the box. This is not the case in every location where a fan could be installed. The rule applies only where special wiring provisions for a fan have been made. Many ceiling fans today do not need special wiring for separate control of the fan and light. Whether or not wiring provisions are made for future fans, if a fan is hung in any location, it must comply with the requirements for separate support or use listed and marked fan support boxes.

14. Verify that all boxes are accessible.
The wiring inside boxes and conduit bodies must be accessible without having to remove any part of a building or structure. Applying the definitions of Article 100 of the *NEC*, wiring in boxes must be **accessible (as applied to wiring methods)** and may not be concealed. Boxes above removable ceiling panels, in attics and crawl spaces, and behind access doors are generally considered to be **accessible** if access is not blocked by other equipment. No specific dimensioned working space is required. Whatever space is necessary for access and any likely servicing must be provided. Wiring in boxes that are covered up by mechanical equipment, luminaires, or the like are often rendered inaccessible. Some underground installations can have boxes covered by light materials such as gravel, but the locations must be marked for excavation.

15. Verify that support means for nonmetallic boxes are outside the box or otherwise isolated from contact with conductors.
Some metal outlet boxes are provided with means of supporting the box by screws or nails that are inside or pass through the box and are effectively connected to the metal box (**FIGURE 3-25**). (Where screws are used as shown at the top of the box in Figure 3-25, the screw threads must be installed or protected so as to prevent conductors from contacting the sharp edges of the threads.) Similar nails or screws in nonmetallic boxes would be effectively isolated but subject to contact with conductors and therefore "likely to become energized." Nonmetallic boxes should not contain such isolated metal parts unless the box is constructed so that contact between conductors and screws or nails is prevented. Most nonmetallic boxes have their mounting means on the outside of the box.

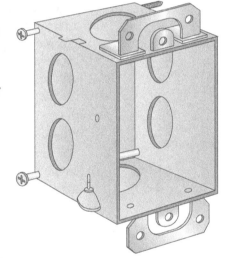

FIGURE 3-25 Support screw or nail passing through interior of device box.

Checklist 3-3: Cabinets and Cutout Boxes

1. Verify that cabinets or cutout boxes are suitable and properly installed in any wet or damp locations.
The definitions of *wet and damp locations* appear in Article 100 of the *NEC* and are also included in this chapter. Cabinets and **cutout boxes** must be identified for use in wet locations if they are to be used in such locations. The enclosure must be arranged so that water will not accumulate inside where installed in a wet or damp location. In addition, many items of equipment have Type ratings that define the specific conditions for which they are suitable. For example, both Type 4 and Type 12 are suitable for some wet locations, but Type 4 is also suitable for hosedown and outdoor locations. A partial description of these ratings can be found in Article 110. These Type ratings are discussed in the General Requirements Inspections chapter of this manual. Common examples of cabinets are the enclosures of panelboards. According to the definition of *panelboard* in Article 100, a panelboard is an assembly that is designed to be installed in a cabinet or cutout box.

Where conduits or cables enter an enclosure in a wet location, the fittings used must be suitable for wet locations unless the entry is below the level of uninsulated live parts inside the enclosure.

2. Verify that cabinets in walls do not have excessive gaps at the edges and are flush with the finished surface or, if surfaces are noncombustible, within ¼ in. (6 mm) of the finished surface.

This rule is the same as the requirement for boxes, which is explained in Item 6 of the Boxes and Conduit Bodies checklist. Similarly, the gaps around a flush-type enclosure in noncombustible surfaces may not exceed ⅛ in. (3 mm), which is the same as the requirement discussed in Item 7 of the Boxes and Conduit Bodies checklist.

3. Check for closure of conductor openings.

This rule is the same as the requirement for boxes, which is explained in Item 5 of the Boxes and Conduit Bodies checklist.

4. Verify that cables are secured to cabinets and cutout boxes or that the conditions for cables with nonmetallic sheaths are met.

This rule is similar to the requirements found in Article 300 and Article 314 and is explained in Item 4 of the Boxes and Conduit Bodies checklist. An exception is provided for cables with nonmetallic sheaths that enter a cabinet through a raceway. The raceway is usually used for physical protection or cable management, and the exception is most commonly applied in dwelling units that are being rewired, since the conditions in the exception require the cabinet and raceway to be surface mounted and, for the most part, exposed. Although in this application the raceway is not a complete system and is often used in short lengths for physical protection, the maximum raceway fill requirements in the General Wiring Methods checklist in this chapter do apply.

5. Check wiring and bending space in cabinets and cutout boxes.

Like boxes and conduit bodies, a certain amount of space is required for conductors in cabinets and cutout boxes. However, conductors are usually terminated in cabinets and cutout boxes, so the space requirements are intended to provide sufficient room for bending and terminating conductors without damaging the conductors, insulation, or terminals. Therefore, the rules in Article 312 of the *NEC* are based on the size, number, and material of conductors connecting to a terminal as well as the way the conductors enter the enclosure in relation to a terminal. With regard to material, compact-stranded AA-8000 aluminum alloy conductors require the same space as copper conductors of about the same ampacity rather than the same size. With regard to the way the conductors enter the enclosure, more space is required for a conductor that enters the enclosure through a wall opposite the terminal than is required if the conductor enters through some other wall.

Reduced bending spaces are permitted in meter enclosures or where removable or lay-in terminals are provided. **FIGURE 3-26** illustrates the two conditions where the wire bending space can be reduced for 350-kcmil and smaller conductors installed in meter enclosures with lay-in terminals. The terminals on the left have an offset not greater than 50 percent of the bending space within the enclosure. The terminals on the right are within a 45-degree angle of the conductor entrance into the enclosure. The bending space requirements in Article 312 also apply to wireways and auxiliary gutters (**FIGURE 3-27**) and to switchboards and panelboards. (According to Article 100, *panelboards* are assemblies of buses, overcurrent devices, and switches that are designed to be placed in a cabinet or cutout box.)

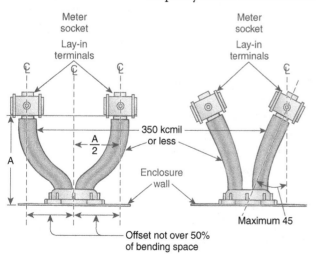

FIGURE 3-26 Application of wire-bending requirements in a meter socket.

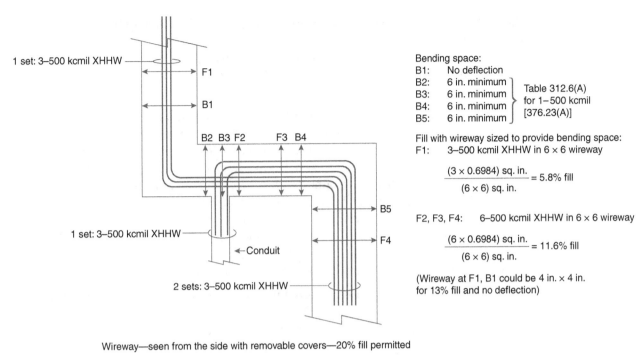

Bending space:
B1:   No deflection
B2:   6 in. minimum ⎫
B3:   6 in. minimum ⎪   Table 312.6(A)
B4:   6 in. minimum ⎬   for 1–500 kcmil
B5:   6 in. minimum ⎭   [376.23(A)]

Fill with wireway sized to provide bending space:
F1:   3–500 kcmil XHHW in 6 × 6 wireway

$$\frac{(3 \times 0.6984) \text{ sq. in.}}{(6 \times 6) \text{ sq. in.}} = 5.8\% \text{ fill}$$

F2, F3, F4:   6–500 kcmil XHHW in 6 × 6 wireway

$$\frac{(6 \times 0.6984) \text{ sq. in.}}{(6 \times 6) \text{ sq. in.}} = 11.6\% \text{ fill}$$

(Wireway at F1, B1 could be 4 in. × 4 in. for 13% fill and no deflection)

Wireway—seen from the side with removable covers—20% fill permitted

FIGURE 3-27   Example of minimum wire-bending space in wireways and auxiliary gutters.

6. Check cabinets and cutout boxes for adequate space for conductors and for splices and taps where they exist.

In addition to the bending space requirements discussed, adequate space for conductors must be provided where the conductors are routed through the cabinet or cutout box. The space requirements are essentially the same as for raceway fill (40 percent), except that splices may fill an area to 75 percent of the wiring space, as in wireways. Generally, cabinets and cutout boxes including panelboards are not intended to be used for through-wiring or for splices, but they may be used for such purposes if the wiring space is adequate. However, where wiring feeds through a cabinet from another source, the enclosure must be marked to identify the closest disconnecting means for the feed-through conductors.

Checklist 3-4: Switches and Receptacles

1. Verify that all switching is done in the ungrounded conductors.
If switching to control electrical equipment is done in all of the ungrounded ("hot") conductors of a circuit, there will be no energized conductors at the equipment when the switch is in the off or open position. Although switching the grounded conductor may serve to control the equipment by causing it to go on or off (**FIGURE 3-28**), switching done in the grounded conductor is prohibited because it creates two hazards:

- The equipment may be energized by an unswitched ungrounded conductor and present a shock hazard even while off.
- Accidental damage to the grounded conductor may cause the equipment to operate with the switch still in the off position, and the circuit overcurrent device will not respond to a damaged neutral or grounded conductor.

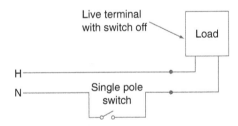

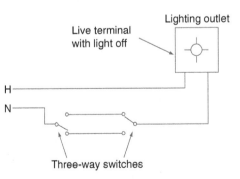

Switching in grounded conductor is *prohibited!*

FIGURE 3-28   Example of problems associated with switching in a grounded (neutral) conductor.

2. Verify that any switches in wet locations are properly installed in weatherproof enclosures.

Like wiring methods, boxes, and enclosures of various types, switches must be suitable for the conditions. A number of types of <u>weatherproof</u> or weather-resistant switch covers and enclosures are available.

3. Verify that switches are located not over 6 ft 7 in. (2.0 m) high and that they can be operated from readily accessible places unless otherwise permitted.

Switches of all kinds are supposed to be mounted in such a way that they can be operated easily. This rule covers not only snap switches like those commonly used for lighting controls but also safety switches, disconnects, and circuit breakers. In the case of overcurrent devices, 240.24(A) also applies, but the rules are identical. The measurement is made from the operating platform to the center of the operating "grip." Where switches are used as disconnects and are mounted adjacent to equipment that is itself elevated above 6 ft 7 in. (2.0 m) high, the switches are permitted to be located more than 6 ft 7 in. (2.0 m) above the floor or working platform and portable means may be used for access.

4. Verify that the voltage between adjacent grouped or ganged devices is not over 300 volts or that barriers are installed.

This rule is most commonly an issue with 277-volt lighting circuits. Where two or more ungrounded conductors of a 480Y/277-volt system are used for lighting and more than one 277-volt lighting circuit is controlled from a single location, the voltage between adjacent switches is 480 volts where the switches are fed from different phases. In such cases, the switches must be separated by barriers that are identified and securely installed or must be placed in more than one box. **FIGURE 3-29** shows two switches that are connected to different phases of a 480Y/277-volt system without a barrier in between the switches. Such a system is prohibited. **FIGURE 3-30** shows a 208Y/120-volt system that does not require a barrier because the voltage between snap switches does not exceed 300 volts. Boxes are available that can be equipped with the permanent barriers required by this rule. Otherwise, separate boxes should be used. This requirement has been expanded to include all wiring devices installed with switches. For instance, where the voltage between a switch and an adjacent receptacle exceeds 300 volts, an identified barrier is required to be securely installed between the two devices.

One solution sometimes used to resolve the problem of an excessive voltage between switches is to switch more than one circuit with the same switch. This is permitted only if the switch is listed and marked as a two-circuit or three-circuit switch.

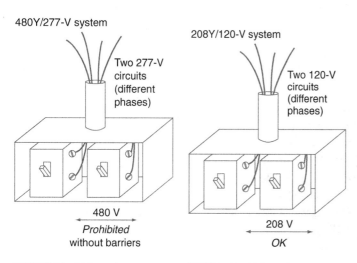

FIGURE 3-29 Voltage between ganged switches more than 300 volts; prohibited without barrier.

FIGURE 3-30 Voltage between ganged switches less than 300 volts; barrier not required.

5. Verify that metal switch boxes, switches, and any metal faceplates are grounded.

Metal switch boxes are required to be connected to an equipment grounding conductor of the types specified in Section 250.118 like most other metal enclosures containing conductors. Metal faceplates also must be grounded but are usually grounded by a mounting connection to a grounded switch. Switches are required to be grounded whether or not a metal faceplate is used. However, in existing installations where there is no grounding means at the switch, a replacement switch is not required to be grounded if all parts within reach of grade or conducting surfaces are nonmetallic or ground-fault circuit interrupter (GFCI) protection is provided. Other exceptions are provided for certain listed nonmetallic assemblies or switches with integral nonmetallic enclosures.

6. Verify that switches or receptacles in boxes have their plaster ears seated against the wall surface or the box.

Switches and receptacles, both also called *wiring devices*, must be firmly seated against a box or wall surface so that they are not free to move when used. Wiring devices that can move in relation to a box can have their terminals shorted to the box, an obvious hazard to a user. **FIGURE 3-31** illustrates a receptacle mounting strap (plaster ears) that is seated securely against the wall.

7. Verify that switches and receptacles are used within their ratings.

Wiring devices have ampere and voltage ratings that must be observed. Receptacles must be used within their voltage ratings and are required to be noninterchangeable, so 125-volt–rated receptacles may be used only on nominal 120-volt circuits, and 250-volt receptacles

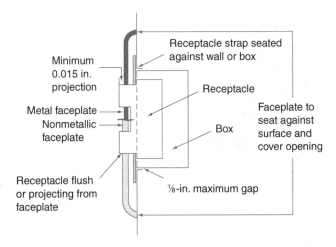

FIGURE 3-31 Receptacle showing proper seating of mounting strap and minimum projection of device from faceplate.

cannot be used on 120-volt circuits. The ratings and configuration of a receptacle should indicate the voltage and current it can supply. Conversely, switches may be applied on circuits rated less than the rating of the switch. So a 30-ampere switch may be used on a 20-ampere circuit, and a 277-volt switch may be used on 120-volt circuit. The rating of the switch must be compatible with the load supplied. Such compatibility is determined by the type and rating of the switch and the type of load supplied.

8. Verify that general-use dimmers are installed only for control of permanently installed incandescent lighting.

General-use dimmer switches are designed and evaluated for use with permanently installed incandescent (resistive) luminaires only. They may be used for controlling other loads only if they are listed and installed in compliance with the listing. Many compact fluorescent lamps can be installed in place of incandescent lamps, but their markings usually say they are "not dimmable." Installing a general-use dimmer for control of a **receptacle outlet** used as a lighting outlet could result in a misapplication of the device if an inductive load is plugged into the receptacle. Additionally, damage to sensitive equipment could result if a dimmer is used to control a receptacle outlet.

9. Check the listing and marking of any switches or receptacles used with aluminum conductors.

Compatibility with terminals is important for any conductor, whether copper or aluminum. The use of aluminum branch-circuit conductors in new installations is uncommon. However, in existing installations or in larger sizes, aluminum is frequently encountered, and larger lugs are often rated for either copper or aluminum. Many of the smaller terminals on wiring devices are rated only for copper. If aluminum conductors are used with wiring devices, the wiring devices must have terminals that are rated for use with aluminum. This suitability is indicated by the marking *CO/ALR*.

10. Check receptacles in wet or damp locations for proper covers and enclosures and weather-resistant ratings.

Receptacles installed in wet or damp locations are a little different from switches installed in the same locations. Many switch covers for wet locations completely enclose the switch and provide for an outside operator. Receptacles, by their nature, require the user to expose the receptacle in order to insert and remove an **attachment plug**. This results in some exposure of the receptacle to dampness or wetness over a period of time, so common straight blade receptacles (15- and 20-ampere, 125- and 250-volt, American National Standards Institute/National Electrical Manufacturers Association [ANSI/NEMA] configurations 5-15, 5-20, 6-15, and 6-20) must be listed as weather-resistant.

FIGURE 3-32   Red Dot® Code Keeper®
universal while-in-use cover.

The *NEC* specifies two basic types of wet-location receptacle covers: weatherproof receptacle covers that are weatherproof only when the cover is closed, and weatherproof covers that allow closing of the cover while a plug is inserted (**FIGURE 3-32**). All 15- and 20-ampere, 125- and 250-volt receptacles installed in wet locations (outdoors or indoors) must have covers that are weatherproof with or without the plug inserted. For other ratings of receptacles, where a plug is to be left inserted in a receptacle while unattended, the cover must be weatherproof with the plug inserted. Otherwise, if the receptacle is to be used only when attended and is not 15- or 20-amperes, 125- or 250-volts, such as one that is used with large portable electric tools or appliances, the receptacle cover is required to be weatherproof only when not in use.

TABLE 3-8 summarizes the required and permitted uses of the two most commonly used weatherproof receptacle covers. One limited exception to the requirements in Table 3-8 was added in 2008: Receptacles in areas subject to "routine high-pressure spray washing" may use covers that are required to be

TABLE 3-8	Requirements for Receptacle Cover (Enclosure) Types

Damp and Wet Receptacle Locations	Cover that *is not* Weatherproof, with Attachment Plug Cap Inserted into Receptacle	Cover that *is* Weatherproof, with Attachment Plug Cap Inserted into Receptacle ("in-use" type)
Outdoor, damp locations 406.9(A)	Minimum type required*	Permitted
Indoor, damp locations 406.9(A)	Minimum type required*	Permitted
Outdoor, wet locations 406.9(B)(1) and (2)	Required for receptacle types other than those rated 15 and 20 amperes, 125 and 250 volts, where the tool, appliance, or other utilization equipment plugged into the receptacle *is* attended while in use.*	(a) Required for receptacles rated 15 and 20 amperes, 125 and 250 volts (b) Required for receptacles other than those rated 15 and 20 amperes, 125 and 250 volts, where the tool, appliance, or other utilization equipment plugged into the receptacle *is not* attended while in use.
Indoor, wet locations 406.9(B)(2)	Required for receptacle types other than those rated 15 and 20 amperes, 125 and 250 volts, where the tool, appliance, or other utilization equipment plugged into the receptacle *is* attended while in use.*	(a) Required for receptacles rated 15 and 20 amperes, 125 and 250 volts (b) Required for receptacles other than those rated 15 and 20 amperes, 125 and 250 volts, where the tool, appliance, or other utilization equipment plugged into the receptacle *is not* attended while in use.

*"In-use" type covers permitted.

Source: *National Electrical Code® Handbook*, NFPA, Quincy, MA, 2014, Exhibit 406.3.

weatherproof only with the attachment plug removed. In applications other than one- or two-family dwellings, the outlet box "hood" or cover must be listed and identified as "extra duty" if the support is from grade on a post or similar support or if the box is supported by a raceway from grade.

11. Verify that isolated ground (IG) receptacles are properly identified and connected to IG conductors.

IG receptacles are required to be identified by an orange triangle on their face. Normally, isolated grounding conductors are extended back to the grounding point of the derived system or service, often through one or more panelboards and intermediate isolated grounding terminals. However, the *NEC* does not say how far back in the system the isolated grounding terminal must go, as long as it terminates directly on an equipment grounding terminal of the system that supplies the branch circuit to which the IG receptacle is connected. The isolated equipment grounding conductor is not permitted to extend beyond the **building** in which the IG receptacle is located. It is important to remember that an equipment grounding conductor, in addition to the IG, is necessary to ground all other metallic non-current-carrying parts such as boxes and faceplates. **FIGURE 3-33** shows an IG receptacle with an insulated equipment grounding conductor and with the device box grounded through the metal raceway. **FIGURE 3-34** shows an IG receptacle with markings and some construction features.

12. Check that receptacles project from metal faceplates or are flush with nonmetallic faceplates and that the faceplates cover openings.

Receptacles must project from, or be flush with, receptacle faceplates so that the faceplate cannot move or rotate in relation to the receptacle. Since nonmetallic faceplates are thicker than metal faceplates, receptacle faces are allowed to be flush with nonmetallic faceplates but are required to project from metal faceplates. Faceplates also must seat firmly against the box or wall surface and completely cover any opening around the receptacle box to complete the enclosure of the receptacle. This requirement, together with the requirements for receptacle installation covered in Item 6, provide the secure receptacle installation needed to withstand the forces imposed by the insertion and removal of attachment plugs.

13. Check receptacles for proper polarity and for grounding and bonding connections.

As may be expected, the most commonly used receptacles are also the ones most often miswired. Simple and inexpensive testers are available for checking the wiring of the most commonly used receptacles. A receptacle polarity tester indicates correct and incorrect

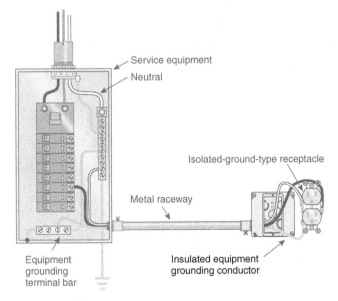

FIGURE 3-33 Equipment grounding connections permitted for installation of IG receptacles.

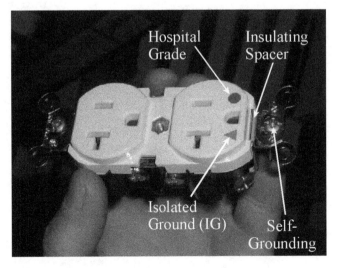

FIGURE 3-34 Isolated grounding receptacle with markings for IG and hospital grade.

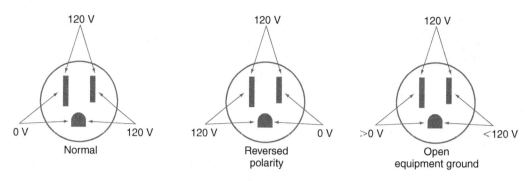

FIGURE 3-35   Various voltage measurements determined through the use of receptacle tester.

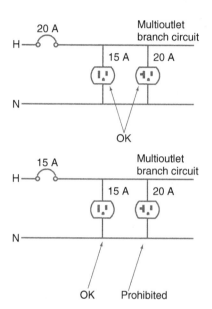

FIGURE 3-36   Rating of receptacles installed in 15- and 20-ampere multioutlet circuits.

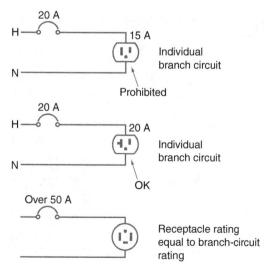

FIGURE 3-37   Rating of receptacles installed in individual branch circuits or rated over 50 amperes.

wiring of a standard 125-volt single-phase, 15- or 20-ampere device. With such simple testers, this check is one of the easiest for inspectors to make, but it must be done after a system has been energized. Voltmeters can be used to check other types of receptacles. Before any test equipment is used, it must be inspected for wear or damage and properly rated for the circuits and environment, and the user must be trained and qualified in its use. These are requirements of the *NEC* and Occupational Safety and Health Administration (OSHA) as well as NFPA 70E®. Checks of other types may be worthwhile, but most inspectors only have time for spot checks of such devices. **FIGURE 3-35** shows expected nominal voltage measurements based on correct and incorrect wiring of a receptacle where there is no load on the branch circuit. GFCI circuit testers are often helpful for this purpose.

14. Verify that receptacle ratings and branch-circuit ratings are compatible. Circuit ratings are determined by the rating of the overcurrent device. Article 210 of the *NEC* restricts the ratings of receptacles that may be installed on circuits of specific ratings. For example, in **multioutlet** circuits, either 15- or 20-ampere receptacles may be installed on 20-ampere circuits, but 20-ampere receptacles may not be installed on 15-ampere circuits (**FIGURE 3-36**). Also, if a circuit has only one outlet and one single receptacle, or if a receptacle is rated over 50 amperes, in general the receptacle rating must not be less than the rating of the circuit (**FIGURE 3-37**). However, receptacles used for welders and some motors may have a rating less than the branch-circuit rating (less than the rating of the overcurrent device protecting the circuit), if the requirements of 210.21 are met. An example of a receptacle rated less than the branch circuit but sized to the conductor rating for a welder is shown in **FIGURE 3-38**. Because circuits rated over 50 amperes can generally be used only as **individual branch circuits**, such circuits may have only one receptacle outlet.

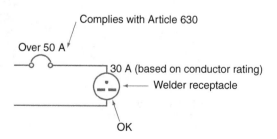

FIGURE 3-38   Receptacle ratings used in welder supply circuits.

KEY TERMS

Article 100 of the *NEC* defines the key terms pertaining to wiring methods as follows:

Accessible (as applied to equipment)   Admitting close approach; not guarded by locked doors, elevation, or other effective means.

Accessible (as applied to wiring methods)   Capable of being removed or exposed without damaging the building structure or finish or not permanently closed in by the structure or finish of the building. [*The terms* concealed *and* exposed *are closely related to this definition of "Accessible."*]

Accessible, Readily (Readily Accessible)   Capable of being reached quickly for operation, renewal, or inspections without requiring those to whom ready access is requisite to use tools, to climb over or remove obstacles, or to resort to portable ladders and so forth. [*Readily accessible* and *accessible* are separate terms. Electrical equipment does not necessarily have to be accessible to be readily accessible. A piece of equipment may be readily accessible and be behind a locked door, for example. Ready access is concerned with access "for those to whom ready access is requisite." Section 110.26(G) states that "Electrical equipment rooms or enclosures housing electrical apparatus that are controlled by lock(s) shall be considered accessible to qualified persons."]

Attachment Plug (Plug Cap) (Plug)   A device that, by insertion in a receptacle, establishes a connection between the conductors of the attached flexible cord and the conductors connected permanently to the receptacle. [**FIGURE 3-39** *illustrates a receptacle, an attachment plug, and a cord connector. The attachment plug or plug cap may be connected to either a receptacle or a cord connector. Cord connectors are also called cord caps.*]

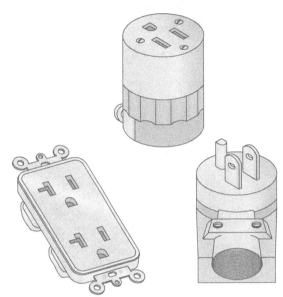

FIGURE 3-39   Example of a receptacle, cord cap, and attachment plug.

Branch Circuit, General-Purpose   A branch circuit that supplies two or more receptacles or outlets for lighting and appliances.

Branch Circuit, Individual   A branch circuit that supplies only one utilization equipment.

Branch Circuit, Multiwire   A branch circuit that consists of two or more ungrounded conductors that have a voltage between them, and a grounded conductor that has equal voltage between it and each ungrounded conductor of the circuit and that is connected to the neutral or grounded conductor of the system.

Building   A structure that stands alone or that is cut off from adjoining structures by fire walls with all openings therein protected by approved fire doors. [*Building codes determine the construction method necessary for a structural assembly to constitute a fire wall.*]

Cabinet   An enclosure that is designed for either surface or flush mounting and provided with a frame, mat, or trim in which a swinging door or doors are or can be hung. [*One of the most commonly encountered types of cabinets is the enclosures of panelboards. According to the definition of a panelboard, the panelboards are the assemblies of buses, switches, and overcurrent devices that are installed in cabinets.*]

Concealed   Rendered inaccessible by the structure or finish of the building.

> *Informational Note:* Wires in concealed raceways are considered concealed, even though they may become accessible by withdrawing them.

Conduit Body   A separate portion of a conduit or tubing system that provides access through a removable cover(s) to the interior of the system at a junction of two or more sections of the system or at a terminal point of the system. Boxes such as FS (shallow)

FIGURE 3-40   Example of a conduit body.

and FD (deep) or larger cast or sheet metal boxes are not classified as conduit bodies. [**FIGURE 3-40** *is a photo of a conduit body in a T configuration. The shape of the body and entries is in the form of a T. Other conduit bodies include LB (L shape with entry in back), LL (L shape with entry on left), LR (entry on right), and X. A conduit body that goes straight through is called a C conduit body.*]

Cutout Box   An enclosure designed for surface mounting that has swinging doors or covers secured directly to and telescoping with the walls of the box proper.

Device   A unit of an electrical system, other than a conductor, that carries or controls electric energy as its principal function. [*Circuit breakers are overcurrent devices, and receptacles and wall switches are wiring devices. Some devices such as switches with lighted handles or control relays may use a small amount of electrical energy, but utilizing energy is not their primary purpose.*]

Enclosed   Surrounded by a case, housing, fence, or wall(s) that prevents persons from accidentally contacting energized parts.

Enclosure   The case or housing of apparatus, or the fence or walls surrounding an installation, to prevent personnel from accidentally contacting energized parts or to protect the equipment from physical damage.

> *Informational Note:* See Table 110.28 for examples of enclosure types.
>
> [*Table 110.28 was Table 430.91 in editions of the NEC prior to 2008 and was actually intended for the selection of motor controller enclosures. Because the table is also useful for selecting enclosures for other purposes, it has been included in Article 110. Table 110.28 is not a complete listing of all available enclosure type numbers, and specifically does not include special types intended primarily for use in hazardous (classified) locations.*]

Equipment   A general term, including fittings, devices, appliances, luminaires, apparatus, machinery, and the like, used as a part of, or in connection with, an electrical installation. [*This is a very broad term that includes virtually everything that is part of an electrical installation. Equipment may have a much narrower meaning as used by installers, but where the word appears in the NEC, it must be used as defined in the NEC.*]

Exposed (as applied to wiring methods)   On or attached to the surface or behind panels designed to allow access. [*According to this definition, the wiring installed above a suspended grid ceiling is still considered to be exposed even though it may not be visible most of the time.*]

Fitting   An accessory such as a locknut, bushing, or other part of a wiring system that is intended primarily to perform a mechanical rather than an electrical function. [*Many fittings such as those used with metallic conduits are also listed and tested for grounding and also perform electrical functions, but the primary function is not electrical.*]

Identified (as applied to equipment)   Recognizable as suitable for the specific purpose, function, use, environment, application, and so forth, where described in a particular *Code* requirement.

> *Informational Note:* Some examples of ways to determine suitability of equipment for a specific purpose, environment, or application include investigations by a qualified testing laboratory (listing and labeling), an inspection agency, or other organizations concerned with product evaluation. [*Enclosure type numbers are a form of identification for a use or environment that may or may not be included in the listing of an enclosure.*]

Location, Damp Locations protected from weather and not subject to saturation with water or other liquids but subject to moderate degrees of moisture.

>*Informational Note:* Examples of such locations include partially protected locations under canopies, marquees, roofed open porches, and like locations, and interior locations subject to moderate degrees of moisture, such as some basements, some barns, and some cold-storage warehouses.

Location, Dry A location not normally subject to dampness or wetness. A location classified as dry may be temporarily subject to dampness or wetness, as in the case of a building under construction.

Location, Wet Installations underground or in concrete slabs or masonry in direct contact with the earth; in locations subject to saturation with water or other liquids, such as vehicle washing areas; and in unprotected locations exposed to weather. [*It is intended that the inside of a raceway in a wet location or the inside of a raceway or enclosure installed underground be considered a wet location. Therefore, any conductors contained therein would be required to be suitable for wet locations. It is the responsibility of the authority having jurisdiction to assess the environmental factors associated with a given installation in order to determine the type of location, either dry, damp, or wet.*]

Multioutlet Assembly A type of surface, flush, or freestanding raceway designed to hold conductors and receptacles, assembled in the field or at the factory.

Outlet A point on the wiring system at which current is taken to supply utilization equipment. [*This definition includes outlets for lighting and appliances where the connection is made without a receptacle. A receptacle outlet is a type of outlet, but not all outlets include receptacles. This definition varies somewhat from the way the word* outlet *is often used in the field.*]

Plenum A compartment or chamber to which one or more air ducts are connected and that forms part of the air distribution system. [*This is not necessarily the same definition used by mechanical codes or people involved in heating, ventilation, and air-conditioning (HVAC) installations. They often refer to a duct as a* plenum. *This is the definition that must be used with the NEC to properly understand rules related to such things as plenum-rated cables and to differentiate between plenums, ducts, and other spaces for environmental air, although other spaces for environmental air are now recognized as a type of plenum.*]

Raceway An enclosed channel of metallic or nonmetallic materials designed expressly for holding wires, cables, or busbars, with additional functions as permitted in the *Code.*

>*Informational Note:* A raceway is identified within specific article definitions.

Rainproof Constructed, protected, or treated so as to prevent rain from interfering with the successful operation of the apparatus under specified test conditions. [*The term* rainproof *is typically used in conjunction with enclosure Types 3R and 3RX. These enclosures do not have to exclude rain completely, so they are often supplied with weep holes to allow small amounts of water to escape.*]

Raintight Constructed or protected so that exposure to a beating rain will not result in the entrance of water under specified test conditions. [*The term* raintight *is typically used in conjunction with enclosure Types 3, 3S, 3SX, 3X, 4, 4X, 6, and 6P.*]

Receptacle A receptacle is a contact device installed at the outlet for the connection of an attachment plug. A single receptacle is a single contact device with no other contact device on the same yoke. A multiple receptacle is two or more contact devices on the same yoke.

Receptacle Outlet An outlet where one or more receptacles are installed. [*Although many people in the electrical industry routinely use the term* outlet *to mean a receptacle outlet, these are separate terms in the NEC, and a receptacle outlet is just one type of outlet. Lighting outlets, appliance outlets, and power outlets are other types.*]

Remote-Control Circuit Any electric circuit that controls any other circuit through a relay or an equivalent device.

Switch, General-Use A switch intended for use in general distribution and branch circuits. It is rated in amperes, and it is capable of interrupting its rated current at its rated voltage.

Switch, General-Use Snap A form of general use switch constructed so that it can be installed in device boxes or on box covers or otherwise used in conjunction with wiring systems recognized by the *Code*.

Utilization Equipment Equipment that utilizes electric energy for electronic, electromechanical, chemical, heating, lighting, or similar purposes.

Weatherproof Constructed or protected so that exposure to the weather will not interfere with successful operation.

> *Informational Note*: Rainproof, raintight, or watertight equipment can fulfill the requirements for weatherproof where varying weather conditions other than wetness, such as snow, ice, dust, or temperature extremes, are not a factor.

KEY QUESTIONS

1. What wiring methods are being used?
2. What is the occupancy type?
3. Are there any fire-rated walls, floors, ceilings, or roofs, and if so, where are they located?
4. Are there any suspended (grid) ceilings, and if so, are suspended ceilings being used for environmental air?
5. Is any electrical equipment installed in wet or damp areas or other areas requiring special protection?
6. Are multiwire branch circuits being used?
7. Is any part of the wiring system being installed underground?

CHECKLISTS

		Checklist 3-1: General Wiring Methods		
✔	Item	Inspection Activity	*NEC* Reference	Comments
	1.	Identify the wiring methods in use and verify their suitability for the occupancy and conditions.	Various Chapter 3 articles	
	2.	Verify that all conductors of a circuit are grouped together.	300.3(B), 210.4(D) and 404.2(C)	
	3.	Check insulation values where conductors of different systems share common enclosures.	300.3(C)(1) and (2)	
	4.	Check wiring methods for spacing from edges of framing and for protection from nails and screws.	300.4(A), (B), (D), (E), and (F)	
	5.	Check for insulating bushings or grommets where NM cable is installed through metal studs or where insulated conductors 4 AWG or larger enter enclosures.	300.4(B)(1) and (G)	
	6.	Check cover, fill, protection, and allowances for ground movement on underground conductors and raceways.	300.5 and Table 300.5	
	7.	Verify that electrical raceways and cable trays are used exclusively for electrical conductors.	300.8	
	8.	Check for continuity and completeness in metal raceways and enclosures.	300.10	
	9.	Verify that wiring methods are securely fastened in place, supported independently of suspended ceilings, and not used as supports.	300.11 and applicable Chapter 3 article(s)	
	10.	Check for continuity of grounded conductors in multiwire branch circuits.	300.13(B)	
	11.	Check for adequate length of free conductors in boxes.	300.14	
	12.	Verify that boxes are installed at junction, splice, outlet, switch, and pull points.	300.15	
	13.	Check conductor fill in raceways.	300.17	
	14.	Verify that raceway systems are complete prior to installation of conductors.	300.18(A)	

(continues)

Checklist 3-1: General Wiring Methods

✔	Item	Inspection Activity	*NEC* Reference	Comments
	15.	Check vertical raceways for adequate conductor supports.	300.19	
	16.	Verify that fire ratings have been restored at electrical penetrations.	300.21	
	17.	Check installations of wiring in ducts, plenums, and other air-handling spaces for proper methods and materials.	300.22	
	18.	Verify that access to equipment behind removable panels is not compromised by cables, raceways, or equipment.	300.23	

Checklist 3-2: Boxes and Conduit Bodies

✔	Item	Inspection Activity	*NEC* Reference	Comments
	1.	Identify wet and damp locations and the suitability of boxes and fittings.	314.15	
	2.	Check boxes and conduit bodies for adequate space for conductors.	314.16	
	3.	Verify that boxes and conduit bodies for conductors 4 AWG and larger are adequately sized.	314.28	
	4.	Verify that raceways and cables are secured to boxes.	314.17(B) and (C)	
	5.	Check for closure of unused openings other than those permitted as part of the design of listed equipment.	314.17(A), 110.12(A)	
	6.	Verify that boxes in walls and ceilings are flush with the finished surface or, if surfaces are noncombustible, within ¼ in. (6 mm) of the finished surface.	314.20	
	7.	Check for excessive gaps between edges of boxes and plaster, plasterboard, or drywall surfaces.	314.21	
	8.	Verify that boxes are securely fastened and supported.	314.23	
	9.	Check boxes for adequate depth for the equipment that will be installed within them.	314.24	
	10.	Check for covers or canopies on boxes.	314.25, 314.28(C)	

CHECKLISTS

Checklist 3-2: Boxes and Conduit Bodies

✔	Item	Inspection Activity	*NEC* Reference	Comments
	11.	Check lighting fixture outlet boxes for suitability.	314.27(A), (B), and (C)	
	12.	Check floor boxes and receptacle/cover assemblies for listing.	314.27(C)	
	13.	Check listing and installation of boxes used for support of ceiling-suspended (paddle) fans.	314.27(C), 422.18	
	14.	Verify that all boxes are accessible.	314.29	
	15.	Verify that support means for nonmetallic boxes are outside the box or otherwise isolated from contact with conductors.	314.43	

Checklist 3-3: Cabinets and Cutout Boxes

✔	Item	Inspection Activity	*NEC* Reference	Comments
	1.	Verify that cabinets or cutout boxes are suitable and properly installed in any wet or damp locations.	312.2	
	2.	Verify that cabinets in walls do not have excessive gaps at the edges and are flush with the finished surface or, if surfaces are noncombustible, within ¼ in. (6 mm) of the finished surface.	312.3, 312.4	
	3.	Check for closure of conductor openings.	312.5(A)	
	4.	Verify that cables are secured to cabinets and cutout boxes or that the conditions for cables with nonmetallic sheaths are met.	312.5(C)	
	5.	Check wiring and bending space in cabinets and cutout boxes.	312.6	
	6.	Check cabinets and cutout boxes for adequate space for conductors and for splices and taps where they exist.	312.7, 312.8	

CHECKLISTS

	Item	Inspection Activity	*NEC* Reference	Comments
		Checklist 3-4: Switches and Receptacles		
	1.	Verify that all switching is done in the ungrounded conductors.	404.2	
	2.	Verify that any switches in wet locations are properly installed in weatherproof enclosures.	404.4	
	3.	Verify that switches are located not over 6 ft 7 in. (2.0 m) high and that they can be operated from readily accessible places unless otherwise permitted.	404.8(A), 240.24(A)	
	4.	Verify that the voltage between adjacent grouped or ganged devices is not over 300 volts or that barriers are installed.	404.8(B) and (C)	
	5.	Verify that metal switch boxes, switches, and any metal faceplates are grounded.	404.9(B), 404.12	
	6.	Verify that switches or receptacles in boxes have their plaster ears seated against the wall surface or the box.	404.10(B), 406.5(A), and (B)	
	7.	Verify that switches and receptacles are used within their ratings.	404.14, 406.3(A), 430.109	
	8.	Verify that general-use dimmers are installed only for control of permanently installed incandescent lighting.	404.14(E)	
	9.	Check the listing and marking of any switches or receptacles used with aluminum conductors.	404.14(C), 406.3(C), 110.14	
	10.	Check receptacles in wet or damp locations for proper covers and enclosures and weather-resistant ratings.	406.9	
	11.	Verify that isolated ground receptacles are properly identified and connected to isolated grounding conductors.	406.3(D)	
	12.	Check that receptacles project from metal faceplates or are flush with nonmetallic faceplates and that the faceplates cover openings.	406.5(D), 406.6	
	13.	Check receptacles for proper polarity and for grounding and bonding connections.	406.4, 250.146, 200.11	
	14.	Verify that receptacle ratings and branch-circuit ratings are compatible.	210.21, 210.24	

CHECKLISTS

CHAPTER 4

Services, Feeders, and Branch Circuits

OVERVIEW

This chapter focuses primarily on electrical installations in occupancies other than dwelling units, though some of the differences between residential and nonresidential requirements also are covered. A detailed discussion of residential inspections appears in the Dwelling Units and Mobile/Manufactured Home Sites chapter of this manual.

Services, **feeders**, and **branch circuits** are the means for distributing electrical power from a utility source to the **luminaires**, **receptacles**, and other **outlets** where the power is used (**FIGURE 4-1**). The rules governing services, feeders, and branch circuits are therefore critical to all electrical installations. The requirements for **services** include minimum sizes for conductors and equipment, requirements for service disconnects and overcurrent devices, and installation requirements for **service conductors** of various types.

Many similar but sometimes less restrictive rules are provided for feeders. The *National Electrical Code®* (*NEC®*) also provides very specific rules regarding the installation and permitted uses of different types and **ratings** of branch circuits. There are instances, such as **buildings** with emergency systems, where feeders and branch circuits are supplied from a utility source and also by the on-site emergency power supply. In some cases, a premises' wiring system may not have any service conductors, as would be the case where the only supply is an on-site generator, solar photovoltaic supply, wind electric system, or fuel cell.

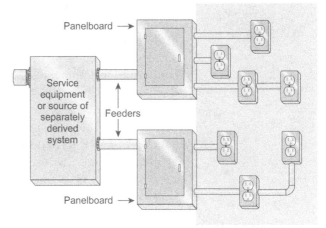

FIGURE 4-1 Distinguishing service, feeder, and branch-circuit conductors and equipment.

SPECIFIC FACTORS

Services, feeders, and branch circuits are the three basic categories of conductor types used in delivering power to **utilization equipment**.

Services stand apart from other parts of a wiring system because they represent the usual source of power to a building. **Service equipment** provides the primary means of disconnecting a building from its power source. In addition, the service defines the location for the main connection of a system to a grounding reference, which in turn defines the origin of the equipment grounding conductors in a system as well as the point where ground-fault current is directed to the grounded service conductor to complete the ground-fault current return path.

Service conductors are treated differently from most other conductors in or on a building because their overcurrent devices are on the load end of the conductors. Therefore, the overcurrent devices at the service provide only overload protection and do not provide short-circuit or ground-fault protection for the upstream service conductors. Feeder and branch-circuit conductors typically have overcurrent protection at or near the point where they receive their supply, so that the entire conductor is protected from all three types of overcurrent. Because service conductors do not have short-circuit or ground-fault protection at the point of supply, the rules about the placement of service equipment in buildings, and about the length of service conductors allowed in buildings, are critical to safe electrical installations. These rules minimize the physical damage exposure of service conductors within buildings.

Feeders and branch circuits complete the distribution of power to the outlets of a system. Feeders provide the link between services and branch circuits. Because branch circuits can originate at service equipment, not all wiring systems have feeders, but feeders are necessary where the number or distribution of branch circuits throughout a building or other **structure** is greater than can be reasonably supplied from a single location.

An important feature of both feeders and services is that they both usually supply power to multiple branch circuits. This means that feeders and services can often be smaller than the sum of the loads connected to all the branch circuits.

Generally, not all the branch circuits are likely to be fully loaded at any one time. This fact justifies the reduction in feeder and service capacity. The ratio between the maximum actual load in amperes and the total connected load in amperes is the **demand factor**. Demand factors may be used in feeder and service calculations. Demand factors are percentages, or decimals less than or equal to 1.00. Thus, when a large number of devices or outlets are combined on a feeder or service, the total connected load can be multiplied by the demand factor to determine the actual load that is likely to be imposed on a conductor at any one time.

Branch circuits, on the other hand, are generally required to be sized for the total connected load. The smallest rated branch-circuit-type overcurrent device in the circuit defines the size or rating of the branch circuit with which it is associated. Branch circuits are the final link in the power supply to the outlets, which, by definition, are the supply points to utilization equipment. Ideally, the overcurrent devices allow small portions of a total wiring system to be isolated either manually or by the automatic action of circuit breakers and fuses. This system allows for isolation of faults and disconnection of individual or small groups of outlets and loads without interrupting the main power to a building.

The requirements that apply to services, feeders, and branch circuits are intended to address the various safety issues that relate to each of these basic elements of a wiring system.

KEY QUESTIONS

1. How many services are there, and where is the service equipment located?

The answer to this question provides the primary location of most service inspections and also provides a logical starting point for inspections of the rest of the distribution system, including feeders and branch circuits and their related distribution equipment and overcurrent devices. The general requirement limits a building or structure to a single service, but there are conditions specified in Section 230.2 that permit additional services. Each service must be inspected individually where more than one service is justified and installed. The service locations can usually be found on electrical floor plans or site plans. The **service point** may also be a useful starting point, but since the service point may be defined by utility ownership, the service point may be relocated during the course of a project.

2. Are the service conductors overhead or underground?

Many of the checklist items are specific to either overhead or underground services. For example, overhead clearances and underground burial depths are not both likely to be applicable to the same service, so some checklist items can be easily eliminated. The answer to this question is usually found on the electrical plans but is also evident on most job sites.

3. What is the required current and voltage rating of service equipment and service conductors?

The answer to this question is usually found on the electrical plans, along with calculations or panel schedules that justify the current ratings. This answer is critical to many of the checklist items regarding service disconnects and overcurrent devices. It also provides the information needed to determine whether ground-fault protection for equipment (GFPE) is required for the service disconnecting means.

4. What are the required conductor and overcurrent device sizes for feeders?

Again, this information must be found from riser or one-line drawings on the electrical plans. Calculations or other evidence of the proper current ratings also should be provided on the plans. The answer provides the criteria needed to complete many parts of the feeder checklist.

5. Is more than one building or structure served from the same supply source, and, if so, what are the required conductor and disconnect sizes?

The answer to this question also may be found on plans, but should be verified on the job site. Disconnects are generally required at a separate building or structure, whether it is served by feeders or by branch circuits.

6. Are multiwire branch circuits being used?

A number of checklist items involve the use of multiwire branch circuits. Most electrical drawings will indicate the use of multiwire circuits by showing a series of circuit numbers sharing a single neutral conductor. Multiwire circuits require disconnecting means that simultaneously disconnect all ungrounded conductors. Multiwire circuits also require identification by grouping the conductors in branch-circuit panelboards or by other means to clearly identify the grounded conductor that is associated with any specific ungrounded conductors of a multiwire circuit. Fluorescent luminaires that utilize double-ended lamps and contain ballasts and that are supplied power by multiwire circuit(s) require disconnecting means for all the supply conductors to that luminaire, including the grounded (neutral) conductor. See 410.130(G).

PLANNING FROM START TO FINISH

Some aspects of the inspection of services must be done at an early stage to determine whether inspections of underground raceways or buried conductors are necessary. Once any applicable underground inspections are taken care of, the other aspects of both underground and overhead services can usually be checked in a single inspection. Often, this inspection is made when the project is ready for power (often called a "power-to-panel inspection" or "service inspection"). A request from the installer usually determines the timing of this inspection.

Inspections of feeders can be made at the same time as the service inspection because the basic distribution system is usually complete at that time. In fact, a completed feeder distribution system is usually desirable before power is supplied to the service, as it eliminates or reduces the inherent exposure to shock and arc flash hazards that are present upon the service being energized. This exposure is reduced for the installers and the inspector as well as for any other occupants or workers.

Branch-circuit inspections include branch-circuit ratings; the proper use of individual, multioutlet ("other than individual"), and multiwire branch circuits; and the use of circuits for various kinds of outlets. Therefore, these inspections can usually be performed at the same time as the service and feeder inspections or in subsequent inspections. Some of the items to be inspected may not be completed and ready for inspection until later inspections are done, perhaps not until the "final" inspection.

Most of the checklists related to services and feeders can be thoroughly covered. However, because of the numbers of branch circuits, most inspectors do not have time to thoroughly inspect each one. Instead, some sampling must be done. Important aspects of many circuits can be checked at one time by starting at the branch-circuit panelboards. From there, the inspector should choose specific branch circuits to inspect more completely. Good targets for investigation are any circuits that have unusual ratings as well as a few samples of the more ordinary lighting and receptacle circuits. (When reference is made to "receptacles" generally in this chapter, the receptacles discussed are 125-volt, 15- or 20-ampere duplex receptacles—the type generally used for convenience receptacles for most cord-and-plug–connected equipment in the United States.)

WORKING THROUGH THE CHECKLISTS

Checklist 4-1: Services

1. Verify that each building or structure has only one service or, if more than one, that additional services are justified and identified.

Generally, each building or structure is allowed only one service. This rule provides for a single centralized location where all power to a building can be disconnected. This central disconnecting means is especially important in emergency situations but is also useful to the occupants of the building or to anyone who has to service the electrical system. Conditions that allow more than one service are divided into four general types:

1. Special conditions, such as for fire pumps or for enhanced reliability
2. Special occupancies, such as large apartment buildings or large commercial and industrial buildings
3. Capacity requirements for large power demands
4. Different characteristics, such as for more than one **voltage** system or rate schedule or for different uses

Some of the rules allowing more than one service require special permission (written consent of the authority having jurisdiction [AHJ]) or may be limited or mandated by the local utility.

2. Verify that each service drop or lateral supplies only one set of service-entrance conductors or, if more than one, that the additional sets are justified and identified.

The basic rule here is that each set of service conductors, whether they are overhead or underground, is permitted to supply only one set of **service-entrance conductors**. The permitted exceptions to this rule are very commonly used. Probably the most commonly applied exception is the one that provides for multiple sets of service-entrance conductors where up to six disconnects in separate enclosures are grouped in one location (**FIGURE 4-2**). **FIGURE 4-3** shows a similar application of multiple sets of service-entrance conductors supplied by a single **service lateral**. **FIGURE 4-4** shows an application similar to Figure 4-3, except that there are no service-entrance conductors, and multiple service laterals or multiple sets of **underground service conductors** are treated as a single service.

3. Check clearances from building openings, grade, roadways, roofs, and swimming pools.

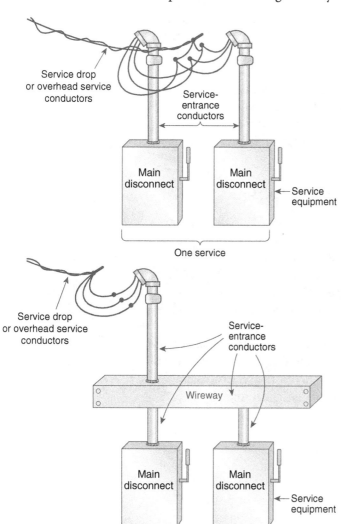

FIGURE 4-2   Two examples of one service consisting of multiple disconnects and sets of service-entrance conductors supplied by a single service drop or a single set of overhead service conductors.

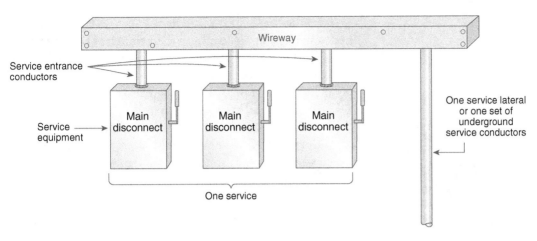

FIGURE 4-3   One service consisting of multiple disconnects and sets of service-entrance conductors supplied by a single service lateral or a single set of underground service conductors.

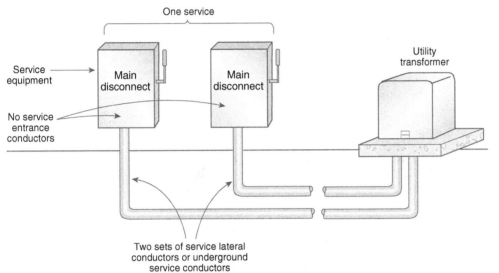

FIGURE 4-4   One service consisting of multiple disconnects grouped together and multiple sets of service lateral conductors or underground service conductors.

Different parts of these rules apply to different kinds of service conductors. For **overhead service conductors** installed above a roof, the main rule specifies an 8-ft (2.4-m) clearance that extends 3 ft (914 mm) horizontally from the roof (**FIGURE 4-5**). Lower clearances are permitted under certain conditions. The rules for clearances from ground (**FIGURE 4-6**) apply only to overhead service

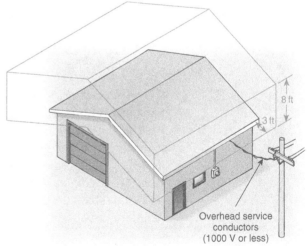

FIGURE 4-5   Basic rule for clearances of overhead service conductors installed above a roof. (Source: *National Electrical Code® Handbook*, NFPA, Quincy, MA, 2011, Exhibit 230.22.)

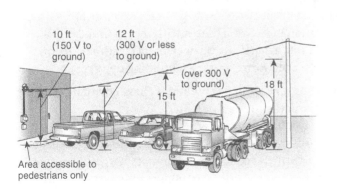

FIGURE 4-6   Minimum clearance above ground for overhead service conductors.

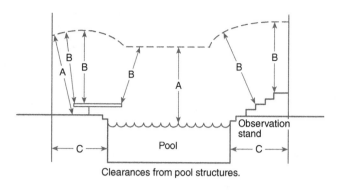

Clearances from pool structures.

FIGURE 4-7 Overhead conductor clearances for swimming pools (see **TABLE 4-1**).

conductors, but the rules for clearances from building openings apply to all service conductors that are not in a raceway or do not have an overall jacket. The rules for swimming pools require greater clearances and apply to all open conductors (**FIGURE 4-7** and **TABLE 4-1**). All of these rules are intended to reduce the likelihood of someone coming into contact with the conductors. In the case of swimming pools, underground clearances between the swimming pool and power conductors are also required.

TABLE 4-1	Overhead Conductor Clearances					
	Insulated Cables, 0–750 Volts to Ground, Supported on and Cabled Together with a Solidly Grounded Bare Messenger or Solidly Grounded Neutral Conductor		All Other Conductors Voltage to Ground			
			0 Through 15 kV		Over 15 Through 50 kV	
Clearance Parameters	m	ft	m	ft	m	ft
A. Clearance in any direction to the water level, edge of water surface, base of diving platform, or permanently anchored raft	6.9	22.5	7.5	25	8.0	27
B. Clearance in any direction to the observation stand, tower, or diving platform	4.4	14.5	5.2	17	5.5	18
C. Horizontal limit of clearance measured from inside wall of the pool	This limit shall extend to the outer edge of the structures listed in A and B of this table but not to less than 3 m (10 ft).					
Source: *NFPA 70, National Electrical Code®*, NFPA, Quincy, MA, 2014, Table 680.8(A).						

4. Verify that the point of attachment for an overhead service drop is adequate and will provide minimum clearances.

Most utilities have a minimum requirement for means of support or mechanical strength of attachment of their service-drop conductors. Their guidelines are usually better than any other general rule because they take into account local conditions such as ice or wind loads. The height of the attachment must be located to maintain vertical clearances when sag in the cable is considered. The location of the attachment with regard to the utility facilities must also be considered to maintain vertical and horizontal clearances from earth, pavement, concrete, roofs, or other platforms and from building openings.

5. Verify that masts used as supports for service-drop conductors have adequate strength and are not used to support other conductors or equipment.

Utilities often have specific minimum "worst case" requirements for service masts, as they do for other points of attachment. Minimum trade sizes of perhaps 1½ or 2 (metric designators 38 or 51) rigid metal or intermediate metal conduit are common. Larger sizes may be required by some utilities or for other reasons. Again, the support rules often take into account local conditions. The *NEC* requires only that supports be structurally adequate but provides no rules relative to the minimum size or type of material; these issues are best determined by the serving utility and AHJ. **FIGURE 4-8** shows a service mast installed through the roof overhang with maximum dimensions shown for length of conductors and distance traveled above the roof, and typical minimum clearance of conductors from the roof surface. The attachment as shown in this illustration is not allowed to be between the weatherhead and a coupling located above the roof. If a coupling is required to obtain adequate height or for other reasons, the coupling must be below the last point of attachment of the raceway to the building. In addition to being structurally adequate, the mast must be high enough to provide the proper clearance above the roof for the conductors.

6. Verify that supports for service conductors passing over a roof are adequate and substantial.

Intermediate supports for service conductors passing over roofs are not as common as the main attachment, as many utilities require that the point of attachment be at a location on a building nearest their facilities. Where such additional or intermediate supports are required, the inspector must use some judgment to determine what is adequate, but requirements similar to those for masts or other points of attachment can be used. The important safety concern is to maintain the required clearances over roofs with supports that will withstand the weather and loading conditions. Ideally these supports will be independent of the building and are required to be independent if that can be done in a practicable way.

7. Check underground conductors for adequate burial depth and protection.

These rules are quite clearly stated in Article 300 of the *NEC* for different wiring methods, voltages, and conditions of use. The requirements are intended to minimize the danger from digging into buried cables and reduce the likelihood of damage to conductors. The minimum cover required for underground wiring methods varies based on the specific method and location. **FIGURE 4-9** shows the different minimum cover requirements from *NEC* Table 300.5 (see Chapter 3) for cables or wiring methods up to 1000 volts. Only the minimum burial depths for rigid polyvinyl chloride (PVC) conduit and reinforced thermosetting resin conduit (RTRC) are shown based on different installation conditions. Both of these wiring methods are also sometimes called rigid nonmetallic conduit. In Table 300.5 they are both called nonmetallic raceways.

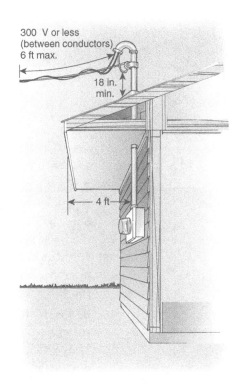

FIGURE 4-8   A through-the-roof service mast.

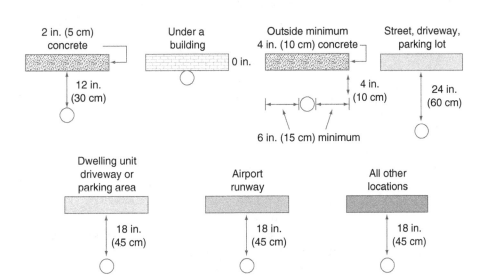

FIGURE 4-9   Minimum cover requirements for PVC or RTRC.

Since minimum depths take into account the conditions and the risk, some conductors that are installed in raceways that are very resistant to damage, that have **ground-fault circuit interrupter (GFCI)** protection, or that are part of low-energy circuits may have less cover than other conductors. However, all conductors, regardless of the circuit type or wiring method, must be buried at least 24 in. (610 mm) under areas that are subject to heavy traffic. Depths are greater for higher voltages. For circuits over 1000 volts, Table 300.50 in the *NEC* provides the minimum cover requirements. A warning ribbon or tape at least 12 in. (305 mm) above the conductors is required for service laterals that are buried 18 in. (457 mm) or more below grade and are not concrete encased.

8. Check above-ground conductors and cables for adequate supports and protection from physical damage.

Some methods of protecting service conductors are specifically listed in the *NEC*. Other methods, such as guards or barriers, also may be acceptable. The judgment of what constitutes a risk of physical damage is intended to be left to the AHJ. Support spacing and clearance for various types of conductors and cables also are specified in Article 230 and the respective articles of Chapter 3 in the *NEC* corresponding to the wiring methods used.

9. Verify that wiring methods and support systems for service-entrance conductors are suitable.

The *NEC* includes a list of the wiring methods that are suitable for service-entrance conductors. Flexible metal conduit and liquid-tight flexible metal conduits are limited to 6 ft (1.8 m) and must be provided with bonding jumpers. A cable tray is also mentioned and allowed for use with specified wiring methods; where used for service-entrance conductors, it must be dedicated to service conductors or the service conductors must be separated from other conductors by permanent barriers. A cable tray that is used for service-entrance conductors must be clearly marked as such. However, a cable tray is not actually a wiring method; rather, it is a support system for other wiring methods.

10. Verify that service raceways are arranged to drain and that service heads are raintight and properly located.

All service raceways exposed to weather are required to be arranged to drain. Service raceways and cables, especially those that enclose service-entrance conductors connected to an overhead **service drop** or overhead service conductors, must be arranged so that rain or other water will be excluded from service or metering equipment. This means that the point of connection of the service drop should be below the service head (also called a weatherhead), and a "drip loop" should be formed so that water would have to run uphill in order to enter the service raceway or cable **(FIGURE 4-10)**. It is also important to ensure that service conductors exposed to direct sunlight are marked as "sunlight-resistant" and that all service conductors located in wet locations are wet-location rated. The service head must be listed for wet locations.

FIGURE 4-10   Service weatherhead, point of attachment, and drip loop.

11. Check service conductors for adequate size and rating.

Service loads are determined under Parts III or IV of Article 220 of the *NEC*. (Loads on farms are calculated under Part V of Article 220.) Service or feeder loads are based on the branch-circuit loads calculated under Part II of Article 220. The service load and the service conductor sizes are usually shown on the electrical project plans. Minimum sizes for certain types of loads or occupancies also are given in Article 230, and these minimum sizes will apply even if the calculated load from Article 220 produces a smaller size.

Service-neutral conductors (grounded service conductors) must be at least large enough to carry the unbalanced load, and they must also be large enough to serve as a fault-return path. The fault-return provision mandates that the service-grounded conductor must always be at least as large as that required for the main bonding jumper, since both will be expected to carry currents of the same magnitude. Either the grounded service conductor or the main bonding jumper may consist of more than one conductor, depending on how the service is arranged. Where the service equipment includes more than one enclosure, there may be more than one main bonding jumper. In any service over 400 amperes, the grounded service conductor is likely to be installed in parallel. The grounded-circuit conductor of a multiwire (single- or 3-phase) service or other circuit is also required to have an **ampacity** of not less than the computed maximum unbalance that may occur in the system (**FIGURES 4-11A** and **B**).

If the loads on the ungrounded conductors are balanced, the normal neutral current is zero. If the loads are not balanced, the neutral load is the normal unbalance in the system. The maximum unbalance, which is one of the criteria for sizing a feeder or service neutral, is based on the largest load that could occur on the neutral when all line-to-neutral loads on any ungrounded conductor are turned off. For 3-phase, 4-wire, wye-connected circuits, the impact of harmonic currents caused by nonlinear loads must be considered. In any case, the larger of the minimum sizes required for the maximum unbalance (220.61) or the fault current [250.24(C)] is the minimum permissible size for the grounded service conductor. The minimum size from 250.24(C) is based on the size of the ungrounded service conductors.

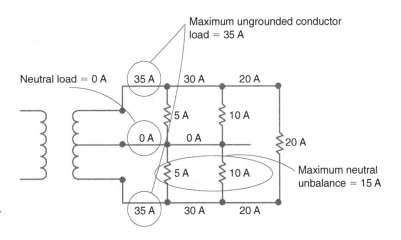

FIGURE 4-11A   Normal neutral load.

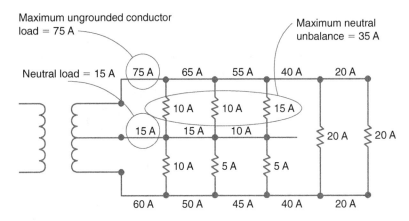

FIGURE 4-11B   Maximum unbalanced load.

12. Verify that service equipment is identified as suitable for that use.

Service equipment must be listed and is usually marked "Suitable for Use as Service Equipment," sometimes abbreviated as "SUSE." The service marking may include other requirements or conditions, such as "when equipped with a main circuit breaker," and may specify locations for grounding electrode conductor connections. Some equipment may be marked "Suitable for Use Only as Service Equipment." According to the listing information on equipment marked "Suitable Only for Use as Service Equipment," such equipment may also be used where the grounded conductor is permitted to be attached to an enclosure at separate buildings (existing installations only) or is used as the disconnecting means and the location of the system bonding conductor for separately derived systems.

One characteristic of equipment marked "Suitable for Use as Service Equipment" is a means (provided or available from the manufacturer) to connect the grounded service conductor to a grounding electrode system. A marking that states "Suitable Only for Use as Service Equipment" indicates that the connection between the grounded conductor terminal and the enclosure is installed or inherent in the design and the connection is not removable.

13. Verify that a service disconnecting means is provided, suitable, and located outside or inside nearest the point of entrance of the service conductors.

A service disconnecting means must be provided. It may consist of up to six switches, which may be in one enclosure or in separate enclosures. The service disconnecting means is required to be located as close as practical to where the service conductors enter a building or structure, or "nearest the point of entrance" **(FIGURE 4-12)**. This placement limits the length of service conductors that are inside the building, as these conductors have no short-circuit or ground-fault protection other than the utility primary overcurrent protection on the supply side. That is, they do not have short-circuit and ground-fault protection that satisfies *NEC* requirements that would apply to branch circuits or feeders.

There is no specified distance stated by the *Code.* The phrase "nearest the point of entrance" is subject to the inspector's interpretation. Some jurisdictions state a specific distance under specific conditions, such as in certain wiring methods. Others simply say it is where the service conductors come through the floor or through an outside wall. The *NEC* leaves the determination of the precise service disconnecting means location up to the discretion of the AHJ. This allows any conditions unique to the service installation to be assessed by the AHJ. The service disconnect must also be suitable for the environmental conditions. (See checklist Item 11 in the General Requirements Inspections chapter of this manual.) Suitability and marking for service use are covered in Item 12 of this checklist. An **intersystem bonding termination** must also be provided. This is often part of or near the service equipment but could be located elsewhere. See the Grounding and Bonding chapter of this manual for more information about requirements for this termination.

14. Verify that service overcurrent protection is provided, properly sized, and part of or adjacent to the disconnecting means.

Service overcurrent protection is sized to accommodate the calculated service load and any **continuous loads** and to provide overload protection at the load end of the service conductors. The rating of the service overcurrent device usually matches the ampacity of the service conductors or is the next larger size. However, where there is more than one service overcurrent device, the sum of the ratings is allowed to exceed the ampacity of the service conductors. A few other exceptions also apply, such as those for motor loads. Generally, the service overcurrent device is part of the service disconnect, as in a circuit breaker or a fusible switch. In older installations, the fuses might be located in a separate cutout box adjacent to the service switch. Such an arrangement is still allowed but is usually impractical in new installations.

15. Verify that service disconnects are grouped together and limited to six in any one location.

The requirements for grouping service disconnects and for limiting the number of disconnects in one location are both aimed at providing a ready means for removing power from a building or structure. Interpretations vary on what constitutes a *location* or a *different location,* and such interpretations are left to the AHJ. Disconnects for certain special purposes, such as for control power for GFPE, or for connection of surge protective devices (SPDs) are not required to be counted as one of the six disconnecting means permitted for a service where the disconnect and associated items are installed as a part of listed equipment. Some additional disconnects, such as a service disconnect/controller for a fire pump, are required to be located remote from the other service disconnects.

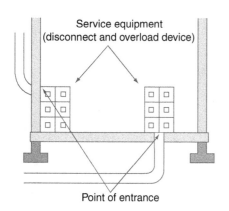

FIGURE 4-12 Location of service disconnecting means.

16. Check ratings of service disconnecting means.

Service disconnecting means must be rated not less than the current rating required for conductors and overcurrent protection, including any adjustment for continuous loads. The service load is calculated in accordance with Article 220 and does not include any factor for continuous loads (**FIGURE 4-13**).

The disconnect rating is not required to be increased for continuous loads, but most service disconnects are also overcurrent devices, and the ratings of overcurrent devices and service conductors must be adjusted for continuous loads [230.42]. Determination of the service load is the first step required in selecting properly sized conductors and disconnecting means.

In addition to the ordinary ratings of overcurrent devices discussed here, all overcurrent devices, panelboards, and similar equipment must have appropriate short-circuit and interrupting ratings based on type, size, and length of conductors, and the impedance of the service transformer(s). As noted in the Introduction chapter of this manual, these ratings are based on a short-circuit study that should be provided by the engineer or designer, but the actual ratings of installed equipment on other than one- or two-family dwellings must be verified in the field from the maximum fault current that is required to be marked on the service equipment.

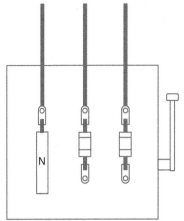

Total noncontinuous loads ....12,200 VA
Total continuous load @ 125%
16,200 × 1.25 ....20,250 VA
　　Total load ....32,450 VA

32,450 VA ÷ 240 V = 135 A

The next higher standard size is 150 A (Section 240.6).

Minimum size disconnect: 135A   230.79
Minimum size ungrounded conductor: 135A   230.42(A)
Standard Switch Rating: 200A
Minimum Standard Fuse: 150A   240.6
Ungrounded Conductor Selection: 150A   230.90(A)

FIGURE 4-13 Example of sizing service equipment and conductors.

17. Check for equipment connected to the supply side of the service disconnecting means and overcurrent protection.

Most equipment is not permitted to be connected to the supply or line side of service disconnects or service overcurrent protection. Certain equipment, such as taps for load management, connections for parallel power production (cogeneration), or metering equipment, among others, may be connected on the line side of service equipment. The rules and exceptions for service disconnects and service overcurrent devices are pretty much the same. However, some exceptions or specific rules recognize the subtle difference that sometimes exists. For example, service switches may be connected ahead of service fuses.

18. Verify that ground-fault protection is supplied where required, and obtain a written record of performance testing.

GFPE is required on services within a specified voltage range for each disconnecting means rated at 1000 amperes or more. The requirements apply to solidly grounded wye systems where the **voltage to ground** is over 150 volts to ground, and the voltage between ungrounded conductors is 1000 volts or less.

Perhaps the most common such system is the 480Y/277 volt 3-phase, 4-wire system. Such systems are especially susceptible to damage from arcing faults. The arcing faults may never reach a value that will cause a 1000-ampere or higher rated overcurrent device to operate quickly but can do substantial damage. (Note: If the neutral conductor is not used as a circuit conductor, the system may not be required to be grounded or may be grounded through an impedance, and ground-fault protection may not be required.)

Where ground-fault protection is required on a service, the system must be tested on site, and a test report must be made available to the AHJ. The test must be conducted in accordance with the manufacturer's instructions. The manufacturer's instructions may require the use of special test equipment or may describe a field-assembled test

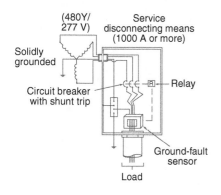

(480Y/ 277 V)

Solidly grounded

Service disconnecting means (1000 A or more)

Circuit breaker with shunt trip

Relay

Ground-fault sensor

Load

FIGURE 4-14 Basic operation of an equipment ground-fault protection device.

apparatus. Most "push-to-test" buttons on GFPE test only the mechanical operation and do not test the sensing equipment; as a result, this test does not meet the testing requirement in the *NEC*.

FIGURE 4-14 illustrates the basic operating scheme of a "zero-sequence"-type ground-fault sensor. The main circuit breaker is operated through a shunt-trip device. The circuit breaker is opened upon the sensing of a load-side ground fault. A ground fault is detected by monitoring the balance (vector summation) of currents on the normal current-carrying conductors in much the same way that a GFCI senses ground-fault currents. However, ground-fault protection is intended to protect equipment from damage due to high currents or arcing faults, and GFCI protection is for personnel protection from relatively low currents. In Figure 4-14, a single current transformer is shown, but multiple current transformers are used in some designs.

Checklist 4-2: Feeders

1. Verify that feeder conductors, including any required neutral conductors, are adequate for the load.

Feeders are required to be adequate for the load determined according to Article 220 of the *NEC*. Feeder loads are determined in the same manner as loads on service-entrance conductors. However, because service-entrance conductors often supply more than one set of feeder conductors, the sizes are not necessarily the same. Feeder-neutral loads are based on the maximum unbalanced load connected to any one ungrounded conductor. (See Item 11 under Services.)

Where there is a small line-to-neutral load in comparison to the line-to-line connected load, the maximum unbalanced load is minimal. A requirement for feeders that include a grounded conductor uses Table 250.122 in the *NEC* and the size or rating of the feeder overcurrent protective device to establish the minimum feeder grounded conductor size. This requirement helps ensure that the feeder neutral has sufficient size to create an effective circuit path to sustain the current level necessary for opening the overcurrent device in the event of a line-to-neutral short circuit. A feeder supplying only line-to-line connected loads is not required to include a grounded or neutral conductor. An equipment grounding conductor must be provided to establish proper grounding and bonding connections for equipment supplied by the feeder. Diagrams or drawing details on the electrical plans showing overcurrent ratings and feeder conductor sizes, including neutrals, are usually shown on project plans. Diagrams showing feeders may be required by the AHJ for review or approval.

2. Check overcurrent device and feeder conductor sizing for continuous and noncontinuous loads.

Feeder conductors and overcurrent devices must be increased in size for continuous loads. This requirement does not apply to the grounded or neutral conductor because it normally does not connect to an overcurrent device. (A grounded feeder conductor normally terminates in a bus or similar termination point on both ends.) For the purpose of establishing a minimum size for overcurrent devices and ungrounded conductors, continuous loads are included at 125 percent unless the overcurrent device and its assembly is listed for operation at 100 percent of its rating.

Feeders must be sized at the calculated value for other (noncontinuous) loads, which may include some reduction in size due to demand factors. The minimum size for a conductor carrying continuous loads need not be further adjusted using correction or adjustment factors from Article 310, or for terminal temperatures. However, the minimum size required for carrying the actual load after correction and adjustment factors are applied (not increased for continuous load) must also be considered. The final minimum size will be the larger of these two sizes. Terminal temperatures and correction and adjustment factors are covered by separate rules in Sections 110.14(C), 310.15, and Table 310.15(B)(16), which must be coordinated. The conductor size chosen to avoid overheating any connected terminal is one factor. The size chosen to accommodate continuous loads is a second factor. A third factor is determined based on the calculated load and conductor sizes that are adjusted or corrected for specific conditions of use (ampacity). These factors must all be addressed, and the *largest* size determined based on any of these three factors gives the minimum-size conductor that must be used. Because these requirements apply to different locations in the circuit run (in terminal enclosures and in the raceway or cable), each of these requirements affecting conductor size may be applied separately. A conductor sized to accommodate continuous loads is a terminal consideration, yet it may also provide the necessary conductor size to address any ampacity correction or adjustment that has to be made. Applying all of the relevant conductor sizing requirements at the same time helps ensure that the requirements are coordinated and nothing affecting conductor sizing is overlooked.

3. Check wiring methods for suitability.

All wiring methods have uses that are permitted, uses that are not permitted, and/or uses that are conditioned on some specific factor, such as the need for protection from corrosion or deterioration or protection from physical damage. Section 230.43 lists wiring methods that may be used for service conductors. Since feeder conductors have overcurrent protection that includes short-circuit and ground-fault protection, they are not subject to exactly the same restrictions on wiring methods that apply to service conductors. Still, in all cases, the wiring methods must be suitable for the occupancy and conditions (detailed in the Wiring Methods and Devices chapter of this manual).

4. Check feeders with disconnecting means rated at 1000 amperes or greater for GFPE if required.

The requirements for **ground-fault protection of equipment (GFPE)** on services are discussed in Item 18 under Services. These same criteria are used to determine ground-fault protection requirements for feeders. If the ground-fault protection is provided at the service, additional protection is not required at the feeder, unless the feeder is supplied by a separately derived system or is part of a health care facility distribution system covered in 517.17 of the *NEC*. If the feeder is supplied by a separately derived system or is otherwise not protected by the service GFPE, then separate protection for the feeder must be provided (**FIGURE 4-15**). If ground-fault protection is provided at the service of a hospital, additional levels of ground-fault protection are required on the next level of distribution downstream from the service disconnect. These additional levels are required to be coordinated to prevent a fault on one branch of the essential electrical system from removing power to another branch.

5. Verify that disconnects are provided at separate structures for feeders running between structures.

When a feeder is run from one structure to another, the downstream structure must be provided with a disconnecting means in much the same manner as a disconnecting means is provided at a service (**FIGURE 4-16**). From the standpoint of the people living or working in any particular building, a feeder to that building is the same as a service. Because overcurrent protection for the feeder is provided at the origin of the circuit, it is

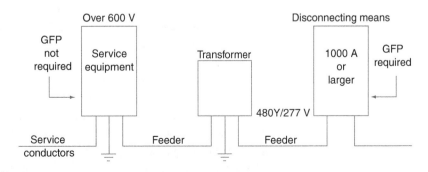

FIGURE 4-15 Ground-fault protection for feeders.

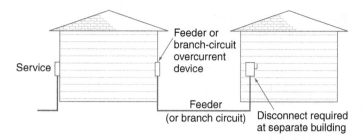

FIGURE 4-16 Disconnecting means at structure supplied by feeder or branch circuit.

not necessary to provide overcurrent protection at the point of termination, unless it is necessary to provide additional overcurrent protection for the equipment, such as a panelboard, that is used as the disconnecting means. Services and feeders have separate definitions in the *NEC,* but they may serve a similar purpose. Under the conditions specified in the Exceptions to 225.32, certain types of structures such as lighting poles and some large facilities with multiple buildings fed from a common service are allowed to have the disconnecting means for the separate building(s) or structure(s) remotely located.

6. Verify that disconnects at separate structures are properly rated, located, grouped, and identified.

Disconnecting means at separate structures are required to be suitable for use as service equipment [225.36] only if they are used where bonding of the grounded conductor to the enclosure is permitted. This is allowed only in existing installations. Otherwise the disconnect may be a circuit breaker or a switch, including, in some cases, a nonfused switch or a snap switch. As noted in Item 12 under Services, equipment that is suitable for use as service equipment is marked accordingly. Disconnecting means must be adequate for the load served and must be located outside or inside the building and nearest the point of entrance of the conductors unless the conditions of the exceptions are met [225.32]. This is different from the requirement for a service disconnect that can be anywhere outside the building and is only required to be nearest the point of entrance when the disconnect is inside the building. Where the service disconnect is outside of and remote from the building, the conductors to the building are usually feeders, and another feeder disconnect must be located nearest the point of entrance of the feeder conductors. The disconnect(s) must be grouped, with no more than six disconnects in any one location [225.33]. In addition, if there is more than one or a combination of services, feeders, or branch circuits supplying a building, a permanent plaque must be provided to indicate the locations of any other supplies and the areas they serve [225.37]. This requirement parallels the requirement of 230.2(E), which refers to 225.37.

7. Verify that any outside feeders use appropriate wiring methods and are properly supported and arranged to drain.

This is the same basic requirement as that for services explained in Item 10 under Services, except that the references are found in Article 225.

8. Check any outside feeders for adequate supports, clearances, and mechanical protection.

This is similar to the requirements for service conductors and wiring methods. The dimensions for clearances from ground or above roofs are the same as for service conductors, and the same considerations regarding supports and mechanical protection also apply.

9. Verify that feeders supplied from transformers or from other feeders are properly protected by overcurrent devices.

Article 240 of the *NEC* includes a number of connection rules for overcurrent protection of feeders tapped from other feeders or supplied through transformers (secondary conductors). The rules vary according to the length of the feeder, the type of occupancy, and whether the feeder is located outside or inside. In limited cases, a feeder can even be protected by the overcurrent device protecting a transformer primary. In fact, feeders are often protected against overload by the transformer secondary overcurrent device. But the rules for transformer protection and the rules for conductor protection are separate, and, although a single device may perform both functions, both rules must be satisfied. Therefore, even though a transformer may not be required to have secondary overcurrent protection for the transformer, protection is usually required at the end of the secondary conductors (**FIGURE 4-17**). As noted, the protective device in the illustration provides only overload protection for the secondary conductors, similar to how the service overcurrent device provides only overload protection for the service conductors. The requirements of Articles 240 and 450 work together in this situation to provide some protection from faults that may occur between the transformer and the first overcurrent device. The conductors on the load side of the secondary overcurrent device then have short-circuit and ground-fault protection in addition to overload protection.

TABLE 4-2 provides a summary of the various "tap rules" that apply to feeders and transformer secondary conductors. Refer to 240.21 for the details of any specific rule.

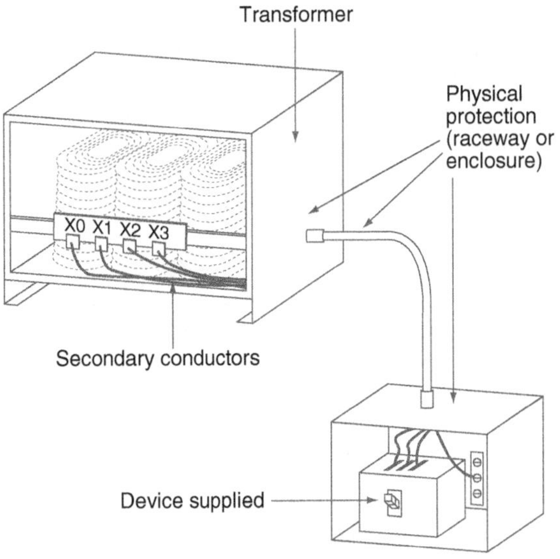

FIGURE 4-17 Overcurrent protection for transformer secondary conductors.

10. Check panelboards supplying or supplied by feeders for overcurrent protection, grounding, and proper enclosures.

Metal parts of panelboards are required to be grounded by connection to an equipment grounding conductor of the types specified in Section 250.118. If separate grounding conductors are used (other than or in addition to raceways), a grounding terminal bar must be provided. Enclosures are required to be dead front (i.e., totally enclosed so that live parts are not exposed at the front with the "dead-front" cover installed) and designed for the purpose. The two previous general categories of panelboards—lighting and appliance panelboards and power panelboards—were eliminated in the 2008 edition of the *NEC*. Since 2008, panelboards are limited to 42 overcurrent devices or spaces only when protected by two main overcurrent devices, a type of installation not as common as it once was but still used by some residential installers and designers. Other panelboards are not restricted in the number of overcurrent devices, but except where used as service equipment, they must be protected by an overcurrent device that does not exceed the panelboard rating. The overcurrent device may be a main device in the panelboard or may be installed in the feeder ahead of the panelboard. These requirements are shown in **FIGURE 4-18**. In general, a panelboard supplied through a transformer may not be protected by the primary overcurrent device unless the secondary is single voltage and supplies only line-to-line connected loads.

11. Verify that an identification scheme for ungrounded feeder conductors has been established and made readily available or posted where the premises wiring system has feeders that are supplied by more than one nominal voltage system.

The ungrounded (hot) conductors in an alternating current (ac) system are required to be identified by system and phase or line where the feeders are supplied by more than one **nominal voltage** system. For instance, where the premises is supplied by a 480Y/277-volt utility service, which is subsequently transformed down to 208Y/120 volts to supply 120-volt outlets and equipment, and feeders are supplied from both systems, the hot (ungrounded) conductors of each system must be identified so that conductors supplied from one system can be distinguished from conductors supplied from the other system(s). Grounded conductors could be identified with the same color in both systems unless the grounded conductors of more than one system share a single enclosure. More commonly, gray is used for one system and white for another, but this requirement in Article 215 is about clearly distinguishing conductors of different systems from each other, and color coding is one technique (probably the most widely used) permitted to meet this rule. Other requirements for color coding of grounded and equipment grounding conductors

| TABLE 4-2 | Summary of Feeder Tap and Transformer Secondary Rule Provisions and Applications |

Section Number	Tap or Secondary Conductor	Maximum Length	Rating Ratio*	Location	Occupancy	Termination	Other Restrictions or Notes
240.21(B)(1)	Tap	10 ft (3 m)	$\frac{1}{10}$	Any	Any	Single device	$\frac{1}{10}$ ratio applies to field taps. Termination may be in device other than overcurrent device.
240.21(B)(2)	Tap	25 ft (7.6 m)	$\frac{1}{3}$	Any	Any	Single OCD†	
240.21(B)(3)	Tap and secondary	25 ft (7.6 m) total	$\frac{1}{3}$	Any	Any	Single OCD	Portion of primary that is protected at its ampacity is not counted in total length, but applies only where primary conductors are tap conductors. Ampacity must be corrected for transformer.
240.21(B)(4)	Tap	100 ft (30.5 m)	$\frac{1}{3}$	Inside	High-bay manufacturing buildings only	Single OCD	Conductors not smaller than 6 AWG cu or 4 AWG al. No penetration of walls, floors, or ceilings permitted.††
240.21(B)(5)	Tap	No limit	N/A	Outside	Any	Single OCD	Must remain outside except at termination.
240.21(C)(1)	Secondary	No limit	N/A	Any	Any	Not specified	Ampacity must be corrected for transformer. Limited to 2-wire secondary and 3-wire delta-delta transformers where secondary conductor may be protected by primary overcurrent device. See 240.4(F).
240.21(C)(2)	Secondary	10 ft (3 m)	$\frac{1}{10}$	Any	Any	Single device	$\frac{1}{10}$ ratio applies to conductors that leave enclosure or vault where connection is made. Termination may be in device other than overcurrent device.
240.21(C)(3)	Secondary	25 ft (7.6 m)	N/A	Any	Industrial only supplying switchboards or switchgear	Multiple OCD grouped together	Conductor ampacity must be at least equal to transformer secondary rating. Sum of overcurrent device ratings cannot exceed conductor ampacity.
240.21(C)(4)	Secondary	No limit	N/A	Outside	Any	Single OCD	Must remain outside except at termination.
240.21(C)(5)	Secondary	25 ft (7.6 m)	$\frac{1}{3}$	Any	Any	Single OCD	Same as 240.21(B)(3), which applies only where primary conductors are tap conductors. Ampacity must be corrected for transformer.
240.21(C)(6)	Secondary	25 ft (7.6 m)	$\frac{1}{3}$	Any	Any	Single OCD	Applies where primary conductors comply with 240.4. Ampacity must be corrected for transformer.

*Rating ratio is the least permitted ratio between the ampacity of the conductor and the rating of the overcurrent device, corrected for transformers where applicable. N/A, not applicable, means that no ratio applies in this rule.
†Single OCD, single overcurrent device with rating not exceeding ampacity of conductors. Rounding up [using 240.4(B)] is not permitted.
††cu, copper; al, aluminum.

still apply. A requirement for identifying the conductors of direct current (dc) systems over 50 volts was established in the 2014 *NEC*, and a specific color code is prescribed for small conductors (6 AWG and smaller) as one acceptable method to identify the positive and negative dc conductors.

Unlike the requirement for identifying the feeder grounded conductor, the requirement for identifying the ungrounded conductor is contingent on the different system conductors being located at the same premises and not on both systems being in the same wiring enclosure or raceway. The identification method and scheme can be unique to that particular premises. Taping, alphanumeric tagging, color coding, or a combination of these methods are all acceptable. The identification scheme is required to be readily available or permanently posted at all feeder distribution equipment—i.e., panelboards and switchboards. There is no specific color code established in the *NEC* for this purpose. One method often used when more than two systems are used in a facility is to identify each system by a color and each phase or line by alphanumeric tagging. The identification is required on feeder conductors at all termination, connection, and splice points.

Checklist 4-3: Branch Circuits

1. Verify that wiring methods used are appropriate for the conditions and occupancy.

All wiring methods have uses that are permitted and not permitted or uses that require some added feature such as mechanical protection or protection from corrosion and deterioration. In addition, some occupancies impose additional restrictions on the use of certain wiring methods or require specific wiring methods. See the Wiring Methods and Devices and the Special Occupancies chapters of this manual for a more complete discussion of this subject.

2. Check panelboards for proper overcurrent protection.

As explained in Item 11 under Feeders, most panelboards feeding branch circuits for lighting and **appliances** are required to have main overcurrent protection. Service panelboards with more than one overcurrent device are not required to have overcurrent protection ahead of the panelboard, but they are limited to a maximum of six overcurrent devices. These devices could supply branch circuits or feeders or both.

3. Check individual and multioutlet branch circuits for proper ratings.

<u>Individual branch circuits</u> may have any rating as long as the rating (the overcurrent device) is appropriate and adequate for the connected conductors and equipment. Such individual branch circuits are permitted to supply loads greater than 50 amperes within any occupancy. **Multioutlet branch circuits** ("other than individual branch circuits") are limited to 15-, 20-, 30-, 40-, and 50-ampere ratings. However, multioutlet branch circuits may be permitted on circuits larger than 50 amperes for special equipment loads in industrial buildings. **FIGURE 4-19** shows examples of both types of branch circuits.

4. Check conductors and overcurrent protection for consideration of continuous and noncontinuous loads, multioutlet loads, and minimum ampacity and size.

All branch-circuit conductors must be at least equal in ampacity to the connected or calculated load. Unlike feeders, there are no demand factors that can be applied to branch circuits. (An exception to this rule is electric ranges where demand values can be used for both feeders and branch circuits, even for a single electric range.) When branch circuits supply more than one receptacle for cord-and-plug–connected portable loads, the conductor ampacity must be no less than the circuit rating; that is, the overcurrent device cannot be "rounded up" to the next standard overcurrent device size. Circuits supplying continuous loads must have ungrounded conductors and overcurrent devices that are increased in rating for the portion of the load that is continuous. This rule is the same as the continuous

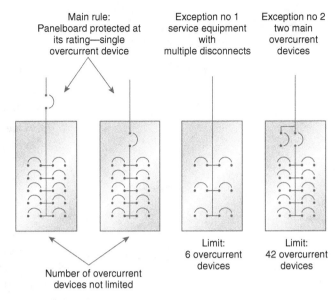

FIGURE 4-18   Overcurrent protection and circuit limitations for panelboards.

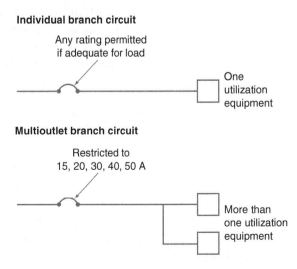

FIGURE 4-19   Individual and multioutlet branch circuits.

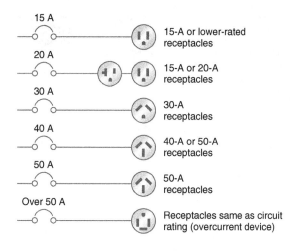

FIGURE 4-20 Maximum overcurrent device rating based on rating of receptacle.

load rule for feeders, which is explained in Item 3 under Feeders. Notice that, like feeders, more than one rule applies to the size, ampacity, and overcurrent protection for branch-circuit conductors. Whichever requires the largest conductor becomes the overriding rule in a specific case.

5. Check branch circuits supplying receptacles and other outlet devices for permitted ratings of circuits and receptacles.
Receptacles and other outlet devices are restricted to circuits of certain ratings, depending on the ratings of the outlet devices. For example, fixtures with lampholders cannot be connected to circuits rated over 20 amperes unless the lampholders are of the heavy-duty type. Many high-intensity discharge luminaires include heavy-duty lampholders, but because few (if any) fluorescent luminaires have heavy-duty lampholders, fluorescent luminaires are usually limited to 15- or 20-ampere circuits. Similar limitations apply to circuits supplying receptacles.

FIGURE 4-20 illustrates the maximum current rating of the branch circuit in relation to the permitted current rating of the receptacle(s).

Note that 15- and 20-ampere receptacles are permitted on a 20-ampere branch circuit, but 20-ampere receptacles may not be connected to 15-ampere circuits. A listed appliance rated over 15 amperes will be equipped with a 20-ampere cord cap, thus preventing it from being used with a 15-ampere receptacle. Also, if a single receptacle is installed on an individual branch circuit, the receptacle must have a rating not less than the branch-circuit rating. For example, a single receptacle on an individual 20-ampere branch circuit must be at least a 20-ampere receptacle. (Note: By definition, a circuit supplying a duplex receptacle is not an individual branch circuit.)

FIGURE 4-21 shows an example of calculating and selecting branch-circuit ratings, conductors, and overcurrent protection for lighting and receptacle outlet circuits.

6. Verify that branch-circuit loads do not exceed maximum permitted loads.
The total load on a branch circuit cannot exceed the rating of the branch circuit. This is simply a restatement of the rules in 201.22 and 210.23 that branch-circuit overcurrent devices (the branch-circuit ratings) must be sized so they are not smaller than the load. However, this rule is also found in Article 220 of the *NEC* because it concerns the branch-circuit load calculation rather than the branch-circuit conductors or overcurrent protection.

7. Verify that branch circuits supplying motors are sized according to Article 430 or 440 and that inductive lighting loads are based on ballast ratings.
Motor loads are generally calculated at 125 percent of the largest motor on a circuit in accordance with Article 430. If there is more than one motor on the same circuit (service, feeder, or branch circuit), the additional smaller motors are calculated at 100 percent. Air-conditioning and refrigeration loads are similarly calculated, except that nameplate currents from the hermetic refrigeration motors are used instead of values from the tables in Article 430. For combination loads, the total load on a branch circuit is calculated from the sum of all the calculated loads plus 125 percent of the largest motor. Inductive lighting loads, such as fluorescent or high-intensity discharge types, are supplied through ballasts. The input ratings of the ballasts must be used in calculating such loads, and not the output ratings of the lamps. For example, a metal halide fixture will have an input ballast current rating about 10 percent to 25 percent higher than the current rating implied by the output rating of the lamp.

Application and Coordination of Conductor Sizing, Overcurrent Device Sizing, and Termination Temperature Requirements

Given:

Load A is a continuous multioutlet electric discharge lighting load. The load on each ungrounded conductor of the multiwire branch circuit is 15 amperes. Because of the ballasts used, the load is considered to be nonlinear (neutral is current-carrying). The lampholders are not heavy duty.

Load B is a noncontinuous multioutlet receptacle load for portable appliances and tools. Each ungrounded conductor of the multiwire branch circuit supplies ten 20-ampere, 125-volt duplex receptacles. The neutral carries only the unbalanced current of the circuit, so it is not counted as a current-carrying conductor.

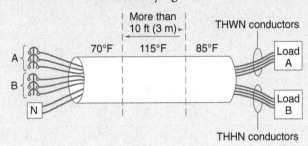

Find:

What is the rating of the overcurrent devices and what are the minimum conductor sizes?

Load A Solution

Minimum rating of overcurrent protection and minimum ampacity of conductors for the continuous load [210.19, 210.20]:

125 percent × 15 amperes = 18.75 – minimum 20-ampere overcurrent device and minimum 18.75-ampere ampacity.

Based on 60°C terminals, the minimum conductor size for a 15-ampere load is 14 American Wire Gauge (AWG) from the 60°C column of Table 310.15(B)(16)[110.14(C)]. Generally, however, 14 AWG must be protected at 15 amperes, so the minimum size is 12 AWG to coordinate the conductor size with the minimum overcurrent device rating [240.4(D)].

Maximum rating of overcurrent protection: Maximum 20 amperes, based on load restrictions on higher-rated circuits [210.23].

Actual calculated load from Article 220: 15-ampere load (continuous/noncontinuous adjustment is not part of Article 220 calculation).

Minimum conductor size for load:

Temperature correction factor for 75°C insulation in a 115°F ambient is 0.75.

Adjustment factor for seven current-carrying conductors in the raceway is 0.70.

Adjusted ampacity = Table ampacity × 0.75 × 0.70.

Minimum adjusted ampacity = 15 amperes from Article 220.

Therefore the Table ampacity × 0.70 × 0.75 must be at least equal to 15 amperes, and

Minimum table ampacity (75°C column before correction or adjustment) = 15 amperes/0.70 × 0.75 = 15 amperes/0.525 = 28.57 amperes.

Minimum conductor size based on conditions of use: 10 AWG THWN [310.15 and Table 310.15(B)(16)].

(*continues*)

FIGURE 4-21   Application and coordination of conductor sizing, overcurrent device sizing, and termination temperature requirements.

Overcurrent device: 20 amperes to comply with both the minimum and the maximum sizes determined above.

Conductor size: No. 10 AWG THWN, the largest of the minimum sizes determined above [210.19(A)].

Minimum conductor size based on conditions of use: 10 AWG THWN [310.15 and Table 310.15(B)(16)].

Overcurrent device: 20 amperes to comply with both the minimum and the maximum sizes determined above.

Conductor size: No. 10 AWG THWN, the largest of the minimum sizes determined above.

Load B Solution

Load calculated in accordance with Article 220:

Load = 10 × 180 VA per outlet = 1800 VA [220.14(I)].

1800 VA/120 volts = 15-ampere noncontinuous load.

Minimum rating of overcurrent protection and minimum ampacity of conductors for the noncontinuous load:

100 percent × 15 A = 15 – minimum 15-ampere overcurrent device and minimum 15-ampere ampacity [210.19, 210.20].

Based on 60°C terminals, the minimum conductor size is 14 AWG from the 60°C column [110.14(C)].

Minimum overcurrent device ratings and conductor sizes for type of load:

Only 20-ampere overcurrent device may be used for 20-ampere receptacles [210.21(B)(3)]. (This is both the maximum and minimum rating of the overcurrent device in this example.)

Because 14 AWG must generally be protected at 15 amperes, the minimum conductor size is 12 AWG to coordinate the conductor size with the 20-ampere overcurrent device rating [240.4 and 240.4(D)].

Minimum conductor size for load:

THHN conductors are 90°C conductors in dry or damp locations. However, since they are associated with 75°C conductors in the same raceway, they are limited to 75°C operating temperatures [310.15(A)(3)].

Temperature correction factor for 75°C insulation in a 115°F ambient is 0.75.

Adjustment factor for seven current-carrying conductors in the raceway is 0.70.

Adjusted ampacity = Table ampacity × 0.75 × 0.70.

Minimum adjusted ampacity = 15 amperes from Article 220.

Therefore, the Table ampacity × 0.70 × 0.75 must be at least equal to 15 amperes, and

Minimum table ampacity (75°C column before adjustment or correction) = 15 amperes/0.70 × 0.75 = 15 amperes/0.525 = 28.57 amperes.

Minimum conductor size based on conditions of use: 10 AWG THHN [310.15 and Table 310.15(B)(16)].

Overcurrent device: 20 amperes to comply with all the minimum and maximum sizes determined above.

Conductor size: 10 AWG THHN, the largest of the minimum sizes determined above.

In the example of **FIGURE 4-22**, a ballast draws 0.75 amperes at 120 volts, or 90 VA, to operate two F32T8 lamps for a light output of 64 watts. Part of the difference is due to power factor, but the rest is due to the power consumed by the ballast. Assuming a power factor of 80 percent (many ballasts of this type are higher power factor, but the power factor of this ballast is not stated), the input of 90 VA is about 72 watts for a rated output of 64 watts, about a 12 percent difference in this case.

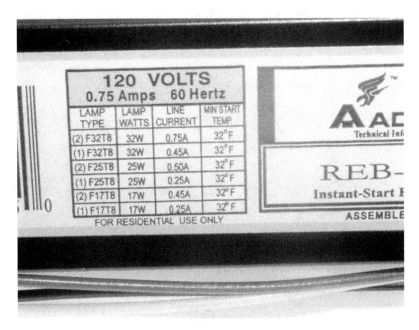

FIGURE 4-22   Example of the nameplate input and output ratings of a fluorescent ballast.

8. Verify that branch circuits are used to supply only permissible loads based on their ratings.

As explained in Item 3 of this manual, an individual branch circuit may supply any load for which it is rated, but multioutlet branch circuits are limited to specific ratings. The use of multioutlet branch circuits is further restricted in terms of the types of loads that may be supplied. The load restrictions are based on the ratings of the branch circuits and the occupancies in which they are used. Essentially, 15- and 20-ampere branch circuits can supply any types of loads in any types of occupancies, as long as the loads (including the increase for any continuous loads) do not exceed the circuit ratings. However, if a 15- or 20-ampere circuit is used for both fixed equipment (other than lighting) and other equipment (such as lighting or receptacle loads), not more than 50 percent of the circuit may be used for the fixed equipment. No individual cord-and-plug–connected load may exceed 80 percent of the circuit rating on 15-, 20-, and 30-ampere circuits. Circuits rated for 30, 40, and 50 amperes may be used only for certain types of loads in dwelling units and may be used for fixed lighting in other occupancies only if the lighting units have heavy-duty lampholders. Branch circuits larger than 50 amperes cannot be used for lighting in any occupancy. Large installations of luminaires (such as the banks of high-output floodlights used for outdoor sports lighting) require multiple branch circuits.

9. Verify that the number of branch circuits is adequate and that the load is evenly proportioned among the branch circuits.

Once the permitted circuit ratings are determined, the required number of branch circuits can be found by dividing the total load by the size of the circuits. The total load is to be divided as evenly as possible between the branch circuits so that no circuits are overloaded and so that the load is reasonably balanced on the system. If the calculated load is more than the actual connected load, branch circuits are required to be supplied only for the connected load. This often happens with lighting loads in commercial occupancies, especially where energy use restrictions or "energy codes" require the use of highly efficient lighting sources. Feeder capacity, however, is generally required for the entire calculated load even though the maximum load that actually exists may be less and may be required to be less.

FIGURE 4-23 illustrates the general requirement. An exception was added to 220.12 in the 2014 *NEC* that permits the calculated load to be reduced to the values specified in an adopted energy code. See 220.12 Exception for the additional conditions that must be met

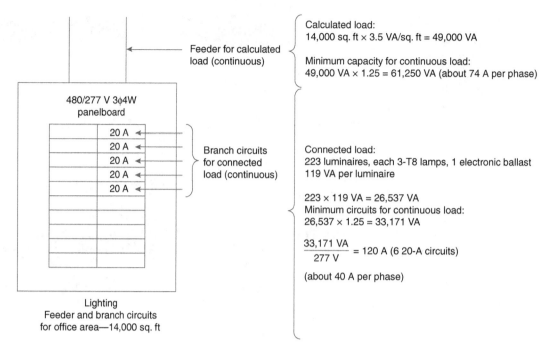

Feeder for calculated load (continuous)

Calculated load:
14,000 sq. ft × 3.5 VA/sq. ft = 49,000 VA

Minimum capacity for continuous load:
49,000 VA × 1.25 = 61,250 VA (about 74 A per phase)

480/277 V 3ϕ4W panelboard

20 A
20 A
20 A
20 A
20 A

Branch circuits for connected load (continuous)

Connected load:
223 luminaires, each 3-T8 lamps, 1 electronic ballast 119 VA per luminaire

223 × 119 VA = 26,537 VA
Minimum circuits for continuous load:
26,537 × 1.25 = 33,171 VA

$$\frac{33,171 \text{ VA}}{277 \text{ V}} = 120 \text{ A (6 20-A circuits)}$$

(about 40 A per phase)

Lighting
Feeder and branch circuits
for office area—14,000 sq. ft

FIGURE 4-23   Example of connected load less than calculated load.

to use these reduced values prescribed by the locally adopted energy code. Load calculations for additions to existing occupancies may be based on the actual measured maximum load.

Note that unlike feeders and service conductors, the neutral conductors in branch circuits are required to be increased in size due to continuous loads because although they do not connect to overcurrent devices, the items they are connected to on the load end may require larger conductors. This is different for branch circuits than for feeder or service conductors because grounded feeder or service conductors are likely to connect to busbars or equivalent terminations on each end. Common practice is to size neutral (grounded) conductors the same as the ungrounded conductors.

10. Check for compliance with branch-circuit voltage limitations.
Dwelling units and similar occupancies, such as hotel and motel guest rooms, are the primary places where actual limits are imposed on voltages. In such occupancies, voltages at the terminals of lighting fixtures or small cord-and-plug–connected appliances are limited to 120 volts. Most of the other "voltage limitations" are actually permissive rules that allow specific types of equipment to be supplied by specific ranges of voltages. Other types of equipment may be supplied in accordance with their listings. However, other voltage limitations may apply for specific equipment or occupancies. For example, according to Article 410 of the *NEC*, electric discharge lighting systems, such as neon signs or art forms, with open circuit voltages exceeding 1000 volts are not permitted in or on dwelling units.

11. Verify that branch circuits for specific loads meet the requirements of the applicable articles.
Definite rules apply to the branch circuits that supply certain types of equipment. The rules for specific equipment must be followed. For example, central heating equipment is generally required to be supplied by individual branch circuits. Many types of equipment, such as electric water heaters and some space heating, are required to be treated as continuous loads, at least at the branch circuit level, even though the equipment may

not always be operating in a continuous mode. Sizing branch-circuit overcurrent protective devices and conductors for continuous loads is covered in Item 4 of this checklist.

12. Check for proper use and identification of multiwire branch circuits.

Multiwire branch circuits supplying multiple outlets and protected by single-pole overcurrent devices are restricted to supplying only line-to-neutral loads. Prior to 2008, multiwire circuits and multiple circuits that supply more than one device on the same yoke, such as split-wired receptacles, were required to have multipole disconnecting means to manually disconnect all ungrounded conductors (**FIGURE 4-24**). This requirement now applies to all multiwire branch circuits. Generally, an overcurrent device must disconnect all ungrounded conductors of a circuit both manually and automatically, but 240.15(B) provides special rules for multiwire branch circuits and a few other special cases.

The requirement for simultaneous manual disconnection of all

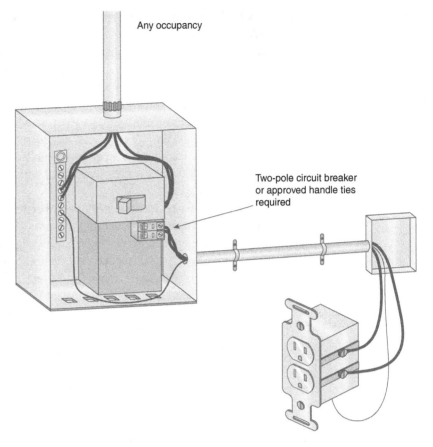

FIGURE 4-24   Simultaneous disconnection for "split-wired" receptacles; now applies to all multiwire branch circuits.

ungrounded conductors can be accomplished either with identified handle ties or multipole circuit breakers. Where multipole circuit breakers are used to disconnect automatically all ungrounded conductors when the circuit breaker operates on an overcurrent, the circuits may also supply line-to-line loads. The previous rule was concerned only with multiple ungrounded conductors on the same device and required a device that would disconnect all the ungrounded conductors. The current rule recognizes that similar hazards may be present at any location where more than one ungrounded conductor of a multiwire branch circuit is present. A similar requirement applies to certain luminaires connected to multiwire branch circuits. Fluorescent luminaires that utilize double-ended lamps and contain ballasts and are supplied power by multiwire circuit(s) require disconnecting means for both the ungrounded and neutral conductor [see 410.130(G)]. However, this disconnecting means is to be provided at each luminaire rather than at the origin of the branch circuit, and a similar disconnecting means that disconnects the ungrounded conductors is required for all such luminaires, even those not supplied from multiwire branch circuits. This rule does not apply in dwelling units.

Another provision that was discussed in the General Requirements Inspections chapter of this manual is the requirement to group all the conductors of a multiwire branch circuit in at least one location where the branch circuit originates. This can be accomplished through the use of wire ties, but such additional means is not required where the grouping is obvious by entry in a raceway or cable that is unique to the circuit. Other methods of identifying which grounded conductors are associated with which ungrounded conductors may be used as long as the method is acceptable to the AHJ. These rules apply to all occupancies. Multiwire circuits are permitted to be considered as single circuits or as multiple circuits (or both), depending on the rule that is being applied.

13. Check for required receptacles and lighting at mechanical equipment and service equipment.

Receptacle outlets must be provided for the servicing of all mechanical equipment for heating, air-conditioning, and refrigeration (HACR). These receptacle outlets must be within 25 ft (7.6 m) and on the same level as the equipment. Such receptacles are also often required for ventilation equipment by mechanical or building codes, but are not necessarily required by the *NEC*, and the *NEC* specifically exempts evaporative coolers on one- and two-family dwelling units. The outside receptacles required for dwelling units may satisfy this requirement if properly located. Receptacles must also be provided within 50 ft (15 m) of the electrical service equipment, but this requirement does not apply in one- and two-family dwellings.

Lighting outlets must be provided in attics and crawl spaces that contain mechanical equipment or other equipment that requires servicing. The lighting outlet(s) must be controlled by a wall switch or by a switch at the lighting outlet. A control must be located at the point of entry to the space, and the lighting outlet must be located near the equipment that is to be serviced. In dwelling units, lighting outlets must also be located in attics, crawl spaces, utility rooms, and basements that are either used for storage or contain equipment that requires servicing. Mechanical codes may contain requirements for lighting other types of spaces that contain mechanical equipment.

14. Check for required outlets or receptacles for show windows and signs.

In commercial buildings with **show windows**, receptacles must be provided directly above the windows for show window lighting. One receptacle is required for every 12 ft (3.7 m) of show window or major fraction. The effect is that any show window up to 18 ft (5.5 m) in width must have at least one receptacle outlet. Windows 18 ft to 30 ft (5.5 to 9.1 m) wide must have two receptacle outlets, and an additional outlet is required for each additional 12 ft (3.7 m) of window. The receptacle outlet must be located within 18 in. (457 mm) of the top of the window. Additional receptacle outlets may be installed in floors or walls of show windows. The calculated load for show window receptacles is based on either the number of required receptacles or the horizontal dimension of the window(s).

At least one outlet (not necessarily a receptacle outlet) must be installed on a dedicated 20-ampere circuit for **electric sign** loads in all commercial buildings or occupancies. These outlets are required at each pedestrian customer entrance to each tenant space. The load on these outlets is calculated at a minimum of 1200 VA or the actual load if it is known to be greater and is considered to be a continuous load.

15. Verify that receptacles are provided for all cord-and-plug–connected appliances and where other flexible cords are used.

Receptacles must be installed where flexible cords with attachment plugs are used, such as with appliances or portable equipment. This will prevent the unnecessary use of extension cords by requiring receptacles where they are most likely to be used. Where the receptacle is for a specific appliance in a dwelling unit, it must be within 6 ft (1.8 m) of the intended location of the appliance. The attachment plugs are not required if the flexible cords are permanently connected, where permitted in Article 400. However, Article 400 places significant restrictions on the use of flexible cord. Any use of flexible cord must be permitted by 400.7 and not prohibited by 400.8.

16. Verify that GFCI protection is provided for receptacles in bathrooms, in kitchens, near sinks, outdoors, on rooftops, in indoor wet areas, in locker rooms with showers, and in garages.

Receptacles (125-volt, 15- and 20-ampere) are not required to be installed in nonresidential **bathrooms,** but where they are installed, they must have GFCI protection. Similarly, where such receptacles are installed in **kitchens**, on rooftops, outdoors, or within 6 ft (1.8 m) of a sink, they must be GFCI protected. A GFCI-protected receptacle is required to be installed within 25 ft (7.6 m) of HACR equipment located on rooftops or other wet locations. For an example, see **FIGURE 4-25** and Item 13 in this checklist. This receptacle is for service personnel and must be located on the same level as the equipment that requires servicing.

Receptacles must also be provided within 25 ft (7.6 m) of HACR equipment in other locations per 210.63, but GFCI protection is required in accordance with Section 210.8. Receptacles for certain types of equipment and in certain occupancies are exempted from the GFCI requirements. The requirement for GFCI protection in nonresidential kitchens applies to all 125-volt, 15- and 20-ampere receptacles, not just those that serve countertop areas. Receptacles in locker rooms with associated showering facilities now require

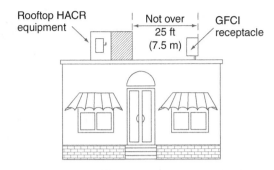

FIGURE 4-25   GFCI-protected receptacles for servicing HACR equipment.

GFCI protection although the *NEC* does not require the receptacles themselves. The requirement for GFCI protection in garages and service bays is based on the type of equipment to be used and is also found in 511.12. GFCI protection is required in other areas or occupancies, and those additional requirements are found in other articles, such as Article 590 for temporary installations, Article 513 for aircraft hangars, Article 517 for patient care areas, and Article 620 for elevator and escalator machine or control rooms or spaces and pits. GFCI requirements for dwelling units are covered in the Dwelling Units and Mobile/Manufactured Home Sites chapter of this manual.

17. Verify that disconnects are provided at separate structures for branch circuits running between structures.

As with feeders used to supply separate structures, branch circuits supplying a separate structure are required to be provided with disconnects at the separate structure. The general rule is that only one feeder or one branch circuit is permitted to supply a separate building, but there are conditions permitting multiple supplies. Disconnects are not required at some structures such as light poles. (See the discussion in Item 6 of the Feeders checklist.)

18. Check for adequate size and clearances for outside branch circuits.

Outside branch circuits that run overhead between buildings are required to have adequate mechanical strength or be supported by a messenger wire. Minimum sizes are specified based on the voltage and the length of the span. Such branch circuits also must be installed so that the same clearances required for overhead feeders or service conductors are maintained.

19. Verify that an identification scheme for ungrounded branch-circuit conductors has been established and made readily available or posted where the premises wiring system has branch circuits that are supplied by more than one nominal voltage system.

The ungrounded (hot) conductors are required to be identified by system and phase or line where the branch circuits are supplied by more than one nominal voltage system. For instance, where the premises is supplied by a 480Y/277-volt utility service, which is subsequently transformed down to 208Y/120 volts to supply 120-volt outlets and equipment, and branch circuits are supplied from both systems, the hot (ungrounded) conductors of each system must be identified at all termination, connection, or splice points so that conductors supplied from one system can be distinguished from conductors supplied from the other system(s).

Unlike the requirement for identifying the branch-circuit grounded conductor, the requirement for identifying the ungrounded conductor is contingent on the different system conductors being located at the same premises and not on both systems being in the same wiring enclosure or raceway. Grounded conductors could be identified with the same color in both systems unless the grounded conductors of more than one system share a single enclosure. More commonly, gray is used for one system and white for another, but this requirement in Article 210 is not about color coding and does not require that color coding be used. Other requirements for color coding grounded and equipment grounding conductors still apply. The identification method and scheme can be unique to a particular premises, and taping, alphanumeric tagging, and color coding, or a combination of these, are all acceptable methods. The 2014 *NEC* does establish a specific color code for 6 AWG and smaller conductors of dc systems above 50 volts. The color code is specified where colors are used, but other identification methods are permitted.

The identification scheme is required to be permanently posted at all branch-circuit distribution equipment (i.e., panelboards and, in some cases, switchboards) or to be otherwise readily available. One method often used when more than two systems are used in a facility is to identify each system by a color and each phase or line by alphanumeric tagging. The identification is required on branch-circuit conductors at all termination, connection, and splice points.

KEY TERMS

Article 100 of the *NEC* defines the key terms pertaining to services, feeders, and branch circuits listed below as follows:

Ampacity The maximum current, in amperes, that a conductor can carry continuously under the conditions of use without exceeding its temperature rating.

Appliance Utilization equipment, generally other than industrial, normally built in standardized sizes or types, that is installed or connected as a unit to perform one or more functions, such as clothes washing, air-conditioning, food mixing, deep frying, and so forth.

Bathroom An area including a basin with one or more of the following: a toilet, a urinal, a tub, a shower, a bidet, or similar plumbing fixtures.

Branch Circuit The circuit conductors between the final overcurrent device protecting the circuit and the outlet(s). [*The* outlets *may be receptacle outlets, or in the case of lighting outlets or appliance outlets, may be the junction box where the connection to the leads of the luminaire or appliance is made.*]

Branch Circuit, Individual A branch circuit that supplies only one utilization equipment. [*These circuits are often called "dedicated circuits," but this term does not specify that a circuit supplies only one utilization equipment. Circuits that supply more than one utilization equipment are called "other than individual branch circuits" in the* NEC, *but they are also commonly called "multioutlet branch circuits" elsewhere, including in the* NEC *and in this manual.*]

Branch Circuit, Multiwire A branch circuit consisting of two or more ungrounded conductors that have a voltage between them and a grounded conductor that has equal voltage between it and each ungrounded conductor of the circuit and that is connected to the neutral or grounded conductor of the system. [*Most circuits have more than one wire. This definition is specific to circuits that have more than one ungrounded wire sharing the same grounded conductor, which is usually a neutral.*]

Building A structure that stands alone or that is cut off from adjoining structures by fire walls with all openings therein protected by approved fire doors. [*See* Structure.]

Continuous Load A load where the maximum current is expected to continue for 3 hours or more.

Demand Factor The ratio of the maximum demand of a system, or part of a system, to the total connected load of a system or the part of the system under consideration.

Electric Sign A fixed, stationary, or portable self-contained, electrically illuminated utilization equipment with words or symbols designed to convey information or attract attention.

Feeder All circuit conductors between the service equipment, the source of a separately derived system, or other power supply source and the final branch-circuit overcurrent device.

Ground-Fault Circuit Interrupter (GFCI) A device intended for the protection of personnel that functions to deenergize a circuit or portion thereof within an established period of time when a current to ground exceeds the values established for a Class A device.

> *Informational Note:* Class A ground-fault circuit interrupters trip when the current to ground is 6 MA or higher and do not trip when the current to ground is less than 4 MA. For further information, see UL 943, *Standard for Ground-Fault Circuit Interrupters.*

Ground-Fault Protection of Equipment (GFPE) A system intended to provide protection of equipment from damaging line-to-ground fault currents by operating to cause a disconnecting means to open all ungrounded conductors of the faulted circuit. This protection is provided at current levels less than those required to protect conductors from damage through the operation of a supply circuit overcurrent device.

Intersystem Bonding Termination A device that provides a means for connecting intersystem bonding conductors for communication systems to the grounding electrode system.

Kitchen An area with a sink and permanent provisions for food preparation and cooking.

Lighting Outlet An outlet intended for the direct connection of a lampholder or luminaire.

Luminaire A complete lighting unit consisting of a light source such as a lamp or lamps, together with the parts designed to position the light source and connect it to the power supply. It may also include parts to protect the light source or the ballast or to distribute the light. A lampholder itself is not a luminaire.

Outlet A point on the wiring system at which current is taken to supply utilization equipment.

Rating (of a branch circuit) [*Derived from Section 210.3*] The rating of the branch-circuit overcurrent device or, in the case of a motor circuit, the rating of the branch-circuit short-circuit and ground-fault protective device.

Receptacle A receptacle is a contact device installed at the outlet for the connection of an attachment plug. A single receptacle is a single contact device with no other contact device on the same yoke. A multiple receptacle is two or more contact devices on the same yoke.

Receptacle Outlet An outlet where one or more receptacles are installed.

Service The conductors and equipment for delivering electric energy from the serving utility to the wiring system of the premises served. [*A building that is not supplied by an electric utility but is supplied only by a customer-owned generator or similar system has no service as defined. The power supply is usually a separately derived system in such cases, and the supply to the building consists of one or more feeders or branch circuits.*]

Service Conductors The conductors from the service point to the service disconnecting means.

Service Conductors, Overhead The overhead conductors between the service point and the first point of connection to the service-entrance conductors at the building or other structure.

Service Conductors, Underground The underground conductors between the service point and the first point of connection to the service-entrance conductors in a terminal box, meter, or other enclosure, inside or outside the building wall.

> *Informational Note:* Where there is no terminal box, meter, or other enclosure, the point of connection is considered to be the point of entrance of the service conductors into the building.

Service Drop The overhead conductors between the utility electric supply system and the service point.

Service-Entrance Conductors, Overhead System The service conductors between the terminals of the service equipment and a point usually outside the building, clear of building walls, where joined by tap or splice to the service drop or overhead service conductors.

Service-Entrance Conductors, Underground System The service conductors between the terminals of the service equipment and the point of connection to the service lateral or underground service conductors.

> *Informational Note:* Where service equipment is located outside the building walls, there may be no service-entrance conductors, or they may be entirely outside the building.

Service Equipment The necessary equipment, usually consisting of a circuit breaker(s) or switch(es) and fuse(s), and their accessories, connected to the load end of service conductors to a building or other structure, or an otherwise designated area, and intended to constitute the main control and means of cutoff of the supply.

Service Lateral The underground conductors between the utility electric supply system and the service point.

Service Point The point of connection between the facilities of the serving utility and the premises wiring.

Show Window Any window used or designed to be used for the display of goods or advertising material, whether it is fully or partly enclosed or entirely open at the rear and whether or not it has a platform raised higher than the street floor level.

Structure That which is built or constructed.

Utilization Equipment Equipment that utilizes electric energy for electronic, electromechanical, chemical, heating, lighting, or similar purposes.

Voltage (of a circuit) The greatest root-mean-square (rms) (effective) difference of potential between any two conductors of the circuit concerned.

> *Informational Note:* Some systems, such as 3-phase 4-wire, single-phase 3-wire, and 3-wire direct current, may have various circuits of various voltages.

Voltage, Nominal A nominal value assigned to a circuit or system for the purpose of conveniently designating its voltage class (e.g., 120/240, 480Y/277 volts, or 600 volts).

> *Informational Note No. 1:* The actual voltage at which a circuit operates can vary from the nominal within a range that permits satisfactory operation of equipment.

> *Informational Note No. 2:* See American National Standards Institute (ANSI) C84.1-2006, *Voltage Ratings for Electric Power Systems and Equipment (60 Hz).*

[*Although the* NEC *designates nominal system values for the purposes of load calculations in Article 220, other nominal values may also be used for the same systems for other purposes. For example, motors may be designated 115/230 volts on their nameplates, but this refers to the same system as the 120/240-volt values given in Article 220, so a 115-volt rated motor may be properly used on a 120-volt circuit. In fact, 110, 115, 118, and 120 volts are all different classifications used for the same nominal voltage.*]

Voltage to Ground For grounded circuits, the voltage between the given conductor and that point or conductor of the circuit that is grounded; for ungrounded circuits, the greatest voltage between the given conductor and any other conductor of the circuit.

[*The* NEC *provides a definition of the voltage to ground in ungrounded systems so that rules that refer to the voltage to ground, such as working space requirements, can be easily applied with such systems. Under ground-fault conditions in an ungrounded system, the voltage to ground is the line-to-line voltage. The actual measured voltage to ground in an ungrounded system is not quite so predictable during normal operations and may in fact be greater than the maximum voltage between conductors and may fluctuate over time. Voltage to ground is stabilized in grounded systems only under normal conditions [see 250.4(A)(1)].*]

KEY QUESTIONS

1. How many services are there, and where is the service equipment located?
2. Are the service conductors overhead or underground?
3. What is the required current and voltage rating of service equipment and service conductors?
4. What are the required conductor and overcurrent device sizes for feeders?
5. Is more than one building or structure served from the same supply source, and, if so, what are the required conductor and disconnect sizes?
6. Are multiwire branch circuits being used?

CHECKLISTS

Checklist 4-1: Services				
✔	Item	Inspection Activity	*NEC* Reference	Comments
	1.	Verify that each building or structure has only one service or, if more than one, that additional services are justified and identified.	230.2	
	2.	Verify that each service drop or lateral supplies only one set of service-entrance conductors or, if more than one, that the additional sets are justified and identified.	230.40, 230.40 Exception No. 1, 230.2	
	3.	Check clearances from building openings, grade, roadway, roofs, and swimming pools.	230.9, 230.24, 680.8	
	4.	Verify that the point of attachment for an overhead service drop is adequate and will provide minimum clearances.	230.26, 230.27	
	5.	Verify that masts used as supports for service-drop conductors have adequate strength and are not used to support other conductors or equipment.	230.28	
	6.	Verify that supports for service conductors passing over a roof are adequate and substantial.	230.29	
	7.	Check underground conductors for adequate burial depth and protection.	230.32, 230.50, 300.5	
	8.	Check above-ground conductors and cables for adequate supports and protection from physical damage.	230.50, 230.51	
	9.	Verify that wiring methods and support systems for service-entrance conductors are suitable.	230.43, 230.44	
	10.	Verify that service raceways are arranged to drain and that service heads are raintight and properly located.	230.53, 230.54	
	11.	Check service conductors for adequate size and rating.	230.23, 230.31	

CHECKLISTS

Checklist 4-1: Services

✔	Item	Inspection Activity	*NEC* Reference	Comments
	12.	Verify that service equipment is identified as suitable for that use.	230.66	
	13.	Verify that a service disconnecting means is provided, suitable, and located outside or inside nearest the point of entrance of the service conductors.	230.70	
	14.	Verify that service overcurrent protection is provided, properly sized, and part of or adjacent to the disconnecting means.	230.90, 230.91	
	15.	Verify that service disconnects are grouped together and limited to six in any one location.	230.71, 230.72	
	16.	Check ratings of service disconnecting means.	110.24, 230.79, 230.80	
	17.	Check for equipment connected to the supply side of the service disconnecting means and overcurrent protection.	230.82, 230.94	
	18.	Verify that ground-fault protection is supplied where required, and obtain a written record of performance testing.	230.95	

Checklist 4-2: Feeders

✔	Item	Inspection Activity	*NEC* Reference	Comments
	1.	Verify that feeder conductors, including any required neutral conductors, are adequate for the load.	215.2, 220.40, 220.61, 215.5	
	2.	Check overcurrent device and feeder conductor sizing for continuous and noncontinuous loads.	220.61, 215.2, 215.3	
	3.	Check wiring methods for suitability.	*NEC* Chapter 3	
	4.	Check feeders with disconnecting means rated at 1000 amperes or greater for ground-fault protection for equipment if required.	215.10, 230.95	
	5.	Verify that disconnects are provided at separate structures for feeders running between structures.	Article 225, Part II, 225.32	
	6.	Verify that disconnects at separate structures are properly rated, located, grouped, and identified.	Article 225, Part II	

(continues)

CHECKLISTS

Checklist 4-2: Feeders

✔	Item	Inspection Activity	*NEC* Reference	Comments
	7.	Verify that any outside feeders use appropriate wiring methods and are properly supported and arranged to drain.	225.10, 225.20 through 225.22	
	8.	Check any outside feeders for adequate supports, clearances, and mechanical protection.	225.15 through 225.20	
	9.	Verify that feeders supplied from transformers or from other feeders are properly protected by overcurrent devices.	240.4(E) and (F), 240.21(B) and (C)	
	10.	Check panelboards supplying or supplied by feeders for overcurrent protection, grounding, and proper enclosures.	408.36 through 408.40	
	11.	Verify that an identification scheme for ungrounded feeder conductors has been established and made readily available or posted where the premises wiring system has feeders that are supplied by more than one nominal voltage system.	215.12(C)	

Checklist 4-3: Branch Circuits

✔	Item	Inspection Activity	*NEC* Reference	Comments
	1.	Verify that wiring methods used are appropriate for the conditions and occupancy.	*NEC* Chapter 3	
	2.	Check panelboards for proper overcurrent protection.	408.36 through 408.40	
	3.	Check individual and multioutlet branch circuits for proper ratings.	210.3	
	4.	Check conductors and overcurrent protection for consideration of continuous and noncontinuous loads, multioutlet loads, and minimum ampacity and size.	210.19, 210.20, Article 220, 310.15	
	5.	Check branch circuits supplying receptacles and other outlet devices for permitted ratings of circuits and receptacles.	210.21, 210.24	
	6.	Verify that branch-circuit loads do not exceed maximum permitted loads.	201.22, 210.23, 220.10 through 220.14, 220.18	
	7.	Verify that branch circuits supplying motors are sized according to Article 430 or 440 and that inductive lighting loads are based on ballast ratings.	220.14(C), 220.18	
	8.	Verify that branch circuits are used to supply only permissible loads based on their ratings.	201.22, 210.23	
	9.	Verify that the number of branch circuits is adequate and that the load is evenly proportioned among the branch circuits.	210.11, 220.12	
	10.	Check for compliance with branch-circuit voltage limitations.	210.6	

Checklist 4-3: Branch Circuits

✔	Item	Inspection Activity	*NEC* Reference	Comments
	11.	Verify that branch circuits for specific loads meet the requirements of the applicable articles.	210.2, 422.12, 422.13, 424.3, 424.22	
	12.	Check for proper use and identification of multiwire branch circuits.	210.4	
	13.	Check for required receptacles and lighting at mechanical equipment and service equipment.	210.63, 210.64, 210.70(A)(3), 210.70(C)	
	14.	Check for required outlets or receptacles for show windows and signs.	210.62, 220.14(G), 600.5(A), 220.14(F)	
	15.	Verify that receptacles are provided for all cord-and-plug–connected appliances and where other flexible cords are used.	210.50, 400.7, 400.8	
	16.	Verify that GFCI protection is provided for receptacles in bathrooms, in kitchens, near sinks, outdoors, on rooftops, in indoor wet areas, in locker rooms with showers, and in garages.	210.8(B)	
	17.	Verify that disconnects are provided at separate structures for branch circuits running between structures.	Article 225, Part II	
	18.	Check for adequate size and clearances for outside branch circuits.	225.6, 225.18, 225.19	
	19.	Verify that an identification scheme for ungrounded branch-circuit conductors has been established and made readily available or posted where the premises wiring system has branch circuits that are supplied by more than one nominal voltage system.	210.5(C)	

CHECKLISTS

CHAPTER 5

Grounding and Bonding

OVERVIEW

This chapter covers inspections of grounding and bonding of electrical systems and equipment. The reasons for grounding and bonding fall into two general categories:

1. Limiting voltages to ground
2. Providing low-impedance paths for ground-fault current

Both of these reasons support the primary purpose of the *National Electrical Code®* (*NEC®*) as stated in Article 90: the protection of persons and property from hazards arising from the use of electricity. Electrical systems, circuit conductors, and electrical **equipment** are **grounded** to provide safety to the users of electrical equipment during normal and fault conditions. These reasons for grounding can be further subdivided according to whether systems, circuit conductors, or equipment is being considered.

Limiting the voltage to **ground** is important because the users of electrical equipment are at or near ground potential simply because they are often directly or indirectly in contact with ground. Thus, the voltage to ground is basically the same as the voltage to which the users of the system are most likely to be exposed. *Ground* is essentially the environment in which users of electrical equipment are located (e.g., the earth). In fact, in some cases, such as on the upper floors of metal frame buildings, the frame of the building defines the grounded environment and serves as a conducting extension from the earth.

Systems refer to supply systems such as common 120/240-volt, single-phase; 208Y/120-volt, 3-phase; or 480Y/277-volt, 3-phase systems, among others. Such systems are grounded to limit voltages imposed by lightning, line surges, and unintentional contact with higher-voltage lines. A system that is grounded also has a limited and stabilized voltage to ground under normal conditions. The voltage to ground due to lightning, for example, is more controlled and the path to ground for lightning current is more predictable in a grounded system than it is in an ungrounded system. A grounded system, such as a single-phase, 120/240-volt system, has a stable line-to-ground voltage of 120 volts during normal operation. If the system neutral is ungrounded, the voltage to ground of either line conductor is unpredictable during both normal and abnormal conditions. "Normal conditions" do not include fault conditions.

Equipment with a conductive enclosure, such as an appliance housing or a metal raceway, is also grounded to limit the voltage to ground during both normal and abnormal conditions. In this case, the voltage to ground may be thought of as the voltage to which a person is exposed when in contact with electrical equipment. If both the people and the equipment are connected to the same conductive reference, then the people will be exposed to little or no voltage difference (touch-potential) when using such equipment. This type of grounding is sometimes called equipotential grounding because it provides for equal potentials between all exposed parts of electrical equipment with which people may normally come into contact.

Equipment grounding conductors (EGCs) are **bonded** to the system-**grounded conductor** to provide a low-impedance path for fault current that will facilitate the automatic operation of overcurrent devices under ground-fault conditions or facilitate the operation of ground detectors on ungrounded systems. When an energized (ungrounded) conductor makes electrical contact with the ground or a grounded part, the connection is called a **ground fault**. Low-impedance paths between a grounded part and the grounded service conductor are necessary to allow fault currents to rise to high enough values to cause circuit breakers or fuses to operate quickly. While the same consideration applies to short circuits, short circuits alone do not involve equipment grounding circuit pathways, although they may involve grounded conductors.

Whether the circuit is intentional or accidental, a voltage difference across a complete circuit pathway results in the circuit conductors carrying current. When the circuit voltage is fixed, such as the nominal system voltage, the magnitude of the current is inversely proportional to the impedance of the circuit path. Thus, a low-impedance path for fault current results in a high current, which facilitates the operation of the overcurrent device and also reduces the voltage difference to which people may be exposed while there is fault current present. This is the purpose of bonding: to connect non-current-carrying metal parts and metallic systems together through reliable, low-impedance pathways with adequate current-carrying capacity. Furthermore, bonding EGCs to grounded system conductors completes the fault-current path to the utility transformer or other voltage source.

Because the current that occurs on grounding and **bonding conductors** is a relatively short-time event, the minimum size conductor specified is smaller than conductor sizes required to carry long-term currents for the normal operation of electrical equipment. Short-time current-carrying capacities such as those established by the sizes for EGCs in Table 250.122 in the *NEC* do not have the same heating effect on conductors as does the continuous current ampacity for the same size conductor specified in an allowable ampacity table such as Table 310.15(B)(16) in the *NEC*.

FIGURES 5-1, 5-2, and **5-3** illustrate the function of the grounded and grounding/bonding conductors during normal and fault conditions. It is important to understand that the operation of the overcurrent device is not facilitated by the grounding (earth) connections. The bonding of the EGCs to the grounded (neutral) conductor creates and completes the fault-current circuit back to the source. Connections to earth through a **grounding electrode** establish a voltage reference to ground for shock protection and provide a path for those circuits such as lightning in which the earth is part of the circuit.

Whether systems are grounded or not, non-current-carrying metal parts are bonded together to provide good pathways in the event of accidental contact or insulation failures. The same parts are grounded to limit the voltage between parts and between such parts and the environment.

In addition to these general objectives, Article 250 of the *NEC* provides specific requirements related to the following:

- Systems, circuits, and equipment required, permitted, or not permitted to be grounded
- The circuit conductor to be grounded on grounded systems

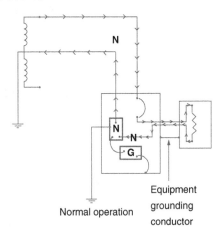

FIGURE 5-1 Normal circuit current path in a grounded system. (Reprinted with permission from *National Electrical Code®* Seminar Module 4, Slide 7, page 186, as modified, Copyright © 2011, National Fire Protection Association, Quincy, MA 02169. This reprinted material is not the complete and official position of the NFPA or the referenced subject, which is represented only by the standard in its entirety.)

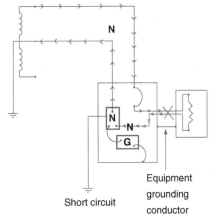

FIGURE 5-2 Short-circuit current path in a grounded system. (Source: *National Electrical Code®* Seminar, NFPA, Quincy, MA, 2011, Module 4, Slide 8, page 186, as modified.)

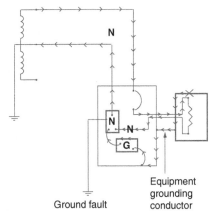

FIGURE 5-3 Ground-fault circuit path in a grounded system. (Source: *National Electrical Code®* Seminar, NFPA, Quincy, MA, 2011, Module 4, Slide 9, page 186, as modified.)

- The location of grounding connections
- The types and sizes of grounding and bonding conductors and electrodes
- Permissible methods of grounding and bonding

Significant changes in many terms related to grounding and bonding were made in the 2008 edition of the *NEC* in an ongoing effort to provide more consistent terminology and foster better understanding of the requirements and objectives of Article 250. Often, in previous editions of the *NEC*, equipment was required to be "effectively grounded." This somewhat subjective definition was eliminated in 2008, and throughout the *Code*, rules that required things to be effectively grounded or that otherwise required grounding have been replaced with specific language that describes how grounding is to be accomplished in each specific case. In most cases, equipment is grounded by connecting it to an EGC. In some cases, a system or certain items of equipment may be grounded by connecting them to a grounding electrode and **grounding electrode conductor**. In other cases, some equipment may be grounded by connecting it to a grounded service conductor or other grounded conductor.

SPECIFIC FACTORS

Usually, service grounding and bonding is inspected first, followed by grounding and bonding of specific equipment. In most cases, the inspection of system and equipment grounding coincides with the inspection of other electrical system components.

Service grounding is concerned primarily with grounding of the system and bonding of conductive parts on the line side of the **service equipment**. Typically, the electrical service utility establishes the system type (single-phase, 3-phase wye, or 3-phase delta), the nominal voltage, and grounding requirements for the source-side supply system. However, a grounded reference for the **premises wiring** is also needed in the building or structure to meet *Code* requirements.

Service grounding is different from the grounding of other parts of a wiring system because the service is the place where a reference to the local wiring system environment (the earth) is established. Therefore, a reliable connection to ground is very important. A reliable connection depends on proper selection and installation of grounding electrodes, connections to electrodes, grounding electrode conductors, and, in grounded systems, connections to grounded circuit conductors. This pathway to earth is typically not called on to carry significant amounts of fault current from the premises. However, it may be required to carry lightning currents or other transient currents from inside or outside the premises, so conductors and connections must be reliable and adequate for and routed to accommodate such currents.

Service grounding and bonding differ from equipment grounding and bonding in at least the following three respects:

- Because the grounded service conductor completes the electrical path to the source for both normal circuit currents and fault currents, the grounded conductor must be adequate for both types of currents. Downstream from the **service**, EGCs provide circuit pathways for fault currents, and grounded circuit conductors provide circuit pathways for normal currents; these conductors are not interchangeable.
- Overcurrent protection for the service conductors is generally limited to overload protection. Because the overcurrent device is located at the load end of service conductors, no ground-fault or short-circuit protection that meets *NEC* requirements is provided for the upstream service conductors. Therefore, the impedance, capacity, and reliability of bonding connections are especially vital to safe operation of the service conductors and equipment. Feeders and branch circuits and their associated EGCs are provided with short-circuit and ground-fault protection in addition to overload protection.

- The magnitude of fault currents at the service equipment is limited primarily by the capacity of the source, the impedance of the supply transformer, and the impedance of the service conductors. Both the total circuit impedance and the rating and type of the overcurrent device determine the magnitude and duration of faults on EGCs. Since feeder and branch-circuit conductors are usually longer and smaller than service conductors are, the total impedance is higher. In addition, some of the available fault current has been reduced by the impedance of the service conductors and equipment. Therefore, assuming we are still looking at the same system, the fault currents available from feeders and branch circuits are usually lower than those at services, and the branch circuits and feeders have overcurrent devices ahead of them that can clear most fault currents. However, when a transformer is added to step the voltage down, the fault current, while subject to the added impedance of the transformer, is still often much higher on the secondary side than it was at the higher voltage on the primary.

EGCs are both different from and similar to other circuit conductors in important ways. The difference is that they are not expected to carry current typically. The important similarity is that EGCs and **bonding jumpers** become circuit conductors in the event of a ground fault. At those times, they are vital to the safety features of an electrical system. EGCs are not expected to carry current for very long—only until the overcurrent device operates, typically a few cycles at most. Therefore, the sizing requirements are different from those for other normal circuit conductors, although their size is still related to the rating of the overcurrent device. Circuit conductors are sized based on their ampacity as defined—that is, for the calculated (long-term or continuous) load currents—while EGCs are sized for ground-fault currents that occur only for very short periods of time.

The primary practical difference between service grounding and bonding and equipment grounding and bonding is in the method used to size the grounding and bonding conductors. The size of grounding and bonding conductors on the supply side of the first overcurrent device in a system is based on the size of the ungrounded conductors. The minimum size for grounding and bonding conductors on the load side of overcurrent devices is based on the rating of the upstream overcurrent device but in all cases must be of sufficiently low impedance to provide an effective path for ground-fault current. Also, because EGCs have short-circuit and ground-fault protective devices ahead of them, the requirements for **equipment bonding jumpers** and bonding connections are in some cases less stringent than the requirements on the supply side of services.

KEY QUESTIONS

1. What size are the ungrounded (hot) service-entrance conductors?

Generally, this question can be answered by examining the electrical design plans. Otherwise, the size of the service conductors can be determined by examining the conductors in the field. The answer to this question is the basis for sizing many supply-side elements of the grounding and bonding system as well as the grounded conductor of the service.

2. What types of grounding electrodes are (or will be) present on the premises?

The answer to this question can usually be determined by examining the electrical design plans, but in some cases, structural and mechanical plans may have to be consulted. Often, more than one inspection, at various stages of construction, may be required to verify that all electrodes present are used and bonded together to form the grounding electrode system.

3. On the basis of all available grounding electrodes present, what size grounding electrode conductor or conductors and bonding jumpers are required?

The arrangement of grounding electrodes, grounding electrode conductors, and bonding jumpers, along with the size of the service conductors, provides the answers to this question.

4. Is the equipment required to be grounded?

Most metallic equipment or enclosures that contain insulated conductors are required to be grounded. Certain types of equipment, such as double-insulated equipment and some "low-voltage" equipment that could not create a shock hazard, are not required to be grounded.

5. What type of EGC is used?

EGCs may be individual conductors, metallic raceways, cable trays, metallic cable sheaths, or a combination of a raceway or cable tray and an individual conductor. EGCs of the wire type must be sized on the basis of the circuit overcurrent device rating, but those sizing requirements do not apply to raceways.

6. If a separate EGC is used, what is the conductor material?

The conductor material, copper or aluminum, is a factor in the determination of the proper size of conductor. Some restrictions that apply to the use of aluminum conductors do not apply to copper conductors.

7. What is the rating of the overcurrent device protecting the circuit?

EGC sizing is based primarily on the rating of the overcurrent device protecting the circuit.

PLANNING FROM START TO FINISH

Service grounding and bonding may involve one or many separate inspections. If the job is only the changing of a service, for example, all inspections can likely be done at one time. But in new construction, the necessary inspections may occur throughout the progress of a job. Inspection of a concrete-encased electrode may be one of the first inspections done, probably at the same time as the inspections of the footings and prior to any concrete pours. The last inspection of service grounding and bonding may occur as late as the final inspection, or perhaps a "power-to-panel" (energizing) inspection. Usually, no more than two inspections are needed, but some flexibility may be called for. Grounding and bonding are so critical to some of the safety functions of the electrical system that, in a way, grounding and bonding inspections are an inherent part of all inspections.

Other than specific parts of an electrical installation that may be inspected individually, most inspections can logically begin with the service equipment. The plans or electrical drawings may provide the answers to the key questions, but, if not, answers to such questions as the size of the service-entrance conductors can be determined by starting at the service equipment. A more complete examination of the job, perhaps during underground and rough inspections, will reveal which electrodes are present. In any case, the size of grounding electrode conductor(s) and the grounded service conductor and the size and type of the **main bonding jumper** can be checked at the service equipment. Bonding of the service enclosures and raceways can usually be inspected at the same time and place. From here, the installation of the grounding electrode conductor(s) can be checked, and the inspection can be finished at the grounding electrode(s). A logical and orderly progression like this will ensure that nothing important is missed. Unlike many other types of inspections, the inspection of the service grounding and bonding can be quite thorough without an extraordinary expenditure of time and without resorting to sampling techniques. Also, as with some other items of inspection, because access to everything except some electrodes is required by the *NEC*, accessibility is not generally a problem.

Separately derived systems, if any, can be inspected using a similar approach, starting at the system disconnecting means or grounding point and ending at the electrodes.

A complete inspection of the equipment grounding on a job would require the key questions to be answered for each feeder and branch circuit. Such an inspection is impossible in most cases. Instead, the questions can usually be answered in a general way. For example, the predominant wiring method can be determined, and spot checks can be performed. Other wiring methods that may exist can also be sampled for code compliance, such as EGCs connected to the grounded conductor via the main bonding jumper (**FIGURE 5-4**). This is the path necessary to complete the fault-current circuit. The EGC also connects exposed non-current-carrying metal parts of the electrical system to earth or to an extension of the grounding connection to eliminate voltages between exposed parts and the associated shock hazards.

In many cases, the service grounding inspection and equipment grounding inspections can overlap for purposes of efficiency. For example, while the service grounding at the service equipment is being inspected, equipment grounding connections for feeders or branch circuits that originate at the service equipment can also be examined. A number of circuits or feeders can be inspected at their origins. Then individual items of equipment such as boxes, receptacles, and the like can be checked as they are encountered. In fact, branch-circuit and feeder overcurrent devices are the best places to start equipment grounding inspections, because they are in the same place that EGCs originate, and they determine the sizes of separate EGCs and load-side bonding jumpers. From there, equipment grounding and bonding can be checked as part of any other inspections that are being done.

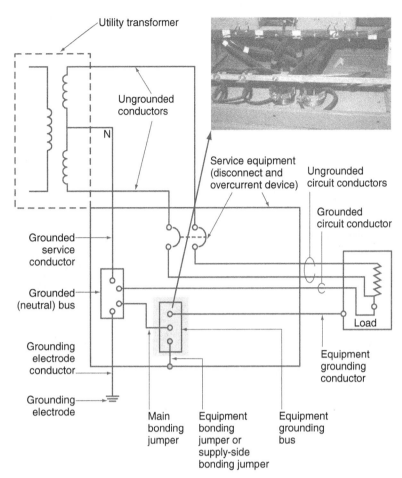

FIGURE 5-4   EGCs connected to grounded conductor via main bonding jumper.

In most cases, the EGC either is the wiring method itself, as in the case of metallic raceways, or is included in the wiring method, as in nonmetallic-sheathed (NM) cable or metal-clad (MC) cable. In other cases, as in rigid nonmetallic conduit (polyvinyl chloride [PVC] conduit or reinforced thermosetting resin conduit [RTRC]) or, unless specific conditions are met, flexible metal conduit (FMC) and liquidtight flexible metal conduit (LFMC), the wiring must be examined to be sure that separate EGCs are supplied. In any case, check grounding connections, including metallic raceway connections for adequacy and integrity. Where an EGC in the form of a wire is used to supplement an EGC in the form of a raceway, neither can be disregarded. In fact, in the event of a ground fault in such an installation, most of the fault current will be imposed on the conduit, even if there is an equipment grounding wire in the assembly. **FIGURE 5-5** is a simplified comparison between the method for selecting grounding conductors on the supply side of the service and the method for selecting grounding conductors on the load side of the service. The supply-side versus load-side distinction also applies to separately derived systems derived from a transformer, where the transformer shown is customer-owned rather than a utility transformer.

The EGC shown in Figure 5-5 is also labeled as a **supply-side bonding jumper**. Consider the similar diagram in **FIGURE 5-6**. In this illustration, assume the EGC as shown is a nonflexible metallic conduit. Note the two **ground-fault current paths** shown by arrows.

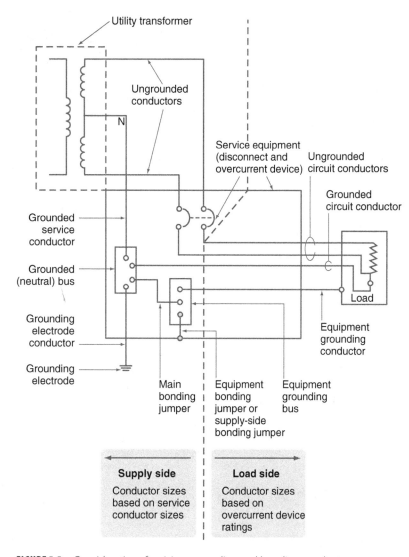

Supply side

Conductor sizes based on service conductor sizes

Load side

Conductor sizes based on overcurrent device ratings

FIGURE 5-5   Considerations for sizing grounding and bonding conductors.

The path shown by large arrows would carry current from a supply-side fault, through the service equipment enclosure and through the jumper to the equipment grounding bus. In this case the jumper would be acting as a supply-side bonding jumper. The path shown by the small arrows is from a load-side fault in the load equipment. This path is through the conduit, the service equipment enclosure and through the jumper or through the EGC to the equipment grounding bus. In either case the ground-fault current path is completed through the main bonding jumper, to the grounded service conductor and back to the source transformer.

The main bonding jumper is required to carry ground-fault current from either supply-side or load-side ground faults, so it is sized according to the rule that gives the larger size, which is the size from Table 250.102(C). (Theoretically the size based on the overcurrent device could be larger in some special case, but that is highly unlikely.) The only way to determine which size should be used is to examine the specific situation. The equipment bonding jumper would also be a supply-side bonding jumper only in situations similar to that shown in Figure 5-6—that is, service equipment—or in the enclosure for the first overcurrent device in a separately derived system. In fact, the bonding jumper shown in this illustration is often just one or more bolts or machine screws that directly connect the equipment grounding bus to the enclosure, and no sizing in the field is required.

WORKING THROUGH THE CHECKLISTS

Checklist 5-1: Service Grounding and Bonding

1. Determine what grounding electrodes are present on the premises.
The types of electrodes present at the premises will determine how the grounding electrode system is formed and what, if any, additional electrodes will be required. The type of electrode also affects the size of the grounding electrode conductor. In some cases, electrodes are present as components of nonelectrical installations. For instance, where the building plumbing system is supplied by 10 ft (3 m) or more of underground metal water piping, a grounding electrode for the building electrical system is present and must be integrated into the building's grounding electrode system. Similarly, if the structural frame of a building is formed by metal that has been installed or connected to be considered a grounding electrode, a grounding electrode is present and, like the water pipe, is required to be incorporated into the building grounding electrode system. In addition to underground metal water piping and some metal building frames, other recognized grounding electrodes

include concrete-encased electrodes, ground rings, rod and pipe electrodes, plate electrodes, underground metal systems or structures, and other listed electrodes.

Where the building construction does not provide a grounding electrode(s), it is necessary that one or more of these electrodes be installed. It should be noted that buildings or structures with footings or foundations containing 20 ft (6.1 m) or more of reinforcement steel (rebar) sized ½ in. (13 mm) and larger in diameter, located horizontally near the bottom or vertically and within that portion of a concrete foundation or footing that is in direct contact with the earth and encased in at least 2 in. (51 mm) of concrete, are required to have this concrete-encased electrode incorporated as a component of the grounding electrode system. The exception to this provision exempts existing buildings or structures where access to the rebar can be accomplished only by disturbing some portion of the concrete encasing the rebar. Section 250.52(A)(3) specifically states that usual steel tie wires can be used to bond reinforcing bars, so there is no requirement that a single 20 ft (6.1 m) piece be located entirely near the bottom or vertically in concrete that is in contact with the soil. Multiple pieces of rebar can be used to satisfy the length requirement. Where multiple separate sections or footings contain electrodes, only one is required to be used. None of these electrodes is required to be existing on the premises, but if any are present through build-

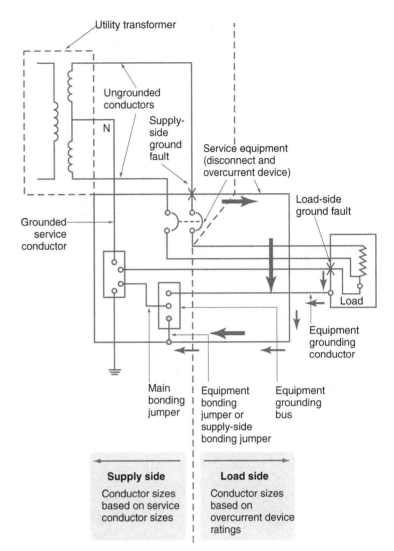

FIGURE 5-6   Supply-side and load-side ground-fault current paths.

ing construction or by installation (except for underground structures and systems), they all must be bonded together to form the grounding electrode system. Concrete-encased electrodes are likely to be present in most new construction. However, if the rebar is encapsulated in a nonconductive, corrosion-resistant coating or the footing is installed on top of a vapor barrier or insulation, the concrete-encased electrode will not be present. One could still be "made" by adding uncoated rebar or using 4 American Wire Gauge (AWG) or larger bare copper wire. Some jurisdictions and designers/engineers require a concrete-encased ("Ufer") electrode in all cases, often because it is the best electrode for the area or conditions, but providing this type of electrode where it does not exist as part of the structure is not mandated by the *NEC*.

The water pipe must be supplemented with another electrode [250.53(D)(2)], and the building steel must be intentionally grounded to be usable as an electrode. Section 250.52(A)(2) describes specific methods that may be used to establish that a metal building frame is effectively connected to earth. An earth connection to a structural metal building frame may be established through a structural member that is in direct contact with the earth or encased in concrete that is in direct contact with the earth for 10 ft (3.0 m) or more. Otherwise, a connection must be made between the hold-down bolts and a concrete-encased electrode.

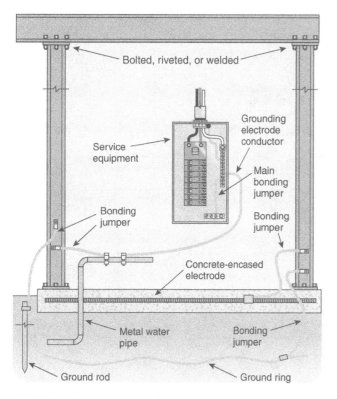

Bolted, riveted, or welded

Service equipment

Grounding electrode conductor

Main bonding jumper

Bonding jumper

Bonding jumper

Concrete-encased electrode

Metal water pipe

Bonding jumper

Ground rod

Ground ring

FIGURE 5-7   Grounding electrode system.

A metal water pipe that is not buried or exposed building steel frame that does not satisfy the requirements to be an electrode will still have to be bonded (see Items 10 and 11 of this checklist). **FIGURE 5-7** illustrates the bonding together of electrodes to form the grounding electrode system. In this example, a single grounding electrode conductor has been run to the water pipe, and bonding jumpers are used to interconnect all of the electrodes. A connection between the hold-down bolts and the concrete-encased electrode is not shown in this illustration. If such a connection is not made, the metal frame is not an electrode in this illustration, but some of the jumpers are required by 250.104(C) to bond the metal frame to the electrical system.

2. Determine which other electrodes are required or used.

Other electrodes are required if none of the building construction-type electrodes (underground metal water piping, structural steel, or concrete-encased) are present or if the only electrode present is a water pipe. The electrodes that may be used include a ground ring, rod or pipe electrodes, other listed electrodes, plates, and other metal underground structures or systems such as isolated metal well casings. Except for the ground ring, these were formerly referred to in the *Code* as "made electrodes." Driven ground rods are the most common type of other electrode. Buried rods or pipes may be used instead of driven rods or pipes, but only if the rods or pipes cannot be driven because of rock. The *Code* requires that rods be driven first perpendicular to the ground, then, if rock bottom is encountered, at up to a 45-degree angle from vertical. Only if the rod cannot be driven can it be buried in a trench. Plate electrodes must be buried not less than 30 in. (762 mm) deep. If other listed electrodes, such as chemical "rods" (actually tubes filled with a chemical that leaches into the surrounding soil over a period of time), are used, they must be used and installed in accordance with their listings.

3. Verify that the grounding electrode conductor or conductors and bonding jumpers are properly sized.

Grounding electrode conductor sizes and the sizes of bonding jumpers used to interconnect electrodes are based on the size of the service-entrance conductors and the types of electrodes used. One or more grounding electrode conductors may be used, and where more than one is used, each may be a different size, depending on the electrode type and the size of the service-entrance conductors related to each grounding electrode conductor. The installation of the grounding electrode conductor and bonding jumpers to interconnect electrodes based on ungrounded conductor size and type of electrode is shown in **FIGURE 5-8. TABLE 5-1** is used to size grounding electrode conductors. This table, copied from *NEC* Table 250.66, must be used in conjunction with the text of 250.66.

The table sizes apply where grounding electrode conductors connect to water pipes or building steel, but smaller sizes may be used with other types of electrodes. For example, the grounding electrode conductor or the bonding jumper that is the sole connection to the driven ground rod in Figure 5-8 is not required to be larger than 6 AWG copper regardless of the size of the service-entrance conductors.

Figure 5-8 shows two different possible ways of forming a grounding electrode system with multiple electrodes. The solid lines show a single grounding electrode conductor to

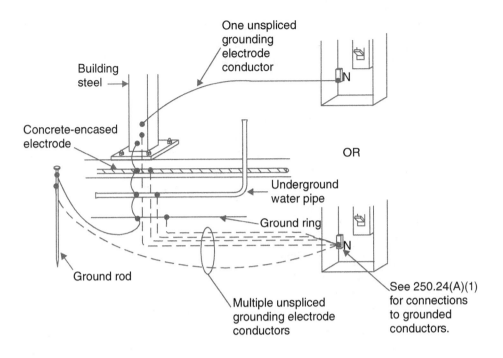

FIGURE 5-8 Permitted connections for grounding electrode conductors and bonding jumpers.

one electrode (the building steel) with the other electrodes bonded together with bonding jumpers. Because the electrodes include water pipe and building steel, the single grounding electrode conductor must be sized from *NEC* Table 250.66, and the bonding jumpers are sized based on the electrode(s) to which they connect, which may or may not allow smaller jumpers. The dotted lines show separate grounding electrode conductors run to each electrode and bonded together at the service equipment. These would each be sized based on the electrodes to which they connect (or Table 250.66, whichever is smaller),

TABLE 5-1   Grounding Electrode Conductor for Alternating-Current Systems

Size of Largest Ungrounded Service-Entrance Conductor or Equivalent Area for Parallel Conductors[a] (AWG/kcmil)		Size of Grounding Electrode Conductor (AWG/kcmil)	
Copper	Aluminum or Copper-Clad Aluminum	Copper	Aluminum or Copper-Clad Aluminum[b]
2 or smaller	1/0 or smaller	8	6
1 or 1/0	2/0 or 3/0	6	4
2/0 or 3/0	4/0 or 250	4	2
Over 3/0 through 350	Over 250 through 500	2	1/0
Over 350 through 600	Over 500 through 900	1/0	3/0
Over 600 through 1100	Over 900 through 1750	2/0	4/0
Over 1100	Over 1750	3/0	250

Notes:
1. If multiple sets of service-entrance conductors connect directly to a service drop, set of overhead service conductors, set of underground service conductors, or service lateral, the equivalent size of the largest service-entrance conductor shall be determined by the largest sum of the areas of the corresponding conductors of each set.
2. Where there are no service-entrance conductors, the grounding electrode conductor size shall be determined by the equivalent size of the largest service-entrance conductor required for the load to be served.
[a]This table also applies to the derived conductors of separately derived ac systems.
[b]See installation restrictions in 250.64(A).

Source: *NFPA 70, National Electrical Code®*, NFPA, Quincy, MA, 2014, Table 250.66.

except the conductors to building steel and water pipe, which would have a minimum size determined strictly from *NEC* Table 250.66. Another option would be to bring all the bonding jumpers together on a busbar and run one grounding electrode conductor from the busbar to the service equipment. The requirements for a busbar assembly can be found in the *NEC* in 250.64(F)(3). (See Item 6 in this checklist.)

4. Verify that the grounding electrode conductors are protected and secured.
All exposed grounding electrode conductors must be securely fastened to the surface on which they are carried. In addition, conductors smaller than 6 AWG must be installed in a raceway for protection. Conductors (6 AWG or smaller) must be protected by a raceway if exposed to physical damage, and 4 AWG or larger conductors must also be protected if exposed to physical damage, but not necessarily by a raceway. In addition to cable armor for smaller sizes, the raceways that may be used for protection are rigid metal conduit (RMC), intermediate metal conduit (IMC), rigid PVC (commonly Schedule 80 PVC), RTRC, and electrical metallic tubing (EMT). Conductors concealed in walls, installed through joists or trusses, or isolated by elevation are generally not considered to be exposed to damage. Those portions of grounding electrode conductors and the bonding jumpers used to interconnect grounding electrodes are not subject to the minimum burial depth requirements of 300.5.

5. Verify that grounding electrode conductor enclosures are properly bonded.
Ferrous metal enclosures for grounding electrode conductors must be made electrically continuous and be connected in parallel with the conductor from the enclosure or cabinet to grounding electrodes. Raceways enclosing grounding electrode conductors may be made electrically continuous by bonding each end of the raceway to the conductor. **FIGURE 5-9** shows the use of bonding jumpers to make the ferrous metal raceway enclosing the grounding electrode conductor electrically continuous. Listed pipe electrode connectors are available with hubs that connect the raceway directly to the pipe, eliminating the need for jumpers on the electrode end. A clamp of this type is shown in **FIGURE 5-10**. Bonding methods (fittings, jumpers, etc.) must comply with 250.92(B)(1) through (4) at service locations or 250.92(B)(2) through (4) at other locations. If bonding jumpers are used, they must not be smaller than the grounding electrode conductor. The use of rigid nonmetallic conduit or nonferrous metal raceways such as RMC of red brass, aluminum, or nonmagnetic stainless steel is permitted as noted in Item 4 of this checklist, and such use eliminates the need for bonding of the raceway.

6. Verify that the grounding electrode conductor is either unspliced or spliced using appropriate methods.
Generally, grounding electrode conductors are required to be installed in continuous lengths without splices. Splices or joints are permitted in grounding electrode conductors where necessary. Such splices are permitted in grounding electrodes of the wire type only if spliced by exothermic welding or by irreversible compression-type connectors that are listed as grounding and bonding equipment. Although many compression connectors are listed, not all are listed for use with grounding electrode conductors. Connectors also need to be suitable for the conditions to which a grounding electrode conductor is exposed, such as in the case of directly buried grounding electrode conductors. Where busbars are used as grounding electrode conductors, they are permitted to be spliced. Accessible busbars may also be used for a common bonding point

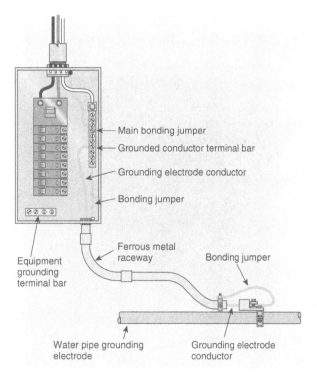

Main bonding jumper

Grounded conductor terminal bar

Grounding electrode conductor

Bonding jumper

Equipment grounding terminal bar

Ferrous metal raceway

Bonding jumper

Water pipe grounding electrode

Grounding electrode conductor

FIGURE 5-9   Required bonding connections for metal raceway enclosing grounding electrode conductor.

FIGURE 5-10 Grounding electrode connector equipped with a hub for connection of a metal raceway.

for a group of electrodes (**FIGURE 5-11**), but the connections of the bonding jumpers to the busbar must still be made with listed connectors or exothermic welding. In this case, the compression connectors shown are required to be listed, but not necessarily listed as grounding and bonding equipment. Other items such as structural metal frames of buildings and metal water piping are sometimes used as extensions of the grounding electrode conductor from the end of the conductor to the earth itself. The bolted, riveted, or welded connections in structural metal or the threaded, welded, brazed, soldered, or bolted-flange connections in metal water piping are recognized as providing electrical continuity through the respective systems. Additionally, reinforcement steel or a copper conductor extended out of the footing or foundation to a convenient point at which the grounding electrode conductor may be connected to a concrete-encased electrode is specifically permitted by the 2014 *NEC*.

7. Check for correct size and installation of any rod, pipe, or plate electrodes.

The requirements for rod and pipe electrodes are clearly detailed in the *NEC*. Rod and pipe electrodes must be at least 8 ft (2.4 m) long, and at least 8 ft (2.4 m) must be in contact with the soil. If the connection to the electrode is above ground, a rod longer than 8 ft (2.4 m) must be used. Where buried in the earth, rod, pipe, and plate electrodes must be installed at least 2½ ft (762 mm) below grade. **FIGURE 5-12** shows three correct installation techniques for ground rods, but these techniques must be utilized in a specific order. First, they should be driven in vertically, and if they cannot be driven vertically, they can be driven at an angle at up to 45 degrees from vertical. Rod or pipe

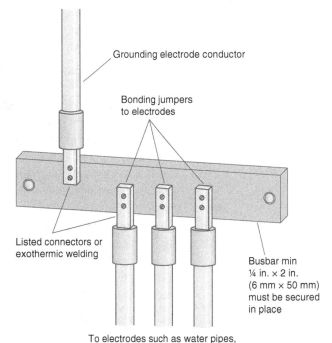

Grounding electrode conductor

Bonding jumpers to electrodes

Listed connectors or exothermic welding

Busbar min ¼ in. × 2 in. (6 mm × 50 mm) must be secured in place

To electrodes such as water pipes, concrete-encased, driven rod

FIGURE 5-11 Busbar used for connecting bonding jumpers together and to a grounding electrode conductor.

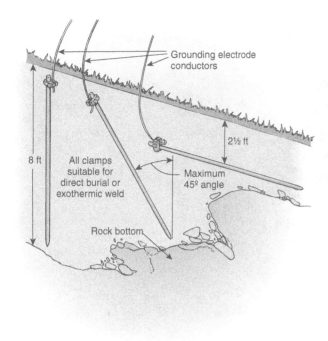

FIGURE 5-12 Acceptable installation techniques for rod and pipe electrodes.

electrodes are permitted to be buried in a trench only if they cannot be driven. Plate electrodes must be at least 1 ft² (0.1 m²) on each side to provide a total of at least 2 ft² (0.2 m²) of surface in contact with the earth. Larger plates may be desirable but are not required. Ideally, electrodes will be installed below permanent moisture level, but this is not possible in some climates. Multiple electrodes must be spaced at least 6 ft (1.8 m) apart, and where the only electrode is a rod, pipe, or plate, it must be supplemented by another electrode unless the resistance of one rod, pipe, or plate electrode is 25 ohms or less. A greater spacing may provide better efficiency in establishing a good connection to the earth. A common recommendation or rule of thumb is to drive multiple electrodes as far apart as they are long, but this is not a *Code* requirement. When there are multiple grounding systems that use rods, pipes, or plates as electrodes, those electrodes must also be at least 6 ft (1.8 m) apart. For example, the rods, pipes, or plates used as ground terminals connected to strike termination devices in a lightning protection system must be established independently from the grounding electrode system used for power systems.

8. Verify the accessibility of grounding electrode conductor connections.

The connection between each grounding electrode and the attached grounding electrode conductor must be **accessible** unless the connection is directly buried, concrete encased, or encased in structural metal fireproofing using listed fittings or clamps. Where connections are covered by fireproofing, the connection is required to be made by exothermic welding or irreversible compression connectors. If irreversible compression connectors are used, the mechanical means (such as nuts and bolts) used to attach the connectors to structural metal members are also permitted to be covered by fireproofing. Access, but not ready access, is required for other types of connections, so the connection may be concealed behind removable panels or covers or elevated as long as access is provided.

9. Check for proper grounding electrode conductor connections, including buried connections.

Grounding electrodes are unlikely to be called on to carry significant currents from faults in premises wiring because the earth is not a normal part of such circuits. However, other currents, such as lightning and some currents that are due to utility faults and surges, may result in circuits that include the earth. Such currents usually have very short duration but may be very large in magnitude. Thus, all connections to grounding electrodes must have very low impedance and be highly reliable. To ensure the integrity of such connections, they must be tested for the conditions to which they may be exposed. Exothermic welding and irreversible compression connections that are listed for grounding and bonding will provide such reliability (**FIGURE 5-13**). Other connections, including buried connections, should be made only with devices listed and identified for the purpose. Connectors designed for direct burial are available and are marked to indicate such suitability (**FIGURE 5-14**). Listed connectors for direct burial must be marked "direct burial" or "DB."

10. Verify that metal water pipe installed in or attached to a structure is bonded.

Whether or not interior metal water pipe or metal water pipe attached to a structure extends to buried underground water pipe, it must be bonded to avoid possible shock hazards due to ground faults that may energize the piping. Reliable connections between hot and cold water piping are also necessary. For a residence, this connection is usually easily

FIGURE 5-13   Variety of exothermic welding examples, some of which are considered adequate for splicing grounding electrode conductors. Note: The examples in the upper row are typically used for signal reference grids rather than for establishing electrode systems for power.

FIGURE 5-14   Direct burial connectors for rod- and pipe-type electrodes.

done at a water heater. Such interconnections may be provided by mixing valves in some cases, but many modern plumbing fixtures use nonmetallic connections even with metal water piping. Bonding conductors to interior water piping are sized in the same manner as grounding electrode conductors to buried underground piping using Table 250.66. Bonding and grounding connections to water piping (interior and exterior) are shown in **FIGURE 5-15**. Where electrical continuity between the hot and cold water pipes through mechanical connections can be confirmed, the jumper is not required. In Figure 5-15, the electrode connection is made within 5 ft (1.5 m) of the water entry, and the bonding jumpers serve to bond the interior portions of the piping as required by 250.104(A).

11. Verify that exposed structural building frames are bonded.

Usually, structural metal building frames are used as grounding electrodes. However, if the frames are not effectively grounded or bonded by one of the methods listed in 250.52(A)(2), are exposed, and may become energized, they must be bonded by using the same methods and conductor sizes as if they were suitable for use as electrodes. Building steel may become energized if any conductive electrical equipment that contains insulated conductors is attached to the steel. Keep in mind that *exposed* in this case (as applied to wiring methods) does not necessarily mean *visible*. Also, only structural metal that

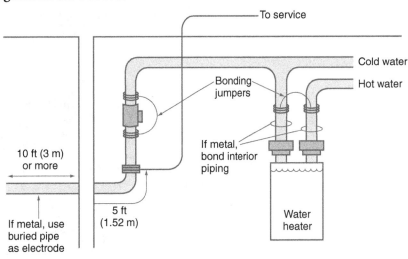

FIGURE 5-15   Bonding and grounding connections to water piping.

forms a building framework is required to be bonded. Isolated metal members or parts such as metal trusses in concrete or masonry construction and metal wall studs are not required to be deliberately bonded under this rule, although they usually are incidentally bonded by connections to raceways and equipment. Section 250.104(C) lists four items to which the metal frame may be bonded, but typical choices are the grounding electrode conductor or one of the grounding electrodes. In previous editions of the *NEC*, once the structural metal was bonded to any of the electrodes, it effectively became an electrode, but since the 2011 edition, there are only two ways to qualify the structural metal frame as an electrode: by connection through the anchor bolts (hold-down bolts) to a concrete-encased electrode or by direct contact with the earth. However, even if the metal framework does not qualify as an electrode, in many cases it may still be used as a common interconnection point for the grounding electrode system, as a bonding jumper to interconnect electrodes, or even as a grounding electrode conductor [250.68(C)(2)].

12. Check for proper size and length of bonding jumpers around water meters and the like.

Where both interior metal piping and buried metal piping exist, the grounding electrode conductor to the underground piping may also be used to bond the interior piping. Jumpers between the two must be provided to ensure that the connections are not dependent on removable plumbing equipment such as pressure-reducing valves, meters, or filters (**FIGURE 5-16**). Such jumpers must be long enough to allow servicing or replacement of the plumbing devices without compromising the integrity of the bonding connection.

13. Check the size, type, and installation of the main bonding jumper.

The main bonding jumper is a critical part of the **effective ground-fault current path** between premises wiring and the utility source. The main bonding jumper connects EGCs and equipment bonding jumpers (including some supply-side bonding jumpers) to the grounded service conductor to complete the fault-current path. To ensure its adequacy as a fault-current path, it must be sized in the same manner as for grounding electrode

FIGURE 5-16 Bonding around removable devices (a water filter) and between hard, soft, and hot water piping.

conductors to primary electrodes. However, a new Table 250.102(C) was added in the 2014 edition to be used for this purpose rather than Table 250.66. The difference between the tables is that Table 250.66 is for sizing grounding electrode conductors and jumpers that are not expected to carry a significant amount of ground-fault current. Grounding electrode conductors are never required to be larger than 3/0 AWG copper or 4/0 AWG aluminum. Table 250.102(C) is specific to elements of the effective ground-fault current path, including the main bonding jumper. Elements of the fault-current path continue to increase in size as the supply conductors increase in size. As the table heading says, it is also used to determine the minimum size of the grounded service conductor, which is part of the same fault-current path. If the cumulative size of the service-entrance conductors is larger than 1100-kcmil copper or 1750-kcmil aluminum, the bonding jumper must be sized at 12.5 percent of the total or equivalent area of the service-entrance conductors. The notes to the table establish what is often referred to as the "twelve and a half percent rule" (**TABLE 5-2**). Generally, the main bonding jumper will be factory installed (or supplied) in the service equipment panelboard (**FIGURE 5-17**). The "equipment bonding jumper" shown could be a "supply-side bonding jumper," depending on the origin of the fault current it carries.

TABLE 5-3 shows a number of examples of main bonding jumpers sized for various combinations of service-entrance conductors. The bonding jumper may also be a screw if it is colored green and furnished with the service equipment. In this case or when other types of main bonding jumpers are supplied with equipment that is "suitable for use as service equipment," it is not necessary to calculate the bonding jumper size because the

TABLE 5-2 Grounded Conductor, Main Bonding Jumper, System Bonding Jumper, and Supply-Side Bonding Jumper for Alternating-Current Systems

Size of Largest Ungrounded Service-Entrance Conductor or Equivalent Area for Parallel Conductors (AWG/kcmil)		Size of Grounded Conductor or Bonding Jumper (AWG/kcmil)	
Copper	Aluminum or Copper-Clad Aluminum	Copper	Aluminum or Copper-Clad Aluminum
2 or smaller	1/0 or smaller	8	6
1 or 1/0	2/0 or 3/0	6	4
2/0 or 3/0	4/0 or 250	4	2
Over 3/0 through 350	Over 250 through 500	2	1/0
Over 350 through 600	Over 500 through 900	1/0	3/0
Over 600 through 1100	Over 900 through 1750	2/0	4/0
Over 1100	Over 1750	See Notes	

Informational Note: See *NEC* Chapter 9, Table 8, for the circular mil area of conductors 18 AWG through 4/0 AWG.
Notes:
1. If the ungrounded supply conductors are larger than 1100 kcmil copper or 1750 kcmil aluminum, the grounded conductor or bonding jumper shall have an area not less than 12½ percent of the area of the largest ungrounded supply conductor or equivalent area for parallel supply conductors. The grounded conductor or bonding jumper shall not be required to be larger than the largest ungrounded conductor or set of ungrounded conductors.
2. If the ungrounded supply conductors and the bonding jumper are of different materials (copper, aluminum, or copper-clad aluminum), the minimum size of the grounded conductor or bonding jumper shall be based on the assumed use of ungrounded supply conductors of the same material as the grounded conductor or bonding jumper and will have an ampacity equivalent to that of the installed ungrounded supply conductors.
3. If multiple sets of service-entrance conductors are used as permitted in 230.40, Exception No. 2, or if multiple sets of ungrounded supply conductors are installed for a separately derived system, the equivalent size of the largest ungrounded supply conductor(s) shall be determined by the largest sum of the areas of the corresponding conductors of each set.
4. If there are no service-entrance conductors, the supply conductor size shall be determined by the equivalent size of the largest service entrance conductor required for the load to be served. For the purposes of this table, the term bonding jumper refers to main bonding jumpers, system bonding jumpers, and supply-side bonding jumpers.

Source: *NFPA 70, National Electrical Code®*, NFPA, Quincy, MA, 2014, Table 250.102(C)

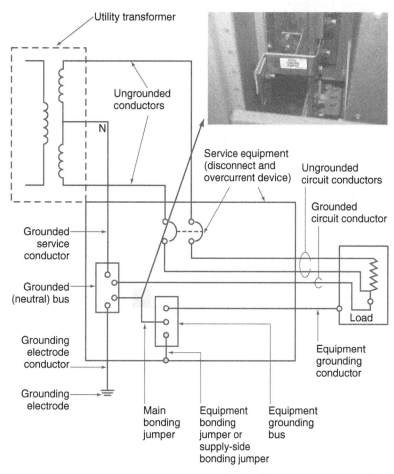

FIGURE 5-17 Main bonding jumper factory installed in the service equipment panelboard.

bonding jumper is covered by the listing and labeling of the service equipment. Reliable, low-impedance connections are vital in this part of the electrical installation. Table 5-3 provides example sizes for main or supply-side bonding jumpers. These are the bonding jumpers used to connect all metallic non-current-carrying parts into a complete electrically conductive path for fault current from any point in the wiring system to the supply system source.

14. Verify that service raceways and enclosures are correctly bonded.

Because the overcurrent protection for service equipment and conductors provides overload protection only and does not provide short-circuit or ground-fault protection, the possible fault-current pathways at service equipment need special attention. Specific methods of ensuring adequate bonding at services include threaded connections, threadless fittings, threaded hubs, bonding jumpers, and special bonding fittings. Supplementary bonding jumpers or fittings must be used with ordinary locknuts and around oversized, concentric, or eccentric knockouts.

There are several methods of bonding service equipment and raceways (**FIGURES 5-18A, B,** and **C**). Note that the ordinary

| TABLE 5-3 | Sizing Main Bonding Jumpers—Copper* |

Service-Entrance Conductor Sizes	Number of Conductors (Parallels)	Area of Conductor (kcmil)	Equivalent Area for Parallel Conductors	Size from Table 250.66 of the NEC	12½% Area (kcmil)	Equivalent Bonding Jumper (kcmil)	Equivalent Area of Busbar (in.²)
1/0 AWG	1	106	106	6 AWG	NA	26	0.020
1/0 AWG	4	106	422	2 AWG	NA	66	0.052
3/0 AWG	1	168	168	2 AWG	NA	66	0.052
3/0 AWG	4	168	671	2/0 AWG	NA	133	0.104
250 kcmil	1	250	250	2 AWG	NA	66	0.052
250 kcmil	4	250	1000	2/0 AWG	NA	133	0.104
500 kcmil	1	500	500	1/0 AWG	NA	106	0.083
500 kcmil	4	500	2000	NA	250	250	0.196
500 kcmil	6	500	3000	NA	375	375	0.295
750 kcmil	1	750	750	2/0 AWG	NA	133	0.104
750 kcmil	5	750	3750	NA	469	469	0.368
750 kcmil	8	750	6000	NA	750	750	0.589
*Also applies to grounded service conductors and supply-side bonding jumpers.							

FIGURE 5-18A   Threaded hubs (in conduit body) used to bond service raceways.

FIGURE 5-18B   Threadless connector used to bond service raceway.

FIGURE 5-18C   Grounding bushing with connected supply-side bonding jumper used for bonding of service equipment.

locknut shown in Figure 5-18C and the similar locknuts that are probably on the inside of the enclosures in Figures 5-18A and B are not adequate bonding means for service equipment and must be supplemented with other devices such as the grounding bushing shown in Figure 5-18C.

15. Check the size of service-equipment supply-side bonding jumpers.

Like the main bonding jumper and the grounded service conductor, equipment-bonding jumpers on the supply side of service equipment are exposed to fault currents whose magnitude and duration are not limited by overcurrent protection similar to that required for feeders and branch circuits. The magnitude of these fault currents is limited primarily by the impedance of the circuit between the utility and the service equipment. Therefore, components that are intended to carry these fault currents must be sized on the basis of the size of the service-entrance conductors using the same table from the *NEC* that is used to size the grounding electrode conductor. Where service-entrance conductors are larger than 1100-kcmil copper or 1750-kcmil aluminum, supply-side (equipment) bonding jumpers must be increased in size in the same manner as the main bonding jumper. Service equipment and raceways on the supply side of the service may also be bonded by a direct connection to the grounded service conductor. **FIGURE 5-19** is a photograph of the grounded conductor bus in a meter base that is bonded directly to the meter enclosure by

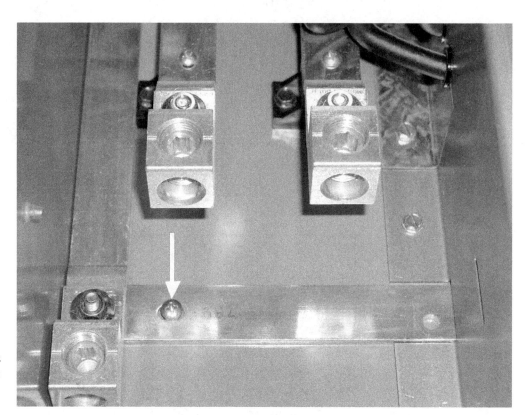

FIGURE 5-19 Bonding of service equipment enclosures by inherent direct connection to grounded service conductor.

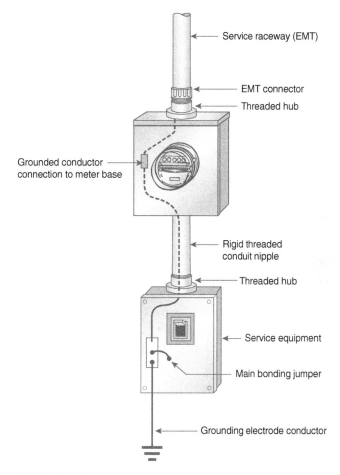

Service raceway (EMT)

EMT connector

Threaded hub

Grounded conductor connection to meter base

Rigid threaded conduit nipple

Threaded hub

Service equipment

Main bonding jumper

Grounding electrode conductor

FIGURE 5-20 Bonding of service equipment and service raceways using various methods from 250.92(B).

construction. **FIGURE 5-20** shows an example of various parts of service equipment that are bonded together using the methods required by 250.92(B). Note the grounded conductor connection to the meter base.

Section 250.102(C) addresses three specific cases: where the conductors are in a single raceway, where they are paralleled in more than one raceway, and where the supply conductors and equipment bonding jumper are of different materials. Sizing is determined from Table 250.102(C). For a single raceway, the size of the supply-side bonding jumper is based on the total area of the ungrounded service conductors in any one phase or line in that raceway. If there are multiple raceways, each with a separate supply-side bonding jumper, each jumper is sized based on the circular mil area of the ungrounded conductors in that raceway. If multiple raceways are bonded with a single supply-side bonding jumper, the jumper is sized as if all the conductors are in one raceway. If the supply-side bonding jumper and service conductors are different materials, the sizing is based on the equivalent size of conductors that have the same ampacity as the installed ungrounded conductors and the same material as the bonding jumper.

16. Verify that the grounded service conductor size is adequate.

Generally, the grounded service conductor (often a **neutral conductor**) is sized based on the load likely to be imposed. In fact, most commonly designers and

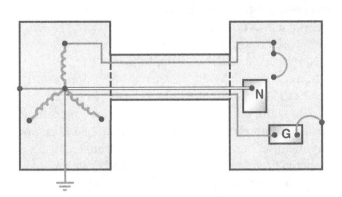

FIGURE 5-21A   Grounding electrode conductor connection made at the derived system source. (Source: *National Electrical Code®* Seminar, NFPA, Quincy, MA, 2011, Module 4, Slide 112, page 233.)

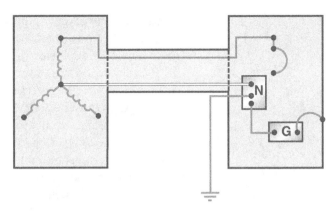

FIGURE 5-21B   Grounding electrode conductor connection made at the disconnecting means. (Source: *National Electrical Code®* Seminar, NFPA, Quincy, MA, 2011, Module 4, Slide 113, page 233.)

installers simply specify the same size as the ungrounded service conductors. However, the grounded conductor must always be capable of carrying fault current, so the same rule that is used to determine the minimum size of the main bonding jumper and supply-side bonding jumpers at service equipment also applies to grounded service conductors. To determine the minimum size of a grounded service conductor, the two functions of this conductor must be considered. The grounded conductor is required to have an ampacity not less than the maximum computed unbalance of the system (220.61), and in addition, it must be sized to carry fault current [250.24(C) and Table 250.102(C)]. Whichever of these conditions produces the largest conductor determines the minimum size for the grounded (neutral) conductor.

17. Check separately derived systems for proper grounding electrodes, grounding electrode conductors, and system bonding jumpers.

Separately derived systems are treated in much the same way as services. A grounding electrode must be selected and connected to the grounded conductor with a grounding electrode conductor. A **system bonding jumper**, which functions like a main bonding jumper, must be installed to complete the fault-current path from equipment and enclosures to the source. The grounding electrode conductor connections to the system-grounded conductor must generally be located at the same point where the system bonding jumper is also connected, which, in most cases, may be anywhere between the derived system source and the disconnecting means (**FIGURES 5-21A** and **B**). If the separately derived system is located outdoors, the grounding and bonding connections must be at the source, and in some limited cases the system bonding jumper may be installed at both the source and the first disconnecting means. Grounding electrodes are selected based on proximity to the derived system (**FIGURE 5-22**). Note that water pipe and building steel must be used if present, and the other electrodes, as shown in (c) of Figure 5-22, are permitted only if building steel or

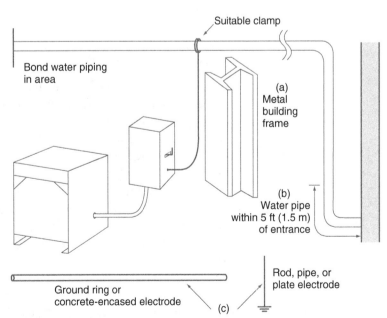

Use *nearest* of (a) or (b), or use (c) if (a) or (b) are not available

FIGURE 5-22   Grounding electrode selection based on proximity to the derived source.

underground metal water piping is not present. Metal building frames may be used as grounding electrode conductors if the frame does not meet the requirements to be an electrode. Any of the other electrodes may be used as supplemental electrodes.

The connection to a water pipe electrode must be done within the first 5 ft (1.5 m) of where the water piping enters the building. Water piping in the area served by a separately derived system must be bonded as shown in Figure 5-22. See Item 18 below. In other respects, sizing and installation of bonding jumpers and grounding electrode conductors follow the same conventions as the similar components at services. Grounding and bonding connections must be made in a manner that prevents neutral current from being carried on conductors other than the insulated grounded (neutral) conductor. Connections resulting in parallel neutral current paths are prohibited because they can cause neutral current to be imposed on metal raceways or other electrically conductive equipment that is normally supposed to be non-current-carrying and can present a serious shock hazard to persons, especially if the pathways are not part of the electrical installation.

18. Verify that water pipe and structural metal building frames in the area served by each separately derived system are bonded.
Where the water pipe in the area of a separately derived system is not suitable for use as an electrode (for example, where only the interior piping is metal), and where structural metal building frames are not considered electrodes, they must still be bonded in the same manner required as if they had been used as an electrode. Bonding to the water piping and metal structure are requirements based on the same considerations that require bonding of exposed structural metal frames or interior metal piping to services (i.e., eliminating shock hazards). However, if the water pipe is bonded to building steel somewhere in the area served by the separately derived system and the building steel is used as the electrode, the piping need not have a separate jumper from the separately derived system.

19. Verify that an <u>intersystem bonding termination</u> has been provided.
A provision for the connection of at least three bonding terminations for systems other than the power system, such as cable or satellite TV and communications, must be provided. This may be a set of terminals mounted to the meter enclosure; a bonding terminal bar near the service equipment, meter, or raceway; or a bonding terminal bar near the grounding electrode conductor, and the terminals must be listed for grounding and bonding. This termination provision must be accessible for connection by other system installers and for inspection. Methods that were considered adequate for this purpose prior to the 2008 *NEC* may still be used in existing buildings.

20. Verify that where a wire-type EGC is also used as a grounding electrode conductor, it meets all applicable requirements for both grounding electrode conductors and EGCs.
Wire-type EGCs are permitted to also be used as a grounding electrode conductor. In order to use a single conductor to perform the two functions, its installation must meet the requirements for both EGCs and grounding electrode conductors. Sizing of the conductor would have to be the larger of that determined by 250.66 (grounding electrode conductor) or 250.122 (EGC). The requirements covering the continuity and connection of grounding electrode conductors are more restrictive than those for EGCs, thus making the typical equipment grounding connection in electrical equipment can be a more complicated installation. For instance, if the EGC of a feeder supplying a panelboard is also used as the grounding electrode conductor, the typical connection of the EGC to the panelboard equipment grounding terminal has to be coordinated with the requirement that grounding electrode conductors be installed without a splice or joint. Terminating the EGC to the panelboard grounding terminal and then running a separate conductor from the grounding terminal to the grounding electrode would not meet the requirement for the grounding electrode conductor to be installed without a splice or joint. Practical application of this requirement requires careful adherence to the multiple *NEC* rules covering the installation of EGCs and GECs.

Checklist 5-2: Equipment Grounding and Bonding

1. Identify equipment that is required to be grounded.

Electrical equipment that includes exposed conductive parts and contains energized conductors is **likely to become energized** and must be grounded by connection to an EGC. This is generally true whether the equipment is fastened in place, movable, or portable. Equipment grounding is usually required if the equipment is connected by permanent wiring methods or by cord and plug. The most common exception is double-insulated equipment. Some low-voltage equipment and some equipment that is guarded or otherwise not subject to contact by personnel are also excluded. In most cases, except for equipment covered by Chapter 8 of the *NEC,* it can be assumed that conductive equipment is required to be grounded unless there is a specific exemption in Article 250. Listed appliances that are supplied by the manufacturer with 2-wire cords need not be investigated further.

2. Verify appropriate grounding methods for equipment fastened in place or connected by permanent wiring methods.

Section 250.118 provides a list of wiring methods that may be used as EGCs. The list includes certain metallic wiring methods. Equipment that is fastened in place or connected by permanent wiring methods may be grounded by such wiring methods or by an EGC that is contained in, or included with, the same raceway or cable. Some equipment may be considered to be grounded because it is attached to, or mounted on, other grounded equipment [250.136].

3. Verify appropriate types of EGCs.

Article 250 provides a list of wiring methods that may be used as EGCs [250.118]. These wiring methods may include an EGC or may themselves be an acceptable EGC (**FIGURE 5-23**). Some wiring methods, such as flexible metal conduit, are permitted for grounding but

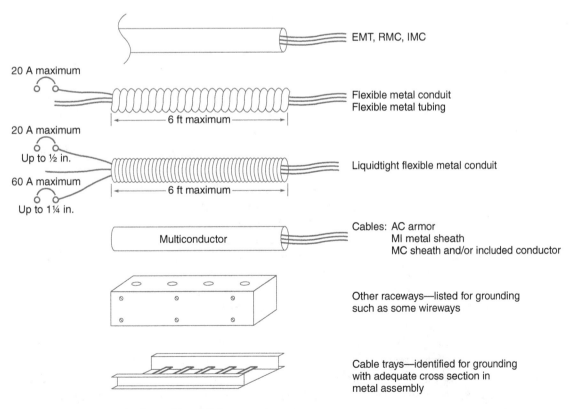

FIGURE 5-23 Wiring methods suitable as EGCs.

TABLE 5-4	Minimum Size Equipment Grounding Conductors for Grounding Raceways and Equipment		
		Size (AWG or kcmil)	
Rating or Setting of Automatic Overcurrent Device in Circuit Ahead of Equipment, Conduit, etc., Not Exceeding (Amperes)		Copper	Aluminum or Copper-Clad Aluminum*
15		14	12
20		12	10
60		10	8
100		8	6
200		6	4
300		4	2
400		3	1
500		2	1/0
600		1	2/0
800		1/0	3/0
1000		2/0	4/0
1200		3/0	250
1600		4/0	350
2000		250	400
2500		350	600
3000		400	600
4000		500	750
5000		700	1200
6000		800	1200

Note: Where necessary to comply with 250.4(A)(5) or (B)(4), the EGC shall be sized larger than given in this table.
*See installation restrictions in 250.120.

Source: *NFPA 70, National Electrical Code*®, NFPA, Quincy, MA, 2014, Table 250.122.

with restrictions on length and circuit ratings. Others, such as Type NM cable, are provided with separate EGCs. Wiring methods that are themselves EGCs may be supplemented with separate conductors, but such supplementary EGCs are not required except in certain specific applications such as in patient care areas of health care facilities.

4. Check separate EGCs for proper sizing and identification.

Separate EGCs installed in raceways or as a part of cables are required to be sized based on the rating of the largest overcurrent device protecting the associated branch-circuit or feeder conductors. The *NEC* provides a table for sizing copper and aluminum EGCs (**TABLE 5-4**). Larger EGCs are required where the circuit conductors are increased in size (circular mil area) above a conductor of adequate ampacity. The size increase might be for voltage drop or for other reasons. Full-sized EGCs based on the rating of the overcurrent device are required where the circuit is run in parallel in multiple raceways or cables, but are not required to be larger than the ungrounded conductors.

The notes to the table point out that in some cases larger sizes may be necessary to provide the required low-impedance pathway for fault current. As a general rule, the sizes shown in the table were originally based on circuit lengths up to 100 ft (30.5 m), but they are frequently adequate for much longer lengths. Smaller conductors are permitted in flexible cords or where the general rule would result in an EGC that is larger than the circuit conductors. **FIGURE 5-24** shows three examples of EGC sizing. Unlike supply-side grounded conductors and bonding jumpers, it is the rating of the overcurrent device and not the ungrounded conductor size that drives the selection.

EGCs of the wire type are required to be clearly identified. Identification is usually in the form of color coding. Sizes up to 6 AWG are required to be bare, covered, or insulated. Where they are covered or insulated, the insulation or covering must be green or green with one or more yellow stripes for its entire length. Sizes larger than 6 AWG are permitted to be reidentified by stripping the insulation from the conductor or coloring the conductor green at each end and most places where the conductor is accessible. Green tape is the method most commonly used for this purpose. In most cases, green or green with yellow stripe(s) is reserved for EGCs, although this identification is also often used for bonding jumpers or grounding electrode conductors. The *NEC* does not prohibit or require the use of green identification for bonding jumpers or grounding electrode conductors.

5. Check connections of EGCs within outlet boxes.

Metallic outlet, device, and junction boxes are required to be grounded. Where more than one EGC enters a box, all the EGCs, except isolated grounding conductors, must be joined together and to the box. The connections between grounding conductors and metal boxes are not permitted to depend on fixtures or devices. The connection(s) to the box must be made with listed devices or with screws that are used for no other purpose and that engage at least two threads or are secured with nuts. Sheet metal screws essentially form a single thread and are not permitted for making grounding or bonding connections. **FIGURE 5-25A** shows listed devices that may be used for making grounding connections. **FIGURE 5-25B** shows types of screws that may be used for making grounding connections. The screws on the left form machine screw threads (one has a nut installed to illustrate this point) and may be used for grounding connections as long as at least two threads are engaged in the metal of the enclosure. **FIGURE 5-25C** shows examples of sheet metal screws that cannot be used for grounding connections because they are not included in

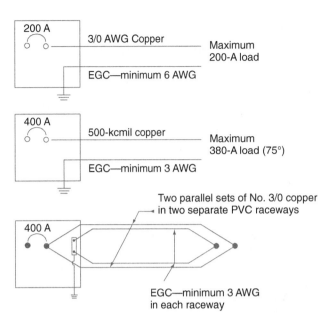

FIGURE 5-24   Three examples of sizing EGCs. "Load" includes any required allowance for continuous load.

FIGURE 5-25A   Listed connectors that may be used for grounding connections.

FIGURE 5-25B   Screws with machine screw threads that may be used for grounding connections.

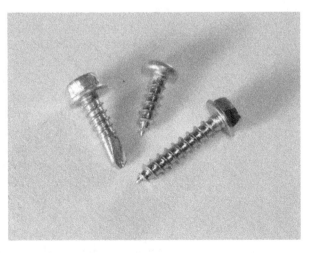

FIGURE 5-25C   Screws that are not allowed for grounding connections.

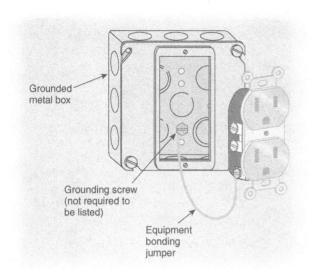

FIGURE 5-26   Bonding of metal box and grounding-type receptacle.

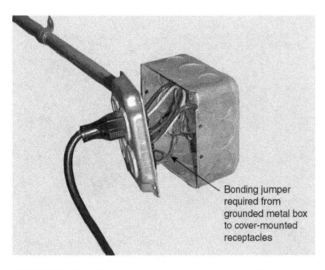

FIGURE 5-27   Bonding jumper to receptacle required for this type of cover-mounted device.

the list of permitted connection methods. One has a self-drilling point, but it has threads that essentially create one thread in sheet metal. EGCs in nonmetallic boxes must be joined together in a way that will permit a grounding connection to be made to devices such as receptacles or switches that are installed in the box. These requirements do not apply to EGCs in boxes that are used only for pulling points if the circuit conductors associated with the EGCs are not spliced or terminated in or at the box and the box is otherwise grounded through a connection to a wiring-method-type EGC such as a metal conduit.

6. Verify that proper methods are used to bond receptacles to boxes.

Except for some isolated grounding receptacles, all grounding-type receptacles installed in metal boxes must be connected to grounded boxes with a bonding jumper or EGC or the equivalent (**FIGURE 5-26**). Where direct metal-to-metal contact is provided between receptacle yokes and surface-mounted grounded boxes or metal extension rings with a flat contact surface to a grounded box or where the receptacle is a listed self-grounding device, the separate bonding jumper may be omitted because the direct contact or self-grounding screw provides the bonding connection. Cover-mounted receptacles must be provided with jumpers unless the cover is listed for the purpose (**FIGURE 5-27**).

7. Check installation of equipment bonding jumpers, especially where flexible connections or cords are used.

Where the continuity or capacity to conduct fault current is suspect or subject to damage, as at loose connections, at flexible connections, at movable or portable equipment, or where connections are impaired by paint or other nonconductive coatings, bonding jumpers or equivalent methods must be used to ensure such continuity and capacity. (Note: Some fittings, especially those with locknuts, are designed to cut through paint to establish adequate connections if they are installed properly.)

To ensure continuity and capacity, EGCs and grounding-type receptacles, plugs, and connectors must generally be used with flexible cords that include an EGC. Similarly, equipment bonding jumpers must be supplied around flexible conduit if the conduit connects to equipment that may be moved (requiring flexibility after installation), or the flexible conduit is used to minimize transmission of vibration from the equipment, or the flexible connection is made from a box cover (subject to interruption). Bonding jumpers up to 6 ft (1.8 m) long may be installed on the outside of flexible conduits, but they must be routed with the conduits. Longer external jumpers are permitted at outside pole locations for bonding isolated sections of metal raceways. Bonding jumpers for equipment supplied by feeders or branch circuits are sized on the basis of the overcurrent device protecting the applicable circuit. Where cords are used, an EGC is required where the cord supplies equipment that must be grounded, and where the circuit conductors are 10 AWG or smaller, the EGC must be the same size as the circuit conductors and not smaller than 18 AWG. Where separate flexible wires or straps are used, protection from physical damage also must be provided.

8. Verify grounding of panelboard enclosures and connections of EGCs to panelboard metallic enclosures.

Panelboard enclosures and panelboard frames (often collectively referred to as "panels") are required to be grounded. Raceways permitted as EGCs in 250.118 of the *NEC* may be used to ground panels and equipment supplied from panels. If separate EGCs of the wire type are used, terminal bars must be provided and bonded to the panel enclosure for connecting EGCs. Where grounded conductors are connected to grounding electrode conductors, such as at services, separately derived systems, and some existing separate structures, a single terminal bar may be used for both grounded (white or gray) and grounding (green or bare) conductors. However, at other locations, such as feeder-supplied panelboards or "subpanels," separate terminal bars must be used for grounded and equipment grounding conductors, and the terminals for grounded conductors must be isolated from the panel enclosure (**FIGURE 5-28**).

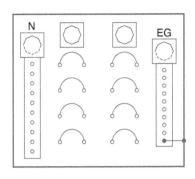

FIGURE 5-28 Neutral conductor and equipment grounding separation in feeder-supplied panelboard (subpanel).

Isolated grounding conductors, such as those permitted for isolated ground receptacles, are permitted to pass through one or several enclosures without terminating or connecting to the enclosures in route to the selected termination point on an equipment grounding bus of that particular supply system. A separate terminal for isolated grounding conductors is permitted in any panelboard enclosure for which an isolated equipment grounding circuit is supplied. Isolated equipment grounding terminals or isolating kits are available from panelboard manufacturers.

9. Verify proper grounding at separate buildings or structures.

In general, the power supplies to separate structures supplied by feeders or branch circuits must be treated much like services. However, in most cases an EGC must be run with the feeder or branch circuit. A grounding electrode or grounding electrode system must be established at each separate structure unless there is only one branch circuit and an EGC is supplied with the branch circuit [250.32(A)]. (Note: The one branch circuit may be a multiwire branch circuit.) *In existing installations only that were installed and remain in compliance with previous editions of the* NEC, where no EGC is included with the supply to a separate building, no ground-fault protection for equipment is installed at the service, and no other metallic paths exist between grounding systems in the two buildings, the grounded conductor must be connected to the grounding electrode and the disconnect enclosure at the disconnect for the separate building in order to serve as a ground-fault return path [250.32(B)(1), Exception]. Otherwise (for all new installations), an EGC must be included with the supply [250.32(B)(1)]. In either case, the grounding electrode(s) at the separate building must be bonded to both the disconnect and the EGCs at the separate building, but may not be bonded to the grounded conductor.

These conditions for grounding and bonding at separate buildings are shown in **FIGURE 5-29**. The illustration of 250.32(A), Exception, shows no bonding jumper at Building 2, which is consistent with 250.32(B)(1), because there is an EGC run with the circuit conductors. Note the emphasis on parallel paths in 250.32(B)(1), Exception, where there is no EGC run with the circuit conductors to Building 2 and a bonding jumper is installed at Building 2. There may well be another metallic path in parallel with the grounded (neutral) conductor, and the main rule of 250.32(B) is the only option where such other paths exist or in new installations. The other paths could be water, gas, process, or other piping; racks or other structures; metallic conduits; or common grounding grids; even reinforcing steel. Where any of these exist or are likely, and in all new or existing installations, the main rule of 250.32(B)(1) is the intended grounding method.

10. Check equipment grounding of electric ranges and clothes dryers.

Grounded (usually neutral) conductors are not permitted to be used to ground the frames of ranges and dryers except in existing installations. New installations are required to

FIGURE 5-29 Grounding and bonding at separate buildings.

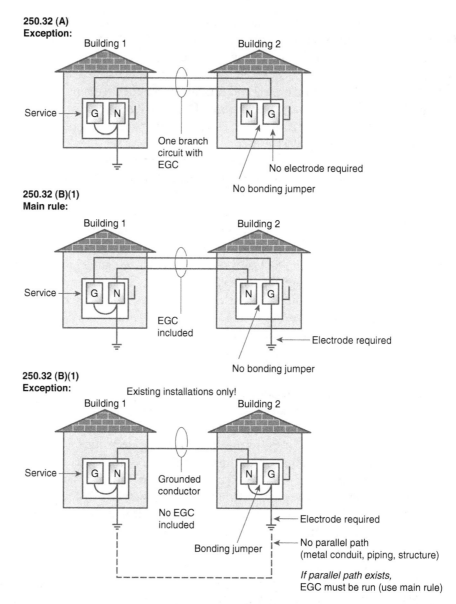

250.32 (A)
Exception:

Building 1 · Building 2

Service → G N

One branch circuit with EGC

No electrode required

No bonding jumper

250.32 (B)(1)
Main rule:

Building 1 · Building 2

Service → G N

EGC included

Electrode required

No bonding jumper

250.32 (B)(1)
Exception:

Existing installations only!

Building 1 · Building 2

Service → G N

Grounded conductor

No EGC included

Electrode required

Bonding jumper

No parallel path (metal conduit, piping, structure)

If parallel path exists, EGC must be run (use main rule)

include EGCs, along with appropriate cords, plugs, and receptacles for grounding of such appliances. The electrical connections within the appliance must be done so as to electrically isolate the EGC connection from the neutral connection. The requirements and connections for new and existing installations are shown in **FIGURES 5-30A** and **B**.

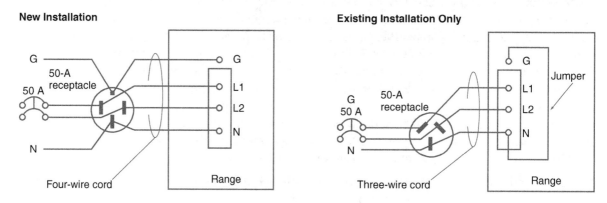

New Installation

G

50-A receptacle

50 A

N

Four-wire cord

G
L1
L2
N

Range

Existing Installation Only

G

50-A receptacle

G
50 A

N

Three-wire cord

G
L1
L2
N

Jumper

Range

FIGURE 5-30 (A) Grounding of electric ranges and clothes dryers (new installation). (B) Grounding of electric ranges and clothes dryers (existing installation only).

11. Verify bonding of equipment operating at over 250 volts to ground.

Equipment supplied by circuits of over 250 volts to ground (most commonly 277 volts to ground) is required to meet special bonding requirements. These special rules are basically the same as the rules for bonding at services, except that connection to the grounded conductor is not permitted. However, as **FIGURE 5-31** illustrates, ordinary locknuts are permitted for bonding in this case if concentric, eccentric, or oversized knockouts are not encountered, or if such knockouts are encountered but they are listed for the purpose, and if (1) the locknuts are supplied with threadless fittings, such as EMT connectors or MC cable connectors; (2) double locknuts are used with threaded IMC or RMC; or (3) other fittings listed for the purpose are used. **FIGURE 5-32** shows a metal box with eccentric knockouts that are listed for grounding above or below 250 volts to ground.

According to the *UL Guide Information for Electrical Equipment* ("The White Book"), *Product Category QCIT, Metallic Outlet Boxes,* "all metallic outlet boxes with concentric or eccentric knockouts have been investigated for bonding and are suitable for bonding without any additional bonding means around concentric (or eccentric) knockouts where used in circuits above or below 250 volts, and may be marked as such." (The marking is not required.)

12. Check installations with isolated grounding conductors for proper connections and for grounding of the associated enclosures and wiring methods.

Isolated grounding conductors must originate at a grounding terminal of the applicable service or derived system, and the terminal must be within the same building or structure. Isolated grounding conductors must be run with the other conductors of the circuit and may not be run to separate grounding electrodes. Usually, isolated grounding conductors originate at the grounding point of the service, separate structure, or derived system, but they may be terminated in some intermediate panelboard if that point of connection is deemed by the designer to provide the needed reduction of electrical noise. Although isolated grounding conductors are not required to be connected to intervening panels, boxes, or other enclosures, such enclosures are still required to be grounded. If, for example, an isolated grounding receptacle is installed in a metal box supplied by a long length of flexible metal conduit or electrical nonmetallic tubing, there must be an isolated grounding conductor for the receptacle and another EGC or equipment bonding jumper for the metal box. Isolated grounding receptacles are constructed and identified as shown in **FIGURE 5-33**.

Supplemental grounding electrodes (called "auxiliary grounding electrodes" in 250.54) are permitted; however, earth cannot be used as a return path for fault current. The grounding connections for an isolated ground receptacle are described and illustrated in **FIGURE 5-34**. All references in the *NEC* to isolated grounding refer to 250.146(D) for these primary requirements. The purpose of these rules in the *Code* is to accommodate special

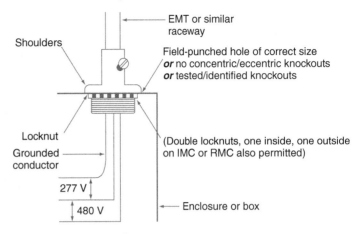

Bonding over 250 V to ground

FIGURE 5-31   Permitted bonding connection between raceway and box in a circuit over 250 volts to ground.

FIGURE 5-32   Listed box with knockouts for suitable grounding in circuits over 250 volts to ground.

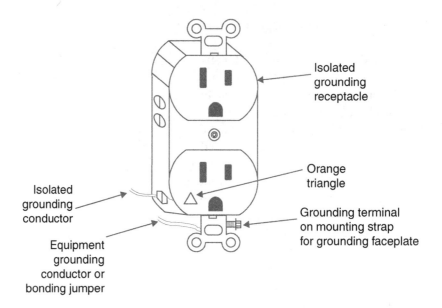

FIGURE 5-33   Isolated grounding receptacle.

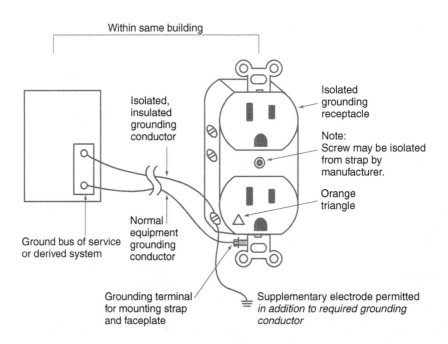

FIGURE 5-34   Grounding connections for an isolated ground receptacle.

needs or desires for isolation of electrical noise without any sacrifice of the essential safety objectives of equipment grounding and bonding and without any endorsement of the effectiveness of the method.

13. Check for occupancies or equipment with special grounding or bonding requirements.

Many occupancies and equipment involve specific additional hazards and requirements. Special bonding and grounding requirements apply in hazardous locations, patient care areas in health care facilities, swimming pools, spas and hot tubs, agricultural buildings, and marinas and boatyards, to mention a few locations. In some cases, bonding is for local equipotential only and a connection to earth directly or indirectly through an EGC is not always required. Such connections may be incidental because of other requirements. In other cases the rules establish a higher level of reliability but the issues are the same as in Article 250. A more complete list of additional or special requirements is given in *NEC* Table 250.3.

KEY TERMS

Article 100 of the *NEC* provides the following definitions for the key grounding and bonding terms. The arrangement of many of the items represented by these terms in a typical electrical system is illustrated in **FIGURE 5-35**.

Accessible (as applied to wiring methods) Capable of being removed or exposed without damaging the building structure or finish or not permanently closed in by the structure or finish of the building. [*This term is used in the context of grounding and bonding in requirements for access to things such as connections to electrodes or to bonding jumpers.*]

Bonded (Bonding) Connected to establish electrical continuity and conductivity. [*This revised definition emphasizes the fact that bonding is about connecting things together. Usually, at least in Article 250, this is about establishing an effective ground-fault path, but it also provides for equal potential by connecting equipment and EGCs together. Outside of Article 250, bonding is often, sometimes strictly, about establishing an equipotential plane or reference.*]

Bonding Conductor or Jumper A reliable conductor to ensure the required electrical conductivity between metal parts required to be electrically connected.

Bonding Jumper, Equipment The connection between two or more portions of the EGC. [*Often, a single conductor serves as both an EGC and an equipment bonding jumper. One of the more obvious examples of this is the bonding jumper that connects a receptacle to a box and thereby completes the connection between the EGC of an appliance cord to the EGC of the fixed wiring system. Since EGCs are sized based on the upstream overcurrent device, this definition does not quite fit when a bonding jumper is used on the supply side of any overcurrent devices, as at services and separately derived systems. The bonding jumper in these situations is called a "supply-side bonding jumper." See Bonding Jumper, Supply Side*]

Bonding Jumper, Main The connection between the grounded circuit conductor and the EGC at the service. [*This definition is a precise description of the purpose of a main bonding jumper. Similar devices or conductors are used where separately derived systems are grounded and sometimes at the disconnect for an existing separate structure, but only the jumper at the service is called the "main" bonding jumper. See the definition for Bonding Jumper, System.*]

Bonding Jumper, Supply Side A conductor installed on the supply side of a service or within a service equipment enclosure(s), or for a separately derived system, that ensures the required electrical conductivity between metal parts required to be electrically connected. [*This definition comes from Section 250.2.*]

Bonding Jumper, System The connection between the grounded circuit conductor and the supply-side bonding jumper, or the EGC, or both, at a separately derived system. [*This is essentially the same as a main bonding jumper except it applies to a separately derived system rather than a service.*]

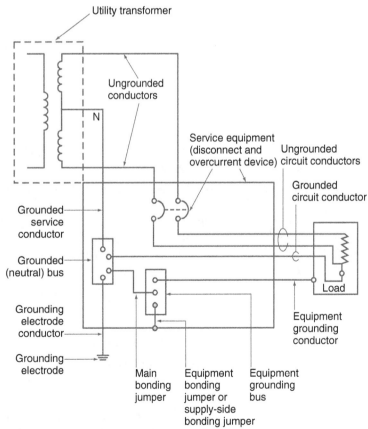

FIGURE 5-35 Components of a typical grounded alternating current (ac) system.

Effective Ground-Fault Current Path An intentionally constructed, low-impedance electrically conductive path designed and intended to carry current under ground-fault conditions from the point of a ground fault on a wiring system to the electrical supply source and that facilitates the operation of the overcurrent protective device or ground fault detectors. [*This definition should be reviewed along with the performance requirements for the Effective Ground-Fault Current Path in 250.4(A)(5). The concept is somewhat different in an ungrounded system because two faults on different ungrounded conductors are needed to produce the high currents required to open overcurrent devices quickly. See 250.4(B)(4).*]

Equipment A general term including fittings, devices, appliances, luminaires, apparatus, machinery, and the like used as a part of, or in connection with, an electrical installation.

Ground The earth. [*This very straightforward definition was revised in 2008 to eliminate the language that said a conducting body could serve in place of earth. Instead, a conducting body can serve to extend the ground connection from the earth.*]

Grounded (Grounding) Connected (connecting) to ground or to a conductive body that extends the ground connection. [*See Ground.*]

Grounded Conductor A system or circuit conductor that is intentionally grounded. [*The grounded conductor is normally a current-carrying conductor operating near ground potential. In most cases, this conductor is the neutral conductor in the electrical system. Often, the grounded conductor is the neutral of a 4-wire, wye-connected electrical system or the neutral of a 120/240 single-phase system. However, the grounded conductor also can be a phase conductor of a corner-grounded, 3-phase, 3-wire delta system, and in that case it is not a neutral. See* Neutral Conductor *and* Neutral Point.]

Grounded, Solidly Connected to ground without inserting any resistor or impedance device. [*Most grounded systems are solidly grounded. High-impedance grounded neutral systems use a resistor or impedance device in the ground-fault return path so that fault currents are limited to a predetermined value, usually to increase the reliability of the system and reduce automatic operation of overcurrent devices, especially on large feeders.*]

Ground Fault An unintentional, electrically conductive connection between an ungrounded conductor of an electrical circuit and the normally non-current-carrying conductors, metallic enclosures, metallic raceways, metallic equipment, or earth. [*See* Effective Ground-Fault Current Path *and* Ground-Fault Current Path. *Ground faults are distinguished from short circuits by the fact that only one insulation failure is necessary to create a ground fault, while short circuits involve two circuit conductors and usually two insulation failures. For this reason, most faults at least begin as ground faults. Although a single fault to ground in an ungrounded system meets the definition of a ground fault, a second fault to ground involving a second ungrounded conductor is necessary to create the high current levels typically associated with ground faults. Ground-fault circuit interrupters (GFCIs) and ground-fault protection for equipment (GFPE) are devices that protect against damage due to ground faults of various magnitudes. GFCIs are intended to sense and protect personnel against low levels of ground-fault current that would otherwise be a shock hazard. GFPE protects against damage to property due to faults in equipment, such as heat tracing and deicing cables. Another type of GFPE protects larger equipment, such as feeders and service equipment, from extensive damage due to high-level ground faults that might otherwise not be automatically interrupted by large overcurrent devices.*]

Ground-Fault Current Path An electrically conductive path from the point of a ground fault on a wiring system through normally non-current-carrying conductors, equipment, or the earth to the electrical supply source.

> *Informational Note:* Examples of ground-fault current paths could consist of any combination of EGCs, metallic raceways, metallic cable sheaths, electrical equipment, and any other electrically conductive material such as metal water and gas piping, steel framing members, stucco mesh, metal ducting, reinforcing steel, shields of communications cables, and the earth itself.

[*Note the difference between this definition and the definition of* Effective Ground-Fault Current Path, *which requires the path to be intentionally constructed, permanent, and low impedance, while this term considers all paths for fault current, effective or otherwise.*]

Grounding Conductor, Equipment (EGC)   The conductive path(s) that provide a ground-fault current path and connects normally non-current-carrying metal parts of equipment together and to the system grounded conductor or to the grounding electrode conductor, or both.

> *Informational Note No. 1:* It is recognized that the equipment grounding conductor also performs bonding.
>
> *Informational Note No. 2:* See 250.118 for a list of acceptable equipment grounding conductors. [*The EGC is not used to carry normal load current, but is expected to complete the bonding function of providing a low-impedance fault-current path. Under normal conditions, its grounding function is to reduce shock hazards by maintaining ground potential on the exposed, non-current-carrying metal parts of electrical equipment.*]

Grounding Electrode   A conducting object through which a direct connection to earth is established.

Grounding Electrode Conductor   A conductor used to connect the system grounded conductor or the equipment to a grounding electrode or to a point on the grounding electrode system. [*Although not used in the* NEC, *this is sometimes abbreviated as GEC similar to the acronym EGC that appears in the definition of* Grounding Conductor, Equipment. *A grounding electrode conductor will connect to both equipment and a grounded conductor in a grounded system whether directly or indirectly, but it will connect only to equipment in an ungrounded system. The grounding electrode conductor is not used to carry normal load current or high levels of ground-fault current. It is intended to carry lighting current and other transient currents that originate outside a building. Under normal conditions, it establishes an earth reference for the grounded system and grounded equipment.*]

Intersystem Bonding Termination   A device that provides a means for connecting intersystem bonding conductors for communications systems to the grounding electrode system. [*This was a new term but not a new concept in the 2008* NEC. *This connection has been required in the past, but the specific termination methods had not been addressed previously.*]

Likely to Become Energized   Failure of insulation on energized parts or conductors. [*This term is used but not defined in the* NEC. *The definition comes from the* NEC *Style Manual. Conductive metal parts of equipment that could become energized if the insulation on enclosed parts were to fail are "likely to become energized." Therefore, virtually any metal parts in contact with or enclosing insulated energized parts are "likely to become energized" and, therefore, usually required to be connected to an EGC. "Likely" in this sense does not address the probability that an insulation failure will occur.*]

Neutral Conductor   The conductor connected to the neutral point of a system that is intended to carry current under normal conditions.

Premises Wiring (System)   Interior and exterior wiring, including power, lighting, control, and signal circuit wiring together with all their associated hardware, fittings, and wiring devices, both permanently and temporarily installed. This includes (a) wiring from the service point or power source to the outlets, or (b) wiring from and including the power source to the outlets where there is no service point. Such wiring does not include wiring internal to appliances, luminaires, motors, controllers, motor control centers, and similar equipment.

> *Informational Note:* Power sources include, but are not limited to, interconnected or stand-alone batteries, solar photovoltaic systems, other distributed generation systems, or generators.

Separately Derived System   An electrical source, other than a service, having no direct connection(s) to circuit conductors of any other electrical source other than those established by grounding and bonding connections. [*"Separately derived" does not require isolation of EGCs or other grounded equipment. The definition only considers normal current-carrying conductors and recognizes that bonding and grounded conductors of different systems are electrically connected. The following additional explanation is found in 250.30, Informational Note No. 1: "An alternate ac power source such as an on-site generator is not a separately derived system if the grounded conductor is solidly interconnected to a service supplied system grounded conductor. An example of such a situation is where alternate source transfer equipment does not include a switching action in the grounded conductor and allows it to remain solidly connected to the service-supplied grounded conductor when the alternate source is operational and supplying the load served."*]

Service   The conductors and equipment for delivering electric energy from the serving utility to the wiring system of the premises served.

Service Equipment   The necessary equipment, usually consisting of a circuit breaker(s) or switch(es) and fuse(s) and their accessories, connected to the load end of service conductors to a building or other structure, or an otherwise designated area, and intended to constitute the main control and cutoff of the supply.

KEY QUESTIONS

1. What size are the ungrounded (hot) service-entrance conductors?
2. What types of grounding electrodes are (or will be) present on the premises?
3. On the basis of all available grounding electrodes present, what size grounding electrode conductor or conductors and bonding jumpers are required?
4. Is the equipment required to be grounded?
5. What type of EGC is used?
6. If a separate EGC is used, what is the conductor material?
7. What is the rating of the overcurrent device protecting the circuit?

CHECKLISTS

✔	Item	Inspection Activity	*NEC* Reference	Comments
		Checklist 5-1: Service Grounding and Bonding		
	1.	Determine what grounding electrodes are present on the premises.	250.50, 250.52(A)(1) through (7)	
	2.	Determine which other electrodes are required or used.	250.52(A)(4) through (8)	
	3.	Verify that the grounding electrode conductor or conductors and bonding jumpers are properly sized.	250.66, 250.64(F), 250.53(C)	
	4.	Verify that the grounding electrode conductors are protected and secured.	250.64(A) and (B)	
	5.	Verify that grounding electrode conductor enclosures are properly bonded.	250.64(E)	
	6.	Verify that the grounding electrode conductor is either unspliced or spliced using appropriate methods.	250.64(C), 250.64(F)	
	7.	Check for correct size and installation of any rod, pipe, or plate electrodes.	250.52, 250.53(A), (B), and (G)	
	8.	Verify the accessibility of grounding electrode conductor connections.	250.68(A)	
	9.	Check for proper grounding electrode conductor connections, including buried connections.	250.70	
	10.	Verify that metal water pipe installed in or attached to a structure is bonded.	250.104(A)	
	11.	Verify that exposed structural building frames are bonded.	250.104(C)	
	12.	Check for proper size and length of bonding jumpers around water meters and the like.	250.66, 250.68(B), 250.52(A)(1)	

(continues)

✔	Item	Inspection Activity	*NEC* Reference	Comments
		Checklist 5-1: Service Grounding and Bonding		
	13.	Check the size, type, and installation of the main bonding jumper.	250.24(B), 250.28	
	14.	Verify that service raceways and enclosures are correctly bonded.	250.92(A), 250.92(B), 250.102	
	15.	Check the size of service-equipment supply-side bonding jumpers.	250.102(C)	
	16.	Verify that the grounded service conductor size is adequate.	220.61, 250.24(C)	
	17.	Check separately derived systems for proper grounding electrodes, grounding electrode conductors, and system bonding jumpers.	250.30(A)	
	18.	Verify that water pipe and structural metal building frames in the area served by each separately derived system are bonded.	250.104(D)(1), 250.104(D)(2)	
	19.	Verify that an intersystem bonding termination has been provided.	250.94	
	20.	Verify that where a wire-type EGC is also used as a grounding electrode conductor, it meets all applicable requirements for both grounding electrode conductors and EGCs.	250.121 Exception	

✔	Item	Inspection Activity	*NEC* Reference	Comments
		Checklist 5-2: Equipment Grounding and Bonding		
	1.	Identify equipment that is required to be grounded.	250.110, 250.112, 250.114, 250.116	
	2.	Verify appropriate grounding methods for equipment fastened in place or connected by permanent wiring methods.	250.134, 250.136	
	3.	Verify appropriate types of EGCs.	250.118	
	4.	Check separate EGCs for proper sizing and identification.	250.122, 250.119	
	5.	Check connections of EGCs within outlet boxes.	250.146, 250.148, 250.8	
	6.	Verify that proper methods are used to bond receptacles to boxes.	250.146, 250.8	

✔	Item	Inspection Activity	*NEC* Reference	Comments
		Checklist 5-2: Equipment Grounding and Bonding		
	7.	Check installation of equipment bonding jumpers, especially where flexible connections or cords are used.	250.96, 250.102, 250.118, 314.22, 350.60, 348.60	
	8.	Verify grounding of panelboard enclosures and connections of EGCs to panelboard metallic enclosures.	408.40	
	9.	Verify proper grounding at separate buildings or structures.	250.32	
	10.	Check equipment grounding of electric ranges and clothes dryers.	250.140, 250.142	
	11.	Verify bonding of equipment operating at over 250 volts to ground.	250.97	
	12.	Check installations with isolated grounding conductors for proper connections and for grounding of the associated enclosures and wiring methods.	250.146(D), 250.96(B), 406.3(D), 408.40, Exception	
	13.	Check for occupancies or equipment with special grounding or bonding requirements.	250.3	

CHAPTER 6

Dwelling Units and Mobile/ Manufactured Home Sites

OVERVIEW

This chapter covers the general aspects of dwelling unit and mobile/manufactured home site inspections. Special systems and **equipment** such as swimming pools, electric deicing and snow melting, motors, air conditioners, dumbwaiters, wheelchair and stairway lifts, and communications wiring are not covered in this chapter but are discussed in other chapters of this manual and in specific articles of the *National Electrical Code®* (*NEC®*).

The *NEC* rules for **dwelling units** are more detailed than the rules for most other occupancy types. The detailed rules covering dwelling units make inspecting them both easier and more difficult than inspecting other types of occupancies. On the one hand, inspections are easier because the rules are more specific, more issues are covered, and fewer matters are left to design discretion. On the other hand, inspections can be more difficult because there are many more rules to remember and apply. Because many dwelling unit electrical installations are laid out by the electrician or the electrical contractor, it is necessary to provide specific prescriptive requirements that provide for a safe electrical plan and installation.

The inspection of the mobile or manufactured home site supply equipment is typically the extent of a field inspector's involvement with these types of dwelling units. Because their construction, including the electrical wiring, is performed within the manufacturing facility, these units are subject to inspection during the manufacturing process. In addition, manufactured homes are built in accordance with U.S. Department of Housing and Urban Development (HUD) construction standards, which supersede locally adopted codes. Thus, only the site supply equipment for these new homes is within the jurisdiction of local code enforcement officials.

The checklists provided at the end of this chapter are intended to help readers keep the most important and most general requirements in mind. Similarly, by expanding on the intent of the rules, the explanations of the individual checklist items found in the manual are intended to help readers make reasoned and uniform interpretations.

SPECIFIC FACTORS

The *NEC* provides more specific requirements for dwelling units than for any other occupancy type. Although many special rules are provided for parts of some special occupancies, such as motor fuel dispensing stations and hospitals, the *NEC* provides requirements for almost every room in a dwelling unit. Included are spacing requirements for **receptacle outlets**, required locations for **lighting outlets** and, in some cases, wall **switches**, and many rules for circuits intended for specific loads, such as **kitchen appliances** and laundry equipment. In contrast, the only **receptacles**, lighting **outlets**, or wall switches required in commercial and industrial occupancies are for specific equipment, such as heating, air-conditioning, and other equipment requiring servicing, and for a few specific **locations**, such as elevator pits and machine rooms. The electrical layout in structures other than dwelling units is left far more to the discretion of the user, installer, designer, or engineer.

The fact that dwelling units are so numerous and have so many similarities makes the adoption of uniform minimum requirements possible and reasonable. Other occupancies are much more dependent on design and intended use. Some large or custom homes may also have special requirements that are dictated by design, and the *Code* does not necessarily provide all the requirements for such installations. For example, the *NEC* requires only two small-appliance **branch circuits** for kitchens and one circuit for **bathrooms**. While such rules provide for a basic level of safety, they do not always provide for convenience or good design. Still, there is a certain amount of convenience contained in the requirements for dwelling units. Receptacle spacing requirements provide receptacle placements that are reasonably convenient in order to deter occupants from resorting to the use of extension cords or other unsafe substandard installations. These types of considerations help to reduce electrical fire and shock hazards within the dwelling unit.

Mobile homes and **manufactured homes** meet the *NEC* definition of *dwelling unit*; however, because these units are not constructed onsite, there are additional considerations regarding their electrical systems that relate to their transportability and the fact that they are often delivered to the site in more than one section on a steel chassis. Thus, the requirements for this type of dwelling unit are contained in Article 550 of the *NEC*. Articles within Chapter 5 of the *NEC* contain requirements for special occupancies.

Although many of the wiring requirements for these dwelling units are similar to those contained in Chapters 1 through 4 of the *NEC* for onsite constructed units, the necessary provisions to address the unique construction features of mobile/manufactured homes amend or modify the general code requirements, and therefore, in accordance with the structure of the *NEC*, the requirements are contained within Chapter 5 of the *NEC*. If the requirements of Article 550 do not amend or modify the general rules of the *Code*, the general rules apply to the installation.

For the most part, the permanent wiring of the mobile/manufactured home is performed at the manufacturing facility and is not subject to inspection by the local **authority having jurisdiction (AHJ)**. However, the site supply equipment is field installed and is subject to local inspection. Additionally, modifications to the wiring systems of these units after they have been initially located onsite are subject to the requirements of the *NEC* and specifically Article 550. Mobile and manufactured homes meeting the definitions in Article 550 are not "modular homes." The modular home type of construction is covered in *NEC* Article 545, "Manufactured Buildings."

KEY QUESTIONS

Dwelling Units

1. What wiring methods are used and are the wiring methods suitable for the conditions?

Type NM (nonmetallic-sheathed) cable is one of the most commonly used wiring methods for branch circuits and feeders in dwelling units in the United States, and except for some large multifamily dwellings where the installation rules for various construction types may preclude its use, it is a suitable wiring method for most dwelling unit electrical systems. However, many dwelling units have areas where NM cable is not suitable for the circuits installed. For example, circuits installed outdoors, underground, or in other **wet locations** cannot be installed using Type NM cable; therefore, other suitable wiring methods must be chosen.

2. Are multiwire branch circuits used?

Disconnecting means, likely circuit breakers, that disconnect both ungrounded (hot) conductors simultaneously must be used where **multiwire branch circuits** are used. Since dwelling units are commonly wired with NM cable or other cable-type wiring methods, multiwire branch circuits can be easily **identified** by the configuration of the cables

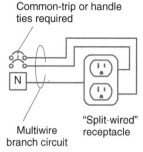

Common-trip or handle ties required

N

Multiwire branch circuit

"Split-wired" receptacle

FIGURE 6-1 Multiwire branch circuits in dwelling units.

used. **FIGURE 6-1** shows a duplex receptacle that has been "split wired" using two separate ungrounded conductors of a multiwire branch circuit. Simultaneous opening of the ungrounded conductors is required when the disconnect is manually operated for this type of application or any multiwire branch circuit, even if separate ungrounded conductors are not in the same box or connected to the same **device**.

3. What rooms require small-appliance branch circuits?

Permanent provisions for cooking are a part of the definition of a *dwelling unit,* so a kitchen is required. A kitchen is an area that has more concentrated use of electrical appliances than in other rooms. Not all dwelling units have dining rooms, breakfast rooms, pantries, or similar areas that are also required to be served by small-appliance branch circuits. Where multipurpose rooms take the place of formal dining rooms and similar areas where small appliances are likely to be used, small-appliance branch circuits should be provided for receptacle outlets. Such spaces should be identified prior to inspections to ensure appropriate application of *Code* requirements.

4. What sizes of service equipment, service-entrance conductors, and feeder conductors are required?

The answer to this question is critical to the inspection of services and feeders and is based on the load calculation and power distribution scheme.

5. Is the service overhead or underground?

The applicability and dimensions of clearances, burial depths, and points of attachment and the permissible types of service **raceways** are all determined by the means of delivering electrical service to a dwelling unit. Also, the applicability of the *NEC* and the timing and types of inspections may vary according to whether the service is underground or overhead.

6. What types of grounding electrodes are present?

Where a concrete-encased electrode or water pipe electrode is present at a dwelling unit, the *Code* requires that it be incorporated as part of the premises grounding electrode system. The use of concrete-encased electrodes is mandatory if the footing or foundation contains 20 ft (6.1 m) or more of electrically conductive reinforcing steel with a diameter of ½ in. (13 mm) or greater, encased by at least 2 in. (50 mm) of concrete, and located horizontally near the bottom or vertically and within that portion of a concrete foundation or footing that is in direct contact with the earth. The 20-ft (6.1-m) requirement does not require a single length of 20 ft (6.1 m) or more, because the usual steel tie wires are specifically permitted to serve to bond reinforcing steel bars together. The use of the concrete-encased electrode is not mandatory for existing structures where access to the rebar requires disturbing (i.e., jack-hammering or chiseling) the concrete. Where electrodes that are part of structural elements or mechanical (plumbing) systems are not present at the dwelling, driven rods are often used to create the grounding electrode system. It should be noted that a single 8-ft (2.4-m) rod is permitted if its resistance to earth is 25 ohms or less. Otherwise, a second 8-ft (2.4-m) rod, driven at least 6 ft (1.8 m) away from the first rod, must be installed. The earth resistance requirement applies to the installation of ground rods, pipes, and plates where used as the only grounding electrode at the premises and where used to supplement an underground metal water-pipe-type electrode. Other listed electrodes must be used in accordance with their listing and labeling instructions, but are not subject to the 25-ohm requirement.

7. What sizes of grounding electrode conductors are required?

The sizes of the grounding electrode conductor(s) or bonding jumpers are dependent on the size of the **ungrounded service-entrance conductors** and the type(s) of grounding electrode(s) used to form the grounding electrode system.

8. Where is mechanical equipment located, and what types of equipment will be installed?

Lighting and receptacle outlets are required for mechanical equipment such as furnaces, boilers, fan coils, and air-conditioning equipment. In some cases, the receptacles for such equipment are required to have **ground-fault circuit interrupter (GFCI)** protection.

Mechanical equipment for heating, air-conditioning, and other functions is subject to the requirements of Articles 210, 422, 424, 430, and 440. Systems for charging electric vehicles may also be found as part of dwelling unit installations. These systems (along with motor loads, including air-conditioning loads) are covered by the Commercial and Industrial Inspections chapter of this manual.

Mobile/Manufactured Home Sites

1. Does the unit meet the definition of a mobile home or of a manufactured home?

Although the interior installation requirements for mobile and manufactured homes are the same, the distinction makes a significant difference relative to the installation of service equipment. Units that meet the definition of manufactured homes are permitted to have the service equipment installed on the structure.

2. If the unit is on-site, what is its electrical rating in amperes?

The minimum rating for the service equipment is 100 amperes. Mobile and manufactured homes are required to have their electrical capacity requirements marked on a nameplate, or, in the case of manufactured homes, the capacity requirement is permitted to be included with the installation instructions provided by the manufacturer. Where the capacity requirement is greater than 100 amperes, the service equipment must be upgraded accordingly.

3. Will any accessory structures be installed?

Necessary wiring provisions and load calculations will be required if there will be any electrical loads in addition to the mobile or manufactured home. This will affect the determination of the minimum size for the site service equipment.

4. Are electrical installations being made to an existing mobile or manufactured home?

Where existing mobile or manufactured home wiring systems are added to or otherwise modified, the requirements of the *NEC* that are in effect within a jurisdiction apply to this "field" installation. For instance, the current (2013) HUD requirements do not require **arc-fault circuit interrupter (AFCI)** protection for branch circuits supplying 125-volt, 15- and 20-ampere outlets installed in manufactured home bedrooms or elsewhere. However, if a 125-volt, 15-ampere branch circuit is field installed in a manufactured home, AFCI protection may be required, depending on which edition of the *NEC* has been adopted within the jurisdiction in which the unit is located.

The scope of the HUD Standard, 24 CFR 3280.801, was revised in 2007 and references the requirements of the 2005 edition of the *NEC*. HUD has endorsed the use of AFCIs, and manufacturers may install them if they wish, but the HUD standard specifically states in 24 CFR 3280.801(b) that AFCIs are not required. The rules for GFCI protection do apply, but they are based on the 2005 *NEC*. In accordance with 24 CFR 3280.806(b), "All 120 volt single phase, 15 and 20 ampere receptacle outlets, including receptacles in light fixtures, installed outdoors, in compartments accessible from the outdoors, in bathrooms, and within 6 feet of a kitchen sink to serve counter top surfaces shall have ground-fault circuit protection for personnel." Additions should comply with the edition of the *NEC* adopted and enforced by the local jurisdiction.

PLANNING FROM START TO FINISH

The checklists in this chapter are broken into four groups:
- Residential rough inspections
- Residential service, feeder, and grounding inspections
- Residential finish inspections
- Mobile/manufactured home inspections

These groupings provide a good general outline of the electrical inspections necessary for dwelling units. Often, all necessary inspections can be done in two visits, with part of the service, feeder, and grounding inspection done with the rough inspection, and the rest completed with the finish inspection. Additional inspections may be necessary for underground installations, exterior wiring, or wiring embedded in masonry or concrete. Special systems such as those relating to snow melting or swimming pools may also require additional inspections.

Many jurisdictions have specific schedules or plans for making inspections on dwelling units and may, for example, require that rough electrical inspections take place at the same time as the inspections for rough framing, plumbing, and mechanical installations. This is sometimes called a four-way inspection, and it covers all items necessary before the work is <u>concealed</u> by drywall or other wall finishes.

The location of the service equipment, whether or not it is completely installed, is a good place to start both rough and finish inspections. Many of the checklist items for services, feeders, and grounding can be covered from this location. From the service, the feeders and other power distribution equipment such as a panelboard (generally referred to as "subpanels"), if any, can be followed to the origination of the branch circuits. Service panels or "subpanels" provide a good place to examine many aspects of the branch circuits. Small-appliance branch circuits, laundry branch circuits, bathroom circuits, and general lighting circuits can all be identified here. The general arrangement of the rough and finish checklists is by room or area. There is no specific order that is best for the room-to-room inspections, but the order of the checklists provides a good path to follow in most cases. The general requirements that apply to all areas can then be checked as a part of the room-to-room inspections.

"Rough" inspections are really inspections of equipment and wiring that are going to be concealed by building finishes. Thus, the actual items inspected on a rough inspection may vary from job to job. For example, if the basement is to remain unfinished, it might not be wired at the time of the rough inspection, because the contractor is primarily interested in having an inspection of the wiring and equipment that will be concealed on future inspections. In some cases, the entire basement wiring may be inspected at the finish stage. In other cases, or in some jurisdictions, all of the wiring must be installed at the rough or "four-way" inspection.

Many of the items checked at the rough stage are rechecked at the finish stage. For this reason, many items in the rough inspection checklists are repeated in the finish inspection checklists, but the inspector will be looking for different aspects of that item. For example, the spacing of receptacle outlet boxes can be checked at the rough stage, but this will have to be rechecked at the finish to be sure that the correct devices were installed in the boxes, or that a box thought to be a required countertop receptacle at the rough stage did not turn out to be a switch on the finish inspection.

WORKING THROUGH THE CHECKLISTS

RESIDENTIAL ROUGH INSPECTIONS

Checklist 6-1: General Requirements (All Areas; Residential Rough Inspection)

1. Check wiring methods (usually cable assemblies) for support and suitability for the conditions.

All wiring methods have specific support requirements and restrictions on their application. The most common type of cable used in residential wiring, Type NM cable, is required to be supported every 4½ ft (1.4 m) and within 12 in. (300 mm) of boxes and other enclosures (334.30). If the NM cable is not attached directly to a box (as is the case with some nonmetallic device boxes), support must be within 8 in. (200 mm) of the box [314.17(C), Exception].

Where NM cable runs through holes in framing members at intervals not to exceed 4½ ft (1.4 m), it is considered to be secured and supported if it is adequately protected in accordance with 300.4 (see Item 2 that follows). Type NM cable may be "unsupported" (installed without any support hardware) where fished through concealed spaces, and up to 4½ ft (1.4 m) may be unsupported between a luminaire and the last point of support in an **accessible** ceiling. (Type NM cable is permitted in dropped or suspended ceilings only in one- and two-family dwelling units.)

Other support restrictions also apply. For example, flat cables may not be stapled on edge, and staples used to secure and support cables must be used in a manner that will not otherwise damage the cable and in accordance with any identification included with the staples. This identification usually includes size and number of cables with which the staples are intended to be used. Generally, staples should secure the NM cable to wooden framing members without pinching the cable. An NM cable with staple damage is a future trouble spot and a potential hazard. NM is generally limited to normally dry locations. Wiring methods for remote control, signaling, fire alarm, communications, and cable television (CATV) or satellite television circuits are subject to the requirements of Articles 725, 760, 800, and 820.

FIGURE 6-2   Steel plate used to protect a Type NM cable within 1¼ in. (32 mm) of the edge of a wood stud. (Courtesy of RACO.)

2. Check cable installation through or parallel to framing members and furring strips for 1¼ in. (32 mm) clearance or protective steel plates.

Type NM cable and other cables must be spaced at least 1¼ in. (32 mm) from the face of framing members to reduce the likelihood of penetration by nails or screws. Protective steel plates or sleeves must be provided where the spacing cannot be maintained (**FIGURE 6-2**). These steel plates are required to be at least ¹⁄₁₆ in. (1.6 mm) thick unless they are listed and marked for the purpose. Where wiring methods are attached to framing members and run parallel to a member, a 1¼-in. (32-mm) space from the edge of the framing member to the wiring is required. Type NM cable installed in metal framing must be protected by grommets that are listed for the purpose and remain in place after installation (**FIGURE 6-3**). Where smaller NM cables (smaller than cables with two, 6 American Wire Gauge [AWG] or three

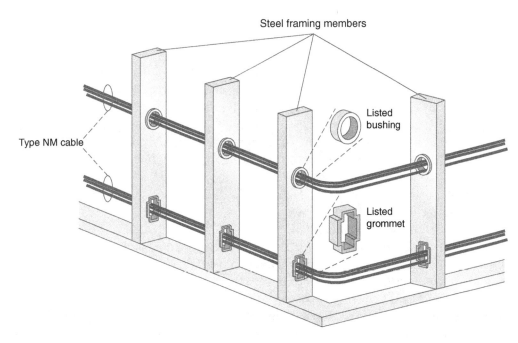

FIGURE 6-3   Listed grommets or bushings for protection of Type NM cable.

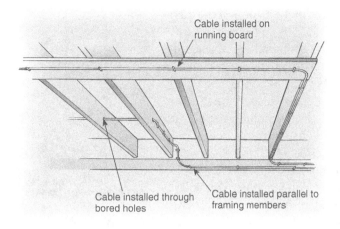

Cable installed on running board

Cable installed through bored holes

Cable installed parallel to framing members

FIGURE 6-4   Protection of Type NM cable installed exposed in basement.

8 AWG conductors) are not run through bored holes but are installed at angles to and on the underside of joists in unfinished basements or crawl spaces, they must be protected by installing the cables on running boards (**FIGURE 6-4**). Larger cables are permitted to run at angles to and across the underside of joists without running boards.

3. Check boxes for suitability for the use.

The boxes most commonly used in residential wiring are suitable only for **dry** or **damp locations**. All boxes must be arranged to avoid the accumulation of moisture in the box. Boxes used in wet locations must be listed for the purpose. Floor boxes must be listed as floor boxes. Similarly, boxes used for supporting paddle fans must be listed for that use. (This specific listing requirement does not apply to boxes used to supply but not support paddle fans if the fan is supported independently of the box.)

Except for device boxes supporting some small wall-mounted luminaires, boxes used for supporting luminaires must be designed for support of luminaires.

4. Verify that boxes are installed in accessible locations for all junctions and outlets and pull points.

Because raceways are not usually needed for Type NM cable and other cables, pull boxes are not usually required unless raceways also are used. However, boxes are required at all junction and outlet points. Boxes must be installed so that contained wiring is accessible and they may not be completely covered by drywall, plaster, concrete, or other building finishes. However, because boxes are not required to be readily accessible, they may be installed in attics, in crawl spaces, above grid ceilings, or behind access doors and panels.

5. Check that cables are secured to boxes.

Generally, cables must be fastened to boxes. Type NM cable need not be secured to one-gang nonmetallic boxes if the cable is fastened within 8 in. (203 mm) of the box. The 8 in. (203 mm) is measured along the sheath of the cable, and the cable sheath or jacket is required to extend into the box no less than ¼ in. (6 mm). Nonmetallic boxes other than one-gang are provided with integral clamps that help prevent the cables from being pulled out.

6. Check boxes for conductor fill.

A certain amount of space is required in a box for each conductor. The general rule is that each enclosed conductor must have free space within a box. Some fittings and devices that occupy space in a box also count as conductors for the purpose of determining the required space in a box. The amount of space required for each conductor varies according to the size of the conductor or, in the case of fittings, by the size of the largest conductor in the box and, in the case of devices, by the size of the largest conductor connected to the device. To verify compliance:

A. Multiply the total number of conductors by their applicable required volumes in cubic inches or cubic millimeters.

B. Add the sum of the volumes.

C. Compare the total volume with the volume of the box in question.

The volumes of common metal boxes are provided in the *Code*. Nonmetallic and metal boxes not covered in the *Code* are required to have the volumes marked on the boxes (see the Wiring Methods and Devices chapter of this manual).

FIGURE 6-5 shows an example of an extension box being used for increasing the available space in the box and providing additional raceway entries from the box in the desired orientation. In this example, there are two 12 AWG conductors in each NM cable and in each flexible metal conduit (FMC) raceway for a total of eight grounded and ungrounded conductors. There are no fittings or clamps within the box, but the combination of four additional 12 AWG equipment grounding conductors (one from each raceway or cable) count as one more conductor, for a total of nine conductors. Each 12 AWG conductor requires 2.25 in.^{3} (36.9 cm^3) of volume, so the total volume required is 2.25 in.^{3} (36.9 cm^3) × 9 conductors = 20.25 in.^{3} (331.9 cm^3). Since the volumes marked on the box and the extension box are each 15.5 in.^{3} (254 cm^3) for a total of 31 in.^{3} (508.1 cm^3), the total volume is adequate for the number of conductors, but the box without the exten-

FIGURE 6-5 Round/octagonal box and extension box, each 4 × 1½ in. (102 × 38 mm), and each marked 15.5 in.^{3} (39.4 cm^3).

sion would be overfilled. An alternative approach to this question is to use *NEC* Table 314.16(A) directly. According to Table 314.16(A), a 4 × 1½ in. (102 × 38 mm) round/octagonal box has adequate space for only six conductors. However, with the addition of a 4 × 1½ in. (102 × 38 mm) round/octagonal extension box, the allowable number of conductors is 12, and the installation as shown complies with the *Code*.

7. Check positioning of boxes that are intended to be flush with combustible and noncombustible finished surfaces.

Boxes mounted in walls should have their edges flush with the wall surface. Up to ¼ in. (6 mm) recess from the wall surface is allowed with wall finishes that are noncombustible, such as drywall (gypsum board) or plaster. Boxes in combustible wall finishes, such as wood paneling, are required to be flush with or extend from the surface (see the Wiring Methods and Devices chapter of this manual).

8. Check for splicing devices on all equipment grounding conductors within boxes and bonding connections to metal boxes.

<u>Listed</u> means, usually listed pressure connectors, must be used to splice equipment grounding conductors in boxes. All equipment grounding conductors in a box must be spliced together and connected to the box if the box is metal. The connection to the box must be made by a screw dedicated to the purpose as shown by the arrow in **FIGURE 6-6** or by a listed device. Screws and clips used for this purpose are often green, but a green color is not required for these EGC connection methods. These requirements do not apply to boxes used only for pulling points in conduit-and-wire installations, where the conductors are not spliced or terminated in or at the box, and the box is grounded through its connection to a grounded metal raceway. Isolated grounding conductors are

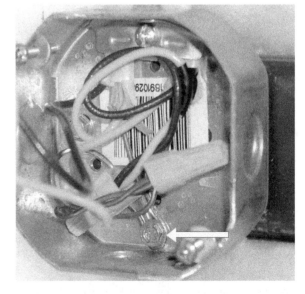

FIGURE 6-6 Equipment grounding conductors connected together and to a metal box with a screw used for no other purpose.

FIGURE 6-7   Listed floor box for use in a wood floor.

not required to be connected to a box, but isolated grounding conductors are not often used in dwelling units.

9. Check equipment grounding conductors for suitability and size.

Metal raceways and conduits such as electrical metallic tubing (EMT) and rigid metal conduit (RMC) may be used as equipment grounding conductors for the circuits they enclose. Some metallic conduits, such as FMC, are permitted to be used as equipment grounding conductors only on small circuit ratings and in limited lengths [250.118]. Otherwise, the sizes of separate equipment grounding conductors are based on the ratings of the circuit overcurrent devices.

Type NM cable is required to contain an equipment grounding conductor. The conductor included is the same size as the other conductors in smaller cable sizes (14 AWG through 10 AWG) or, in larger sizes, is based on the appropriate overcurrent device rating for 60°C-rated conductors of the size included in the cable. For example, Type NM cable designated as 10/2 w/g (two 10 AWG conductors with equipment ground) includes a 10 AWG copper equipment grounding conductor based on the 60°C 30-ampere ampacity of the 10 AWG conductors. An 8/3 w/g (three 8 AWG conductors with equipment ground) also includes a 10 AWG copper equipment ground based on the 40-ampere ampacity of the 8 AWG conductors.

10. Check boxes used in floors or for support of ceiling fans for listing.

As is mentioned in Item 3 of this checklist, boxes used as floor boxes or for the support of ceiling-suspended paddle fans must be listed for the purpose. Note, however, that if a fan is supported directly from the structure, rather than from the box, the box need not be listed for fan support. Many ceiling fans provide support straps that can be attached to a box, but are supported directly from the structure. **FIGURE 6-7** shows an example of a listed floor box suitable for use in a wood floor. The box itself in this case is constructed of cast iron and is also suitable for embedment in concrete.

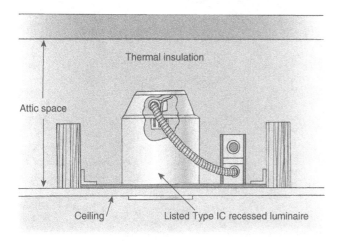

FIGURE 6-8   A listed Type IC recessed luminaire suitable for use in insulated ceilings and installed in direct contact with thermal insulation. (Courtesy of Thomas Lighting, a Philips group brand.)

11. Check recessed luminaires for clearances from combustibles and insulation.

Unless a recessed luminaire is listed for contact with insulation, recessed parts, including power supplies such as ballasts and transformers, must be kept at least ½ in. (13 mm) from combustible materials and 3 in. (76 mm) from thermal insulation. Thermal insulation may not be installed above or in contact with recessed luminaires unless the luminaires are rated for contact with insulation. Recessed luminaires suitable for contact with insulation are designated as Type IC (insulation contact). This marking will be found on the luminaire. A luminaire with the IC marking can be in direct contact with the insulation (**FIGURE 6-8**). Additionally, where recessed luminaires are installed in sloped ceilings, the luminaire is required to be listed for that specific application.

12. Check cables installed in contact with thermal insulation or without maintaining spacing (fire- or draft-stopped, etc.) for possible adjustment factors.

Where more than two NM cables containing two or more current-carrying conductors are installed, without maintaining spacing between the cables, through the same opening in wood framing that is fire- or draft-stopped using thermal insulation, caulk, or sealing foam, the adjustment factors for more than three current-carrying conductors of Table 310.15(B)(3)(a) must be applied. This same rule applies to NM cables installed in contact with thermal insulation in walls or attics without maintaining spacing between the cables.

13. Check wall switch locations for the presence of grounded conductors.

A grounded conductor of the controlled lighting circuit must be provided at most wall switch locations. If type NM cable is used, this requirement applies unless the wall is only partially finished—that is, the wall is open at the top or bottom on the same level or is unfinished on one side so that an additional or replacement cable can be added without disturbing the finish materials. Other switch locations that do not require a grounded conductor were added in 2014. They include switches in nonhabitable rooms such as bathrooms or laundry rooms and may include switches that are part of three-way or four-way switch arrangements. Where NM cable is used, a change in many common wiring practices is required in order to provide the grounded conductor, and different cabling arrangements than have been commonly used under older editions of the *NEC* may be required. Four-conductor NM cable is available that has two pairs that typically consist of one black and one white with a black stripe, and one red and one white with a red stripe (called 12/2 plus 2 or 14/2 plus 2). Another available construction contains one white conductor with three other differently colored conductors (called 12/4 or 14/4).

Checklist 6-2: Kitchen (Residential Rough Inspection)

Note: The receptacles referred to in the following discussions and checklists are 125-volt, 15- or 20-ampere-rated receptacles unless otherwise specified.

1. Check spacing of receptacles for walls and countertops, including islands and peninsulas.

Receptacles must be installed to serve kitchen countertops and be installed so that every countertop space that is 12 in. (300 mm) or more in width and runs along a wall has a receptacle within 24 in. (600 mm) of any point along the wall. The receptacles must be located not greater than 20 in. (500 mm) above the surface of the countertop. If assemblies that are listed for the use are used, the receptacle(s) may be installed in the countertop. If this option is used, the receptacle outlet will not be in place at the rough stage, but the circuit wiring should be present and, if necessary, protected from damage during construction. Peninsula and island countertop spaces must be provided with at least one receptacle outlet unless such spaces are less than 12 × 24 in. (300 × 600 mm) in size. A kitchen with wall counter space, an island, and a peninsula is shown in **FIGURE 6-9.** Receptacles on walls without countertops are required to comply with the standard rule for receptacle wall spacing.

One frequently encountered interpretation issue involves the definition of a countertop. The term is not defined in the *NEC* and does not appear in most dictionaries. The requirements for countertop receptacles are

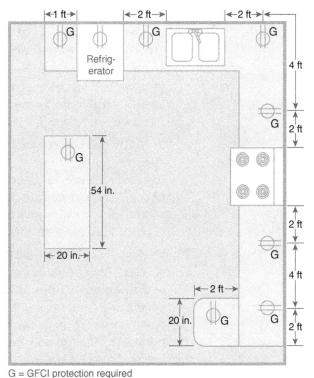

G = GFCI protection required

FIGURE 6-9 Dwelling unit receptacles serving countertop spaces in a kitchen.

FIGURE 6-10 Peninsula from a countertop not treated as countertop space by the AHJ.

based on the idea that such spaces are used as work areas and for small appliances, and the intended or likely use of the spaces should be considered. Areas that are not constructed or designed as kitchen workspaces are probably not appropriately considered "countertops" any more than a table is a countertop. In fact, some portions or extensions of countertop areas are designed and installed to be used as eating areas or tabletops rather than kitchen working spaces. Because of the many variations of kitchen design, it is the responsibility of the AHJ to determine which spaces must be considered as countertop space. **FIGURE 6-10** shows an example of a peninsula extension of a "countertop" that was, in this case, constructed for use as a table and not considered as countertop space by the AHJ for the purposes of receptacle spacing.

Other issues of interpretation concern spaces behind sinks and receptacles in appliance garages. Two diagrams in Figure 210.52(C)(1) in the *NEC* define maximum dimensions of spaces behind sinks or ranges that are not required to have receptacles. If the depth of the countertop behind a sink, range, or cooktop that is installed in an island or peninsula countertop is less than 12 in. (300 mm), that countertop is considered to be divided into two separate spaces and the spaces on each side must be treated as separate countertop spaces. Receptacles in appliance garages, although used for the appliances stored there, are not counted as receptacles serving countertop spaces because they are usually not readily accessible for general countertop use.

2. Verify that a minimum of two 20-ampere small-appliance branch circuits are used for kitchen receptacles.

The required receptacles installed in kitchens, pantries, dining rooms, breakfast rooms, and similar rooms must be connected only to 20-ampere small-appliance branch circuits **(FIGURE 6-11)**. At least two such circuits must be supplied to each kitchen. Some specific appliances, such as some microwave ovens and some refrigerators, may require additional separate circuits (see Item 4 of this checklist).

3. Verify that a wall-switched lighting outlet is provided and wired on a general lighting circuit.

Lighting outlets in kitchens must be connected to general lighting circuits and may not be connected to the small-appliance circuits. At least one such lighting outlet must be controlled by a wall switch as shown in Figure 6-11. Switched receptacles are not permitted to be used to satisfy this requirement in kitchens but could be used in dining rooms as shown. Either 15- or 20-ampere circuits may be used for general lighting.

4. Verify that properly sized circuits are provided for specific kitchen appliances, such as dishwashers, disposals, ranges, <u>counter-mounted cooking units</u> (cooktops), trash compactors, and the like.

Small-appliance branch circuits are intended only for use with cord-and-plug–connected portable appliances and refrigerators. Additional circuits are required for appliances that are fixed in place, are permanently installed, or require dedicated circuits as part of their labeling or because of their ratings. All such circuits must be adequate for the load they serve. The load of any one appliance that is fastened in place may not exceed 50 percent of the rating of the circuit if other appliances are to be served by the same circuit.

5. Check for additional small-appliance branch circuits where there is more than one kitchen.

Each kitchen is required to be served by at least two small-appliance branch circuits, and no such circuit may serve more than one kitchen. **FIGURE 6-12** illustrates the small-appliance branch-circuit requirements for a dwelling unit with more than one kitchen.

6. Check for other outlets or appliances on small-appliance branch circuits.

Small-appliance branch circuits may serve only receptacles in kitchens and related rooms and may not serve other receptacle or lighting loads. Additional circuits must be provided for other types of outlets or for outlets in other areas.

Checklist 6-3: Dining Room (Residential Rough Inspection)

1. Check receptacle outlets for proper spacing.

The locations of dining room receptacles are based on the same spacing rule that applies to most other rooms in a dwelling. That is, a receptacle must be installed in each wall space over 2 ft (600 mm) in width, and additional receptacles must be installed so that no point on any unbroken wall space is over 6 ft (1.8 m), measured horizontally along the wall, from a receptacle (**FIGURE 6-13**). See Item 1 in the Rough Inspection checklist for Other Habitable Rooms.

2. Verify that all required receptacle outlets are supplied by small-appliance branch circuits.

All receptacles in dining rooms should be on small-appliance branch circuits except for those that are used for general lighting and are controlled by a wall switch. Wall switch-controlled receptacles are sometimes installed to supply cord-and-plug–connected luminaires, and such receptacles are considered to satisfy the requirement for a lighting outlet in a dining room.

3. Check for a wall switch-controlled lighting outlet on a general lighting circuit.

A wall switch controlling either a fixed lighting outlet or a receptacle on a general lighting circuit must be installed in a dining room.

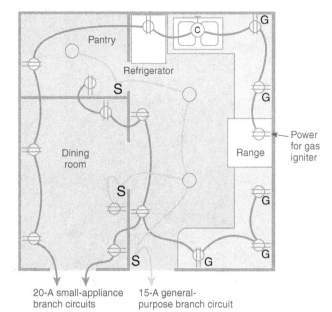

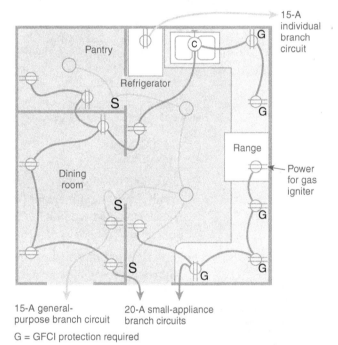

FIGURE 6-11 Two arrangements for small-appliance branch circuits serving kitchen and dining areas.

Checklist 6-4: Bathrooms (Residential Rough Inspection)

1. Verify that receptacle outlets are installed adjacent to and within 36 in. (900 mm) of each basin.

Each basin in a bathroom must have a receptacle within 36 in. (900 mm) of the outside edge of the basin. The receptacle must be located on a wall or partition adjacent to the basin or basin countertop, or it may be installed on the face or side of the basin cabinet if it is not more than 12 in. (300 mm) below the countertop. A receptacle outlet may serve

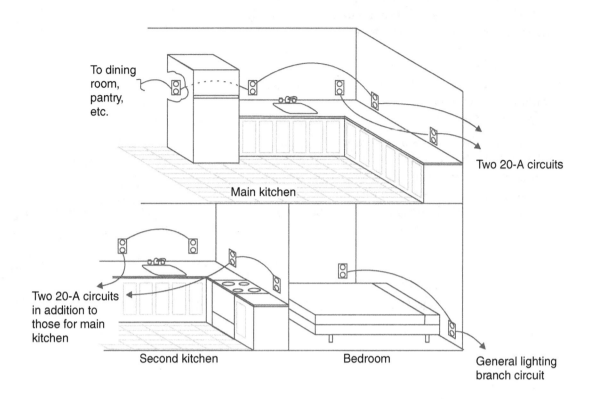

To dining room, pantry, etc.

Main kitchen

Two 20-A circuits

Two 20-A circuits in addition to those for main kitchen

Second kitchen

Bedroom

General lighting branch circuit

FIGURE 6-12   Required small-appliance branch circuits for dwellings with more than one kitchen.

more than one basin if it is within 36 in. (900 mm) of each basin. A receptacle may also be installed in the countertop if it is part of a listed outlet assembly. If this option is used, the receptacle outlet will not be in place at the rough stage, but the circuit wiring should be present and, if necessary, protected from damage during construction.

2. Verify that receptacles are supplied by dedicated 20-ampere branch circuits.

Bathroom receptacles must be supplied by 20-ampere branch circuits that either are dedicated to bathroom receptacle outlets in one or more bathrooms or are dedicated to individual bathrooms. In other words, the required 20-ampere circuit(s) may supply only bathroom receptacles in multiple bathrooms, or they may supply receptacles and other loads but only in one bathroom. The other loads, such as an exhaust fan, cannot exceed 50 percent of the branch-circuit rating. In the example of **FIGURE 6-14**, which depicts two permitted circuit arrangements for supplying bathroom receptacles, the load of the exhaust fan could not be more than 10 amperes.

3. Check for a wall switch-controlled lighting outlet on a general lighting circuit.

Unless the 20-ampere receptacle circuit for a bathroom is dedicated only to that bathroom, the lighting in the bathroom must be supplied by a general lighting circuit. A switched receptacle does not satisfy the requirement for a lighting outlet in a bathroom. A permanently installed luminaire is required.

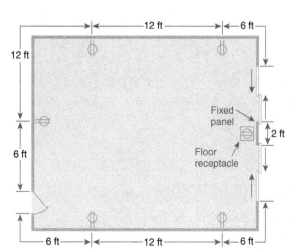

FIGURE 6-13   Required receptacle outlets based on room configuration. (Source: *National Electrical Code® Handbook*, NFPA, Quincy, MA, 2011, Exhibit 210.23.)

Checklist 6-5: Other Habitable Rooms (Bedrooms, Family Rooms, Parlors, and Dens; Residential Rough Inspection)

1. Check receptacle outlets for proper spacing.
Spacing of wall receptacles is based on the same 6-ft (1.8-m) rule that applies to wall spaces in kitchens and dining rooms (that is, a receptacle must be installed in each wall space over 2 ft [600 mm] in width, and additional receptacles must be installed so that no point on any unbroken wall space is over 6 ft [1.8 m] from a receptacle). Figure 6-13 illustrates the required wall receptacle arrangement. A floor receptacle that is located within 18 in. (450 mm) of the wall can be used as a required receptacle. Floor receptacles are often needed when wall spaces are created by glass or railings that do not provide a vertical surface in which to install a wall outlet as shown in **FIGURE 6-15**. The next receptacle outlet is installed within 12 ft (3.6 m) away in the column in this illustration.

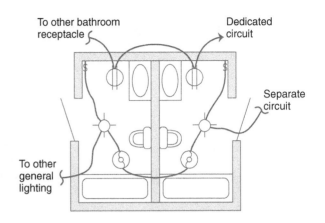

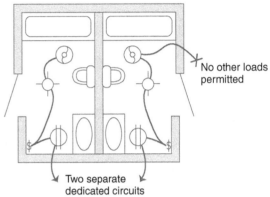

FIGURE 6-14   Permitted circuit arrangements for supplying bathroom receptacle outlets.

FIGURE 6-15   Floor receptacle outlet used to comply with spacing requirements at a railing.

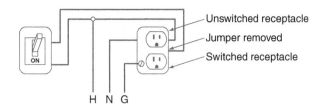

FIGURE 6-16 Wiring diagram and plan symbol for switched receptacle outlet.

2. Check for wall switch-controlled lighting outlets (including switched receptacles).

At least one wall switch-controlled lighting outlet is required in each habitable room. The outlet may be a fixed lighting outlet. Except in bathrooms, hallways, stairways, **garages**, storage or equipment spaces, and kitchens, none of which would typically be considered a habitable room, the lighting outlet can be a switched receptacle. Switching one-half of a duplex receptacle meets the requirement for providing a wall switch-controlled lighting outlet at a location where a convenience receptacle is also required (**FIGURE 6-16**). If both receptacles in the duplex receptacle shown were controlled by the switch, the receptacle outlet could only be counted as a lighting outlet and could not be counted as meeting the requirements for a receptacle outlet.

Checklist 6-6: Hallways and Foyers (Residential Rough Inspection)

1. Check for at least one wall switch-controlled (or automatic-, remote-, or centrally controlled) lighting outlet.

Switched receptacles are not permitted as the required lighting outlets in hallways. Local switches are not required if central, automatic, or remote controls, such as home automation systems or motion sensors, control hallway lighting outlets.

2. Verify that hallways that are continuous for 10 ft (3.0 m) or more have at least one receptacle outlet.

Hallway length is measured along the centerline of the hallway. Hallways that are broken into sections by doorways are considered separate hallways (**FIGURE 6-17**).

3. Verify that foyers that are not part of a hallway have receptacle outlets installed as required.

If a foyer is not part of a hallway and is over 60 ft² (5.6 m²) in area, receptacles are required in each unbroken wall space that is 3 ft (900 mm) or more in width. The wall space is considered unbroken if there are no doorways, floor-to-ceiling windows, or similar openings.

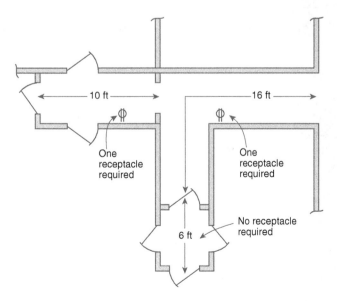

FIGURE 6-17 Receptacle requirements for hallways.

Checklist 6-7: Stairways (Residential Rough Inspection)

1. Check for at least one wall switch-controlled (or automatic-, remote-, or centrally controlled) lighting outlet.

Switched receptacles are not permitted as the required lighting outlets in stairways. Local switches are not required if central, automatic, or remote controls, such as home automation systems or motion sensors, control stairway lighting outlets.

2. Verify that wall switches are provided at each floor level where there are six or more steps between levels.

Remote, automatic, or central controls may replace local three-way switches in stairways. This rule is applicable to the interior stairways within a dwelling unit. If an intermediate landing includes an entryway, another switch must be located at the landing level. Since the rule actually counts risers, a landing may otherwise count as a step for the purposes of this rule.

Checklist 6-8: Closets (Residential Rough Inspection)

1. Check clearances between luminaires and storage spaces if luminaires are to be installed.

Lighting outlets are not required in clothes closets. Where they are installed, clearances between the defined storage spaces in a closet must be maintained. The highest part of the storage area shown in **FIGURE 6-18** is the ceiling of the closet. Incandescent luminaires with open lamps are not permitted in clothes closets. Incandescent lamps must be totally enclosed. A clearance of 12 in. (300 mm) is required from surface-mounted incandescent or light-emitting diode (LED) luminaires, and a clearance of 6 in. (150 mm) is required from recessed incandescent or LED luminaires with completely enclosed lamps or from surface- or recessed-mounted fluorescent luminaires. The type of luminaire that may be installed in a clothes closet depends on the size of the closet, in order to provide the minimum clearances (Figure 6-18). Some fluorescent or LED luminaires may be installed within the defined storage area if they are **identified** for the use.

Checklist 6-9: Laundry Area (Residential Rough Inspection)

1. Verify that at least one receptacle outlet is installed for the laundry.

At least one receptacle outlet must be installed in an area designated for laundry equipment, and this outlet typically includes a duplex receptacle. More than one receptacle or outlet in the laundry area is permitted to be supplied by the laundry branch circuit. This

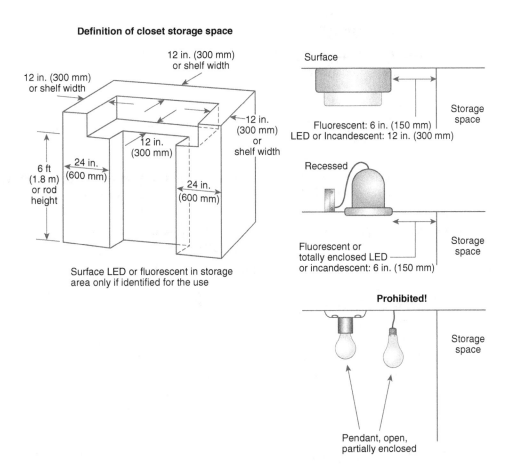

Definition of closet storage space

Surface LED or fluorescent in storage area only if identified for the use

Surface

Fluorescent: 6 in. (150 mm)
LED or Incandescent: 12 in. (300 mm)

Storage space

Recessed

Fluorescent or totally enclosed LED or incandescent: 6 in. (150 mm)

Storage space

Prohibited!

Storage space

Pendant, open, partially enclosed

FIGURE 6-18 Permitted locations for luminaires in clothes closets.

outlet is not required in single dwelling units of a multifamily dwelling with shared laundry facilities or if laundry equipment will not be installed or is not permitted.

2. Verify that a dedicated 20-ampere circuit supplies the laundry outlet(s) and no other outlets.

All receptacle outlets in the laundry area may be supplied by a single laundry branch circuit. However, lighting in the laundry area and receptacles in other areas must be supplied by a general lighting branch circuit or other circuits.

3. Check for a laundry receptacle outlet within 6 ft (1.8 m) of the intended appliance location.

The intended location for a washing machine can usually be easily determined from the location of the plumbing serving that appliance.

4. Check for proper branch-circuit conductors, including equipment grounding conductors, for 240-volt dryers (if used).

Generally, a 30-ampere 4-wire branch circuit that includes two ungrounded conductors, a grounded conductor, and an equipment grounding conductor is required for new 240-volt dryer installations.

The use of the neutral (grounded) conductor for grounding the frames of dryers is no longer permitted in new installations. The 30-ampere, 240-volt rating corresponds to the minimum feeder load of 5000 VA for a dryer in a dwelling unit. At 5000 VA, a 240-volt dryer is a 20.8-ampere load, which requires a 10 AWG copper or 8 AWG aluminum conductor if Type NM cable is used. A 30-ampere set of fuses or a 30-ampere circuit breaker supplying a 30-ampere receptacle is normally used for household electric clothes dryers. Some dryers in washer–dryer combination units may operate on lower amperage or 120-volt circuits, in which case the nameplate of the appliance should be used to determine the required circuit characteristics. The 30-ampere circuits normally installed for residential dryers can be easily modified to accommodate lower-rated appliances by changing the overcurrent device and the receptacle to lower-rated devices.

5. Verify that lighting outlets for the area are supplied from general lighting circuits.

The lighting for a laundry area must be supplied by circuits other than the laundry branch circuit.

Checklist 6-10: Basements and Attics (Residential Rough Inspection)

1. Verify that at least one receptacle outlet is provided in unfinished basement areas in addition to any receptacles installed for laundry equipment or other specific equipment.

An unfinished basement must be furnished with one or more GFCI-protected receptacles for the use of portable appliances and tools. This requirement applies to each separate unfinished portion of a basement and only applies to one-family dwellings.

2. Verify that a receptacle outlet is provided for servicing mechanical equipment, if any.

In any location that contains a furnace or other heating, air-conditioning, or refrigeration equipment, at least one receptacle must be provided within 25 ft (7.5 m) of the equipment for servicing the equipment. Mechanical codes may require receptacles for other types of equipment. Some previous editions of the *NEC* required these outlets only on rooftops or in attics or crawl spaces. This receptacle must have GFCI protection if this receptacle is in an unfinished area of a basement or other area that requires GFCI-protected receptacle outlets. Multifamily dwellings, other than one- or two-family dwellings, also require a receptacle within 50 ft (15 m) of the electrical service equipment (210.64).

3. Verify that individual branch circuits are supplied for central heating equipment, if any.

Central heating equipment, such as gas or electric furnaces or heat pumps, is required to be supplied by individual (dedicated) branch circuits. These circuits may supply other equipment that is directly related to the central heating equipment, such as humidifiers and electrostatic air cleaners. The purpose of this rule is to increase the reliability of heating equipment circuits and reduce the possibility of freezing damage due to a loss of heating. An exception allows permanently connected air-conditioning equipment to be installed on these circuits as well. However, other requirements of Article 210 also apply. For example, on 15- or 20-ampere branch circuits, the load of any one permanent appliance cannot exceed 50 percent of the rating of the circuit, and the total load cannot exceed the rating of the branch circuit (210.23), so the circuit would have to be rated for all connected loads (210.20 and 210.23). Also, the concept of a noncoincident load cannot be applied to the branch circuit because noncoincident load calculations apply only to feeders (220.60).

4. Verify that a wall switch-controlled lighting outlet or a lighting outlet containing a switch is provided at the entrance to areas for storage or equipment requiring servicing.

Illumination controlled by wall switches or a pull chain or other switches integral to the luminaire or lampholder must be provided for mechanical equipment and other equipment that requires servicing. At least one lighting outlet must be located near the equipment requiring servicing, and the switch for that lighting outlet must be located at the point of entrance to the room or area containing the equipment.

5. Check basements, accessible attics, attic entrances, and scuttle holes for clearances from or protection of cable assemblies.

Type NM cable must be protected where subject to physical damage. (The requirements for Type AC cable [armored] and Type MC cable [metal-clad] are similar but not identical.) Exposed work in unfinished basements and crawl spaces should follow building surfaces or be installed on running boards (**FIGURE 6-19**). Larger NM cables (two-conductor 6 AWG or three-conductor 8 AWG or larger) may be installed across the bottoms of joists without running boards, and any size cables may be run through bored holes without running boards. All cables must comply with 300.4 for protection where run through bored holes or parallel to framing.

Cables in attics and around attic entrances must be protected from damage by persons entering or working in such spaces. Similarly, the cables must be supported and secured so that people do not trip over or otherwise become entangled in the cables. Where protection is required, guard strips or raceways should be used to protect the cable. **FIGURE 6-20** shows the areas within an accessible attic or roof space where physical protection is required for cable assemblies that are run across the top of floor joists or across rafters or studding. (These rules do not apply to cables run through bored holes or parallel to and on the sides of joists or rafters.) The area where protection is required is the entire accessible space from the top of the joists to the bottom of any rafters or studding within 7 ft (2.1 m) of the joists if the space is

FIGURE 6-19   Type NM cable installations on running board across joists in basement.

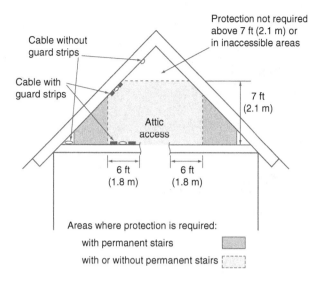

Cable without guard strips

Cable with guard strips

Protection not required above 7 ft (2.1 m) or in inaccessible areas

Attic access

7 ft (2.1 m)

6 ft (1.8 m)   6 ft (1.8 m)

Areas where protection is required:

with permanent stairs

with or without permanent stairs

FIGURE 6-20 Protection for cable assemblies that run across joists or rafters.

accessible by permanent stairs. The same type of protection is required in a smaller space surrounding the access area where not accessible by permanent stairs. Although only guard strips are mentioned in Section 320.23, 334.15(B) specifically allows certain raceways to be used to provide physical protection for Type NM cable. Section 334.23 refers to 320.23 for protection of NM cable, so these rules apply to both NM and AC cable as well as MC cable.

Checklist 6-11: Attached Garages and Detached Garages or Accessory Buildings with Electric Power (Residential Rough Inspection)

1. Verify that at least one receptacle outlet is provided.
Attached garages are required to have at least one receptacle outlet. Detached garages and some accessory buildings are not required to have electrical power, but those that do are required to have at least one receptacle outlet. In garages there must be at least one outlet for each car space. The circuits supplying the receptacles for a garage may not supply outlets elsewhere. These rules apply only to one-family dwellings. If a specific outlet(s) is provided for the purpose of supplying *electric vehicle supply equipment* (EVSE), the branch circuit supplying this outlet(s) cannot be used to supply other outlets. This requirement does not mandate the installation of such circuits; nor does it require that it be an **individual branch circuit** (210.17).

2. Verify that a wall switch-controlled lighting outlet is provided.
A lighting outlet controlled by a wall switch is required in attached garages and in detached garages that have electrical power. Switched receptacles do not satisfy this rule. The lighting included with some garage door openers does not meet this requirement. This requirement does not specifically apply to accessory buildings.

Checklist 6-12: Outdoors (Residential Rough Inspection)

1. Check for at least two receptacle outlets, one each at the front and back of a dwelling.
Outdoor receptacles that are readily accessible while standing at grade level and not over 6½ ft (2.0 m) above grade level must be provided at the front and back of **one-family** and

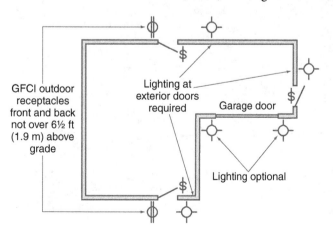

GFCI outdoor receptacles front and back not over 6½ ft (1.9 m) above grade

Lighting at exterior doors required

Garage door

Lighting optional

FIGURE 6-21 Outdoor receptacles and lighting outlets for one- and two-family dwellings.

two-family dwellings (FIGURE 6-21). In addition, GFCI-protected receptacles must be provided within 25 ft (7.5 m) of any heating, air-conditioning, or refrigeration equipment located outdoors, but the receptacles shown in Figure 6-21 could be located to meet both requirements. This requirement also applies to individual units of two-family dwellings located at grade level. In addition, one outdoor receptacle is required for individual dwelling units that are located at grade level and are part of a multifamily dwelling

2. Check for receptacle outlets on balconies, decks, and porches.
One receptacle is required on balconies, decks, and porches that are attached to and accessible from inside *any* dwelling unit. This receptacle may not

be over 6.5 ft (2.0 m) above the surface of the balcony, deck, or porch. For decks or porches at or accessible from grade level, the receptacle supplied to meet the general requirement for outdoor receptacles may also satisfy this requirement, and could meet the requirements for servicing of mechanical equipment, depending on the location of the mechanical equipment.

3. Check for wall switch-controlled (or remote-, central-, or automatic-controlled) exterior lighting outlets at outdoor entrances or exits with grade-level access.
Exterior lighting outlets are required at all outdoor entrances with grade-level access other than garage doors for vehicles. Remote, central, or automatic controls may be substituted for wall switch control of such lighting outlets. Figure 6-21 illustrates the required outdoor receptacle and lighting outlets for a one-family dwelling. In lieu of a lighting outlet at each door, area or flood lighting can be used to illuminate the exterior of the dwelling, because the rule requires illumination of specific areas rather than specific locations of outlets. This requirement also applies to detached garages if they have electrical power. Other detached accessory buildings are not covered by this rule. Lighting outlets at vehicle doors are optional, because these doors are not considered to be outdoor entrances or exits for people. Because exterior lighting outlets may be in wet or damp locations, NM cable may not be permitted if the cable is also in the wet or damp location. Type NM cable is only permitted in "normally dry" locations. Typically, the cable is concealed within the normally dry interior of the wall assembly and within the interior of the outlet box supporting the exterior luminaire.

RESIDENTIAL SERVICE FEEDER AND GROUNDING INSPECTIONS

Checklist 6-13: Services and Feeders and System Grounding (Residential Rough Inspection)

1. Review the calculation of service load and determine the minimum size of service conductors.
Service load calculations are usually provided for plan review. If not, the calculation should be provided by the electrical designer, contractor, or installer at the request of the inspector. (The *NEC* does not specifically require a load calculation to be submitted, but the inspector may require one as a condition of approval of the plans or the installation.) Article 220 provides the method for performing this calculation. For single-family dwellings, the minimum size corresponds to the minimum rating of the service disconnect, which is 100 amperes. **BOX 6-1** shows an example of a load calculation for a one-family dwelling. Conductor sizing may be based on a reduced value according to 310.15(B)(7) if the conductors supply the entire load and the rating of the service is between 100 and 400 amperes.

2. Verify that service disconnects and overcurrent devices are located outside or inside nearest the point of entrance of the service conductors.
The *NEC* requires the service disconnect (and overload devices) to be located "outside . . . or inside nearest the point of entrance of the service conductors." Because **service conductors** are usually not protected by short-circuit and ground-fault protective devices, at least not to *NEC* requirements, the length of service conductors inside of a building must be minimized. Local requirements vary as to how far or under what conditions service conductors may be run into a dwelling. The AHJ must decide what is meant by "nearest."

Box 6-1   Optional Calculation for One-Family Dwelling, Heating Larger Than Air Conditioning

The dwelling has a floor area of 1500 ft² (139 m²), exclusive of an unfinished cellar not adaptable for future use, unfinished attic, and open porches. It has a 12-kW range, a 2.5-kW water heater, a 1.2-kW dishwasher, 9 kW of electric space heating installed in five rooms, a 5-kW clothes dryer, and a 6-ampere, 230-volt room air-conditioning unit. Assume range, water heater, dishwasher, space heating, and clothes dryer kW ratings equivalent to kilovolt-amperes (kVA). [*See* NEC *220.82 and Annex D, Example D2(a).*]

Air-Conditioner Kilovolt-Ampere Calculation

   6 A × 230 ÷ 1000 = 1.38 kVA

This 1.38 kVA (*item 1 from 220.82[C]*) is less than 40 percent of 9 kVA of separately controlled electric heat (*item 6 from 220.82[C]*), so the 1.38 kVA need not be included in the service calculation.

General Load

1500 ft² (139 m²) at 3 volt-amperes (VA)/ft² (33 VA/m²)	4,500 VA
Two 20-ampere appliance outlet circuits at 1500 VA each	3,000 VA
Laundry circuit	1,500 VA
Range (at nameplate rating)	12,000 VA
Water heater	2,500 VA
Dishwasher	1,200 VA
Clothes dryer	5,000 VA
Total	29,700 VA

Application of Demand Factor (*see Section 220.82[B]*)

First 10 kVA of other load at 100%	10,000 VA
Remainder of general load at 40% (19.7 kVA × 0.4)	7,880 VA
Total of general load	17,880 VA
9 kVA of heat at 40% (9000 VA × 0.4) =	3,600 VA
Total	21,480 VA

Calculated Load for Service Size

21,480 VA ÷ 240 V = 89.5 amperes

The minimum service size would be 100 amperes in accordance with 230.42 and 230.79.

Feeder Neutral Load, in Accordance with 220.61

1500 ft² (139 m²) at 3 VA	4,500 VA
Three 20-ampere circuits at 1500 VA	4,500 VA
Total	9,000 VA
3000 VA at 100%	3,000 VA
9000 VA – 3000 VA = 6000 VA at 35%	2,100 VA
Subtotal	5,100 VA
Range: 8 kVA at 70%	5,600 VA
Clothes dryer: 5 kVA at 70%	3,500 VA
Dishwasher	1,200 VA
Total	15,400 VA

Calculated Load for Neutral

15,400 VA ÷ 240 V = 64.2 or 64 amperes

3. Verify that service disconnects are grouped together, with no more than six in any one location.

Most dwelling units have a single disconnect or a group of disconnects in a single enclosure. The *NEC* permits up to six separate circuit breakers or switches to be used as the "**disconnecting means**," but they must all be grouped in one location or be part of the same panelboard. Service panelboards are the only panelboards that do not require overcurrent protection for the panelboard itself in new construction.

4. Check for proper accessibility, working clearances, and dedicated spaces around service equipment.

Service overcurrent devices must be readily accessible to the occupants of a **dwelling unit**. Clear working space must be provided in front of the service equipment (**FIGURE 6-22**). The working space must be at least 30 in. (762 mm) wide or the width of the equipment if the equipment is wider than 30 in. (762 mm). The working space must be at least 6½ ft (2 m) high and 3 ft (914 mm) deep in front of the equipment. (The depth varies according to the conditions and voltages, but 3 ft [914 mm] is adequate for all 120/240-volt residential applications.) In addition to the working space, a space equal to the width and depth of the service equipment must be dedicated to electrical equipment, cables, and raceways, not just the panelboard. As of the 2014 edition, this requirement applies to both indoor and outdoor panelboards. This dedicated space extends from the floor to the structural ceiling or up to 6 ft (1.8 m) above panelboards. Foreign equipment such as plumbing pipes and ducts cannot run through the dedicated space. Even leak-protection apparatus must be located outside (above) the dedicated space. Dedicated space is not required above and below all electrical equipment. For example, separate switch or circuit breaker enclosures are not covered by this requirement. The other types of equipment that require dedicated space (switchboards, switchgear, and motor control centers) are not often used in one- or two-family dwelling units. All such electrical equipment, indoors or out, must be protected from accidental contact, spillage, leakage, or damage by vehicles.

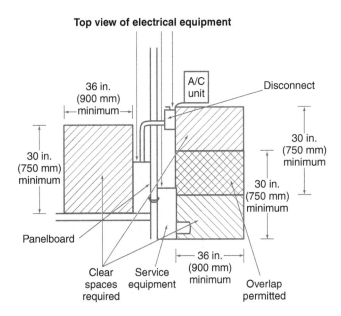

Top view of electrical equipment

FIGURE 6-22   Required working space for 120/240-volt electrical equipment likely to be serviced while energized.

5. Check service-entrance wiring methods for suitability, support, and protection from damage.

Article 230 includes a list of the wiring methods that may be used for service-entrance conductors. In addition to being included in the list, wiring methods for service-entrance conductors must also be suitable for the specific conditions. The article covering each wiring method lists uses that are permitted and not permitted. The wiring method articles also include support and protection requirements. However, where cables are used as service conductors, additional protection and support may be required in accordance with Article 230. For example, **service cables** such as Type SE (service-entrance) cable must be protected by raceways such as EMT (electrical metallic tubing), IMC (intermediate metal conduit), RMC (rigid metal conduit), PVC (polyvinyl chloride conduit), or RTRC (reinforced thermosetting resin conduit) suitable for the location where they are exposed to physical damage. While Section 230.43 does not limit PVC or RTRC to special types, Articles 352 and 355 require special identification for PVC (Schedule 80) or RTRC (AG, XW if above ground) if they are to be installed where subject to physical damage. Some of the wiring methods permitted for service-entrance conductors (such as HDPE conduit) are not permitted to enter buildings, so they must terminate outside the building; however, the wiring methods that

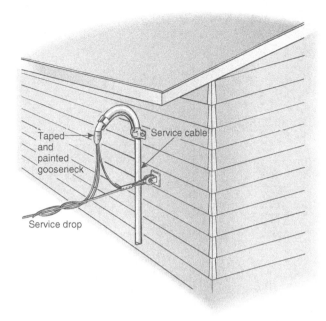

FIGURE 6-23   Example of SE cable formed into a gooseneck. (Source: *National Electrical Code® Handbook,* NFPA, Quincy, MA, 2011, Exhibit 230.24.)

are specifically mentioned for protection of service cables from physical damage in 230.50(A)(1) may all be installed inside buildings.

6. Check for a proper drip loop and weatherhead on overhead services.

Overhead services must be arranged so that water will not enter SE cables or raceways. Drip loops provide a low point in the service-drop cable or conductors to keep water from running along the cable or conductors and entering a service raceway or weatherhead. Weatherheads (called service heads in Article 230) shed water and protect the opening to a service raceway or cable. Service cables used without raceways may be equipped with weatherhead fittings or formed into a gooseneck (**FIGURE 6-23**). The service head or weatherhead, where used, must be listed for use in wet locations.

7. Verify that the point of attachment for overhead service is adequate and will provide required support and clearances above roofs and grade.

Most utilities have some minimum requirement for the means of support of their service-drop conductors. Their guidelines are in addition to the general rules and take into account local conditions such as ice or wind loads. The height of the attachment may never be less than 10 ft (3.0 m) above finished grade and must be located so as to maintain clearances even when sag in the cable is considered. In most cases, the point of attachment is also required to be below the weatherhead or gooseneck to help exclude water from the weatherhead or cable. The point of attachment must also be selected to maintain clearances from communications conductors as required by Section 800.44(A)(4).

8. Check service masts for adequate strength and support.

Utilities often have specific minimum worst-case requirements for service masts, as they do for other points of attachment. Minimum requirements of trade size 1½ or 2 RMC or intermediate metal conduit (IMC) are common. Larger sizes may be required. Again, many utility rules take into account local conditions. The *NEC* requires that supports be structurally adequate but relies on the AHJ to determine precisely what is adequate based on local conditions. If the point of attachment is to a service mast, it may not be made between a conduit coupling located above the roof and the weatherhead.

9. Check for proper clearances of service conductors from building openings.

Service conductors that are not in a raceway or a cable with an overall jacket must be installed so that a minimum clearance of 3 ft (900 mm) from most building openings is maintained. This restriction applies primarily to the open conductors of a service-drop and the drip loop of an overhead service. The 3-ft (900-mm) clearance does not apply to service conductors in a raceway or SE cable. The conductors that emerge out of the weatherhead in **FIGURE 6-24** are the open conductors that require 3 ft (900 mm) of clearance from windows and other building openings.

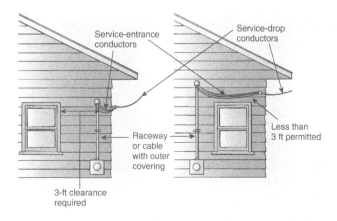

FIGURE 6-24   Required dimensions for service conductors located alongside a window and above the top level of a window designed to be opened.

10. Check underground service conductors for proper depth, fill, protection, marking, and allowances for ground movement.

Article 300 provides minimum cover requirements for different conditions and wiring methods. Depths of 18 or 24 in. (450 or 600 mm) are required for most service conductors, depending on conditions and wiring method. The minimum cover requirements are intended to minimize the danger from digging into buried cables and minimize the likelihood of damage to conductors. Backfill of trenches should use materials that will not damage the raceways or direct-buried conductors. Unless the service conductors are encased in concrete, a warning tape is required to be placed at least 12 in. (300 mm) above conductors that are 18 in. (450 mm) or more deep (**FIGURE 6-25**). Tapes that are too close to the conductors do not provide adequate "warning" to someone digging in the area. Service conductors installed in, for example, RMC or IMC may be permitted to be as little as 6 in. (150 mm) below the surface or even less if under concrete, where a warning tape would not be effective. In many cases, electric utilities will mandate the minimum burial depth for service raceways or direct-buried service cables.

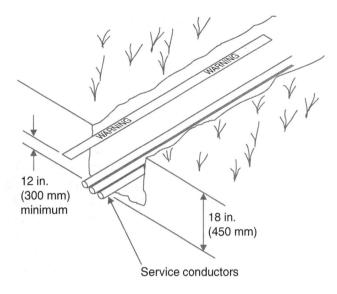

12 in. (300 mm) minimum

18 in. (450 mm)

Service conductors

FIGURE 6-25   Warning ribbon for direct-buried service conductors.

11. Determine which grounding electrodes are available and verify that they are bonded together to form a grounding electrode system.

The types of electrodes present will determine how the grounding electrode system is formed and what, if any, additional electrodes are required. The type of electrode also affects the size of the grounding electrode conductor. The possible electrodes that may be present as part of the structure include the metal underground water pipe, the metal frame of a building or structure, the concrete-encased electrode, and the ground ring. None of these electrodes is required to be installed, but water pipe and concrete-encased grounding electrodes are often present and, if so, must be incorporated into the grounding electrode systems with any other grounding electrodes that are present or installed. Ground rings are not usually used at dwelling units. The water pipe must be supplemented with at least one other electrode, and the metal building frame must meet specific requirements to be usable as an electrode. The ways to ensure that a metal building frame can be used as an electrode are specified in the *NEC* [250.52(A)(2)]. In any case, a metal building frame will have to be bonded. Metal frame construction has not been widely used in residential construction in the past but has become somewhat more common in recent years. Ground terminals (strike termination devices) of lightning protection systems are required to be independently established and bonded to the power supply grounding electrode system and may not be used in place of electrodes for the power system.

12. Check any rod, pipe, plate, or other listed electrodes for proper size, type, and installation.

Rod, pipe, or plate electrodes are required if none of the other electrodes is present or if the only electrode is a water pipe. Driven ground rods are the most common type of these other electrodes. Other types that may be used are other listed electrodes such as so-called "chemical ground rods," local metal underground pipes or structures, or plate electrodes. Buried rods or pipes may be used instead of driven rods or pipes, but only where the rods or pipes could not be driven because of shallow rock. Rod and pipe electrodes must be at least 8 ft (2.44 m) long, and at least 8 ft (2.44 m) must be in contact with the soil. If the connection to the electrode is above ground, a rod longer than 8 ft (2.44 m) must be used. Ten-foot (3.0-m) ground rods that are often used for this purpose allow for a connection to the rod to be above ground or otherwise not completely buried. Rod and pipe electrodes that cannot be driven and plate electrodes must be buried at least 30 in. (750 mm) deep in the earth.

The resistance of a single rod, pipe, or plate electrode must be not over 25 ohms to ground, or an additional electrode must be used. The additional electrode must be at least 6 ft (1.83 m) from the first one. Many installers just put two electrodes in without testing, because installing the second electrode is often less expensive and easier than testing the first one. Some jurisdictions simply require two rod or pipe electrodes for the same reason.

13. Verify that grounding electrode conductors are unspliced and protected and that any ferrous metal enclosures are bonded and electrically continuous.

Generally, grounding electrode conductors are supposed to be continuous without splices or joints. Splices or joints are permitted only if made by exothermic welding or by irreversible compression-type connectors that are listed as grounding and bonding equipment. Many compression connectors are listed but not listed for use with grounding electrode conductors, because in many cases, such splices will be made in embedded locations. Multiple bonding jumpers from electrodes may be bonded together on an accessible busbar, and the continuous grounding electrode conductor may be run from the bonding point. Irreversible compression connectors on busbars are not required. Connectors to busbars are required to be listed, but not necessarily as grounding and bonding equipment. This method is more likely to be used in nondwelling occupancies.

Grounding electrode conductors must be securely fastened to the surface on which they are carried. In addition, conductors smaller than 6 AWG must be installed in a raceway for protection. Grounding electrode conductors 6 AWG and larger must be protected by a raceway if exposed to physical damage. Conductors concealed in walls, installed through and attached to joists or trusses, or otherwise isolated by elevation are generally not considered to be exposed to damage. Where exposed to physical damage, RMC, IMC, Schedule 80 PVC conduit, RTRC, EMT, or cable armor must be used for protection for 6 AWG or smaller conductors.

Ferrous metal enclosures for grounding electrode conductors must be made electrically continuous from the enclosure or cabinet to grounding electrodes. Ferrous metal raceways enclosing grounding electrode conductors may be made electrically continuous by bonding each end of the raceway to the conductor.

The use of PVC or RTRC (Schedule 80 PVC or RTRC-XW if used where subject to physical damage) is specifically permitted as well as the use of nonferrous RMC, and these types eliminate the need for bonding of the raceway that contains the grounding electrode conductor because they are non-magnetic.

14. Check grounding electrode conductor(s) and bonding jumpers for proper sizing.

Sizing of grounding electrode conductors is based on the size of the service-entrance conductors and the electrodes used. One or more grounding electrode conductors may be used, and each may be a different size, depending on the electrode type and the size of the service-entrance conductors related to each grounding electrode conductor. **FIGURE 6-26** shows three permissible sizing and connection schemes (there could be others) for connecting multiple grounding electrodes together

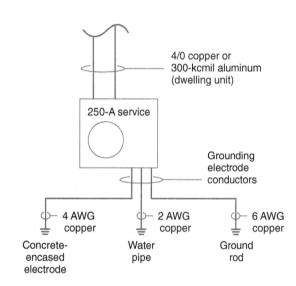

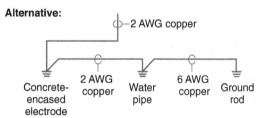

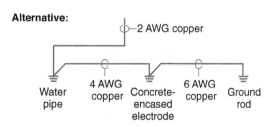

FIGURE 6-26 Three methods for connecting multiple grounding electrodes together.

to form the grounding electrode system. The two alternative methods in Figure 6-26 could use a busbar as the common connection point from which the 2 AWG grounding electrode conductor could be run to the service equipment (see Item 13 of this checklist). Only the conductors running to the service equipment are grounding electrode conductors. The others used to interconnect electrodes in the alternative illustrations are bonding jumpers.

15. Check grounding electrode connections for proper type, protection, and accessibility.
All connections to grounding electrodes must have very low impedance and be highly reliable. To ensure the integrity of such connections, the connecting devices or methods must be tested for the conditions. Irreversible compression connections that are listed for grounding and bonding and exothermic welding will provide such reliability. Other connections, including buried connections, should be made only with devices listed and identified for the purpose. Devices that are intended for direct burial are marked "Direct Burial" or "DB."

The connection between each grounding electrode and the attached grounding electrode conductor must be accessible unless the connection is concrete encased or buried (encasing in fireproofing of structural steel is also permitted, but such fireproofing is typically not used in one- and two-family construction). Because ready access is not required, other connections may be covered or elevated as long as access is provided. Concrete-encased electrodes (rebar or copper wire) may be extended above the footing, perhaps stubbed up into a wall, to an accessible location so the connection is not embedded. Unless identified for general use without protection, grounding connections must be protected from physical damage by enclosure, elevation, guarding, or other means.

16. Verify that the main bonding jumper is installed and is of the proper size and type.
Often, the main bonding jumper is an integral part of the service panelboard, especially in service equipment intended for residential use. The main bonding jumper is a critical part of the fault-current path between **premises wiring** and the utility source. To ensure its adequacy as a fault-current path, it must be sized in the same manner as other portions of the supply-side fault-current path using Table 250.102(C). This table includes the requirement that if the size of the SE conductors is larger than 1100-kcmil copper or 1750-kcmil aluminum, the bonding jumper must be sized at a minimum of 12.5 percent of the area of the service-entrance conductors. This rule is not commonly applied in dwelling units unless the service rating is over about 800 amperes. Bonding jumpers may be busbars or other conductors. The bonding jumper may be a screw if it is colored green and furnished with the service equipment (**FIGURE 6-27**). Where screws or other bonding devices are provided with equipment that is listed for use as service equipment, the diagrams or other instructions included with such equipment must be followed, but a calculation for the minimum size is not required. Reliable, low-impedance connections are vital.

17. Verify that interior metal piping systems are bonded, that bonding jumpers are properly sized, and that continuity around removable devices is assured.
Whether or not interior metal water pipe extends to buried underground metal water pipe, it must be bonded to avoid possible shock hazards due to ground faults that may energize the piping. Reliable connections between hot- and cold-water piping are also necessary. Such interconnections may be provided by mixing valves or other plumbing fittings in some cases, but many modern plumbing fixtures use nonmetallic connections even with metal water piping. Bonding conductors to interior water piping are sized in the same manner as grounding electrode conductors to buried underground piping using Table 250.66.

Where both interior metal piping and buried metal piping exist, the grounding electrode conductor to the underground piping may

FIGURE 6-27 Screw-type main bonding jumper in panelboard.

FIGURE 6-28 Bonding jumper around removable valve in a metallic water piping system.

also be used to bond the interior piping. Jumpers between the two must be provided to ensure that the connections are not dependent on removable plumbing equipment such as pressure-reducing valves, meters, or filters. Such jumpers must be long enough to allow servicing or replacement of the plumbing devices without compromising the integrity of the bonding connection. **FIGURE 6-28** shows an example of a bonding jumper installed around a pressure-reducing valve in an incoming water line.

18. Verify that service raceways and enclosures are properly grounded and bonded.

Because the overcurrent protection for service equipment and conductors provides overload protection only and does not provide short-circuit or ground-fault protection for the service conductors, the possible fault-current pathways at service equipment need special attention. Specific methods of ensuring adequate bonding at or ahead of services include direct connections to the grounded service conductor, threaded connections, threadless fittings, hubs, bonding jumpers, and special bonding fittings. Supplementary bonding jumpers or fittings must be used with ordinary locknuts and around oversized, concentric, or eccentric knockouts (see the Grounding and Bonding chapter of this manual). Proper bonding of the service raceways and enclosures will ensure that they are also grounded to prevent shock hazards.

19. Verify that an <u>intersystem bonding termination</u> has been provided.

A provision for the connection of at least three bonding terminations for systems other than the power system, such as cable or satellite TV and communications, must be provided. This may be a set of terminals mounted to the meter enclosure; a bonding bar near the service equipment, meter, or raceway; or a bonding bar near the grounding electrode conductor, and the terminals must be listed as grounding and bonding equipment. This termination provision must be accessible for connection by other system installers and for inspection. Methods that were considered adequate for this purpose prior to the 2008 *NEC* may continue to be used in existing buildings.

Checklist 6-14: Feeders and Panelboards (Residential Rough Inspection)

1. Review the calculation of feeder loads, and verify that conductors are properly sized and rated.

As with service load calculations, feeder load calculations are also usually provided for plan review. If not, the calculation and feeder diagrams should be provided by the electrical designer, contractor, or installer at the request of the AHJ in accordance with 215.5. The minimum feeder size is simply based on the calculated load after considering any **continuous loads**. A grounded conductor need not be increased in size for continuous loads because it usually does not connect to an overcurrent device. Article 220 provides the method for calculating the load. The service conductors and service equipment for a one-family dwelling are required to be rated at least 100 amperes (230.42 and 230.79), but this load may be divided among multiple feeders or branch circuits (**FIGURE 6-29**). Note that the service and feeder conductor sizing table for dwelling units has been eliminated in the 2014 *NEC* and has been replaced by a series of rules that allow the conductors to be sized based on 83 percent of the service rating (the rating of the service equipment as determined by the calculated load). This permission applies only to the conductors that carry "the entire load associated with an individual dwelling." In Figure 6-29, these would be the conductors **labeled** "service conductors," and these dwelling unit conductor sizes could not be used for the feeders because the only conductors shown that carry all the loads in this example are the service conductors.

2. Verify that panelboards have proper ratings and protection.

Most panelboards in dwelling units must be protected by no more than two (usually one) overcurrent devices. If they are protected by two overcurrent devices on the supply side, they are limited to 42 overcurrent devices, and a two-pole circuit breaker counts as two overcurrent devices for the purposes of the 42-device rule. Panelboard ratings must be at least equal to the calculated load for the feeder supplying the panelboard. Panelboards used as service equipment may have up to six overcurrent devices on a single bus structure and are not required to be protected at the rating of the panelboard.

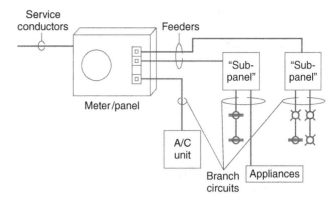

FIGURE 6-29   Service, feeder, and branch-circuit conductors and equipment.

3. Check for proper accessibility, working clearances, and dedicated spaces around panelboards.

Feeder and **branch-circuit overcurrent devices** must be readily accessible to the occupants of a dwelling unit. Clear working space must be provided in front of the panelboards. The working space must be at least 30 in. (762 mm) wide or the width of the panelboard if the panelboard is wider than 30 in. (762 mm). The working space must be at least 6½ ft (2.0 m) high and 3 ft (914 mm) deep in front of the panelboard. (The depth varies according to the conditions and voltages, but 3 ft [914 mm] satisfies the requirement for 120/240-volt residential applications as shown in Figure 6-22.)

In addition to the working space, a space equal to the width and depth of the panelboard must be dedicated to electrical equipment, cables, and raceways. This dedicated space extends from the floor to the structural ceiling (indoors) or up to 6 ft (1.8 m) above the panelboard (indoors or outdoors). Equipment not associated with the electrical installation such as plumbing pipes and ducts may not run through the dedicated space, but can be run above the dedicated space if the panelboard is protected from leaks or condensation that might come from the foreign equipment. The leak protection apparatus, if any, cannot be within the dedicated space. Other electrical equipment is not considered "foreign" to the electrical equipment requiring dedicated space because it is part of the electrical installation, but nothing can compromise the working space required by 110.26(A).

4. Verify that at least the minimum number of overcurrent devices and circuits has been provided.

In dwelling units, the minimum number of circuits can be determined simply by dividing the computed general lighting load in volt-amperes (VA) by the capacity of the branch circuits in volt-amperes and adding the other circuits that are required for specific equipment or areas. For example, 20-ampere branch circuits have a capacity for noncontinuous loads of 2400 VA. To the required number of general lighting circuits, add the number of additional branch circuits, such as circuits for small appliances, laundry, central heating, ranges, dryers, and so on, to arrive at a total minimum number of branch circuits **(FIGURE 6-30)**.

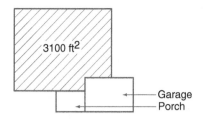

Given: 3100 ft² exclusive of open porches, garages, etc.

Find: How many general lighting branch circuits are required for lighting and general-use receptacles?

3100 ft² × 3 VA/ft² = 9300 VA general lighting load
9300 VA/120 V = 77.5 or 78 A
Using 15-A branch circuits: 78/15 = 5.2   *6 circuits required*
Using 20-A branch circuits: 78/20 = 3.9   *4 circuits required*

Notes: At least one additional 20-A branch circuit is required for bathroom receptacles.
Branch circuits for small appliances, laundry, heating, and special appliances are in addition to the general lighting circuits.

FIGURE 6-30   Calculating the minimum number of general lighting circuits for a dwelling unit.

5. Verify that grounded feeder conductors are insulated and isolated from equipment grounding conductors and grounded enclosures.

Grounding connections to grounded conductors are generally prohibited except at service equipment. The few exceptions to this rule are not applicable to new dwelling units. Grounding (green or bare) conductors must be connected to panelboard cabinets and other metal boxes and enclosures, but grounded (white or gray) conductors must be insulated and isolated from such panels and enclosures.

6. Verify that panelboards are grounded by an appropriate and properly sized equipment grounding conductor (or conductors).

A nonflexible metal conduit enclosing the feeder supplying a panelboard is permitted to serve as the required equipment grounding conductor. Where equipment grounding conductors are installed as separate conductors in raceways or cables, they must be included with the feeder and sized on the basis of the overcurrent device protecting the feeder and panelboard.

RESIDENTIAL FINISH INSPECTIONS

Note: Many of the items listed for inspection at the finish stage have already been checked at the rough stage. For example, locations of boxes for receptacles and switches were checked earlier, but at the finish stage, these items need to be confirmed. They were only boxes with wiring at the rough stage, but at this stage, the assumption that a certain box was for a switch or a receptacle needs to be verified.

Checklist 6-15: General Requirements (All Areas; Residential Finish Inspection)

1. Check for correction of any deficiencies noted on previous inspections.

2. Check positioning of boxes intended to be flush with combustible and noncombustible finished surfaces.

Boxes mounted in walls should have their edges flush with the wall surface. A recess of up to ¼ in. (6 mm) from the wall surface is allowed with wall finishes that are noncombustible, such as drywall (gypsum board) or plaster. Boxes in combustible wall finishes, such as wood paneling, are required to be flush with the surface. However, listed box extenders may be used to effectively extend the enclosure to be flush with the wall surface. Box extenders are especially useful where an original noncombustible wall surface has been covered with additional layers of combustible or noncombustible finish material.

3. Check for proper positioning of receptacles and faceplates on walls.

Receptacle and switch yokes should seat firmly against the box or wall surface so that they do not move when used. Receptacle faces should be flush with nonmetallic faceplates and must protrude a minimum of 0.015 in. (0.4 mm) from metallic faceplates. Faceplates are not intended to hold devices in place, and devices should not be recessed from faceplates, unless the assembly or faceplates are specifically listed for covering the receptacle face.

4. Check for gaps around outlet boxes in walls.

Gaps greater than ⅛ in. (3 mm) around boxes intended for use with flush-type plates must be patched to restore the wall surface around the box. Oversized plates may be used to cover defects in wall finishes at boxes, but should not be used for covering gaps around boxes. Where boxes extend from the wall surface for use with surface plates, gaps around the box are not an issue under the *NEC* unless the wall has a fire rating.

5. Verify that conductor terminations and splicing methods are compatible with conductor materials.

All connections to conductors must be made with devices that are suitable for the conductor materials. Many devices are suitable for both aluminum and copper, but many others are suitable only for copper or only for aluminum. Compatibility is equally important for either type of conductor. Most devices, such as snap switches and receptacles, are intended only for use with copper. Although aluminum is not often used in small branch-circuit sizes, if it is used, devices must be marked CO/ALR to indicate suitability for use with aluminum conductors.

6. Verify that receptacles are bonded to metal boxes and that receptacles, switches, and metal faceplates are grounded.

Receptacles are required to be bonded to metal boxes where such boxes are used. This can be accomplished by direct metal-to-metal contact between device yokes and boxes if the boxes are surface-mounted, by the use of self-grounding devices, or by separate bonding jumpers. In many cases, nonmetallic device boxes are used in dwelling units. Even though bonding to a box is obviously not required in such cases, receptacles and switches are still required to be grounded. If metal faceplates are used, they will be grounded either by a metal screw connection to the device yoke or by a direct connection to a box. The required bonding between a receptacle and a metal box can be accomplished through the use of a self-grounding device **(FIGURE 6-31)**.

7. Check polarity of devices and luminaires.

Luminaires with screw shells should be wired so that the screw shell is connected to the grounded conductor. Most luminaires are furnished with color-coded wiring to ensure correct polarity. Polarity of common receptacles can be most easily checked using a plug-in tester. These testers can also identify a number of other wiring errors. Such testers provide an easy way to make an important check of the primary interface that users have with their wiring systems (see the Wiring Methods and Devices chapter of this manual).

8. Check all luminaires and lampholders for listing.

Listing is required for all luminaires, lampholders, and **retrofit kits**. The construction specifications including requirements for interior wiring, clearances, and suitability for any specific use or location are contained in safety standards. Such judgments should not be made based on the *NEC*, so listing provides an appropriate basis for approval.

9. Check for splicing devices on all equipment grounding conductors within boxes and for bonding connections to metal boxes.

Listed means, usually listed pressure connectors, must be used to splice equipment grounding conductors in boxes. Equipment grounding conductors in a metal outlet or junction box must be spliced together and connected to the box. The connection to the box must be made by a screw dedicated to the purpose or by a listed device such as a metal grounding clip, but may not be dependent on connections to a wiring device or luminaire. Screws must engage at least two threads in the box. Although the grounding connections to a receptacle are required to be identified with a green color or with the letters "G" or "GR," the screw connection to a box is required only to be dedicated to that purpose and is not required to be green. Green screws and green screws with green "pigtail" conductors attached are readily available and commonly used, however.

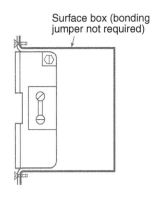

Spring-type grounding strap for holding mounting screw

Surface box (bonding jumper not required)

FIGURE 6-31 Self-grounding receptacle and ordinary receptacle with direct contact to a metal box.

In the infrequent cases where isolated grounding conductors are used in dwelling units, they are not required to be connected to a box. (See Item 8 in the Rough Inspection checklist for General Requirements.)

10. Verify that device ratings are compatible with circuit and equipment ratings.
Switches must be at least equal to the voltage and current rating of the circuits and equipment they supply. See Article 430 for specific rules that apply to switches used to disconnect or control motors. Based on their ratings, receptacles are restricted to certain circuit ratings. For example, 20-ampere, 125-volt receptacles may be used only on 20-ampere circuits, but similar 15-ampere receptacles may be used on either 15- or 20-ampere circuits. Single receptacles that are the only outlet on a circuit (an individual branch circuit) generally must match the rating of the circuit (see the Wiring Methods and Devices chapter of this manual).

A dedicated circuit supplying only one duplex receptacle is not an individual branch circuit because it is designed to supply two pieces of **utilization equipment**, so a 20-ampere branch circuit is permitted to supply one 15-ampere *duplex* receptacle, but cannot be used to supply only one single receptacle rated 15 amperes.

11. Check for proper use of connectors and fittings and for protection of cables.
Fittings used with raceways and cables must be designed and listed for the specific wiring method with which they are used. Suitable fittings must be installed on the ends of conduits or tubing to protect cables from abrasion, where conduits or tubing are used to protect or support cables. For example, where receptacles are installed on exterior basement walls or power is supplied to furnaces, EMT is often used to protect and support NM cable, and a fitting or bushing is required where the cable enters the tubing.

12. Check for bushings or equivalent protection for cables entering boxes and other enclosures.
For 4 AWG or larger conductors, protection by insulated bushings or similar insulating surfaces where they enter cabinets, boxes, enclosures, or raceways is required. Similar requirements apply to conductors leaving some cable assemblies, such as Type AC cable. The requirement to extend the sheath of NM cable at least ¼ in. (6 mm) into a box helps provide such protection as well. Similar requirements apply to flexible cord pendants that pass through covers or canopies.

13. Verify that unused openings in boxes and other enclosures are closed.
Boxes and other enclosures for electrical equipment are used to guard the user from contact with energized parts and to contain the arcs and sparks that may be produced either by normal operation or by accidents or other abnormal conditions. Unused openings not intended for operation of the equipment or for mounting purposes such as knockouts that were mistakenly removed or hubs that are not used must be closed to restore the integrity of the enclosure. Plugs of various types are available for sealing threaded hubs or holes in enclosures. For example, knockout plugs for Type 4 or 4X enclosures can be used to restore the water and corrosion resistance of such enclosures.

14. Verify that appliances, motors, and other equipment are grounded.
Most equipment that has exposed metal parts and contains conductors of power or lighting circuits is required to be grounded by connection to an equipment grounding conductor. Listed cord-and-plug–connected appliances that are required to be grounded will be furnished with grounding-type plugs. Most other types of equipment should be furnished with an equipment grounding conductor, unless the equipment is double-insulated. Electric ranges and dryers may not be in place when a finish inspection takes place, but the receptacles for these appliances can be inspected to ensure that they have equipment grounding terminals separate from the terminals for grounded conductors. This same

separation of the EGC from the grounded (neutral) conductor must also occur within the appliance.

15. Check burial depth of underground raceways and cables for minimum cover.

Minimum cover requirements for various wiring methods and conditions are well defined in Article 300. Inspectors often allow partial backfill before making inspections if sections of the installation are left open to verify compliance with burial depth, backfill, and wiring method requirements.

16. Check installation of listed equipment for compliance with manufacturer's instructions.

Installers should be able to provide the installation instructions if they are not included on the equipment. Such instructions should be available to the occupant as well.

17. Verify that fire rating of building assemblies has been restored at electrical penetrations.

Most one-family dwelling units are not fire-rated construction relative to separation between rooms and levels, but walls between a garage and other areas are usually required to have a 1-hour fire rating. If these or other fire-rated barriers are penetrated, the penetrations must be firestopped using an approved fire-sealing caulk, putty, or other **approved** methods. Firestopping is usually done with intumescent materials that will expand when heated to fill in any gaps where wiring penetrates a fire-rated wall. Building codes usually allow certain maximum sizes of openings in drywall for outlet or device boxes, but may not allow openings on both sides of a wall unless separated by a specified distance. In multifamily construction, fire-rated separations are common and maintaining the integrity of fire-rated construction between units is an extremely important requirement.

18. Check for disconnecting means on both permanently connected and cord-and-plug–connected appliances.

Appliances are required to have a disconnecting means that in some cases may be in addition to the branch-circuit overcurrent device. Appliances not over 300 VA or ⅛ horsepower (hp) are exempted and the branch-circuit overcurrent device is sufficient as a disconnect. Some have integral unit switches that may be used as disconnects. Such switches are required to disconnect all power within the appliance. Generally, an accessible disconnect switch or cord-and-plug connection must be provided unless the branch-circuit overcurrent device is in sight of the appliance or capable of being locked in the off position. The locking means must remain in place with or without a lock installed if the out-of-sight, lockable device is to be used as the required disconnecting means.

Note that an **attachment plug**/receptacle used as a disconnect is required to be accessible, not readily accessible, so a receptacle that is accessible by moving an appliance (that is not fastened in place) meets this requirement. The other disconnecting means in this case, the branch-circuit overcurrent device, is required by 240.24(A) to be readily accessible. **FIGURE 6-32** is an example of a plug and receptacle used as a disconnect for a household range. It is accessible by pulling the range out, and the range is designed with a recess in the back to accommodate the cord and plug while still fitting close to the wall. This receptacle is not accessible from the front of the range and the range is not fastened in place, but if the range were fastened in place and the plug and receptacle

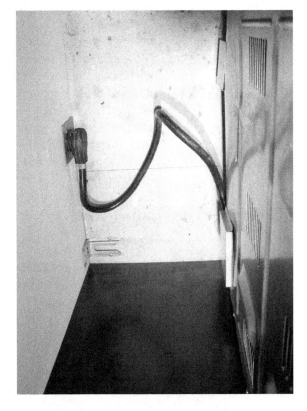

FIGURE 6-32   Cord-and-plug disconnecting means that is accessible but not readily accessible.

were accessible from the front by removing a drawer, that arrangement would also be acceptable.

19. Verify that circuits for mechanical equipment have correct conductor size and overcurrent protection.

In most cases, the nameplate information on appliances and other mechanical equipment provides the necessary information that enables the inspector to determine compliance with the applicable article. Appliances are covered in Article 422, electric heating equipment is covered in Article 424, and air-conditioning equipment is covered in Article 440. Air-conditioning equipment is covered in more detail in the Commercial and Industrial Inspections chapter of this manual.

20. Check for protection on all 120-volt, 15- and 20-ampere branch circuits serving outlets in most areas.

This requirement applies to 120-volt, 15- and 20-ampere branch circuits supplying outlets in most rooms except bathrooms, unfinished basement areas, garages, and the like. This requires a listed combination-type AFCI device to protect branch-circuit wiring in "kitchens, family rooms, dining rooms, living rooms, parlors, libraries, dens, bedrooms, sunrooms, recreation rooms, closets, hallways, laundry areas, or similar rooms or areas." Most of the omitted areas are areas where GFCI protection is required. This is not to say that GFCIs provide equivalent protection, but this division or consideration provided a somewhat limited scope for the expansion of the requirement from the original requirement that covered bedrooms only.

The methods and arrangements for providing this protection have been significantly expanded in the 2014 *NEC*. AFCI protection of the entire branch circuit may be provided through the use of multiple protective devices. Since the requirement for AFCI protection applies to the entire branch circuit, the most straightforward way to accomplish this is through the use of a combination-type AFCI device that is integral to the branch-circuit overcurrent device. Other methods are provided.

Similar protection can be provided by using a listed "branch/feeder–type AFCI" at the origination with a listed "outlet branch-circuit–type AFCI" at the first outlet and marking the outlet to identify it as the first outlet. Alternatively, a portion of the branch-circuit wiring is permitted to be protected only by the ordinary listed branch-circuit overcurrent device if the portion up to the first outlet is protected from physical damage. This protection can be provided by RMC, IMC, EMT, MC cable, or AC cable with steel armor, or by installation in a raceway with 2-in. (50-mm) concrete encasement. The AFCI protection must then be provided by an outlet branch-circuit–type AFCI at the first outlet.

Other alternatives allow for a continuous and limited length of branch-circuit wiring from the branch-circuit overcurrent device to the first outlet. These methods require the AFCI protection to be provided by a combination of devices, one of which is an outlet branch-circuit–type AFCI located at the first outlet. The other device must be located at the origination of the branch circuit and may be either a "listed supplemental arc protection circuit breaker" or a listed branch-circuit overcurrent device that, in combination with the outlet branch-circuit AFCI at the first outlet, is identified and listed as a "system combination AFCI." These options also require that the first outlet be identified. The six methods are summarized in **TABLE 6-1** and illustrated in **FIGURE 6-33**.

Although not explicitly stated in 210.12, the requirement for AFCI protection also applies to outlets for residential smoke alarms. It does not apply to smoke detectors connected to a central fire alarm panel as covered in Article 760. (Such detectors are not connected to "outlets" as defined in Article 100.) Either type of fire detection method could be used to satisfy the requirements of the building code or NFPA 72®, *National Fire Alarm and Signaling Code*®, in most dwelling units. According to Section 760.1, Article 760 does not apply to residential smoke alarms because they are not powered and controlled by a fire alarm panel. The objective of AFCI devices is to disconnect power to circuits before

TABLE 6-1   Summary of the Six Permitted Methods to Provide AFCI Protection for the Entire Branch Circuit

210.12(A) Reference	AFCI Protection Method	Additional Installation Requirements
210.12(A)(1)	Combination type AFCI circuit breaker installed at origin of branch circuit.	No additional requirements
210.12(A)(2)	Branch/feeder type AFCI circuit breaker installed at origin of branch circuit, plus Outlet branch circuit type AFCI device installed at first outlet in branch circuit	Marking of first outlet box in branch circuit
210.12(A)(3)	Supplemental arc protection type circuit breaker installed at origin of branch circuit, plus Outlet branch circuit type AFCI device installed at first outlet in branch circuit	Continuous branch circuit wiring; "Home run" conductor length restrictred (14 AWG–50ft., 12 AWG–70 ft.); Marking of first outlet box in branch circuit
210.12(A)(4)	*Branch circuit overcurrent protective device, plus *Outlet branch circuit type AFCI device installed at first outlet in branch circuit *The combination of devices must be listed and identified to provide *system combination type* arc-fault protection for the "home run" conductors.	Continuous branch circuit wiring; "Home run" conductor length restricted (14 AWG–50ft., 12 AWG–70 ft.); Marking of first outlet box in branch circuit
210.12(A)(5)	Outlet branch circuit type AFCI device installed at first outlet in branch circuit.	Branch circuit conductors installed In specific types of metal raceways or metal cables and metal boxes from origin of branch circuit to the first outlet
210.12(A)(6)	Outlet branch circuit type AFCI device installed at first outlet in branch circuit.	Branch circuit conduit, tubing, or cable Encased in 2" of concrete from origin of branch circuit to the first outlet

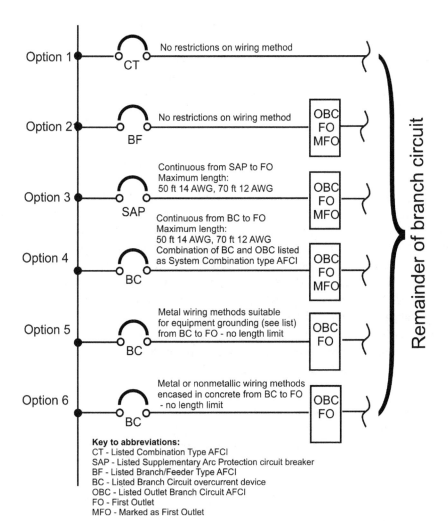

FIGURE 6-33 Diagram showing the six permitted methods to provide AFCI protection for the entire branch circuit from origination of branch circuit to the outlet(s).

FIGURE 6-34 Tamper-resistant 125-volt, 20-ampere receptacle also identified as hospital-grade. (Courtesy of Pass & Seymour/Legrand.)

they start a fire. An AFCI will not interrupt line voltage power to residential smoke alarms by fires from other causes, or at least not until the fire results in damage to the wiring system, and smoke alarms are required to have backup power sources in the event that power is interrupted for any reason, including interruption by an AFCI device. Where fire alarm systems with a central control panel are used in dwellings, 760.41(B) and 760.121(B) prohibit the use of AFCIs on the power supply circuits to the control panel and require the use of circuits that supply no other loads. In that case, AFCI protection is permitted to be omitted, but the wiring of the branch circuit, including any boxes, from its origination to the fire alarm system must be in metallic wiring methods that also qualify as equipment grounding conductors. All AFCI devices are required to be readily accessible.

21. Check for tamper-resistant receptacles in all areas where 125-volt, 15- and 20-ampere receptacle outlets are required.

This is intended to address the large number of injuries to children who insert foreign objects into receptacles in dwellings. The requirement for tamper-resistant receptacles applied only to pediatric areas of health care facilities prior to the 2008 *NEC*. **FIGURE 6-34** shows an example of a tamper-resistant receptacle. This example is one that is also hospital grade, but these receptacles are also available in residential grade and as GFCI receptacles. Tamper-resistant receptacles are identified with the words "Tamper-Resistant" or "TR." Note the marking "TR" on the upper right of the device in Figure 6-34. Tamper-resistant receptacles are required where receptacle outlets are required by 210.52, but would not be required for all receptacles such as those required by 210.63 for servicing mechanical equipment. Tamper-resistant receptacles are not required for receptacles that are located more than 5.5 ft (1.7 m) above the floor, that are part of a luminaire or appliance, that are located in dedicated space where receptacles are used for specific appliances that are not easily moved in normal use (like refrigerators or perhaps washing machines), or that are in remodel applications if nongrounding replacement receptacles are permitted.

Checklist 6-16: Kitchen (Residential Finish Inspection)

1. Check spacing of receptacles for walls and countertops, including islands and peninsulas.

Receptacles must be installed to serve kitchen countertops and be installed so that every countertop space 12 in. (300 mm) or more in width has a receptacle within 24 in. (600 mm) of any point measured along the wall. Peninsula and island countertop spaces must be provided with at least one receptacle outlet unless such spaces are less than 12 × 24 in. (300 × 600 mm) in size. Wall receptacles are required to be in accordance with the basic wall spacing rule. See Item 1 in the Rough Inspection checklist for Kitchens for more about these requirements. Factors that may not have been obvious at the rough inspection may be apparent at the finish stage because cabinets and countertops are usually not installed until the finish stage, and sometimes changes are made in cabinet and countertop layouts. If receptacles are installed in the countertop, they will be inspected at this stage to verify they are part of a listed assembly intended for the use. Receptacles are not permitted to be installed face-up in a countertop.

2. Verify that a minimum of two 20-ampere small-appliance branch circuits are used for kitchen receptacles.

Receptacles installed in kitchens, pantries, dining rooms, breakfast rooms, and similar rooms must be connected only to 20-ampere small-appliance branch circuits. At least two such circuits must be supplied to each kitchen. Circuiting may be more easily inspected at the rough stage, but the presence of at least two identified small-appliance branch circuits can also be verified at the finish stage as part of the inspection of the panelboard.

3. Verify that small-appliance branch circuits are used only for receptacles in kitchen, dining room, pantry, and so forth.

Small-appliance branch circuits may serve only receptacles in kitchens and related rooms and may not serve other loads. Additional circuits must be provided for other types of outlets or for outlets in other areas.

4. Verify that all 125-volt, 15- or 20-ampere receptacles serving countertop surfaces, all receptacles within 6 ft (1.8 m) of a kitchen sink, and outlets for dishwashers are provided with GFCI protection.

Previously this rule required all receptacles serving countertop surfaces in kitchens to be GFCI-protected. Receptacles below countertops, such as those installed for waste disposers, trash compactors, and dishwashers, or receptacles in spaces for specific equipment, such as those installed behind and for refrigerators, were not required to be GFCI-protected. The 2014 *NEC* GFCI requirements apply not only to the receptacles serving the countertops but also to *any* 125-volt, 15- or 20-ampere receptacle located within 6 ft (1.8 m) of a sink in the kitchen as measured from the nearest edge of the sink to the receptacle. Additionally, Section 210.8(D) requires GFCI protection for outlets (receptacle or direct-wired) supplying dishwashers. All GFCI devices are required to be readily accessible.

5. Verify that refrigeration equipment is supplied by a small-appliance branch circuit or an individual branch circuit.

Receptacles for refrigerators in kitchens may be supplied by 15-ampere or higher rated individual branch circuits (with a single receptacle) that are installed in addition to the required small-appliance branch circuits or by 20-ampere small-appliance branch circuits.

6. Verify that a wall-switched lighting outlet is provided and wired on a general lighting circuit.

Lighting outlets in kitchens must be connected to general lighting circuits and may not be connected to the small-appliance circuits. At least one such lighting outlet must be controlled by a wall switch. A switched receptacle may not be used to satisfy this requirement in a kitchen.

7. Verify that properly sized circuits are provided for specific kitchen appliances, such as dishwashers, disposals, ranges (cooktops), trash compactors, and the like.

Small-appliance branch circuits are intended only for use with portable appliances and refrigerators. Additional circuits are required for appliances that are fixed in place, that are permanently installed, or that require dedicated or individual branch circuits. All such circuits must be adequate for the load they serve.

8. Check for proper type, length, and use of flexible cords for appliance connections.

Cords for disposals (in-sink waste disposers) must be between 18 in. (450 mm) and 36 in. (900 mm) in length measured from the cord entry to the disposal (**FIGURE 6-35**). Cords for dishwashers or trash compactors must be between 3 ft (0.9 m) and 4 ft (1.2 m) long measured from the back plane of the appliance. Cord must be of a type identified in the appliance manufacturer's instructions. This rule refers to the type of cord (see Article 400 for examples of cord types) and not to a marking for use

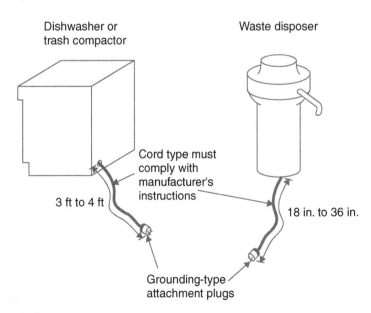

FIGURE 6-35 Maximum and minimum cord lengths for specific kitchen appliances.

with specific appliances. Cords are not necessarily required for such appliances, but they provide a convenient means of connection and disconnection, and a disconnect is usually required unless the branch-circuit overcurrent device is **within sight** of the appliance or is capable of being locked in the open position. To avoid excessive cord length under counters and in cabinets, the length is restricted based on the specific appliance. Kitchen range hoods are also permitted to utilize a cord-and-plug connection where the cord length is between 18 in. (450 mm) and 36 in. (900 mm) and the receptacle supplying the appliance is supplied by an individual branch circuit. The receptacles used as disconnects must be accessible.

Checklist 6-17: Dining Room (Residential Finish Inspection)

1. Check receptacle outlets for proper spacing.
The locations of dining room receptacles are based on the same spacing rule that applies to most other rooms in a dwelling (that is, a receptacle must be installed in each wall space over 2 ft [600 mm] in width, and additional receptacles must be installed so that no point on any unbroken wall space is over 6 ft [1.8 m] from a receptacle located in that wall space). See Item 1 in the Rough Inspection checklist for Other Habitable Rooms.

2. Verify that all required receptacle outlets are supplied by small-appliance branch circuits.
All receptacles in dining rooms should be on small-appliance branch circuits except for those used for general lighting and controlled by a wall switch. Wall switch-controlled receptacles may be used as a method to control decorative luminaires that are cord-and-plug connected.

3. Check for a wall switch-controlled lighting outlet on a general lighting circuit.
A wall switch controlling either a fixed lighting outlet or a receptacle must be installed in a dining room, and may not be connected to the small-appliance circuit.

Checklist 6-18: Bathrooms (Residential Finish Inspection)

1. Verify that receptacle outlets are installed adjacent to and within 36 in. (900 mm) of each basin.
Each basin in a bathroom must have a receptacle located on a wall or partition adjacent to the basin or basin countertop and within 3 ft (900 mm) of the outside edge of the basin. A receptacle outlet may serve more than one basin if it is properly located. Receptacles may also be installed in the face or wall of the basin cabinet if not over 12 in. (300 mm) below the countertop. Receptacles that are part of a listed outlet assembly may be installed in the countertop. Receptacles are not permitted to be installed face-up in a countertop.

2. Verify that receptacles are supplied by dedicated 20-ampere branch circuits.
Bathroom receptacles must be supplied by 20-ampere branch circuits that are either dedicated to bathroom receptacles or dedicated to individual bathrooms. In other words, the required 20-ampere circuits may supply only bathroom receptacles in more than one bathroom, or they may supply receptacles and other loads but only in one bathroom, in which case multiple circuits will be required if there is more than one bathroom. (See Item 2 in the Rough Inspection checklist for Bathrooms.)

3. Verify that bathroom receptacles are GFCI-protected.
The GFCI protection may be either in the receptacle or in the circuit breaker protecting the circuit. GFCI testers may be used, but the manufacturer recommendation for testing the device is to use the test button on the device. If a shower stall or tub is installed separately from a basin so the area does not meet the definition of a bathroom, all receptacles within 6 ft (1.8 m) of the outside edge of the tub or shower stall must be GFCI protected. All GFCI devices are required to be readily accessible.

4. Check for a wall switch-controlled lighting outlet on a general lighting circuit.
Unless the receptacle circuit for a bathroom is dedicated only to that bathroom, a general lighting circuit must supply the lighting in the bathroom.

Checklist 6-19: Other Habitable Rooms (Bedrooms, Family Rooms, Parlors, and Dens) (Residential Finish Inspection)

1. Check receptacle outlets for proper spacing.
Spacing of wall receptacles is based on the same 6-ft (1.8-m) rule that applies to wall spaces in habitable rooms throughout the dwelling unit.

2. Check for wall switch-controlled lighting outlets (including switched receptacles).
This will likely only involve verifying that outlet and switch locations checked during the rough inspection are what they were assumed to be.

Checklist 6-20: Hallways and Foyers (Residential Finish Inspection)

1. Check for at least one wall switch-controlled (or automatic-, remote-, or centrally controlled) lighting outlet.
Switched receptacles are not permitted as the required lighting outlets in hallways. Local switches are not required if central, automatic, or remote controls, such as home automation systems or motion sensors, control hallway lighting outlets.

2. Verify that hallways that are continuous for 10 ft (3.0 m) or more have at least one receptacle outlet.
See Item 2 in the Rough Inspection checklist for Hallways and Foyers.

3. Verify that foyers over 60 ft^2 (5.6 m^2) have receptacles installed as required.
See Item 3 in the Rough Inspection checklist for Hallways and Foyers.

Checklist 6-21: Stairways (Residential Finish Inspection)

1. Check for at least one wall switch-controlled (or automatic-, remote-, or centrally controlled) lighting outlet.
Switched receptacles are not permitted as the required lighting outlets in stairways. Local switches are not required if central, automatic, or remote controls, such as home automation systems or motion sensors, control stairway lighting outlets.

2. Verify that wall switches are provided at each floor level where there are six or more steps.

Checklist 6-22: Closets (Residential Finish Inspection)

1. Check clearances between luminaires and storage spaces if luminaires are installed.
Lighting outlets are not required in clothes closets. Where they are installed, clearances between the defined storage spaces in a closet must be maintained. Incandescent luminaires with open lamps are not permitted in clothes closets. A clearance of 12 in. (300 mm) is required from surface-mounted incandescent or LED luminaires, and a clearance of 6 in. (150 mm) is required from enclosed recessed LED or incandescent luminaires or from surface- or recessed-mounted fluorescent luminaires. Only fluorescent or LED luminaires that are specifically identified for the use can be installed within the defined storage space.

Checklist 6-23: Laundry Area (Residential Finish Inspection)

1. Verify that at least one receptacle outlet is installed for the laundry.

At least one 15- or 20-ampere, 125-volt receptacle outlet must be installed in a laundry area. Additional receptacles, such as a receptacle for ironing, are permitted on the laundry circuit. The circuit must be dedicated to the laundry area, but is not required to be used exclusively for washers or dryers.

2. Verify that a dedicated 20-ampere circuit supplies the laundry outlet(s) and no other outlets.

A laundry branch circuit may supply all receptacle outlets in the laundry area. However, lighting in the laundry area and receptacles in other areas must be supplied by other branch circuits.

3. Check for a laundry receptacle outlet within 6 ft (1.8 m) of the intended appliance location.

All receptacle outlets intended for specific appliances are required to be within 6 ft (1.8 m) of the intended location of the appliance(s).

4. Check for proper receptacle ratings based on branch-circuit ratings, including receptacles for electric dryers (if used).

A 30-ampere, 120/240 volt, 4-wire branch circuit that includes two ungrounded conductors, a grounded conductor, and an equipment grounding conductor is required for most new 240-volt dryers. Compatible receptacles providing for 30-ampere, 4-wire cords and attachment plugs are required where a cord-and-plug connection is used.

5. Verify that lighting outlets for the laundry area are supplied from general lighting circuits.

As is noted in Item 2 of this checklist, the lighting for a laundry area must be supplied by circuits other than the laundry branch circuit.

6. Verify GFCI protection at laundry sink locations and laundry areas in unfinished basements.

GFCI protection for receptacles in laundry areas is likely required under more than one rule. All 125-volt, 15- and 20-ampere receptacles installed within 6 ft (1.8 m) of the outside edge of a sink or in an unfinished basement are required to be GFCI-protected. This measurement is a 6-ft (1.8-m) radius from any edge of the sink. There are no exceptions for receptacles located in dedicated appliance spaces. However, in the 2014 *NEC*, GFCI protection is required for all 125-volt, 15- or 20-ampere receptacles in the laundry area. The AHJ will have to establish the extent of the laundry "area" where it is not obviously defined by its location and construction. The GFCI device(s) affording this protection must be readily accessible. The leakage current requirements for a listed washing machine ensure that the appliance will be compatible with GFCI protection, and there will not be any so-called "nuisance tripping" due to leakage current in the appliance. If a GFCI trips when an appliance is connected, it should be considered to be an indication that a shock hazard exists.

Checklist 6-24: Basements and Attics (Residential Finish Inspection)

1. Verify that at least one receptacle outlet is provided in unfinished basement areas in addition to any receptacles installed for specific equipment.

An unfinished basement must be furnished with one or more receptacles for the use of portable appliances and tools. **FIGURE 6-36** shows the receptacle requirements for the finished and unfinished portions of a dwelling unit basement. ACFI required in finished areas.

2. Verify that a receptacle outlet is provided for servicing mechanical equipment, if any.

If a basement, attic, or crawl space (or any other area) contains a furnace or other heating, air-conditioning, or refrigeration equipment, at least one receptacle must be provided within 25 ft (7.5 m) of the equipment for servicing each item of equipment. Such receptacles in unfinished basements, crawl spaces, or outdoors must be GFCI-protected.

3. Verify that GFCI protection is provided for receptacles in unfinished portions of basements.

A basement that has any portion left unfinished must have a GFCI-protected receptacle installed in each unfinished portion. The 2008 *NEC* removed the exception that said receptacles installed for laundry equipment, sump pumps, or other specific appliances were not required to be GFCI-protected. All such receptacles are now required to be GFCI-protected if they are in unfinished basements. Such receptacles must be installed within 6 ft (1.8 m) of the appliance location. The GFCI device affording this protection is required to be readily accessible.

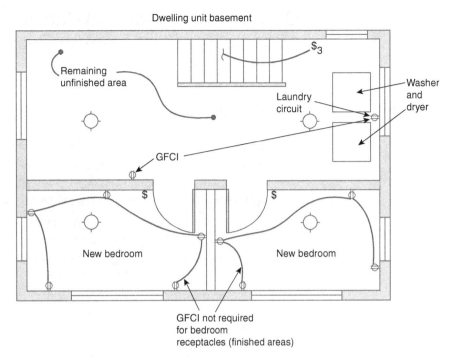

FIGURE 6-36 Required GFCI protection in unfinished basement.

4. Verify that individual branch circuits are supplied for central heating equipment, if any.

Central heating equipment, such as gas or electric furnaces or heat pumps, is required to be supplied by individual branch circuits that supply no other outlets. Equipment directly related to the central heating equipment, such as permanently installed humidifiers or electrostatic air cleaners, may be supplied from the same circuit as the central heating equipment as long as the circuit is adequate for all the connected loads. Receptacles for servicing the heating equipment must be on other branch circuits. (See Item 3 in the Rough Inspection checklist for Basements and Attics for more information regarding central heating equipment and permanently installed air-conditioning equipment supplied from a single branch circuit.)

5. Verify that a wall switch-controlled lighting outlet or a lighting outlet containing a switch is provided at the entrance to equipment requiring servicing.

Illumination controlled by wall switches or by lighting outlets that contain a switch (such as a lampholder with a pull chain) must be provided for mechanical equipment and other equipment that require servicing, and at least one switch must be located at the usual point of entry to the space.

6. Check basements, accessible attics, attic entrances, and scuttle holes for clearances from or protection of cable assemblies.

Cables in attics and around attic entrances must be protected from damage by persons entering or working in such spaces. Similarly, the cables must be supported and secured so that people do not trip over or otherwise become entangled in the cables (see Item 5 in the Rough Inspection checklist for Basements and Attics).

Checklist 6-25: Attached Garages and Detached Garages or Accessory Buildings with Electric Power (Residential Finish Inspection)

1. Verify that at least one receptacle outlet is provided and that the circuit supplying the receptacle outlet(s) does not supply outlets outside of the garage.

At least one receptacle outlet is required in an attached garage. Additional receptacles are required for garages intended for more than one vehicle to provide at least one receptacle outlet for each car space. Detached garages and accessory buildings are not required to have electrical power, but those that do are required to have at least one receptacle outlet, and in detached garages, at least one receptacle outlet for each car space.

2. Verify that GFCI protection is provided for all 125-volt, 15- and 20-ampere receptacles and that GFCI protection device(s) is readily accessible.

Receptacles in garages, including those installed for laundry equipment, sump pumps, or other specific appliances, and those that are not readily accessible are now required to be GFCI-protected (**FIGURE 6-37**). The receptacle outlets in accessory buildings are required to have GFCI protection if they have a floor located at or below grade level, if that floor is used for storage or work areas, and if the floor is not intended to be habitable rooms. All receptacles intended for specific appliances must be installed within 6 ft (1.8 m) of the intended appliance location. The GFCI device affording this protection is required to be readily accessible.

3. Verify that a wall switch-controlled lighting outlet is provided.

A lighting outlet controlled by a wall switch is required in attached garages and in detached garages that have electrical power. The *NEC* does not have a similar requirement for accessory buildings.

Checklist 6-26: Outdoors (Residential Finish Inspection)

1. Check for at least two receptacle outlets, one each at the front and the back of dwelling.

Outdoor receptacles that are accessible from and not over 6½ ft (2.0 m) above grade level must be provided at the front and back of dwelling units. In addition, any heating, air-conditioning, or refrigeration equipment located outdoors must have a GFCI-protected receptacle installed within 25 ft (7.6 m). This includes equipment on rooftops, except for evaporative coolers on one- and two-family dwellings. (Although receptacles are not required for all evaporative coolers, where receptacles are installed outdoors, they must be GFCI-protected. See Item 2 that follows.) The requirement for outside receptacles also applies to individual units of two-family dwellings located at grade level. One outdoor receptacle is required for individual units of multifamily dwellings that are located at grade level. Receptacles are required within 50 ft (15 m) of electrical **service equipment** for multifamily dwellings, but not for one- or two-family dwellings.

2. Verify that outdoor receptacles are GFCI-protected unless they are not readily accessible and are supplied by circuits for deicing or snow-melting equipment. Verify that GFCI protection device(s) is readily accessible.

Article 426 generally requires ground-fault protection of equipment for fixed deicing and snow-melting equipment, but this protection is usually provided by a device that does not provide shock protection for people. Separate, prepackaged, heat-tracing equipment that

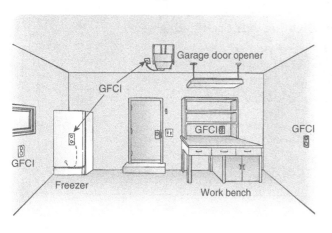

FIGURE 6-37 Required GFCI protection in garages.

comes with flexible cords and attachment plugs is usually listed as an appliance and is covered by Article 422.

3. Check for wall switch-controlled (or remote-, central-, or automatic-controlled) exterior lighting outlets at outdoor entrances or exits with grade-level access.
Exterior lighting outlets are required at all outdoor entrances except garage doors for vehicles. Remote, central, or automatic controls may be substituted for wall switch control of such lighting outlets.

4. Check for boxes at exterior luminaire locations.
Some outdoor luminaires, such as some "wall pack" luminaires used for area or security lighting, are supplied with integral boxes and are intended for direct connection to raceways or cables. Otherwise, a box must be supplied at each lighting outlet. Exterior luminaires are usually in damp or wet locations, and the boxes must be suitable for such conditions. Although illumination of exterior doors is required, the luminaire is not necessarily required to be located at those doors, if a centrally located luminaire provides the required illumination.

5. Check for GFCI-protected receptacle outlets on balconies, decks, and porches.
A receptacle outlet is required to be accessible from balconies, decks, or porches at any dwelling unit if they are attached to the dwelling and directly accessible from inside the dwelling. The receptacle may not be more than 6.5 ft (2.0 m) from the surface of the balcony, deck, or porch surface. Because the receptacle is outdoors, it is required to be GFCI-protected. The GFCI device(s) affording protection of these receptacles is required to be readily accessible.

Checklist 6-27: Service Equipment, Feeders, and Panelboards (Residential Finish Inspection)

1. Review bonding and grounding if not completed during previous inspections.
Some items may not have been completed for inspection at the rough stage, so those items must be revisited and inspected at the finish stage.

2. Check overcurrent devices for compatibility with conductors (terminals, ratings, and ampacities).
Ampacities of conductors must be chosen and coordinated with terminal temperature ratings to prevent overheating of terminals. Overcurrent device ratings and conductor sizes must also be increased for the portion of the load that is continuous in order to conform to the conditions of testing of the overcurrent devices. In general, for dwelling unit loads, this is not a factor except for specific loads that are considered continuous, such as electric water heaters and fixed electric space heating, both of which are considered to be continuous only for purposes of sizing a branch circuit. Minimum conductor sizes and ampacities are based on several factors:
     A. Load calculations and conditions of use
     B. Terminal temperature ratings
     C. Continuous–noncontinuous loads
The largest conductor required by any of these three considerations must be used.

3. Check for proper identification of all overcurrent devices and disconnects.
Disconnects must be identified to indicate their purpose unless the purpose is obvious. For example, on the one hand, the purpose of a single disconnect next to or mounted on a single piece of equipment, such as a switch next to an air conditioner or a switch on a furnace, is obvious. On the other hand, multiple switches next to a group of air-conditioning units should be identified to make it clear which switch is for which unit. Directories on panelboards must be filled out to indicate the purpose or use of each overcurrent device in the

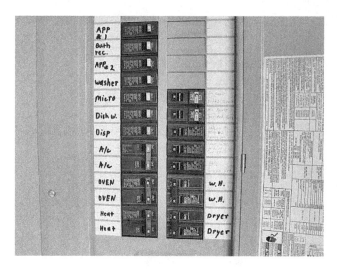

FIGURE 6-38　Panelboard circuit directory. (Courtesy of the International Association of Electrical Inspectors.)

FIGURE 6-39　Class 2 transformer installed on the side of a panelboard to separate Class 2 wiring from power wiring.

panelboard (**FIGURE 6-38**). The directory must be filled out in sufficient detail to differentiate every circuit from every other circuit and may not be based on the name of the person who occupies the space, such as "Joe's bedroom." Spare overcurrent devices (but not unoccupied spaces) must also be designated. These directories facilitate safe operation, and repair and maintenance of the electrical system after the initial installation.

4. Check for open spaces in panelboard fronts or cabinets.

Like unused openings in boxes and other enclosures, open spaces in panelboards must be sealed. Panelboard manufacturers offer blank fillers for this purpose.

5. Verify that doorbell and other Class 2 wiring and transformers are located in appropriate places (not in service equipment or panelboards).

Class 2 wiring must be kept separate from power and lighting circuits. Similarly, power supplies and transformers for Class 2 wiring must be installed in locations or enclosures where the supply- and load-side wiring can be kept separated. The interiors of service equipment and panelboards are not appropriate or permitted locations for Class 2 transformers. While the common 120-volt power supply to such transformers is required to be enclosed in raceways, cables, boxes, and enclosures, the Class 2 wiring is not required to be enclosed. Therefore, signaling transformers can often be mounted on the outside of boxes that contain power wiring or on the outside (end or side) of some panelboards with only the 120-volt supply wiring enclosed (**FIGURE 6-39**). The splice in the panelboard as shown is permitted by 312.8, and the Class 2 wiring and splices are not required to be enclosed. Some manufacturers offer special divided boxes with separate compartments for the power wiring and the Class 2 wiring.

6. Verify that any backfed overcurrent devices are secured in place.

Where main overcurrent devices in panelboards are simply backfed circuit breakers in plug-in spaces normally used for branch-circuit or feeder overcurrent devices, such main devices must be secured in place. Securing devices are available from panelboard manufacturers if the panelboards and main devices are intended for such installations. Unless the connection points of a circuit breaker are specifically marked "line and load," connecting power to a panelboard by backfeeding a secured circuit breaker is permitted.

7. Check for an intersystem grounding termination at the service equipment.

See Item 19 in the Rough Inspection checklist for Services, Feeders, and System Grounding.

Checklist 6-28: Mobile/Manufactured Home Site Inspections (Residential Finish Inspection)

1. Determine whether the unit is a mobile or manufactured home.

The *NEC* defines both mobile and manufactured homes. On the basis of these definitions, the significant difference between the two structures is that a mobile home is built on a permanent chassis and is designed to be used without a permanent foundation (making it truly mobile), whereas a manufactured home is built on a chassis and is designed to be installed with or without a permanent foundation.

From a field inspection standpoint, manufactured homes are built in accordance with standards adopted by the HUD. This standard, CFR Part 3280, *Manufactured Home Construction and Safety Standards,* contains requirements for the installation of electrical conductors and equipment and is not preempted by locally adopted codes. Manufactured homes that are built in accordance with this standard are required to be provided with a label indicating such. These units are subject to inspection during the manufacturing process by a third-party inspection/certifying organization. The *NEC* specifies that the use of the term *mobile home* within Article 550 also includes manufactured homes unless specifically indicated otherwise.

2. Verify that the mobile/manufactured home supply system is rated 120/240 volts nominal, single-phase.

Mobile and manufactured homes are built as standardized units and are equipped with appliances provided by the manufacturer. These standard appliances are designed for connection to supply systems rated at 120/240 volts nominal. The supply system requirement in the *NEC* ensures that these appliances will operate safely and efficiently. Although many appliances with line-to-line connected resistance heating elements will operate safely at 208 volts, they will not perform to their maximum efficiency. Some motor-operated equipment and appliances will operate equally well at either 240 or 208 volts.

3. Review the mobile/manufactured home park load calculations, and verify that demand factors, if used, have been properly applied.

The *NEC* provides **demand factors** for computing the load of a subdivision or **mobile home park** that contains multiple units. These demand factors are similar in concept to those permitted in Article 220 for multifamily dwelling units. Where the load for the actual mobile or manufactured home is not known, the *Code* provides a minimum load requirement of 16,000 VA or the minimum load requirement for the largest unit that can be installed at a given site, whichever is larger.

4. Verify that mobile/manufactured home service is rated not less than 100 amperes.

The rating of the service equipment is required to be not less than 100 amperes, 120/240 volts. In general, modern mobile and manufactured homes have load requirements greater than 50 amperes. A 50-ampere feeder, using a flexible cord connected to a 50-ampere receptacle at the service equipment, supplied many older mobile homes. Mobile and manufactured homes today have larger electrical loads, and this minimum service equipment rating of 100 amperes is necessary to supply these larger loads. It is common with the construction of mobile/manufactured home subdivisions or parks that the sites are provided with electrical supply before the actual unit is installed.

5. Verify location and minimum mounting height of mobile/manufactured home service equipment.

The requirements for the location of the service equipment are based on providing the ability to disconnect the mobile/manufactured home safely, without having to search through the entire park for the disconnecting means. In addition, the relocation of a mobile home from one site to another is facilitated by not permitting the service equipment (and associated utility meter) to be mounted on the mobile home. This requirement specifies that the service equipment must be within sight and not more than 30 ft (9.0 m) from the mobile home. Service equipment may be installed more than 30 ft (9.0 m) from the mobile home site; however, a site-disconnecting means, supplied by a feeder, is required to be located within sight and not more than 30 ft (9.0 m) from the mobile home. **FIGURE 6-40** shows a mobile/manufactured home site supply pedestal.

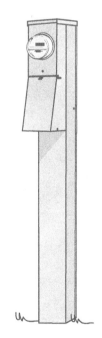

FIGURE 6-40   Site supply equipment for mobile or manufactured home.

The feeder from the service equipment to the site-disconnecting means is permitted to consist of two ungrounded conductors and one grounded conductor as specified in Article 250. The requirement for a feeder with four insulated conductors applies to the conductors installed from the site-disconnecting means to the mobile/manufactured home. This outdoor site-disconnecting means is required to be installed so that the bottom of the disconnecting means enclosure is not less than 2 ft (600 mm) above the finished grade or working platform.

Under conditions specified in 550.32(B), the service equipment for a manufactured home is permitted to be installed on the manufactured home. This is a rule that in its application makes a clear distinction between a mobile home and a manufactured home.

6. Check for proper grounding of mobile/manufactured home service equipment.

The fact that the **mobile home service equipment** is outdoors often results in the use of rod- or pipe-type grounding electrodes for the grounding of the mobile home service. The grounding of this service is no different from other services where a ground rod(s) is the sole grounding electrode. The availability of other types of grounding electrodes is increased where the service equipment is permitted to be installed on a manufactured home. Since the mobile home site is being treated as an occupancy with outside service equipment, additional grounding electrodes at the mobile home supplied by a feeder are not required.

7. Verify that a means to supply other structures or electrical equipment has been provided within the mobile/manufactured home service equipment.

The mobile home service equipment is required to provide means to supply a circuit to an outside **structure** that is in addition to the mobile home or to supply additional electrical equipment that is located outside the mobile home. Although the *Code* does not specify what the specific means is to be, it must be a method that does not require an installer to use or alter the service equipment in a manner that adversely affects the original electrical and/or mechanical integrity of the service equipment.

8. Check for GFCI protection on all 125-volt, 15- and 20-ampere receptacles that are installed in addition to receptacles installed as part of the permanent mobile or manufactured home wiring. Verify that GFCI protection device(s) for "field-installed" receptacles is readily accessible.

Where additional 15- and 20-ampere, 125-volt, single-phase receptacles are provided at the mobile home service equipment or are located elsewhere outdoors at the mobile home site, they are required to be provided with GFCI protection.

9. Verify that the feeder to mobile/manufactured home has four insulated color-coded conductors.

The feeder that is installed from the mobile home site service equipment is required to comprise two ungrounded conductors, a grounded conductor, and an equipment grounding conductor. The identification of the conductors can be done by either the manufacturer or the installer. The connection of the grounded (neutral) conductor and the equipment grounding conductor within the mobile/manufactured home distribution equipment is configured so that these conductors are not electrically connected at their load end. This separation of the grounded and equipment grounding conductors is the same approach that is required for feeder-supplied panelboards in Articles 250 and 408 (other than special conditions permitted in Article 250) and prevents normal neutral current from being introduced on metal raceways, enclosures, or piping systems that are not intended to be current-carrying conductors. Feeder-supplied panelboards are often referred to in the industry as *subpanels*. For mobile and manufactured homes, this wiring configuration and the insulated equipment grounding conductor isolate the steel chassis and any other metal framing or sheathing from neutral current.

10. Verify that feeders installed at <u>mobile/manufactured home lots</u> have a minimum capacity of 100 amperes at 120/240 volts.

This requirement applies to field-installed feeder conductors and correlates with the requirement that the site service equipment be rated not less than 100 amperes. Older mobile homes may be equipped with a 50-ampere, 4-wire power cord; however, that cord is required to be permanently connected to the distribution panelboard within the mobile home and thus is part of the permanent wiring of the unit. Where the load of the mobile/manufactured home is greater than 100 amperes, the feeder is required to be sized not less than the computed load. Conductor ampacities are permitted to be determined according to the alternative method of 310.15(B)(7).

11. Verify that new electrical installations that are added to existing mobile/ manufactured homes comply with applicable requirements of Article 550.

Typically, the electrical inspector is not charged with the responsibility for ensuring *NEC* compliance of the permanent wiring systems installed during the construction of mobile and manufactured homes. This is the responsibility of the third-party inspection/ certification organization and, in the case of manufactured homes, is based on the electrical requirements in the HUD 3280 standard. The electrical inspector is responsible for ensuring that the mobile/manufactured home site and supply wiring complies with the *NEC*, and in addition, where field additions or modifications are made to the wiring system of a mobile or manufactured home, the installation is subject to the requirements of the *NEC* as adopted by that jurisdiction. For these types of field installations, the electrical inspector should require compliance with the applicable rules in the *NEC*, specifically Article 550.

KEY TERMS

Article 100 and Section 550.2 of the *NEC* provide the following definitions for the key dwelling unit terms:

Accessible (as applied to equipment) Admitting close approach; not guarded by locked doors, elevation, or other effective means. [*See* Accessible, Readily.]

Accessible (as applied to wiring methods) Capable of being removed or exposed without damaging the building structure or finish, or not permanently closed in by the structure or finish of the building. [*See* Concealed *and* Exposed.]

Accessible, Readily (Readily Accessible) Capable of being reached quickly for operation, renewal, or inspections, without requiring those to whom ready access is requisite to use tools, climb over or remove obstacles or to resort to portable ladders, and so forth. [*See* Accessible. Readily accessible *and* accessible *are separate terms. Electrical equipment does not necessarily have to be accessible to be readily accessible as defined. A piece of equipment may be readily accessible and be behind a locked door, for example. Ready access is concerned with access "for those to whom ready access is requisite." Section 110.26 (G) states that "Electrical equipment rooms or enclosures housing electrical apparatus that are controlled by a lock(s) shall be considered accessible to qualified persons."*]

Appliance (Utilization equipment), generally other than industrial, that is normally built in standardized sizes or types and is installed or connected as a unit to perform one or more functions, such as clothes washing, air conditioning, food mixing, deep frying, and so forth.

Approved Acceptable to the authority having jurisdiction. [*Everything installed under the NEC is required to be approved. Article 110 provides criteria for an AHJ to determine what should be approved. See also* Authority Having Jurisdiction.]

Arc-Fault Circuit Interrupter (AFCI) A device intended to provide protection from the effects of arc faults by recognizing characteristics unique to arcing and by functioning to de-energize the circuit when an arc fault is detected. [*These devices can be manufactured in several different "types" all of which must pass specific test protocols in order to be a listed AFCI. Up to 2008, the most common type had been the "Branch-Feeder Type," usually used in the form of a circuit breaker. Now a "Combination Type" is required unless the provisions of the exceptions allowing an "Outlet Branch-Circuit Type" or "System Combination Type" are met. The outlet branch-circuit type typically takes the form of a wiring device similar to a GFCI receptacle, and the combination type is typically in the familiar circuit breaker form. The combination type serves to protect the fixed wiring just as the branch-feeder type did, but it also adds protection beyond the fixed outlets for appliance cords and extension cords. The system combination type uses a circuit breaker in conjunction with an outlet branch-circuit type and must be listed together. These devices function by electronically comparing the distortions of the normal waveform of an alternating current (ac) system to the characteristic waveforms created by arcing events. They typically do not trip on a single recorded distortion, but are "programmed" to operate when a pattern is established that would indicate the presence of an arcing fault that is either in series or in parallel with the normal circuit conductors. Since some equipment, such as switches and motors with brushes, produce some arcing in normal operation, the AFCI device must also recognize and not operate when this type of normal arcing occurs. Some objections to these devices claim that they are subject to nuisance tripping. However, in most cases the "nuisance" is actually a detection of a wiring error before it results in damage to the installation or structure.*]

Attachment Plug (Plug Cap) (Plug) A device that, by insertion in a receptacle, establishes a connection between the conductors of the attached flexible cord and the conductors connected permanently to the receptacle. [*See the Wiring Methods and Devices chapter of this manual.*]

Authority Having Jurisdiction (AHJ) An organization, office, or individual responsible for enforcing the requirements of a code or standard, or for approving equipment, materials, an installation, or a procedure.

> *Informational Note:* The phrase *authority having jurisdiction (AHJ)* is used in NFPA documents in a broad manner, since jurisdictions and approval agencies vary, as do their responsibilities. Where public safety is primary, the AHJ may be a federal, state, local, or other regional department or individual such as a fire chief; fire marshal; chief of a fire prevention bureau, labor department, or health department; building official; electrical inspector; or others having statutory authority. For insurance purposes, an insurance inspection department, rating bureau, or other insurance company representative may be the AHJ. In many circumstances, the property owner or his or her designated agent assumes the role of the AHJ; at government installations, the commanding officer or departmental official may be the AHJ.

Bathroom An area including a basin with one or more of the following: a toilet, a urinal, a tub, a shower, a bidet, or similar plumbing fixtures.

Branch Circuit The circuit conductors between the final overcurrent device protecting the circuit and the outlet(s).

Branch Circuit, Individual A branch circuit that supplies only one utilization equipment.

Branch Circuit, Multiwire A branch circuit consisting of two or more ungrounded conductors that have a voltage between them, and a grounded conductor that has equal voltage between it and each ungrounded conductor of the circuit and that is connected to the neutral or grounded conductor of the system.

Branch-Circuit Overcurrent Device A device capable of providing protection for service, feeder, and branch circuits and equipment over the full range of overcurrents between its rated current and its interrupting rating (**FIGURE 6-41**). Such devices are provided with interrupting ratings appropriate for the intended use, but no less than 5000 amperes. [*See* Overcurrent Protective Device, Branch-Circuit *in Article 100. This definition was added in the 2008 NEC partly to distinguish between appropriate overcurrent devices for services, feeders, and branch circuits and supplementary overcurrent devices that are not permitted for such use. Both terms, supplementary overcurrent device and branch-circuit overcurrent device, are used in the NEC. Section 240.10 limits the use of supplementary overcurrent devices and prohibits them from being used in place of branch-circuit overcurrent devices. A definition of supplementary overcurrent device was added in the 2005 NEC. Figure 6-41 shows an assortment of branch-circuit overcurrent devices of the type commonly used in dwelling units. See Supplementary Overcurrent Protective Device.*]

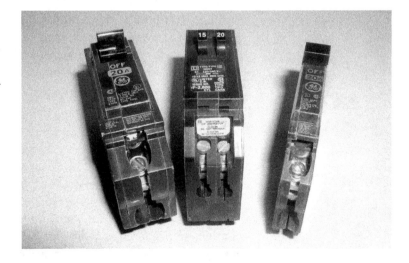

FIGURE 6-41   Branch-circuit overcurrent devices In the form of circuit breakers.

Concealed Rendered inaccessible by the structure or finish of the building.

Informational Note: Wires in concealed raceways are considered concealed, even though they may become accessible by withdrawing them. [*See* Accessible *and* Exposed (as applied to wiring methods).]

Continuous Load A load in which the maximum current is expected to continue for 3 hours or more. [*While lighting in commercial occupancies is generally considered to be continuous, the general lighting load in dwelling units, which includes general-use receptacles, is not considered to be continuous. The load assumed for small-appliance branch circuits is also not considered to be continuous, as the maximum load current is not expected to continue for 3 hours or more. Most large loads in dwelling units will not operate at full load "continuously" because of the normal cycling that occurs in cooking equipment, clothes dryers, and the like. However, some loads, such as storage-type water heaters and fixed electric space heating are treated as continuous loads in accordance with specific rules in their respective articles.*]

Cooking Unit, Counter-Mounted A cooking appliance designed for mounting in or on a counter and consisting of one or more heating elements, internal wiring, and built-in or separately mountable controls. [*These appliances are often called* cooktops.]

Demand Factor The ratio of the maximum demand of a system, or part of a system, to the total connected load of a system or the part of the system under consideration. [*Demand factors are used in calculating feeder and service loads. Except for circuits for some electric cooking equipment, branch circuits in dwelling units are sized on the basis of the connected or calculated load without the application of any demand factors. The optional calculations differ from the standard calculations in Article 220 primarily in the way that demand factors are applied.*]

Device A unit of an electrical system, other than a conductor, that carries or controls electric energy as its principal function. [*Receptacles and wall switches used to control lighting outlets are commonly called* wiring devices. *Although a device such as a lighted switch or a GFCI may actually use a small amount of power, the control of the circuit is still the principal function, and this distinguishes devices from "appliances" or other "utilization equipment." Other equipment such as circuit breakers, fuses, panelboards, and enclosed disconnect switches are also devices.*]

Disconnecting Means A device, or group of devices, or other means by which the conductors of a circuit can be disconnected from their source of supply.

Dwelling, Multifamily A building that contains three or more dwelling units. [*For the purposes of applying NEC requirements to dwelling units, rules that are identified as applying only to one- or two-family dwellings do not apply to multifamily dwellings, even though the individual units may be intended for one family.*]

Dwelling, One-Family A building that consists solely of one dwelling unit.

Dwelling, Two-Family A building that consists solely of two dwelling units.

Dwelling Unit A single unit, providing complete and independent living facilities for one or more persons, including permanent provisions for living, sleeping, cooking, and sanitation. [*A microwave oven alone does not meet the intent of "permanent provisions for cooking." For example, hotel rooms with a bar sink, a refrigerator, and a microwave oven are not considered dwelling units under this definition, but some suite or extended-stay hotels and motels with a sink, range, and refrigerator would be considered dwelling units. A primary consideration is that this definition determines the applicability of the requirement for two small-appliance branch circuits for a dwelling unit.*]

Equipment A general term including fittings, devices, appliances, luminaires, apparatus, machinery, and the like, used as a part of, or in connection with, an electrical installation. [*This broad definition includes virtually all items used in an electrical installation, including conductors, cables, raceways, boxes, and fittings.*]

Exposed (as applied to wiring methods) On or attached to the surface or behind panels designed to allow access. [*According to this definition, the wiring installed above a suspended grid ceiling or behind other access panels is still considered to be exposed even though it may not be visible most of the time.*]

Feeder All circuit conductors between the service equipment, the source of a separately derived system, or other power supply source and the final branch-circuit overcurrent device.

Garage A building or portion of a building in which one or more self-propelled vehicles can be kept for use, sale, storage, rental, repair, exhibition, or demonstration purposes.

> *Informational Note:* For commercial garages, repair and storage, see Article 511.

Ground-Fault Circuit Interrupter (GFCI) A device intended for the protection of personnel that functions to deenergize a circuit or portion thereof within an established period of time when a current to ground exceeds the values established for a Class A device.

> *Informational Note:* Class A GFCIs trip when the current to ground is 6 mA or higher and do not trip when the current to ground is less than 4 mA. For further information, see UL 943, *Standard for Ground-Fault Circuit Interrupters.*

[*GFCIs are widely used in dwelling units. GFCIs operate by detecting any difference between the intended current-carrying conductors. An imbalance in the total current leaving from and returning to the GFCI device is detected. Since some current is finding a way back to the source through other than the intended paths, this is interpreted as a ground fault and the device will interrupt the current in as little as ¼ of a cycle when the leakage current reaches a value of about 5 milliamperes. Although this level of current may cause some pain in some people, it will not be a level that is likely to cause serious shock, heart fibrillation, or electrocution.*]

Identified (as applied to equipment) Recognizable as suitable for the specific purpose, function, use, environment, application, and so forth, where described in a particular *Code* requirement. [*See* Equipment.]

> *Informational Note:* Some examples of ways to determine suitability of equipment for a specific purpose, environment, or application include investigations by a qualified testing laboratory (listing and labeling), an inspection agency, or other organizations concerned with product evaluation.

In Sight From (Within Sight From, Within Sight) Where the *Code* specifies that one equipment shall be "in sight from," "within sight from," or "within sight of," and so forth, of another equipment, the specified equipment is to be visible and not more than 50 ft (15.2 m) distant from the other.

Intersystem Bonding Termination A device that provides a means for connecting intersystem bonding conductors for communications systems to the grounding electrode system. [*This was a new term but not a new concept in the 2008 NEC. This connection has been required in the past, but the specific termination methods have not been addressed previously. The definition was revised for the 2014 Code to provide a more direct description of the purpose of this device.*]

Kitchen An area with a sink and permanent provisions for food preparation and cooking. [*This definition was previously found in 210.8(B) and applied only to nondwelling installations. It is now applicable to residential applications as well. Like the definition of* dwelling unit, *this definition refers to "permanent provisions for . . . cooking." This is not intended to apply to locations such as wet bars that have sinks and portable means for cooking such as a hot plate or a microwave oven.*]

Labeled Equipment or materials to which has been attached a label, symbol, or other identifying mark of an organization that is acceptable to the AHJ and concerned with product evaluation, that maintains periodic inspection of production of labeled equipment or materials, and by whose labeling the manufacturer indicates compliance with appropriate standards or performance in a specified manner.

Lighting Outlet An outlet intended for the direct connection of a lampholder or a luminaire.

Listed Equipment, materials, or services included in a list published by an organization that is acceptable to the AHJ and concerned with evaluation of products or services, that maintains periodic inspection of production of listed equipment or materials or periodic evaluation of services, and whose listing states that the equipment, material, or services either meet identified standards or have been tested and found suitable for a specified purpose.

> *Informational Note:* The means for identifying listed equipment may vary for each organization concerned with product evaluation, some of which do not recognize equipment as listed unless it is also labeled. Use of the system employed by the listing organization allows the AHJ to identify a listed product.

Location, Damp Locations protected from weather and not subject to saturation with water or other liquids but subject to moderate degrees of moisture.

> *Informational Note:* Examples of such locations include partially protected locations under canopies, marquees, roofed open porches, and like locations, and interior locations subject to moderate degrees of moisture, such as some basements, some barns, and some cold-storage warehouses.

Location, Dry A location not normally subject to dampness or wetness. A location classified as *dry* may be temporarily subject to dampness or wetness, as in the case of a building under construction.

Location, Wet Installations underground or in concrete slabs or masonry in direct contact with the earth; in locations subject to saturation with water or other liquids, such as vehicle washing areas; and in unprotected locations exposed to weather. [*Locations underground are wet by definition, and the insides of raceways buried underground are also intended to be wet locations. Protected installations in portions of buildings that are underground, such as basements, are not normally wet locations but may be damp locations as described in the definition of* Location, Damp.]

Manufactured Home [from Section 550.2] A structure, transportable in one or more sections, that, in the traveling mode, is 2.4 m (8 body ft) or more in width or 12.2 m (40 body ft) or more in length; when erected on site, is 29.7 m² (320 ft²) or more; built on a chassis and designed to be used as a dwelling, with or without a permanent foundation, when connected therein. The term *manufactured home* includes any structure that meets all the provisions of this paragraph except the size requirements and with respect to which the manufacturer voluntarily files a certification required by the regulatory agency, and except that such term does not include any self-propelled recreational vehicle. Calculations used to determine the number of square meters (square feet) in a structure will be based on the structure's exterior dimensions, measured at the largest horizontal projections when erected on site. These dimensions include all expandable rooms, cabinets, and other projections containing interior space, but do not include bay windows.

For the purpose of the *Code* and unless otherwise indicated, the term *mobile home* includes manufactured homes.

> *Informational Note No. 1:* See the applicable building code for definition of the term *permanent foundation.*
>
> *Informational Note No. 2:* See Part 3280, *Manufactured Home Construction and Safety Standards,* of the U.S. Department of Housing and Urban Development for additional information on the definition.

Mobile Home [from Section 550.2] A factory-assembled structure or structures transportable in one or more sections that are built on a permanent chassis and designed to be used as a dwelling without a permanent foundation where connected to the required utilities and that include the plumbing, heating, air-conditioning, and electric systems contained therein.

For the purpose of the *Code* and unless otherwise indicated, the term *mobile home* includes manufactured homes.

Mobile Home Lot [from Section 550.2] A designated portion of a mobile home park designed for the accommodation of one mobile home and its accessory buildings or structures for the exclusive use of its occupants.

Mobile Home Park [from Section 550.2] A contiguous parcel of land that is used for the accommodation of occupied mobile homes.

Mobile Home Service Equipment [from Section 550.2] The equipment containing the disconnecting means, overcurrent protective devices, and receptacles or other means for connecting a mobile home feeder assembly.

Outlet A point on the wiring system at which current is taken to supply utilization equipment. [*Receptacles represent one kind of outlet. Lighting outlets are another kind of outlet. Other types of outlets may be used to provide power to appliances or other specific items of electrical equipment. The NEC uses these terms much more carefully than is normally done in casual conversation. Therefore, extra care is necessary when using such terms in applying the* Code.]

Premises Wiring (System) Interior and exterior wiring, including power, lighting, control, and signal circuit wiring together with all of their associated hardware, fittings, and wiring devices, both permanently and temporarily installed. This includes (a) wiring from the service power source to the outlets, or (b) wiring from and including the power source to the outlets where there is no service point.

Such wiring does not include wiring internal to appliances, luminaires, motors, controllers, motor control centers, and similar equipment.

> *Informational Note:* Power sources include, but are not limited to, interconnected or stand-alone batteries, solar photovoltaic systems, other distributed generation systems, or generators.

Raceway An enclosed channel of metallic or nonmetallic materials designed expressly for holding wires, cables, or busbars, with additional functions as permitted in the *Code.*

> *Informational Note:* A raceway is identified within specific article definitions.

Receptacle A receptacle is a contact device installed at the outlet for the connection of an attachment plug. A single receptacle is a single contact device with no other contact device on the same yoke. A multiple receptacle is two or more contact devices on the same yoke.

Receptacle Outlet An outlet where one or more receptacles are installed. [*See* Outlet.]

Retrofit Kit. A general term for a complete subassembly of parts and devices for field conversion of utilization equipment. [*These are also called* Retrofit Luminaire Conversions *in listing standards. They include reflector kits for some luminaires and kits to convert a luminaire that uses one light source to a different light source such as incandescent to fluorescent or fluorescent to LED.*]

Service   The conductors and equipment for delivering electric energy from the serving utility to the wiring system of the premises served.

Service Cable   Service conductors made up in the form of a cable.

Service Conductors   The conductors from the service point to the service disconnecting means.

Service Conductors, Overhead   The overhead conductors between the service point and the first point of connection to the service-entrance conductors at the building or other structure.

Service Conductors, Underground   The underground conductors between the service point and the first point of connection to the service-entrance conductors in a terminal box, meter, or other enclosure, inside or outside the building wall.

> *Informational Note:* Where there is no terminal box, meter, or other enclosure, the point of connection is considered to be the point of entrance of the service conductors into the building.

Service Drop   The overhead service conductors between the utility electric supply system and the service point.

Service-Entrance Conductors, Underground System   The service conductors between the terminals of the service equipment and the point of connection to the service lateral or underground service conductors.

> *Informational Note:* Where service equipment is located outside the building walls, there may be no service-entrance conductors, or they may be entirely outside the building.

Service Equipment   The necessary equipment, usually consisting of a circuit breaker(s) or switch(es) and fuse(s) and their accessories, connected to the load end of service conductors to a building or other structure, or an otherwise designated area, and intended to constitute the main control and cutoff of the supply.

Service Point   The point of connection between the facilities of the serving utility and the premises wiring.

> *Informational Note:* The service point can be described as the point of demarcation between where the serving utility ends and the premises wiring begins. The serving utility generally specifies the location of the service point based on the conditions of service.

Structure   That which is built or constructed.

Supplementary Overcurrent Protective Device   A device intended to provide limited overcurrent protection for specific applications and utilization equipment such as luminaires and appliances. This limited protection is in addition to the protection provided in the required branch circuit by the branch-circuit overcurrent protective device. [*See* Overcurrent Protective Device, Supplementary, *in Article 100. Usually supplementary overcurrent devices are seen in the form of small fuses used to protect components of an appliance.* **FIGURE 6-42** *shows an example of a supplementary fuse to protect the electronic circuitry in a listed gas-fired furnace in a dwelling unit.*]

Switch, General-Use   A switch intended for use in general distribution and branch circuits. It is rated in amperes, and it is capable of interrupting its rated current at its rated voltage.

Utilization Equipment   Equipment that utilizes electric energy for electronic, electromechanical, chemical, heating, lighting, or similar purposes.

FIGURE 6-42   Supplementary overcurrent device in a listed appliance.

KEY QUESTIONS

Dwelling Units

1. What wiring methods are used and are the wiring methods suitable for the conditions?
2. Are multiwire branch circuits used?
3. What rooms require small-appliance branch circuits?
4. What sizes of service equipment, service-entrance conductors, and feeder conductors are required?
5. Is the service overhead or underground?
6. What types of grounding electrodes are present?
7. What sizes of grounding electrode conductors are required?
8. Where is mechanical equipment located, and what types of equipment will be installed?

Mobile/Manufactured Home Sites

1. Does the unit meet the definition of a mobile home or of a manufactured home?
2. If the unit is on-site, what is its electrical rating in amperes?
3. Will any accessory structures be installed?
4. Are electrical installations being made to an existing mobile or manufactured home?

CHECKLISTS

	Checklist 6-1: Residential Rough Inspection: General Requirements (All Areas)			
✔	Item	Inspection Activity	*NEC* Reference	Comments
	1.	Check wiring methods (usually cable assemblies) for support and suitability for the conditions.	*NEC* Chapters 3, 7, and 8, Art 334, 314.17	
	2.	Check cable installation through or parallel to framing members and furring strips for 1¼-in. (32-mm) clearance or protective steel plates.	300.4	
	3.	Check boxes for suitability for the use.	314.15, 314.27	
	4.	Verify that boxes are installed in accessible locations for all junctions and outlets and pull points.	300.15, 314.29	
	5.	Check that cables are secured to boxes.	314.17(B) and (C)	
	6.	Check boxes for conductor fill.	314.16	
	7.	Check positioning of boxes that are intended to be flush with combustible and noncombustible finished surfaces.	314.20	
	8.	Check for splicing devices on all equipment grounding conductors within boxes and bonding connections to metal boxes.	250.8, 250.86, 250.146, 250.148	
	9.	Check equipment grounding conductors for suitability and size.	250.118, 250.122	

Checklist 6-1: Residential Rough Inspection: General Requirements (All Areas)

✔	Item	Inspection Activity	*NEC* Reference	Comments
	10.	Check boxes used in floors or for support of ceiling fans for listing.	314.27(B) and (C)	
	11.	Check recessed luminaires for clearances from combustibles and insulation.	410.116(A)(1), 410.116(B)	
	12.	Check cables installed in contact with thermal insulation or without maintaining spacing (fire- or draft-stopped, etc.), for possible adjustment factors.	334.80, 310.15(B)(2) (a)	
	13.	Check wall switch locations for the presence of grounded conductors.	404.2(C)	

Checklist 6-2: Residential Rough Inspection: Kitchen

✔	Item	Inspection Activity	*NEC* Reference	Comments
	1.	Check spacing of receptacles for walls and countertops, including islands and peninsulas.	210.52(A) and (C)	
	2.	Verify that a minimum of two 20-ampere small-appliance branch circuits are used for kitchen receptacles.	210.11(C)(1), 210.52(B)	
	3.	Verify that a wall-switched lighting outlet is provided and wired on a general lighting circuit.	210.70(A), 210.52(B)(2)	
	4.	Verify that properly sized circuits have been provided for specific kitchen appliances, such as dishwashers, disposals, ranges, cooktops, trash compactors, and the like.	210.23, 422.10	
	5.	Check for additional small-appliance branch circuits where there is more than one kitchen.	210.52(B)(3)	
	6.	Check for other outlets or appliances on small-appliance branch circuits.	210.52(B)(2)	

Checklist 6-3: Residential Rough Inspection: Dining Room

✔	Item	Inspection Activity	*NEC* Reference	Comments
	1.	Check receptacle outlets for proper spacing.	210.52(A)	
	2.	Verify that all required receptacle outlets are supplied by small-appliance branch circuits.	210.52(B)(1)	
	3.	Check for wall switch-controlled lighting outlet on a general lighting circuit.	210.70(A), 210.52(B)(2)	

CHECKLISTS

Checklist 6-4: Residential Rough Inspection: Bathrooms

✔	Item	Inspection Activity	*NEC* Reference	Comments
	1.	Verify that receptacle outlets are installed adjacent to and within 36 in. (900 mm) of each basin.	210.52(D)	
	2.	Verify that receptacles are supplied by dedicated 20-ampere branch circuits.	210.11(C)(3), 210.23(A)(2)	
	3.	Check for a wall switch-controlled lighting outlet on a general lighting circuit.	210.70(A)(1)	

Checklist 6-5: Residential Rough Inspection: Other Habitable Rooms (Bedrooms, Family Rooms, Parlors, and Dens)

✔	Item	Inspection Activity	*NEC* Reference	Comments
	1.	Check receptacle outlets for proper spacing.	210.52(A)	
	2.	Check for wall switch-controlled lighting outlets (including switched receptacles).	210.70(A)(1)	

Checklist 6-6: Residential Rough Inspection: Hallways and Foyers

✔	Item	Inspection Activity	*NEC* Reference	Comments
	1.	Check for at least one wall switch-controlled (or automatic-, remote-, or centrally controlled) lighting outlet.	210.70(A)(2)	
	2.	Verify that hallways that are continuous for 10 ft (3.0 m) or more have at least one receptacle outlet.	210.52(H)	
	3.	Verify that foyers that are not part of a hallway have receptacle outlets installed as required.	210.52(I)	

Checklist 6-7: Residential Rough Inspection: Stairways

✔	Item	Inspection Activity	*NEC* Reference	Comments
	1.	Check for at least one wall switch-controlled (or automatic-, remote-, or centrally controlled) lighting outlet.	210.70(A)(2)	
	2.	Verify that wall switches are provided at each floor level where there are six or more steps between levels.	210.70(A)(2)	

Checklist 6-8: Residential Rough Inspection: Closets

✔	Item	Inspection Activity	*NEC* Reference	Comments
	1.	Check clearances between luminaires and storage spaces if luminaires are to be installed.	410.16	

Checklist 6-9: Residential Rough Inspection: Laundry Area

✔	Item	Inspection Activity	*NEC* Reference	Comments
	1.	Verify that at least one receptacle outlet is installed for the laundry.	210.52(F)	
	2.	Verify that a dedicated 20-ampere circuit supplies the laundry outlet(s) and no other outlets.	210.11(C)(2)	
	3.	Check for a laundry receptacle outlet within 6 ft (1.8 m) of the intended appliance location.	210.50(C)	
	4.	Check for proper branch-circuit conductors, including equipment grounding conductors, for 240-volt dryers (if used).	422.10, 250.134, 250.138, 220.54	
	5.	Verify that lighting outlets for the area are supplied from general lighting circuits.	210.11(C)(2)	

Checklist 6-10: Residential Rough Inspection: Basements and Attics

✔	Item	Inspection Activity	*NEC* Reference	Comments
	1.	Verify that at least one receptacle outlet is provided in unfinished basement areas in addition to any receptacles installed for laundry equipment or other specific equipment.	210.52(G)	
	2.	Verify that a receptacle outlet is provided for servicing mechanical equipment, if any.	210.63, 210.64	
	3.	Verify that individual branch circuits are supplied for central heating equipment, if any.	422.12	
	4.	Verify that a wall switch-controlled lighting outlet or a lighting outlet containing a switch is provided at the entrance to areas for storage or equipment requiring servicing.	210.70(A)(3)	
	5.	Check basements, accessible attics, attic entrances, and scuttle holes for clearances from or protection of cable assemblies.	320.23, 330.23, 334.23, 334.15, 320.15	

CHECKLISTS

Checklist 6-11: Residential Rough Inspection: Attached Garages and Detached Garages or Accessory Buildings with Electric Power

✔	Item	Inspection Activity	*NEC* Reference	Comments
	1.	Verify that at least one receptacle outlet is provided. Verify that if installed, branch circuit(s) supplying outlets for electric vehicle supply equipment (EVSE) do not supply any other loads.	210.52(G), 210.17	
	2.	Verify that a wall switch-controlled lighting outlet is provided.	210.70(A)(2)	

Checklist 6-12: Residential Rough Inspection: Outdoors

✔	Item	Inspection Activity	*NEC* Reference	Comments
	1.	Check for at least two receptacle outlets, one each at the front and back of a dwelling.	210.52(E), 210.63	
	2.	Check for receptacle outlets on balconies, decks, and porches.	210.52(E)(3)	
	3.	Check for wall switch-controlled (or remote-, central-, or automatic-controlled) exterior lighting outlets at outdoor entrances or exits with grade-level access.	210.70(A)(2)	

CHECKLISTS

Checklist 6-13: Residential Rough Inspection: Services and Feeders and System Grounding

✔	Item	Inspection Activity	*NEC* Reference	Comments
	1.	Review the calculation of service load and determine the minimum size of service conductors.	Article 220, 230.42(B), 230.79	
	2.	Verify that service disconnects and overcurrent devices are located outside or inside nearest the point of entrance of the service conductors.	230.70(A)(1), 230.91	
	3.	Verify that service disconnects are grouped together, with no more than six in any one location.	230.71, 230.72, 408.36 Exc. No. 1	
	4.	Check for proper accessibility, working clearances, and dedicated spaces around service equipment.	110.26, 230.70(A)(1), 230.91, 240.24	

✔	Item	Inspection Activity	*NEC* Reference	Comments
		Checklist 6-13: Residential Rough Inspection: **Services and Feeders and System Grounding**		
	5.	Check service-entrance wiring methods for suitability, support, and protection from damage.	230.32, 230.43, 230.50, 230.51	
	6.	Check for a proper drip loop and weatherhead on overhead services.	230.54	
	7.	Verify that the point of attachment for overhead service is adequate and will provide required support and clearances above roofs and grade.	230.24, 230.26, 230.54(C), 23028(B)	
	8.	Check service masts for adequate strength and support.	230.28, 225.17	
	9.	Check for proper clearances of service conductors from building openings.	230.9(C)	
	10.	Check underground service conductors for proper depth, fill, protection, marking, and allowances for ground movement.	300.5, 230.32	
	11.	Determine which grounding electrodes are available, and verify that they are bonded together to form a grounding electrode system.	250.50, 250.52	
	12.	Check any rod, pipe, plate, or other listed electrodes for proper size, type, and installation.	250.52, 250.53	
	13.	Verify that grounding electrode conductors are unspliced and protected and that any ferrous metal enclosures are bonded and electrically continuous.	250.64	
	14.	Check grounding electrode conductor(s) and bonding jumpers for proper sizing.	250.66, 250.64(F)	
	15.	Check grounding electrode connections for proper type, protection, and accessibility.	250.8, 250.10, 250.68, 250.70	
	16.	Verify that the main bonding jumper is installed and is of the proper size and type.	250.24(B), 250.28	
	17.	Verify that metal interior piping systems are bonded, that bonding jumpers are properly sized, and that continuity around removable devices is assured.	250.104(A), 250.68(B)	
	18.	Verify that service raceways and enclosures are properly grounded and bonded.	250.80, 250.92, 250.94	
	19.	Verify that an intersystem bonding termination has been provided.	250.94	

CHECKLISTS

Checklist 6-14: Residential Rough Inspection: Feeders and Panelboards

✔	Item	Inspection Activity	*NEC* Reference	Comments
	1.	Review the calculation of feeder loads, and verify that conductors are properly sized and rated.	Article 220, 310.15, 215.2	
	2.	Verify that panelboards have proper ratings and protection.	Article 220, 408.30, 408.36	
	3.	Check for proper accessibility, working clearances, and dedicated spaces around panelboards.	110.26, 240.24	
	4.	Verify that at least the minimum number of overcurrent devices and circuits has been provided.	210.11	
	5.	Verify that the grounded feeder conductors are insulated and isolated from equipment grounding conductors and grounded enclosures.	250.24(A)(5), 250.142(B), 310.2	
	6.	Verify that panelboards are grounded by an appropriate and properly sized equipment grounding conductor (or conductors).	408.40, 215.6, 250.118, 250.122	

CHECKLISTS

Checklist 6-15: Residential Finish Inspection: General Requirements (All Areas)

✔	Item	Inspection Activity	*NEC* Reference	Comments
	1.	Check for correction of any deficiencies noted on previous inspections.		
	2.	Check positioning of boxes intended to be flush with combustible and noncombustible finished surfaces.	314.20	
✔	3.	Check for proper positioning of receptacles and faceplates on walls.	406.5, 406.6	
	4.	Check for gaps around outlet boxes in walls.	314.21	
	5.	Verify that conductor terminations and splicing methods are compatible with conductor materials.	110.14, 404.14(C), 406.3(C)	
	6.	Verify that receptacles are bonded to metal boxes and that receptacles, switches, and metal faceplates are grounded.	250.146, 250.148, 404.9(B), 406.6(B)	

Checklist 6-15: Residential Finish Inspection: General Requirements (All Areas)

✔	Item	Inspection Activity	*NEC* Reference	Comments
	7.	Check polarity of devices and luminaires.	200.10, 200.11, 406.10, 410.50	
	8.	Check all luminaires and lampholders for listing.	410.6	
	9.	Check for splicing devices on all equipment grounding conductors within boxes and for bonding connections to metal boxes.	250.8, 250.86, 250.148	
	10.	Verify that device ratings are compatible with circuit and equipment ratings.	210.21, 210.24	
	11.	Check for proper use of connectors and fittings and for protection of cables.	300.15	
	12.	Check for bushings or equivalent protection for cables entering boxes and other enclosures.	300.4(F), 314.17, 314.42	
	13.	Verify that unused openings in boxes and other enclosures are closed.	110.12(A), 314.17(A), 312.5(A)	
	14.	Verify that appliances, motors, and other equipment are grounded.	250.110, 250.112, 250.114	
	15.	Check burial depth of underground raceways and cables for minimum cover.	300.5	
	16.	Check installation of listed equipment for compliance with manufacturer's instructions.	110.3(B)	
	17.	Verify that fire rating of building assemblies has been restored at electrical penetrations.	300.21	
	18.	Check for disconnecting means on both permanently connected and cord-and-plug–connected appliances.	Article 422, Part III	
	19.	Verify that circuits for mechanical equipment have correct conductor size and overcurrent protection.	Articles 422, 424, 430, and 440	
	20.	Check for AFCI protection on all 120-volt, 15- and 20-ampere branch circuits serving outlets in most areas and that AFCI devices are readily accessible.	210.12	
	21.	Check for tamper-resistant receptacles in all areas where 25-volt, 15- and 20-ampere receptacle outlets are required.	210.52, 406.12	

CHECKLISTS

Checklist 6-16: Residential Finish Inspection: Kitchen

✔	Item	Inspection Activity	NEC Reference	Comments
	1.	Check spacing of receptacles for walls and countertops, including islands and peninsulas.	210.52(A) and (C)	
	2.	Verify that a minimum of two 20-ampere small-appliance branch circuits are used for kitchen receptacles.	210.11(C)(1), 210.52(B)	
	3.	Verify that small-appliance branch circuits are used only for receptacles in kitchen, dining room, pantry, and so forth.	210.52(B)(2)	
	4.	Verify that all 125-volt, 15- and 20-ampere receptacles serving countertops, all receptacles within 6 ft (1.8 m) of a kitchen sink, and outlets for dishwashers are provided with GFCI protection. Verify that GFCI protection devices are readily accessible.	210.8(A), 210.8(A)(6), 210.8(A)(7), 210.8(D)	
	5.	Verify that refrigeration equipment is supplied by a small-appliance branch circuit or an individual branch circuit.	210.52(B)(1), Exc. No. 2	
	6.	Verify that a wall-switched lighting outlet is provided and wired on a general lighting circuit.	210.70(A), 210.52(B)(2)	
	7.	Verify that properly sized circuits are provided for specific kitchen appliances, such as dishwashers, disposals, ranges (cooktops), trash compactors, and the like.	210.23, 422.10	
	8.	Check for proper type, length, and use of flexible cords for appliance connections.	422.16, Table 400.4	

Checklist 6-17: Residential Finish Inspection: Dining Room

✔	Item	Inspection Activity	NEC Reference	Comments
	1.	Check receptacle outlets for proper spacing.	210.52(A)	
	2.	Verify that all required receptacle outlets are supplied by small-appliance branch circuits.	210.52(B)	
	3.	Check for a wall switch-controlled lighting outlet on a general lighting circuit.	210.70(A), 210.52(B)(2)	

Checklist 6-18: Residential Finish Inspection: Bathrooms

✔	Item	Inspection Activity	*NEC* Reference	Comments
	1.	Verify that receptacle outlets are installed adjacent to and within 36 in. (900 mm) of each basin.	210.52(D)	
	2.	Verify that receptacles are supplied by dedicated 20-ampere branch circuits.	210.11(C)(3)	
	3.	Verify that bathroom receptacles are GFCI-protected and the GFCI device is readily accessible	210.8(A), 210.8(A)(1)	
	4.	Check for a wall switch-controlled lighting outlet on a general lighting circuit.	210.70(A)(1)	

Checklist 6-19: Residential Finish Inspection: Other Habitable Rooms (Bedrooms, Family Rooms, Parlors, and Dens)

✔	Item	Inspection Activity	*NEC* Reference	Comments
	1.	Check receptacle outlets for proper spacing.	210.52(A)	
	2.	Check for wall switch-controlled lighting outlets (including switched receptacles).	210.70	

Checklist 6-20: Residential Finish Inspection: Hallways and Foyers

✔	Item	Inspection Activity	*NEC* Reference	Comments
	1.	Check for at least one wall switch-controlled (or automatic-, remote-, or centrally controlled) lighting outlet.	210.70(A)(2)	
	2.	Verify that hallways that are continuous for 10 ft (3.0 m) or more have at least one receptacle outlet.	210.52(H)	
	3.	Verify that foyers over 60 ft^2 (5.6 m^2) have receptacles installed as required.	210.52(I)	

CHECKLISTS

Checklist 6-21: Residential Finish Inspection: Stairways

✔	Item	Inspection Activity	*NEC* Reference	Comments
	1.	Check for at least one wall switch-controlled (or automatic-, remote-, or centrally controlled) lighting outlet.	210.70(A)(2)	
	2.	Verify that wall switches are provided at each floor level where there are six or more steps.	210.70(A)(2)	

Checklist 6-22: Residential Finish Inspection: Closets

✔	Item	Inspection Activity	*NEC* Reference	Comments
	1.	Check clearances between luminaires and storage spaces if luminaires are installed.	410.16	

Checklist 6-23: Residential Finish Inspection: Laundry Area

✔	Item	Inspection Activity	*NEC* Reference	Comments
	1.	Verify that at least one receptacle outlet is installed for the laundry.	210.52(F)	
	2.	Verify that a dedicated 20-ampere circuit supplies the laundry outlet(s) and no other outlets.	210.11(C)(2)	
	3.	Check for a laundry receptacle outlet within 6 ft (1.8 m) of the intended appliance location.	210.50(C)	
	4.	Check for proper receptacle ratings based on branch-circuit ratings, including receptacles for electric dryers (if used).	210.21, 210.24, 250.140	
	5.	Verify that lighting outlets for the laundry area are supplied from general lighting circuits.	210.11(C)(2)	
	6.	Verify GFCI protection for all 125-volt, 15- and 20-ampere receptacles installed in laundry areas and that the GFCI protection device(s) is readily accessible.	210.8(A) 210.8(A)(10)	

Checklist 6-24: Residential Finish Inspection: Basements and Attics

✔	Item	Inspection Activity	*NEC* Reference	Comments
	1.	Verify that at least one receptacle outlet is provided in unfinished basement areas in addition to any receptacles installed for specific equipment.	210.52(G)	
	2.	Verify that a receptacle outlet is provided for servicing mechanical equipment, if any.	210.63	
	3.	Verify that GFCI protection is provided for receptacles in unfinished portions of basements and that the GFCI protection device is readily accessible	210.8(A) 210.8(A)(5)	
	4.	Verify that individual branch circuits are supplied for central heating equipment, if any.	422.12	
	5.	Verify that a wall switch-controlled lighting outlet or a lighting outlet containing a switch is provided at the entrance to equipment requiring servicing.	210.70(A)(3)	
	6.	Check basements, accessible attics, attic entrances, and scuttle holes for clearances from or protection of cable assemblies.	320.23, 330.23, 334.23, 334.15, 320.15	

Checklist 6-25: Residential Finish Inspection: Garages (Attached or with Electric Power)

✔	Item	Inspection Activity	*NEC* Reference	Comments
	1.	Verify that at least one receptacle outlet is provided and that the circuit supplying the receptacle outlet(s) does not supply outlets outside of the garage.	210.52(G)(1)	
	2.	Verify that GFCI protection is provided for all 125-volt, 15- and 20-ampere receptacles and that GFCI protection device(s) is readily accessible	210.8(A)(2)	
	3.	Verify that a wall switch-controlled lighting outlet is provided.	210.70(A)(2)	

Checklist 6-26: Residential Finish Inspection: Outdoors

✔	Item	Inspection Activity	*NEC* Reference	Comments
	1.	Check for at least two receptacle outlets, one each at front and back of dwelling.	210.52(E), 210.63	
	2.	Verify that outdoor receptacles are GFCI-protected unless they are not readily accessible and are supplied by circuits for deicing or snow-melting equipment. Verify that GFCI protection device(s) is readily accessible.	210.8(A)(3)	
	3.	Check for wall switch-controlled (or remote-, central-, or automatic-controlled) exterior lighting outlets at outdoor entrances or exits with grade-level access.	210.70(A)(2)	
	4.	Check for boxes at exterior luminaire locations.	300.15, 314.27(A)	
	5.	Check for GFCI-protected receptacle outlets on balconies, decks, and porches.	210.52(E)(3), 210.8(A)(3)	

Checklist 6-27: Residential Finish Inspection: Service Equipment, Feeders, and Panelboards

✔	Item	Inspection Activity	*NEC* Reference	Comments
	1.	Review bonding and grounding if not completed during previous inspections.	Article 250	
	2.	Check overcurrent devices for compatibility with conductors (terminals, ratings, and ampacities).	240.4, 110.14(C), 210.20, 215.3, 230.42, 310.15	
	3.	Check for proper identification of all overcurrent devices and disconnects.	110.22, 230.70(B), 408.4	
	4.	Check for open spaces in panelboard fronts or cabinets.	110.12, 408.7	
	5.	Verify that doorbell and other Class 2 wiring and transformers are located in appropriate places (not in service equipment or panelboards).	312.8, 725.133	
	6.	Verify that any backfed overcurrent devices are secured in place.	408.36(D)	
	7.	Check for an intersystem grounding termination at the service equipment.	250.94	

Checklist 6-28: Residential Finish Inspection: Mobile/Manufactured Homes

✔	Item	Inspection Activity	*NEC* Reference	Comments
	1.	Determine whether the unit is a mobile or manufactured home.	550.2	
	2.	Verify that the mobile/manufactured home supply system is rated 120/240 volts nominal, single-phase.	550.30, 550.32(C)	
	3.	Review the mobile/manufactured home park load calculations, and verify that demand factors, if used, have been properly applied.	550.31	
	4.	Verify that mobile/manufactured home service is rated not less than 100 amperes.	550.32(C)	
	5.	Verify location and minimum mounting height of mobile/manufactured home service equipment.	550.32(A), (B), and (F)	
	6.	Check for proper grounding of mobile/manufactured home service equipment.	Article 250, 250.32, 550.16, 550.32, 550.33	
	7.	Verify that a means to supply other structures or electrical equipment has been provided within the mobile/manufactured home service equipment.	550.32(D)	
✔	8.	Check for GFCI protection on all 125-volt, 15- and 20-ampere receptacles that are installed in addition to receptacles installed as part of the permanent mobile or manufactured home wiring. Verify that GFCI protection device(s) for "field-installed" receptacles is readily accessible.	210.8(A) 550.32(E)	
	9.	Verify that the feeder to the mobile/manufactured home has four insulated color-coded conductors.	550.33(A)	
	10.	Verify that feeders installed at mobile/manufactured home lots have a minimum capacity of 100 amperes at 120/240 volts.	550.33(B)	
	11.	Verify that new electrical installations that are added to existing mobile/manufactured homes comply with applicable requirements of Article 550.	Article 550	

CHECKLISTS

CHAPTER 7

Commercial and Industrial Inspections

OVERVIEW

This chapter covers the types of electrical **equipment** that appear most commonly in industrial and commercial occupancies, such as motors, heating and air-conditioning equipment, **transformers**, and capacitors. Although all of these types of equipment (with the possible exception of capacitors) are also frequently used in dwelling units, the equipment installed in dwelling units is generally smaller and less numerous than in commercial and industrial occupancies. In fact, in dwelling units, most motors other than air-conditioning motors are less than 1 horsepower (hp), frequently have integral overload protection, and are often installed on general-purpose **branch circuits**. Few dwelling units use more than one voltage system for power and lighting. Therefore, most transformers in dwelling units are small, inherently power-limited, Class 2 power supplies used for controls, alarms, and signals, or are part of other appliances or equipment that are supplied from a 15- or 20-ampere general-purpose branch circuit. Air-conditioning equipment in dwellings is usually single-phase and relatively small, with the selection of circuit elements based only on nameplate information.

This chapter concentrates on those applications of motors, transformers, capacitors, and equipment with refrigeration compressors where special rules for conductor sizing and application of overcurrent devices are frequently used. It also covers some specific types of equipment that may be found in residential occupancies but that are more common in commercial or industrial occupancies: elevators, dumbwaiters, escalators, platform lifts (formerly called wheelchair lifts), stairway chairlifts as covered in Article 620, and electric-vehicle-charging systems as covered by Article 625. In addition, this chapter covers **signs** and **outline lighting** as covered by Article 600, a type of utilization equipment that is very common in commercial occupancies, sometimes used in industrial occupancies, but uncommon in residential occupancies. In fact, each commercial building and occupancy must have at least one sign outlet supplied by a dedicated branch circuit rated not less than 20 amperes to supply a sign or outline lighting, regardless of whether a sign, outline lighting, or both, are actually installed.

SPECIFIC FACTORS

As noted, some of the equipment covered by this chapter may also be found in residential installations (see the Dwelling Units and Mobile/Manufactured Home Sites chapter of this manual). Motors, transformers, air-conditioning equipment, and capacitors are found in larger sizes and greater numbers in commercial and industrial facilities than in dwellings. Whereas elevators and electric-vehicle-charging equipment may also be found in dwellings, they are less common and, usually smaller than those found in commercial installations, except in large, high-rise multifamily dwellings.

Each of these types of equipment is covered by special rules for installing and protecting the equipment and circuits. Load calculations and conductor sizing requirements are different for motors and air conditioners than they are for other types of equipment. Transformers have special rules for their protection and installation and in some cases require specially built dedicated vaults. Conductors in capacitor circuits are sized by a different rule than are most other conductors. Therefore, characteristics unique to capacitors, such as the requirement to discharge the stored energy, are addressed in Article 460 of the *National Electrical Code® (NEC®)*.

Capacitors covered by Article 460 are also unique in another way: They are often not used to do any useful work, but rather to improve the power factor of the power by a facility, thereby reducing the costs of consuming power. Since a low power factor requires larger conductors and more capacity for a given power load, this often results in higher power bills. (Some capacitors are used to store energy for specific functions, typically as part of other equipment.)

Another significant difference in the rules for the equipment covered by this chapter is that the ratings of some **overcurrent** devices are not required to correspond directly with the conductor ampacity or equipment ratings. In the case of motors covered by both Articles 430 and 440, the functions of overcurrent devices are often separated into two devices, one that provides short-circuit and ground-fault protection and a separate device that provides overload protection. Even where a single device is used to provide both types of protection, the rules for each type of protection are found in separate parts of the articles.

The need for separating the functions of overcurrent devices is created by a basic characteristic of motors: They tend to draw a large amount of current for a short time when they are started. The magnitude of this current may be many times the normal operating current of the motor. **FIGURE 7-1** shows the level of current from start to 100 percent running speed for motors with three different locked-rotor indicating letters. The code letters indicate the locked-rotor kilovolt-amperes per horsepower (kVA/hp). By using this information, the approximate starting, or "inrush," current can be determined. This high current, usually called the *locked-rotor current* (in Article 430), the *stalled-rotor current*, or the *starting current* with regard to motors, or the *inrush current* with regard to inductive loads generally, must be allowed to pass without causing the overcurrent device(s) to operate. Therefore, one part of the overcurrent protection must allow the inrush current while still operating very quickly to clear a fault current.

At the same time, the other part of the overcurrent protection must respond to **overload** currents that are over the normal running current. To accomplish this task, two devices are often used—one (the short-circuit and ground-fault device) to operate very quickly on the high levels of current that might occur during a fault, and the other (the overload device) to operate more slowly on levels of current that are only a little more than the normal operating current. Both devices can be selected to ignore the inrush current and allow the motor to start while still performing their required functions. (Transformers and other inductive loads such as relay coils also have an inrush current, but the current in such equipment stabilizes more quickly than motor inrush currents, which only drop as the motor accelerates.)

FIGURE 7-2 is a one-line diagram showing a motor circuit from main distribution to the motor. Note that the overcurrent protection for the motor branch circuit is provided by two devices. The motor branch-circuit short-circuit and ground-fault protective device may be ahead of, part of, or a function of the disconnect or may be part of a combination **controller** that includes a disconnect.

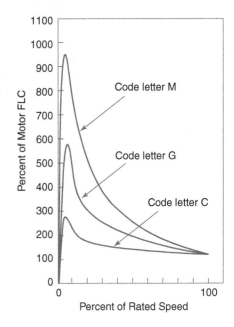

FIGURE 7-1   Approximate starting currents of alternating current (ac) motors.

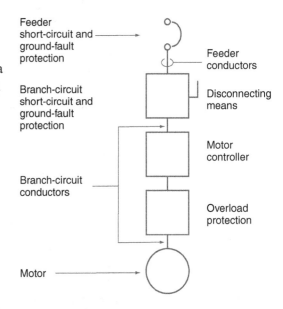

FIGURE 7-2   Typical motor-circuit components.

The overload device is commonly part of the controller. This arrangement allows for the starting current of the motor but also provides protection against short circuits, ground faults, or overloading in the conductors and motor.

Article 450 covers the protection and installation of transformers only. Although it does include footnotes to remind *Code* users of the need to protect conductors in accordance with other articles, Article 450 does not actually address the protection of transformer conductors. Transformer conductor sizing and protection are covered elsewhere in the *Code,* primarily in Articles 215 and 310 for the feeder supplying a transformer and 240 for the overcurrent protection of the primary and secondary conductors.

Elevators, escalators, and the like are specialized applications of motors. The motors themselves are treated much the same as other motors. However, the equipment covered by Article 620 also requires exclusive spaces and has specific rules for equipment in those spaces, and the manufacturer or installer of the elevator or escalator often specifies overcurrent device sizes and sometimes conductor sizes. Electric-vehicle-charging equipment has been around for many years, primarily for charging vehicles such as fork trucks and golf carts. However, with the increased use of other electric vehicles, requirements for the equipment and spaces where this charging takes place have been developed and are included in Article 625. Article 625 covers only those electric vehicles that are intended for road and highway use. This chapter includes checklists for use with both Articles 620 and 625.

This chapter focuses primarily on equipment rated 1000 volts or less, although some checklist items cover equipment that operates at over 1000 volts. Even so, Articles 430, 440, 450, and 460 are broad enough and flexible enough to cover virtually all applications of motors, hermetic refrigeration motor-compressors, power transformers, and capacitors, regardless of size or voltage rating. (Certain types of transformers, such as those used for controls or as parts of other equipment, are covered by other articles or by product listing and certification standards.)

KEY QUESTIONS

1. What are the horsepower ratings of motors, which ones are high-efficiency types, and what are the design letters?

Sizes or ratings of conductors, disconnects, controllers, and branch-circuit and feeder overcurrent devices are all based on the horsepower ratings of motors. The full-load current (FLC) values that are actually used in motor calculations are taken from tables that are based on horsepower. Energy-efficient design B motors may require (or allow) higher-rated overcurrent devices (specifically for instantaneous trip circuit breakers) to accommodate their higher inrush currents. Most new motors are energy-efficient types.

2. What are the nameplate full-load ampere (FLA) ratings of motors?

Generally, only the overload protection for a motor is based on the nameplate rating, because the FLC ratings from the tables are usually somewhat higher than the nameplate rating. In those uncommon cases where nameplate ratings are higher, they must be used instead of the table values.

3. Are there disconnects installed at the controller and at the motor locations?

A disconnect is almost always required in sight of a controller. If it is capable of being locked in the off position, another disconnect at the motor may not be required. Except in some industrial occupancies, disconnects are usually required in both locations. A **disconnecting means** is required to be installed within sight of all air-conditioning and refrigeration equipment. A limited exception to this rule applies to continuous industrial cooling processes.

4. What type of overcurrent device is used for short-circuit and ground-fault protection?

The tables of multipliers used to determine maximum ratings of short-circuit and ground-fault protection are based on the type and design of the motor and the type of overcurrent device used.

5. What are the minimum circuit ampacities and maximum overcurrent device ratings marked on air-conditioning and refrigeration unit nameplates?

These values dictate the minimum conductor size and the maximum overcurrent device rating permitted by the *NEC.* The two values usually do not match. The nameplate values represent the calculation required by the *Code,* but the calculation is already done by the manufacturer and need not be recalculated by the designer or installer.

6. Do the nameplates on hermetic refrigerant motor-compressor equipment specify a certain type of overcurrent device?

If fuses, **circuit breakers**, certain types of fuses, or certain types of circuit breakers are specified on the equipment nameplate or in the manufacturer's installation instructions, they must be provided as specified. The nameplates on some smaller units will usually contain such wording as "maximum fuse or circuit breaker" followed by a standard rating. Some larger units may specify only one type of protective device. Some equipment may indicate a specific type or class of fuse or circuit breaker.

7. What are the impedances, kVA ratings, insulation class, and primary and secondary current and voltage ratings marked on transformer nameplates?

This information is needed to determine maximum overcurrent device ratings for transformers. The impedances are needed only for transformers over 1000 volts. Given this information, the tables in Article 450 can be used to find the appropriate multipliers of the current ratings to determine overcurrent device ratings. The kVA and voltage ratings and the insulation class are used to determine whether clearances from combustible materials apply or whether a fire-resistant room or vault is required.

8. If ventilation clearances are marked on transformers, what are they?

Transformers that are required to have ventilation are also required to be marked with ventilation clearances. Usually, these clearances are roughly the same as the size of the ventilation openings. Encapsulated and nonventilated transformers often do not require any particular clearance and can be mounted directly on or against walls or other structures. The insulation class also helps in determining whether or not clearances are required from combustible construction.

9. Are any transformers liquid-filled, and if so, what type of liquid is used?

Most liquid-filled transformers are installed outdoors or in vaults. Fluid containment areas are sometimes required as well. However, some types of fluid get special treatment under the *Code.*

10. What are the current ratings of capacitors?

Capacitor current ratings are used to determine conductor sizes and disconnect ratings.

11. What means of discharge is provided on capacitors?

Discharge must be **automatic**, either permanently connected or automatically connected, and must bring the voltage down to 50 volts in a specified time. The time varies according to the voltage rating of the capacitor.

12. What are the liquid capacities of capacitors?

If any individual capacitors in an installation contain more than 3 gallons (11.4 liters) of a flammable liquid, they must be installed in vaults or outdoors.

13. Where will elevators or escalators be located?

The machine spaces for **elevators**, escalators, dumbwaiters, and the like are invariably dedicated spaces, and little if any foreign equipment is permitted to be in or pass through such spaces. Also, specific types of equipment must be located in the machine spaces.

14. What is the nameplate information for the elevator or escalator equipment?

Elevators and escalators have specific requirements for conductors, overcurrent devices, and disconnecting means that are dictated by equipment suppliers. Horsepower ratings alone are not usually sufficient for sizing circuit elements.

15. Where will electric-vehicle-charging equipment be located and what types will be used?

Electric-vehicle-charging equipment may be installed in a variety of areas, but the requirements for these spaces, especially ventilation requirements, vary according to the types and numbers of storage batteries.

16. Where will signs or outline lighting be located and what are the electrical ratings?

Commercial occupancies are required to have outlets and circuits installed for the use of signs and outline lighting. Dedicated circuits (circuits that supply no other load) are required for this purpose. The *NEC* does not require a sign to be installed, but certain minimum provisions are required. If the requirements for a sign are greater than the minimum provisions, additional or larger supplies must be provided.

17. What types of signs or outline lighting will be used?

In some cases, outline lighting consists only of ordinary luminaires, which are covered primarily by Article 410 and are required to be listed by those rules. Other signs are self-contained or sectional signs that are required to be listed under Article 600. Neon signs installed as skeleton tubing in the field are not required to be listed as a whole, but many requirements from Article 600 apply, and many elements of such installations are required to be listed. The listing and labeling of the lighting sources is needed to judge such installations.

PLANNING FROM START TO FINISH

The types of equipment covered by this chapter tend to be located throughout commercial and industrial facilities. Therefore, no one recommended starting point will work for all facilities. For each type of equipment, an effective starting point is the origin of the circuit for the specific equipment. The following are suggestions for inspecting the basic types of equipment.

Motors

Once the motor information has been obtained, as outlined in the Key Questions checklist, the **motor control center (MCC)** is a good starting point. However, not all projects have MCCs. In cases where there is no MCC, the individual **motor controller** locations or the panelboards or switchboards that supply the controllers are good places to start. The controllers, which are also included in MCCs, provide easy access to the circuit conductors, the controller itself, **control circuits**, the controller disconnecting means, and the overload protection. Therefore, all of these components of the motor circuit can be inspected in one stop. Some of the key questions concerning disconnects and overcurrent devices can also be answered at this stop. For example, the branch-circuit short-circuit and ground-fault protective device can usually be found in MCCs or combination starters. In other cases, the branch-circuit overcurrent device may be found in a panelboard or switchboard. Where adjustable-speed drives are used, most of the required information is based on the nameplate rating of the drive (the power conversion equipment).

Air-Conditioning and Refrigerating Equipment

The branch-circuit panels supplying air-conditioning and refrigeration equipment are good places to start inspections of such equipment. At this point, note the overcurrent device types and ratings along with the conductor sizes for each item of equipment. This information can then be matched to the nameplate information at each equipment location. While at the equipment locations, also check the disconnecting means location and rating, the locations of receptacles and lighting for servicing, and, where the equipment is not packaged under a single nameplate, other equipment-related items such as controllers and overload devices.

Transformers

Make a note of overcurrent device ratings for circuits feeding transformers while inspecting panels and other distribution equipment. The key questions about transformers can then be answered with the aid of the transformer nameplate, and the noted ratings can be checked for compliance. Other installation issues can also be checked at the transformer location. Some inspection points related to transformer installations are covered in the Services, Feeders, and Branch Circuits chapter and the Grounding and Bonding chapter of this manual. An efficient inspection plan includes an examination of these related requirements while looking at the transformer installation.

Capacitors

Capacitor installations covered by Article 460 of the *NEC* are usually found in two variations: either installed in large sets near service equipment or MCCs (often switched in banks by automatic controls) or installed as small individual capacitors or capacitor sets at the point of use or on the load side of a controller where they are connected to the load at all times. If they are installed in large sets, the entire installation can be inspected at one location, with the possible exception of the overcurrent device, which will likely be nearby. If they are installed individually, they are usually found at individual motor controllers. In that case, they are most easily inspected as part of the motor inspection.

Elevators, Dumbwaiters, Escalators, Platform Lifts, and Stairway Chairlifts

Most of the equipment installed under Article 620 will be located in dedicated spaces of a building, such as the machine room and the shaft or area where the elevator or escalator is installed. Therefore, the inspection of such systems logically starts in the machine room and proceeds from there. The machine room is also the location where equipment controllers and their nameplate data are likely to be found. Frequently, more than one jurisdiction has a say in such installations, as many states or other governing bodies assign the responsibility for elevator and escalator installations and safety to a single specialized agency. Even dumbwaiters that are not intended to carry people are often under the jurisdiction of the "elevator inspector." Usually, these other inspections are done using both the *NEC* and the American National Standards Institute (ANSI) 17.1, *Safety Code for Elevators and Escalators*.

This ANSI standard provides many other requirements that are not within the scope of the *NEC*. For example, the *NEC* provides requirements for lighting outlets and switches in machine spaces, but the required illumination levels are found in ANSI 17.1. Requirements for elevator smoke and heat detectors, sprinklers, and elevator recall in emergencies are covered by various standards, including the building code, the fire alarm code, the sprinkler code, the elevator code, and the *NEC*. These various requirements must be carefully coordinated to ensure compliance with all relevant standards.

Electric-Vehicle-Charging Equipment

Electric-vehicle-charging equipment usually is installed in well-defined areas, typically areas where the vehicles are parked and stored, and most of the aspects of such installations can be found and inspected in those areas.

Signs and Outline Lighting

As with other equipment, inspections of signs can begin at the power source, which may be either feeders for some large signs, such as those advertising multiple retail occupancies at a shopping mall, or branch circuits for signs for individual tenants within the mall. Where the overcurrent devices for the sign are equipped with permanent locking means, they may also be the disconnects for signs and controllers. In any case, the sign itself will be the location for checking compliance with listing requirements, load ratings, and other details of the installation and may also be the location of disconnects and controllers. For skeleton tubing that is not listed as an assembly, many more aspects of the installation, including the load rating, must be checked at the sign. A separate checklist covering those details follows the other checklist for Signs and Outline Lighting.

WORKING THROUGH THE CHECKLISTS
Checklist 7-1: Motors

1. Verify that ampacities and sizing of components other than overload devices are based on table values rather than on nameplate values.

Article 430 provides tables of full-load current (FLC) values for motors of various characteristics. The current values from the tables must be used in calculating values for conductor ampacities, short-circuit and ground-fault devices, disconnect ratings, and so on. The full-load ampere (FLA) rating provided on the nameplate of a motor is used only for sizing overload protective devices for the motor. In the uncommon event that the FLA value from a nameplate is greater than the corresponding FLC value from the table, the nameplate FLA value is used in place of the table value. This is most likely to be true with motors built for low speeds or high torques, such as some direct drive pumps. FLA values will be used for all motors larger than 500 hp as well, simply because the tables do not provide FLC values for motors larger than 500 hp. **FIGURE 7-3** illustrates how the nameplate current value and the table current value for a motor are most commonly used to size circuit components for general applications. For torque motors, the nameplate is used because the motor normally operates in a locked-rotor condition. Calculations for alternating current (ac) adjustable-voltage motors and adjustable-speed drive systems are based on the nameplate of the controller or power conversion equipment. The distinction used in this book between the acronyms FLC for full-load current and FLA for full-load amperes is not a technical distinction, but just a convenient and commonly used way of distinguishing between these values.

2. Verify that conductor ampacities for individual motors are at least 125 percent of table FLC.

Conductors supplying a single motor must have an ampacity not less than 125 percent of the FLC rating from the appropriate table in Article 430. In some cases, where motors are expected to operate in a **short-time** or **varying-duty** application, the conductor size may have to be increased to more than 125 percent of FLC, depending on the duty rating of the motor. If a motor is expected to operate in an intermittent

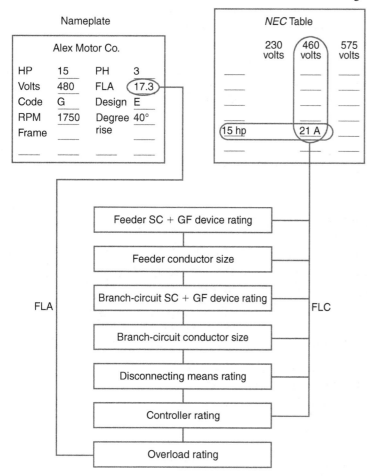

FIGURE 7-3   Use of FLC and FLA.

or **periodic-duty** application, the conductors may be sized at more or less than 125 percent of FLC, depending on the duty rating of the motor. Most motors are considered as **continuous duty**, and 125 percent is the multiplier most commonly applied. Conductors for adjustable-speed drive systems generally must be rated at 125 percent of the rated current input to the power conversion equipment. In some cases, the manufacturer of an adjustable-speed drive will specify a larger minimum conductor size and those instructions must be followed.

3. Check conductors supplying multiple motors or motors and other loads for ampacities equal to at least the sum of other motors or loads plus 125 percent of largest motor.
Conductors supplying more than one motor are required to be sized at not less than 125 percent of the FLC rating of the largest motor plus 100 percent of the FLC rating of any other motors supplied by the same conductor. This rule applies to any conductors of branch circuits, feeders, or services that provide power to more than one motor. **FIGURE 7-4** shows a motor feeder and motor branch-circuit conductors that have been computed in accordance with Article 430. If a conductor supplies both motor and non-motor loads, 100 percent of the noncontinuous non-motor load plus 125 percent of the continuous non-motor load is added to this calculation. This minimum conductor rating is also considered to be the calculated load under Article 220.

4. Verify that motor overload protection does not exceed permitted values.
Motor overload protection is often selected from a table of overload devices based on the FLA rating from the motor nameplate. Where solid-state overload devices are used, such as the solid-state overload relay shown in **FIGURE 7-5**, the setting may be adjusted to the motor nameplate FLA current rating. If fuses are used for overload protection, the usual maximum rating of the fuse is 125 percent of the nameplate FLA current. In some cases, the overload

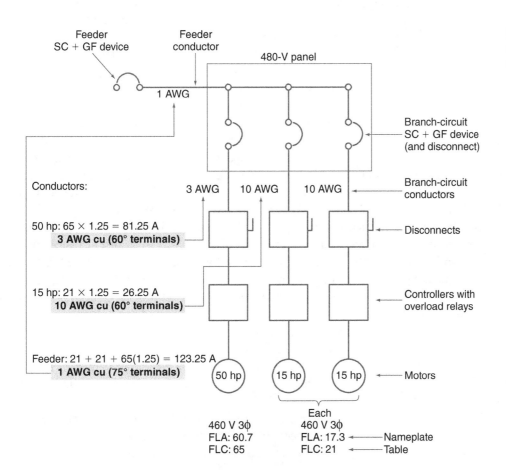

FIGURE 7-4 Calculating motor feeder and branch-circuit conductors.

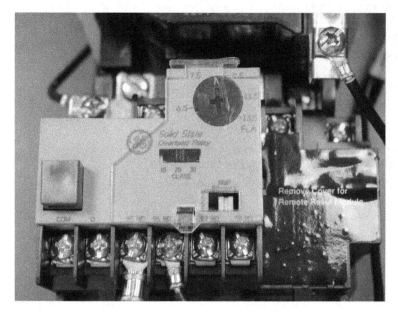

FIGURE 7-5   Solid-state overload relay with adjustable trip settings.

THERMALLY PROTECTED-AUTO				J&B MOTORS
MOD 34560D	HP 1 1/2	RPM 3450		
V 115/230		FLA 18/9		
FR 56J	HZ 60	PH 1		KVA CODE K
INS B	MAX AMB 40 °C	DUTY CONT. DP		
SF 1.3	SFA 22/11	BRGS BALL		
MTR.REF. 12356811139	DATE CODE A87			
TYPE N THERMALLY PROTECTED				

FIGURE 7-6   Motor nameplate marking.

rating may be increased to get a motor started. The device shown in Figure 7-5 will provide protection at 120 percent of the setting and can be adjusted to provide Class 10, Class 20, or Class 30 protection; that is, it can be adjusted to operate in 10, 20, or 30 seconds on an overload of 120 percent of the setting. Motors that are **thermally protected** do not require separate overload devices. As shown in the example nameplate in **FIGURE 7-6**, a motor that is thermally protected must be marked to indicate this feature. Unless the motors are specifically rated for variable-speed operation, motors that are part of adjustable-speed drive systems must be protected from the overtemperature that may be caused by running the motor at reduced speed.

5. Verify that short-circuit and ground-fault protection for motor branch circuits does not exceed permitted values.

Short-circuit and ground-fault protective devices for motor circuits are allowed to exceed the ampacity of the conductors. Whereas the conductors are required to be sized at no less than 125 percent of motor FLC, short-circuit and ground-fault protection in the form of fuses or circuit breakers is allowed to be rated from 150 percent to 1700 percent of motor FLC, depending on the type of motor and the type of overcurrent device that is used. The most commonly used maximum values are 175 percent of FLC for dual-element fuses or 250 percent of FLC for ordinary **inverse-time circuit breakers**. If these calculated maximum ratings do not correspond to the standard ratings listed in 240.6, the next larger standard rating may be used. The next higher size is seldom necessary to allow a motor to start. Circuit breakers may be rated at about twice the ampacity of the conductors they protect under this rule. This increase in rating allows motors to start without the opening of overcurrent devices due to the normal inrush current of the motor. Remember that these devices provide only short-circuit and ground-fault protection for the motor circuit.

Overload protection is provided separately to protect both motors and conductors from overheating due to overload. It is important to recognize that these are maximum settings, and it is acceptable, often desirable, to use lower-rated devices. **FIGURE 7-7** illustrates short-circuit and ground-fault devices that have been selected on the basis of the maximum values permitted by the *NEC* for an inverse-time circuit breaker. It is not required to size the devices at the maximum permitted value. Figure 7-7 also shows that the overload device selection or setting is based on the nameplate current value or FLA, not on the *NEC* table value or FLC. The size of the short-circuit and ground-fault protective device may also be limited by the instructions for overload devices where a motor controller is marked with the maximum short-circuit and ground-fault protective device that can be installed on its supply side.

In the 2014 *NEC* there are new requirements for branch-circuit short-circuit and ground-fault protection where power conversion equipment is used to control a motor(s). The approach is similar to that used for motors with electromechanical or electronic

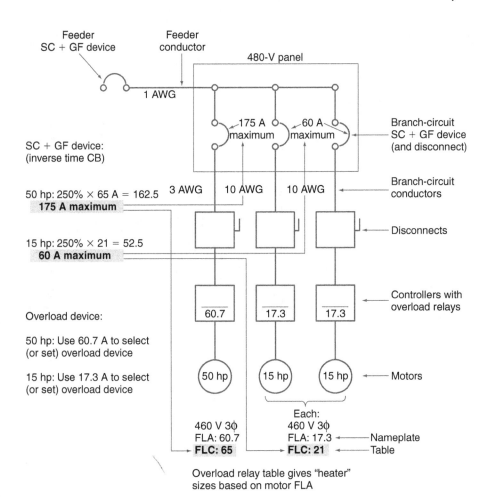

FIGURE 7-7 Calculating motor branch-circuit short-circuit and ground-fault protection and motor overload protection.

controllers that do not involve power conversion. The requirements in 430.52 are to be followed, and those rules for sizing the protective device are based on the type of device used in a specific motor circuit. Additionally, some power conversion units may be marked with a maximum protective device rating; in such cases, the default maximum rating is that marked on the power conversion equipment.

6. Verify that short-circuit and ground-fault protection for motor feeders does not exceed permitted values.

The overcurrent device protecting feeder conductors that supply motor loads is also allowed to have a rating higher than the ampacity of the conductors. In the case of feeders, the motor with the highest-rated branch-circuit short-circuit and ground-fault protective device is separated from the rest of the load. The maximum rating of the feeder overcurrent device is the sum of all of the other loads including motors plus the rating of the largest short-circuit and ground-fault protective device for any motor. This rule allows conductors to be sized in accordance with Article 430, but also allows the feeder overcurrent device to be large enough to start the largest motor while all other loads are connected and running.

It should be noted that this sizing rule does not permit rounding up to the next higher rating or setting for the short-circuit and ground-fault device protecting the feeder circuit. Rounding up may be applied to the branch-circuit device as shown in Figure 7-7, but rounding up from the 217 amperes shown in **FIGURE 7-8** is not permitted. The feeder device may have a higher rating if the conductors are also increased in size so that the conductors are protected at their ampacity. These rules also apply to service conductors according to 230.90(A), Exception No. 1. A feeder supplying several motors is illustrated in Figure 7-8.

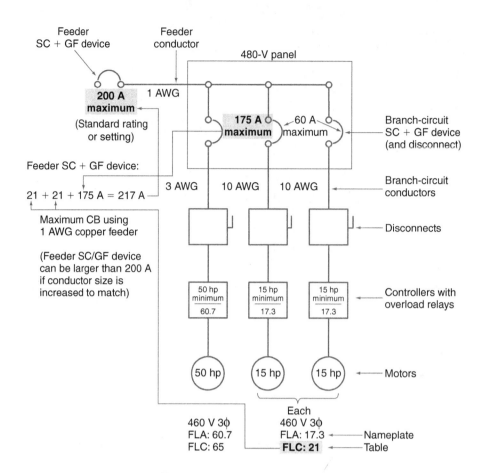

FIGURE 7-8 Calculating motor feeder short-circuit and ground-fault protection.

The rating of the device exceeds the conductor ampacity, but this protection is for short circuits and ground faults only, and the overload protection for the feeder is provided by the individual branch-circuit overload devices provided for each motor supplied by the feeder. Although a motor feeder overcurrent device is permitted to exceed the ampacity of the feeder conductors, a device that protects an MCC supplied by a feeder or used as service equipment is not permitted to exceed the rating of the common power bus of the MCC (see Item 9 in this section).

7. Check motor control circuits for proper overcurrent protection.
The required overcurrent protection for motor control circuits consists primarily of short-circuit and ground-fault protection. The loads on motor control circuits are usually quite limited, and there is little risk of overloads on such circuits. Therefore, in many cases the same device that protects the branch circuit may also serve to protect the control circuit. The maximum rating of an overcurrent device in such a case depends on the size of the control circuit conductors and on whether or not the control circuit conductors leave the controller **enclosure**. Higher ratings are permitted for larger conductors and for conductors that do not extend beyond the controller where they receive their power. Control circuit conductors that do leave the controller are more subject to damage and are likely to form a higher impedance circuit than conductors that remain within the controller. Supplemental fuses can be used to provide the protection when control conductors are too small to be protected by the motor circuit device. Primary fuses on control power transformers with a two-wire secondary may also provide protection for the secondary control conductors.

These rules are based on the assumption that the control circuits are tapped from the load side of the branch-circuit short-circuit and ground-fault protective device. Motor control conductors that are supplied in other ways, such as from a central control panel or from a remote-control power supply, must generally be protected at their ampacities or protected as otherwise permitted in Article 725 for remote-control circuits. Motor control circuits fit into the definition of Class 1 circuits in most cases, but tapped motor control circuits are treated a bit differently with regard to overcurrent protection than other similar control circuits. Other rules for Class 1 circuits from Article 725, such as conductor sizes, insulation types, and separations from other conductors, still apply to motor control circuits. **FIGURE 7-9** shows several approaches to motor control circuit overcurrent protection.

8. Verify that motor controllers are provided for motors and that they are of proper type and have adequate ratings, including short-circuit current ratings.

A means must be provided to start and stop each motor. The usual type of device is a listed controller or "starter" that is rated in horsepower. The horsepower rating of the controller, based on the voltage applied to the motor, must be at least equal to the horsepower rating of the motor it controls. Controllers that do not have horsepower ratings must have adequate current ratings and be rated for the use. Circuit breakers or molded case switches may be used as controllers, in which case their minimum rating would be based on the ampere rating of the switch or circuit breaker. This rating would have to be at least as large as would be required for a disconnect for the same motor, normally at least 115 percent of the motor FLC rating.

Special rules may apply to small portable motors and stationary motors not over 2 hp. Depending on the horsepower rating, such motors may be controlled by a branch-circuit overcurrent device, by a cord-and-plug attachment on a ⅓-hp or smaller portable motor, by a **general-use switch** rated not less than 200 percent of the motor FLC rating, or by a **general-use snap switch** rated not less than 125 percent of the motor FLC rating. The determination of whether a motor controller has been sized properly can easily be accomplished by checking the information that is required to be marked on the motor controller. **FIGURE 7-10** shows the markings that are found on a motor controller, including its horsepower rating. One of the markings required for motor controllers is the short-circuit current rating, which in Figure 7-10 is shown in the lower label (both of these labels are from the same controller). However, in this case, the short-circuit current rating shown applies only to the overload relay, and the short-circuit rating for the contactor is not shown. The short-circuit current, at the point in the premises wiring system where the controller is installed, cannot exceed the rating marked on the controller.

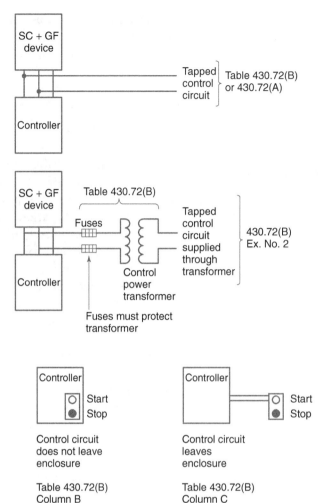

FIGURE 7-9   Motor control circuit overcurrent protection.

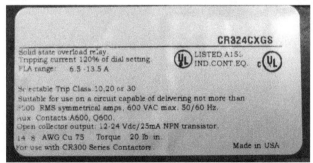

FIGURE 7-10   Motor controller marking.

9. Check MCCs for proper ratings, protection, workspace, and dedicated space.
MCCs must have ratings adequate for the load they supply. Overcurrent protection must be provided that does not exceed the rating of the common power bus in the MCC. The main overcurrent protection may be either ahead of or an integral part of the MCC. Workspace, headroom, and dedicated space requirements for MCCs are the same as those for switchboards, switchgear, and panelboards. Workspace and dedicated space requirements for such equipment are covered in the General Requirements Inspections chapter of this manual.

10. Verify that motor disconnects are of the proper type and rating.
Motor disconnects need to be capable of making and breaking the locked-rotor (inrush) current of a motor. The *NEC* provides for at least 10 different types of devices that may be used as disconnects in various situations. Some, such as isolating switches or general-use switches, are limited to certain types of installations and certain sizes of motors. Because listed manual motor controllers are available only in relatively small sizes, their use as disconnects is similarly limited and is permitted only where they are also marked as suitable for use as motor disconnects. Most manual motor controllers are not listed and tested as disconnecting means, because disconnects are required by product standards to have greater internal clearances than controllers.

Subject to availability, the more commonly used devices that may be used as disconnecting means for motors of any size are listed **motor-circuit switches**, listed molded-case circuit breakers, listed molded-case switches, **instantaneous-trip circuit breakers** that are part of listed combination controllers, listed self-protected combination controllers, and attachment plugs and receptacles. Motor-circuit switches must have horsepower ratings. Except for some motor-operated appliances, room air conditioners, and small portable motors, attachment plugs and receptacles used as disconnects must also have horsepower ratings. The other permitted types of devices are not required to be rated in horsepower but must have ampere ratings at least equal to 115 percent of the motor FLC rating.

Article 430 provides requirements that set maximum ratings and minimum ratings for components in a motor circuit. There may be a number of device ratings that could be used as the short-circuit ground-fault protection and the disconnecting means. Article 430 establishes maximum ratings for motor short-circuit and ground-fault protective devices, and minimum ratings for motor disconnects, but some devices, such as a circuit breaker, could be both. The choice of the device rating, providing it is within the maximum and minimum limits, is up to the system designer or installer. FIGURE 7-11 illustrates circuits with two disconnecting means, one in the form of a circuit breaker in the 480-volt panel and one in the form of an unfused motor circuit switch at the controller and motor, and shows both minimum and maximum ratings for some devices.

If the disconnect is for more than one motor, the equivalent horsepower rating can be determined according to 430.110 using Tables 430.251(A) or (B), which provide a method to convert the total full-load currents and locked-rotor currents to an equivalent horsepower. The equivalent horsepower rating is not just the sum of the horsepower ratings of the motors. For example, consider two 10-hp, 460-volt, 3-phase motors. From Table 430.250, the full-load current for one of these motors is 14 amperes, so the equivalent full-load current is 28 amperes. But according to Table 430.250, 28 amperes corresponds to a motor slightly more than 20 hp. From Table 430.251(B), the locked-rotor current for one motor is 81 amperes, so the total locked-rotor current is 162 amperes for an equivalent horsepower rating from Table 430.251(B) of about 22 hp. The larger value is used, so considering likely available standard ratings, a motor circuit switch rated at 25 hp at 460 volts would be the minimum size.

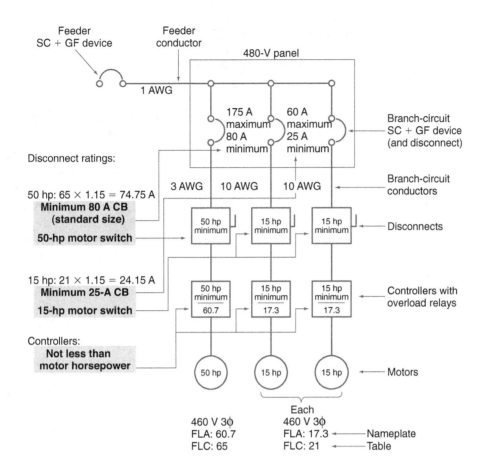

FIGURE 7-11 Calculating motor disconnect and controller ratings.

11. Verify that controller disconnects are in sight of controllers, are readily accessible, and have adequate workspace.

Disconnects must be located **in sight from** all motor controllers. One true exception to this rule that allows the disconnect to be located elsewhere applies to motor circuits over 1000 volts. A single disconnect may be used for a group of coordinated controllers where both the disconnect and controllers are in sight of the multimotor apparatus they serve. This exception modifies the requirement for a separate disconnect for each controller but does not change the requirement about being in sight of each controller. If the disconnect is likely to be worked on or examined while **energized**, *workspace and access to the workspace* must comply with the same rules that apply to MCCs, switchboards, switchgear, and panelboards as discussed earlier and in the General Requirements Inspections chapter of this manual, but the *dedicated space* requirements do not apply to motor-circuit disconnects unless they are part of a switchboard, switchgear, panelboard, or MCC.

Many motors and controllers are located where they are not **readily accessible**, such as above suspended ceilings or in similar spaces. Disconnects that are mounted adjacent to such controllers and motors are not required to be readily accessible, but some disconnect that is readily accessible must be provided. The additional readily accessible disconnect could be the branch-circuit overcurrent device. The working clearance and dedicated space rules for switchboards and panelboards also apply to MCCs (**FIGURE 7-12**). Motor disconnecting means and service switches that are not likely to be examined or serviced while energized must be provided with "sufficient access and working space" in order to facilitate their use.

FIGURE 7-12 Motor control center.

12. Verify that motor disconnects are in sight of motors, are readily accessible, and have adequate workspace.

As with motor controllers, motors themselves are required to have disconnecting means located in sight from the motors. A broad and commonly used exception that once applied to this rule is now significantly restricted. Now, only if locating a disconnect in sight of a motor is impracticable or would cause increased or additional hazards or if the motor is in an industrial setting with written safety (program, training, and lockout/tagout) procedures, a disconnect is not required to be in sight from a motor if the disconnect that is in sight from the controller is equipped with means of locking it in the off position. The means to lock the disconnecting means in the open position must be an associated part of the disconnecting means (factory-installed or manufacturer-sanctioned add-on hardware) at all times and cannot be in place only when a lock is installed (as is the case with many portable locking devices used for compliance with lockout/tagout requirements). This requirement for locking means has been generalized for all equipment requiring lockable disconnecting means and is now located in 110.25.

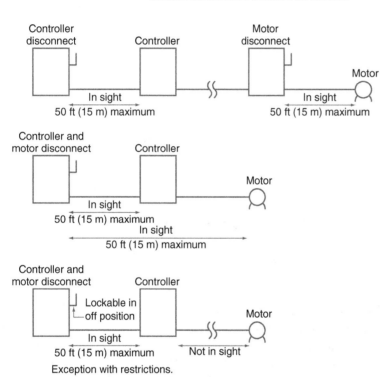

FIGURE 7-13   Location of disconnecting means for motor controllers and motors.

Workspace and access requirements for motor disconnects are generally the same as those for motor controller disconnects. The general rule requires a disconnect within sight of the motor controller in all cases (**FIGURE 7-13**). Where the required controller disconnect can be locked open and the restrictions of the exception are met, a disconnect within sight of the motor is not required. The exception includes Informational Notes that explain what is meant by increased or additional hazards and that refer to NFPA 70E®, *Standard for Electrical Safety in the Workplace*®, 2012 edition, for information on lockout/tagout procedures.

NFPA 70E also covers requirements for safety programs, training, and work practices that are part of the conditions expected in an industrial setting. One aspect of the *NEC* requirements is the expectation that electrical inspections will be performed in an electrically safe work condition, and the disconnect is a critical element in achieving that safe condition.

Checklist 7-2: Air-Conditioning and Refrigerating Equipment

1. Identify equipment subject to Article 440: Equipment Containing Hermetic Refrigerant Motor-Compressor(s).

Article 440 provides rules for a specific class of motors, those consisting of a motor and compressor enclosed in a single housing with no external shaft and with the motor immersed and operating in the refrigerant. Such motors are called "**hermetic refrigerant motor-compressors**" in the *NEC*. Most such motors are part of combination-load or multimotor machinery that includes other motors and sometimes includes electric or other types of heating equipment. Article 440 covers individual hermetic motor-compressors as well as the more common combination-load equipment. Although room air conditioners, water coolers, and household refrigerators and freezers contain hermetic refrigeration motors, they are considered appliances and are covered by Article 422. However, Article 440 does contain some requirements specific to room air conditioners. **FIGURE 7-14** shows the components of a circuit to an air-conditioning unit that are subject to the requirements of Article 440. Where the disconnecting means contains or is a branch-circuit-rated overcurrent device, the supply to the disconnect is classified as a feeder, and the branch circuit begins at the disconnect.

2. Identify the applicable nameplate information for the equipment.

As is mentioned in Item 1, most hermetic refrigerant motor-compressors come in combination-load (multimotor) equipment that includes other motors or other types of loads. The nameplate on such equipment provides circuit sizing and overcurrent selection information. Wire sizes for such equipment are based on what is usually listed as "minimum circuit ampacity" or "MCA." Overcurrent protection, which is usually only short-circuit and ground-fault protection, is specified as "maximum fuse or circuit breaker size" or some variation on those words. For example, the nameplate may specify fuses or may specify a circuit breaker. This nameplate information is based on the calculation methods given in Article 440, and no further calculations are usually necessary. Individual hermetic motor-compressors, where encountered, will list only "**branch-circuit selection current**" and/ or "rated load current." The larger of these two values can be used to determine minimum conductor sizes and maximum overcurrent device ratings in much the same manner as for ordinary motors that use FLC values. **FIGURE 7-15** shows the type of information that is found on the nameplate of packaged air-conditioning equipment. With this information, a supply circuit conductor and overcurrent device can be properly selected. In this example, "minimum circuit amps" is 16.5, and the maximum fuse or circuit breaker size is 25 amperes, so a 12 AWG conductor with a 20- or 25-ampere fuse or circuit breaker is indicated. Larger conductors are permitted, but larger overcurrent devices are not.

FIGURE 7-16 is another example of nameplate information. This example is an illustration from the outdoor unit of a heat pump. The minimum conductor rating in this case is 15 amperes, which could be a 14 AWG copper conductor. The maximum rating of the overcurrent device (fuse or circuit breaker) is 25 amperes. In this example, because the device is a heat pump and must start in cold conditions, the nameplate also provides a minimum rating of 20 amperes for the overcurrent device. Overload protection is provided by individual protection on the motors in the unit. The 15-ampere conductor rating is based on 125 percent of the rated load amps (RLA) value for the compressor, plus the FLC value for the outdoor motor, rounded up.

3. Verify that branch-circuit conductor sizes are adequate on the basis of the applicable nameplate information.

Given a nameplate value for MCA, a conductor can easily be chosen. The only complications would be terminal temperatures and correction or adjustment factors, if any apply. Unless the terminal temperature ratings are known, ampacities for up to 1 AWG conductors can be chosen directly from the 60°C (140°F) column of Table 310.15(B)(16) of the *NEC*. For larger conductors, or where terminal temperature ratings are known to be 75°C

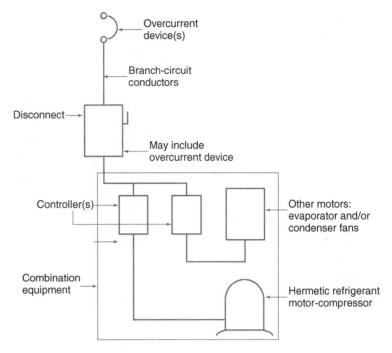

FIGURE 7-14 Components of a circuit to an air-conditioning unit.

FIGURE 7-15 Air-conditioning equipment nameplate.

Courtesy of Trane

FIGURE 7-16   Heat pump equipment nameplate.

(167°F), the ampacities can be chosen directly from the 75°C (167°F) column. The 90°C (194°F) column should be used only if temperature corrections and/or adjustments for more than three conductors are applied. The nameplate may specify the temperature rating of conductors and may specify the temperature column for selection of ampacity. As with other motors, the chosen wire size will not necessarily have an ampacity equal to the maximum overcurrent device rating.

4. Verify that conductors supplying several units are adequately sized.

Article 440 provides requirements for sizing branch-circuit components in a circuit that supplies refrigeration equipment. However, Article 440 does not provide specific rules for sizing feeders; therefore, according to 440.3(A), Article 430 must be used to size feeder circuits that supply multiple air-conditioning or refrigeration units. Most designers will take the sum of the MCAs from the various units to size a feeder. This results in a safe number, because the largest motor in each unit is counted at 125 percent, and the *Code* requires only the single largest motor on a feeder to be counted at 125 percent. Making the calculation according to the *Code* would require obtaining all the information about all the motors in each unit, adding them all up, and then adding 25 percent of the largest motor. Although this information is available on the nameplates, there is no need to gather all this information if the installation passes the simpler test of providing slightly oversized feeders with ampacities at least equal to the sum of the MCAs.

5.   Verify that branch-circuit overload protection is provided and properly sized.

Because most equipment covered by Article 440 is actually combination-load or multimotor equipment under a single nameplate, overload protection is usually provided within the equipment. Some motors are thermally protected, and others have separate overload relays or other devices. Some units even contain supplementary- or branch-circuit-rated fuses. The maximum size overcurrent device specified on the nameplate takes the overload devices into account. Overload protection provided by the electrician or otherwise provided outside the equipment is usually required only for separate hermetic refrigerant motor-compressors that do not have integral thermal overload protection and are not part of packaged, labeled equipment.

6.   Verify that branch-circuit short-circuit and ground-fault protection is provided and properly sized.

The nameplate provides the information needed to determine the maximum allowable size for branch-circuit short-circuit and ground-fault protection. Unless other minimum size information is provided, MCA can be thought of as the minimum size for the branch-circuit overcurrent device, provided that this size is adequate to carry the starting current of the motor. Figures 7-15 and 7-16 show LRA, or locked-rotor amperes (starting current) for each compressor, but overcurrent devices such as inverse-time circuit breakers and dual-element fuses that include some time delay at relatively low currents will allow the motors to start. Remember, these devices are intended to operate very quickly on the high current of a short circuit or ground fault, but that is expected to be many times greater than the starting current. Again, the size of this branch-circuit overcurrent device often does not match the ampacity of the branch-circuit conductors, and such an installation is permitted as long as the nameplate requirements are met.

7. Verify that feeder short-circuit and ground-fault protection is provided and properly sized where applicable.

Article 440 does not cover methods for determining feeder overcurrent device sizes. Therefore, the method given in Article 430 should be used to size feeder short-circuit and ground-fault protection. Section 440.3(A) states Article 430 applies except as amended in Article 440. If conductors are larger than required, the feeder overcurrent device can always be increased to match the ampacity of the feeder. This is the most common method of determining overcurrent protection ratings for feeders supplying multiple items of equipment containing refrigeration compressors, although in many cases, larger overcurrent ratings may be permitted.

8. Verify that controllers have adequate ratings, including short-circuit current ratings, where they are not part of listed multimotor or combination-load equipment.

Checking controller ratings is necessary only for individual hermetic motor-compressors. The internal parts of listed equipment do not have to be reinspected because they are evaluated to a product safety standard, not to the *NEC*; however, it is necessary to ensure that the equipment is not subjected to short-circuit currents higher than the rating of the internal components. In general, controllers for multimotor and combination load equipment are required to have a marked short-circuit current rating. Where controllers are supplied separately, they must have full-load and locked-rotor current ratings at least equal to the motor FLC (full-load current, also called rated load current or rated load amperes, abbreviated RLC or RLA) and locked-rotor current (usually abbreviated LRA) marked on the compressor. Controllers that are horsepower-rated may not be marked with all of the applicable current ratings. In that case, the equivalent current ratings for the horsepower rating can be determined from Article 430.

9. Verify that disconnecting means have ratings adequate for the equipment.

The requirements for sizing disconnecting means based on current values for a single hermetic motor-compressor are basically the same as those for other motors. However, if horsepower-rated switches are used, the equivalent horsepower rating of the compressor is needed. This can be obtained by converting the rated load and locked rotor currents to horsepower using the tables in Part XIV of Article 430.

For example, using the data from the nameplate shown in Figure 7-15, you can see that the unit has two motors. One (the small condenser fan) has a horsepower rating. The other is a hermetic refrigerant motor-compressor rated at 12.8 RLA and 58.3 LRA at 208/230 volts single phase. The equivalent horsepower rating of this motor-compressor is determined by using the RLA with Table 430.248 and the LRA with Table 430.251(A). In this example, the 12.8 RLA is closest to and not more than the full load current for a 2-hp motor from Table 430.248. The 58.3 LRA is closest to and not more than the LRA for a 1½-hp motor from Table 430.251(A). For more accuracy, interpolation may be used. We use the larger of these two values (2 hp) to determine the equivalent minimum horsepower rating for this compressor.

Because this is a combination load, the 1½-hp motor must be added in to get a minimum horsepower rating for the combination. Since the tables in Article 430 do not provide full-load or LRA for such a small motor, 430.110(C)(3) and 440.12(C) provide that the LRA will be considered to be six times the FLC or **rated-load current**. Thus, the total or combined RLA rating would be based on 12.8 RLA for the compressor plus the 0.5 FLA for the fan for a combined total of 13.3 RLA. The total combined RLA would be based on

58.3 LRA for the compressor plus 3 LRA (0.5 times 6) for the fan for a combined total of 61.3 LRA. Consulting the tables with these numbers, we find that on the basis of FLC, the horsepower rating would be just over 2 hp (at 208 volts) or just under 2 hp (at 230 volts) and less than 1½ hp on the basis of LRA. We would use the larger of these two equivalent horsepower ratings. For this example, the 30-ampere 240-volt safety switch that would accept the maximum 25-ampere fuse from the nameplate in Figure 7-15 would likely have at least a 3-hp rating, and the 25-ampere circuit breaker would be more than 115 percent [from 440.12(B)(2)] of the combined 13.3 RLA.

The rating of disconnecting means for the more common combination-load and multimotor equipment can often be found on the nameplate. Disconnecting means and short-circuit and ground-fault protective devices are both sized on the basis of the same motor information. The disconnecting means rating requirements are minimum values that are based on smaller multipliers than the multipliers used for maximum short-circuit and

ground-fault protection. Therefore, a good rule of thumb is that a switch that will accommodate the maximum fuse size or a circuit breaker at least as large as the maximum fuse or circuit breaker size will satisfy the rating requirements for disconnecting means. This rule of thumb would work for the preceding example, but it does not establish the minimum size. Instead, it provides a rating that is high enough to meet *NEC* requirements. Smaller sizes or ratings may be permitted if the rules for sizing disconnects for combination loads are followed precisely. Small air-conditioning units and packaged units that are supplied with disconnecting means often use molded case switches like the one shown in **FIGURE 7-17**. The one in the diagram is marked as providing no overcurrent protection but has a 60-ampere rating. Similar devices are available in larger ratings and in three-pole configurations.

FIGURE 7-17   Molded case switch of the type often used for disconnecting packaged air-conditioning equipment.

10. Verify that the disconnecting means are within sight and readily accessible from the equipment and that working spaces are adequate.

Packaged air-conditioning and refrigeration equipment usually have motors and their controllers in the same enclosure or on the same mounting platform or skid. Therefore, a single disconnect usually serves both motors and controllers. This disconnecting means must be in sight and readily accessible from the equipment. The disconnect may be installed on or within the equipment. The disconnect must not be mounted on access doors, block access doors, or obscure nameplates on the equipment. Clear working spaces about disconnects and control equipment must be provided in accordance with Article 110. The general rule for air-conditioning equipment disconnecting means is that they must be within sight of the packaged equipment. **FIGURE 7-18** shows a disconnect located adjacent to air-conditioning equipment in such a way that the equipment does not interfere with the working space for the disconnect. These disconnects are often the first thing a person servicing or troubleshooting the equipment will open. They provide very easy access to determine if voltage is available and to measure current without removing panels on the equipment.

FIGURE 7-18   Disconnecting means "within sight" of air-conditioning equipment.

11. Verify that conductors, receptacles, cords, and overcurrent devices for room air conditioners are properly sized and that <u>leakage-current detection and interruption (LCDI)</u> **devices or arc-fault circuit interrupter (AFCI) protection is provided for cords.** Only room air conditioners that are single-phase and not over 250 volts are permitted to be cord-and-plug connected, and then only when the manual controls on the unit are readily accessible and within 6 ft (1.8 m) of the floor or another manual disconnecting means is provided within sight. Three-phase equipment and equipment over 250 volts require direct connections. For example, a 277-volt room air conditioner may not be connected by cord and plug. A room air conditioner connected by cord and plug is treated as a single motor for the purpose of sizing branch-circuit components, provided that it is not over 40 amperes and 250 volts, single-phase; it has the total rated load listed on the nameplate; and the rating of the circuit overcurrent device is not greater than the ampacity of the conductors or greater than the rating of the receptacle. This means that the ampacity of the branch-circuit conductors and cords must be at least 125 percent of the total rated load. The receptacle and overcurrent device ratings also must be at least 125 percent of the rated load. If the same branch circuit supplies other loads, the total rated load of the air conditioner may not exceed 50 percent of the overcurrent device rating. This rule repeats the requirements of 210.23(A)(2). Cords are also limited in length: 10 ft (3.0 m) maximum for 120-volt equipment and 6 ft (1.8 m) maximum for 208- or 240-volt equipment.

In addition to the restrictions on the length of cords, the *NEC* also requires special protection for the cords and equipment. Those room air conditioners that are permitted to be connected by cord and plug must have LCDI (leakage current detector-interrupter) or AFCI (arc-fault circuit interrupter) in the **attachment plug** or supply cord. **FIGURE 7-19** shows an example of an LCDI device installed in the supply cord to a through-wall room air conditioner. This requirement is intended to eliminate fires started by air conditioner cords that have been damaged as a result of the annual installation, removal, and storage that is characteristic to the use of this equipment. Since the protection must be in the attachment plug or in the first 12 in. (305 mm) of the cord, special cords or attachment plugs will be required, which may not be supplied with the equipment unless ordered that way. These requirements are in addition to any GFCI protection that may be required by 210.8 for receptacles.

FIGURE 7-19   LCDI device in the supply cord for a room air conditioner.

12. Check for receptacles and adequate lighting for servicing of mechanical equipment. Receptacles for servicing heating, air-conditioning, or refrigeration (HACR) equipment must be located on the same level and within 25 ft (7.6 m) of the equipment. These receptacles are required to be ground-fault circuit interrupter (GFCI) protected if they are located near sinks or outdoors, including rooftops, in any occupancy, or are in garages, crawl spaces, unfinished basements, or boathouses in dwelling units. Weather-resistant outlets are required in some of these areas. (See *NEC* 406.9 and the Wiring Methods and Devices chapter of this manual.) If HACR equipment is located in an attic or underfloor space, utility room, or basement, a lighting outlet must be located near the equipment to comply with 210.70(A)(3).

Checklist 7-3: Transformers

1. Identify transformers that are covered by Article 450.

A number of types of special-purpose transformers are excluded from the scope of Article 450. Most of the excluded types are covered by other articles, such as Article 725 for Class 2 and Class 3 circuits, Article 600 for sign transformers, Article 410 for discharge lighting transformers, and Article 760 for fire alarm circuits. Transformers that are an integral part of other apparatus such as X-ray equipment or transformer-type welders are also excluded from the rules of Article 450. Most transformers used in the distribution of power within a facility are covered by Article 450. Common examples include dry-type transformers used to create 208Y/120-volt systems where the service is at 480 volts or higher. Many industrial installations transform from much higher voltages to produce voltage systems for loads such as 480-volt equipment, 277-volt lights, or even 5000-volt or higher-rated motors. The transformers used for such purposes are covered by Article 450 unless they are owned by a utility. Overcurrent protection for a transformer may be provided in the primary circuit or in the primary and secondary circuits, depending on the voltage level and conditions of use and supervision of the transformer (**FIGURE 7-20**).

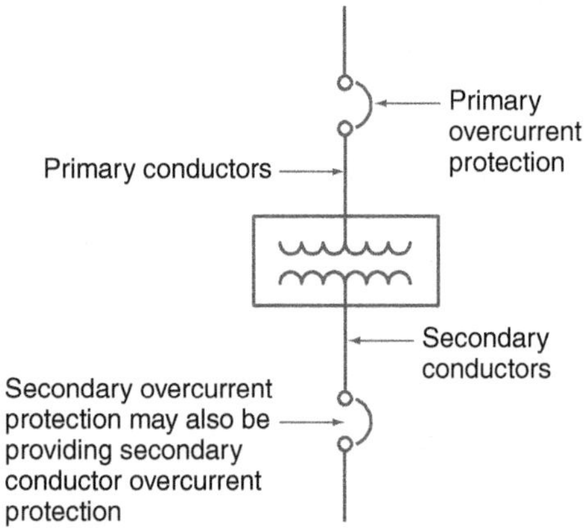

FIGURE 7-20 Transformer circuits and overcurrent protection.

2. Verify that overcurrent protection for transformers over 1000 volts is provided and properly sized.

Transformers over 1000 volts are generally required to have overcurrent protection on both the primary and secondary sides of the transformer. Protection on the secondary side may be omitted only where the installation is supervised and only **qualified persons** will service the installation. The ratings of overcurrent protection can be easily determined from the transformer current ratings. The current rating multiplied by the applicable multiplier gives the maximum overcurrent device rating. The applicable multiplier is based on the type of overcurrent device used and the impedance of the transformer. Generally, the next higher standard size above the calculated size may be used. **FIGURE 7-21** illustrates a transformer with primary and secondary protection where both the primary and secondary ratings are over 1000 volts and the impedance is not over 6 percent. Note that the secondary protection is permitted to consist of multiple devices grouped together in one location, in which case the sum of the ratings of the devices must not exceed the maximum rating permitted for a single device. Depending on the occupancy, supervision, and arrangement of conductors, the multiple devices that protect a transformer may or may not comply with requirements for overcurrent protection of conductors, which must be protected in accordance with 240.100 and 240.101 for conductors over 1000 volts.

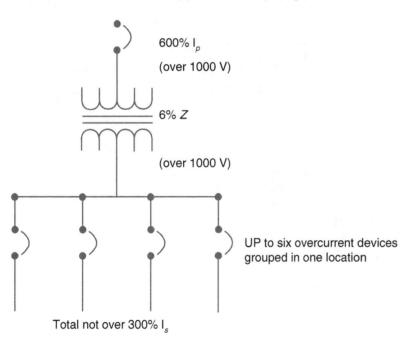

FIGURE 7-21 Overcurrent protection arrangement for transformer rated over 1000 volts.

3. Verify that overcurrent protection for transformers 1000 volts or less is provided and properly sized.

Transformers rated 1000 volts or less are required to have overcurrent protection on the primary side of the transformer only *or* on both the primary and secondary sides. The ratings of overcurrent protection can be easily determined from the transformer current ratings. The current rating multiplied by the applicable multiplier gives the maximum overcurrent device rating. The applicable multiplier is based on ranges of current ratings and whether protection is provided on the primary side only or on both the primary and the secondary sides. Higher multipliers are allowed for smaller transformers. If protection is also provided on the secondary, the allowable rating of the primary protection increases. When the multiplier used is 125 percent and the maximum size calculated does not correspond to a standard overcurrent device size listed in 240.6, the next larger standard size may be used. Unlike transformers over 1000 volts, the use of other multipliers (other than 125 percent) results in absolute maximum ratings, and rounding up is not permitted. **FIGURE 7-22** illustrates two methods of providing overcurrent protection for transformers rated 1000 volts or less.

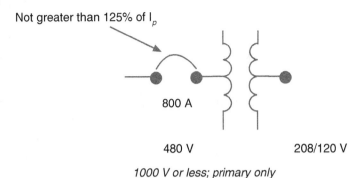

480/208/120 V, 3Ø, 500 kVA

Not greater than 125% of I_p

800 A

480 V   208/120 V

1000 V or less; primary only

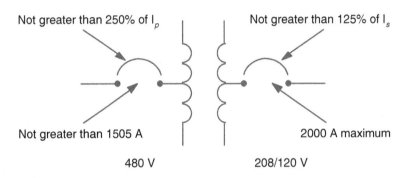

500 kVA, 480/208/120 V, 3Ø

Not greater than 250% of I_p Not greater than 125% of I_s

Not greater than 1505 A   2000 A maximum

480 V   208/120 V

1000 V or less; primary and secondary protection

FIGURE 7-22   Determining overcurrent protection for transformers rated 1000 volts or less. (*Source: National Electrical Code®* Seminar, NFPA, Quincy, MA.)

Like transformers over 1000 volts, the secondary protection is permitted to consist of multiple devices grouped together in one location, in which case the sum of the ratings of the devices must not exceed the maximum rating permitted for a single device (see Item 2 in this checklist). Single or multiple devices that protect a transformer may or may not comply with requirements for overcurrent protection of conductors, which must be protected in accordance with 240.4 and 240.21 for conductors up to 1000 volts (see Item 5 in this checklist). If the devices used to protect transformers do not satisfy the requirements for protection of conductors, additional or different overcurrent devices will be required for the conductors.

4. Verify that overcurrent protection is provided for transformer primary conductors.

Article 450 is concerned primarily with transformer installations and not with conductors. Therefore, the requirements for overcurrent protection in Article 450 do not consider the protection required for the conductors. Conductors must be protected in accordance with the usual rules of Article 240. Conductor protection may be supplied by the same device that protects a transformer if the device is properly sized for both the transformer and the conductors. The "transformer secondary rules" of Article 240 provide other specific rules under which the conductors may be considered to be protected. The rounding that is permitted for transformer protection is not permitted for conductor protection. For transformer secondary conductors, rounding up to the next standard size overcurrent protective device is not permitted.

5. Verify that overcurrent protection is provided for transformer secondary conductors.

Again, because Article 450 is concerned primarily with transformer installations and not with conductors, the requirements for overcurrent protection in Article 450 do not consider the protection required for the secondary conductors. As is stated in Item 4, conductors must be protected in accordance with the usual rules of Article 240, and secondary conductor protection may be supplied by the same device that protects a transformer if the secondary overcurrent device is properly sized for both the transformer and the conductors. Although a primary overcurrent device alone may protect a transformer, primary overcurrent devices seldom provide adequate protection for secondary conductors. Therefore, overcurrent devices to protect secondary conductors are usually required on the secondary side of a transformer even though secondary protection may not be needed for the transformer. Article 240 provides special rules for transformer secondary conductors or certain types of loads, but these rules usually require conductors to terminate in a single overcurrent device. Termination in multiple devices is permitted under some industrial rules, and the 10-ft (3.0-m) rule permits termination in "equipment containing an overcurrent device(s)." This allowance permits, for example, the termination to occur at the main terminals (lugs) of a switchboard.

The requirements of Article 408 require overcurrent protection for all newly installed panelboards supplied by customer-owned transformers, and similar requirements for MCCs are found in Article 430. Transformers may supply more than one set of secondary conductors as long as all sets comply with one of the rules in 240.21(C). Where multiple secondary sets are used with overcurrent protection based on the sum of the overcurrent devices as allowed in *NEC* Tables 450.3(A) and (B), the overcurrent devices must also be grouped in one location. If the transformer were adequately protected by primary-only overcurrent devices, the secondary overcurrent protective devices for the conductors would not be required to be grouped.

FIGURE 7-23 Dry-type transformer rated 600 volts or less.

6. Check transformer installations for adequate ventilation and spacing from walls and obstructions.

Transformers with ventilating openings are required to be marked with the clearances required for proper ventilation. Clearances from walls and other obstructions must be at least equal to the marked dimensions. The transformer shown in **FIGURE 7-23** must be provided with sufficient space to dissipate heat. The transformer manufacturer marks the required clearance on the transformer (**FIGURE 7-24**). Additionally, the 2014 *NEC* prohibits the installation of grounding or bonding connections on or over any ventilation openings in a dry type transformer enclosure.

7. Check transformers for ready access or proper installation in the open or in hollow spaces.

Generally, transformers are required to be readily accessible to qualified personnel. However, ready access is not required for transformers installed in the open, where they are sometimes elevated and mounted on walls, columns, or other structures as shown in **FIGURE 7-25**. Smaller transformers, up to 50 kVA, are allowed to be installed in hollow spaces where they are not readily accessible, provided that they meet the requirements for

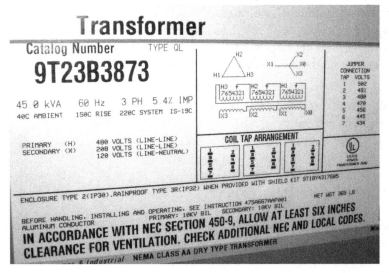

FIGURE 7-24 Transformer nameplate with required ventilation clearances shown.

ventilation and separation from combustibles. Figure 7-25 illustrates another point mentioned in Item 5 in this checklist: A transformer may supply more than one set of secondary conductors.

8. Verify that transformers are supplied with a disconnecting means.

Transformers covered by Article 450 are required to have a disconnecting means. In most cases, the primary overcurrent device for the transformer will satisfy this requirement. The disconnecting means is not required to be in sight of the transformer, but if the disconnect is in a remote location, the location must be indicated on the transformer and the disconnect must be lockable. This location or marking rule was a new requirement in the 2011 *NEC*, but as a practical matter, the disconnecting means usually existed prior to this rule; however, it was not required to be lockable and the location was not required to be marked on the transformer.

9. Check indoor dry-type transformers for separation from combustibles or, based on ratings, installation in fire-resistant rooms or vaults.

Clearances and separations from combustibles depend on the size and voltage rating of the transformer.

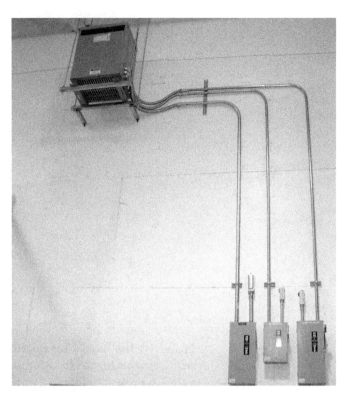

FIGURE 7-25 Transformer mounted in the open not required to be readily accessible.

(These separations are different from the separations for ventilation illustrated in Figure 7-24.) For example, up to 112½ kVA, transformers are supposed to be installed with a minimum 12-in. (305-mm) clearance from combustibles unless fire-resistant heat-insulated barriers are provided, but this rule does not apply to most transformers up to 1000 volts because they are usually totally enclosed except for ventilating openings. Also, transformers larger than 112½ kVA are supposed to be installed in fire-resistant (1-hour-rated) transformer rooms. However, such rooms are not required for transformers that have Class 155 or higher insulation systems and that are completely enclosed except for ventilating openings. Most modern dry-type transformers are totally enclosed except for the ventilating openings, and because Class 155 or higher insulation systems are also common, transformer rooms are usually not required. However, the only way to be sure is to examine the transformer nameplate. (The nameplate in Figure 7-24 shows a 220°C [428°F] insulation system.) Some transformer nameplates or instructions may provide requirements for separations from combustibles that are more stringent than *NEC* requirements. For example, many transformers are not intended for installation on combustible floors, even though the transformer may be enclosed and have a high insulation class rating. Dry-type transformers rated over 35,000 volts and installed indoors must be installed in a vault.

10. Check outdoor dry-type transformers for weatherproof enclosures.

Dry-type transformers are usually not constructed with weatherproof enclosures. Modification kits are usually available from the transformer manufacturer to make the enclosure weatherproof. The kit provides some form of shield to protect the ventilating openings from falling rain or snow. The ventilating openings referred to here can be seen at the top of the transformers in Figures 7-23 and 7-25. Smaller, encapsulated dry-type transformers without ventilating openings often have weatherproof enclosures as part of their standard construction.

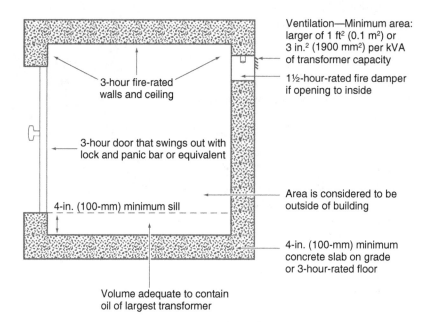

Ventilation—Minimum area: larger of 1 ft² (0.1 m²) or 3 in.² (1900 mm²) per kVA of transformer capacity

1½-hour-rated fire damper if opening to inside

3-hour fire-rated walls and ceiling

3-hour door that swings out with lock and panic bar or equivalent

Area is considered to be outside of building

4-in. (100-mm) minimum sill

4-in. (100-mm) minimum concrete slab on grade or 3-hour-rated floor

Volume adequate to contain oil of largest transformer

FIGURE 7-26   Transformer vault construction.

11. Verify that liquid-insulated transformers are installed in accordance with the requirements for the location and type of insulating liquid.

Vaults are required for most oil-insulated transformers installed indoors. Vaults may not be required or the vault construction requirements may be less stringent for certain types of transformer insulating liquids and certain transformer ratings. For example, a vault may not be required for a "less-flammable liquid-insulated transformer" or for "nonflammable fluid-insulated transformers." All transformers rated over 35,000 volts must be installed outdoors or in a vault. Liquid confinement areas are sometimes required for both indoor and outdoor installations, either by the *NEC* or by other codes or standards.

12. Check transformer vaults for adequate construction, access, ventilation, and drainage and for foreign systems in vaults.

The construction requirements for transformer vaults are provided in Part III of Article 450. Foreign systems and storage are prohibited in transformer vaults. Where the *NEC* requires a transformer vault, its construction must meet the requirements shown in the elevation view in **FIGURE 7-26**. As noted in the drawing and according to 230.6(3), the area inside a vault constructed according to these requirements is considered to be "outside" a building even though the entire vault is inside the building. This consideration is based on the required fire rating of the vault.

Checklist 7-4: Capacitors

1. Check capacitors for proper enclosures and guards.

Capacitors must be located or **guarded** so that unqualified persons cannot accidentally contact **live parts** directly or with conductive objects. Guarding is not required if only qualified persons have access to the capacitor installation. Where individual capacitor units contain more than 3 gallons (11 liters) of flammable liquid, they must be installed outdoors in fenced enclosures or in vaults.

2. Verify that conductors are properly sized on the basis of the current rating of the capacitor(s).

Conductors in capacitor circuits must have an ampacity at least equal to 135 percent of the capacitor current rating. Conductors connecting capacitors to the terminals of a motor must also have at least one-third the ampacity of the motor-circuit conductors.

3. Verify that capacitors other than those connected to the load side of motor overload devices have disconnects and proper overcurrent protection.

Disconnecting means and overcurrent protection must be provided in ungrounded capacitor-circuit conductors. Disconnects are required to be rated at no less than 135 percent of the capacitor current rating. Disconnects and overcurrent protection are not required for capacitor leads connected to the load side of a motor overload device because the overload device provides a limited amount of protection for the capacitor circuit and the motor-circuit disconnect for the motor controller provides for isolation of the capacitor circuit. **FIGURE 7-27** shows two methods of connecting a power factor correction capacitor in a motor circuit. Where the capacitor is connected on the load side of the motor overload device, a separate disconnect for the capacitor is not required and the capacitor can discharge through the motor windings.

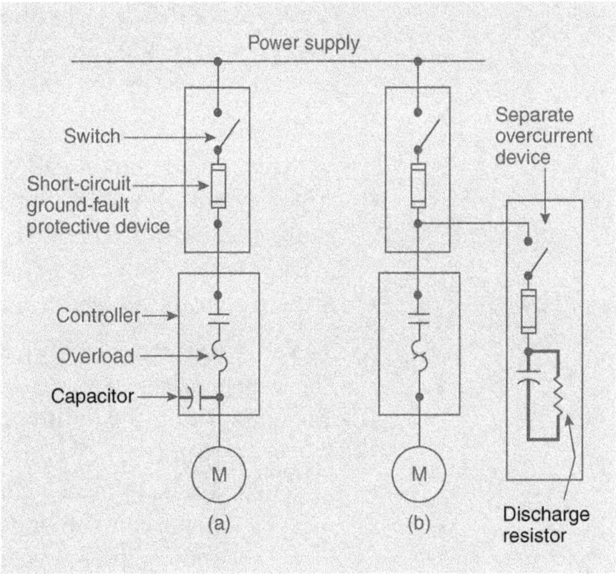

4. Verify that overload device ratings have been corrected where capacitors are connected to the load side of motor overload devices.

Capacitors used with motor circuits to improve the power factor will also reduce the current in parts of the motor circuit. If the capacitors are connected on the load side of the overload units, the overload selection must take into account the improved power factor of the circuit and a lower setting or selection for the overload device must be used.

5. Check capacitors over 1000 volts for proper switching, overcurrent protection, identification, and grounding.

Part II of Article 460 covers the specific requirements for capacitors over 1000 volts, many of which are similar but not identical to the requirements for capacitors not over 1000 volts. See Item 6 for an example of these differences. Part II should be reviewed for specifics when capacitors over 1000 volts are used.

6. Verify that a proper means for discharge has been provided for capacitors.

A means must be provided to discharge the stored energy in a capacitor. The discharge means can be either permanently connected, such as where a capacitor is connected to the load side of a motor overload device and across the windings of a motor, or provided with automatic means to connect the discharge device to the capacitor terminals whenever the capacitor is de-energized. For capacitors rated 1000 volts or less, the residual voltage must be reduced to 50 volts within 1 minute after the capacitor is de-energized. For capacitors rated over 1000 volts, the allowable time is increased to 5 minutes. **FIGURE 7-28** illustrates a group of power factor correction capacitors with discharge resistors contained in an enclosure.

FIGURE 7-27   Two methods of connecting power factor correction capacitors in a motor circuit. (Source: *National Electrical Code®* *Handbook*, NFPA, Quincy, MA, 2011, Exhibit 460.1.)

FIGURE 7-28   Power factor correction capacitors with discharge resistors. (This material and associated copyrights are proprietary to, and used with the permission of, Schneider Electric.)

Checklist 7-5: Elevators, Dumbwaiters, Escalators, Platform Lifts, and Stairway Chairlifts

1. Verify voltage limitations and presence of warning labels where voltage exceeds 600 volts.

Generally, the maximum voltage between conductors is 300 volts. However, voltages to driving machines (power circuits) are often more than 300 volts and may be up to 1000 volts between conductors. Warning signs or labels reading "Danger—High Voltage" complying with 110.21(B) must be posted on the equipment where plainly visible if the voltage exceeds 600 volts. Heating and air-conditioning equipment mounted on elevator cars may also be up to 1000 volts.

2. Verify that all live parts are enclosed.

Virtually all live parts in all parts of the installation must be enclosed. However, for some large elevators, the motors are not of the totally enclosed type and that equipment may be guarded to avoid accidental contact rather than enclosed. Such guarding may be provided in the form of a fence or similar barrier, or it may be accomplished by placing the equipment in a room where access is restricted to qualified persons (see *NEC* 110.27).

3. Verify required working clearances around elevator electrical equipment.

The working clearances for elevator and related electric equipment must generally be the same as required for other equipment of 600 volts or less. The working clearances of 110.26 apply unless:

- A. The equipment is connected by flexible means so the equipment can be repositioned to provide the working clearance.
- B. The equipment is equipped with guards and any necessary work can be accomplished without removing the guards.
- C. The equipment in question does not require servicing to be done while the equipment remains energized [in which case 110.26(A) would not apply anyway].
- D. The equipment operates at less than 30 volts root mean square (rms), 42-volt peak, or 60-volt direct current (dc) so that it does not present a significant shock hazard.

Article 620 is permitted to modify Article 110 in these ways in accordance with Section 90.3 of the *NEC*.

4. Check conductors for proper insulation type and minimum size.

Smaller conductors are permitted and smaller conductors may be paralleled in elevator traveling cables than would be permitted generally in the *Code*. This permission is based on the need to keep the traveling cables to a manageable size, partly to help prevent twisting and looping in the traveling cables. The types of conductors that may be used depend on the use. **Hoistway** cables, other insulated conductors, and all conductors in raceways must have flame-retardant insulations. This is not explicitly stated except for hoistway door interlock wiring, but it is a listing requirement for the types of cables and conductor insulations that are allowed. Permitted types of traveling cables are Types E, EO, ET, ETLB, ETP, and ETT, which are listed as elevator cables in Table 400.4 of the *NEC*. These cables may have control, signaling, or communications circuits in the same cable with power conductors. The application of a traveling cable within the elevator shaft is illustrated in **FIGURE 7-29**. Conductors used for door interlocks must be Type SF (silicone insulated fixture wire) (see *NEC* Table 402.3) or an equivalent 200°C (392°F) insulation type to help maintain the operation of the doors in a fire-rated hoistway. Suitable guards must be provided where necessary to protect the cables from physical damage [620.43, 620.44]. Flexible cords and cables are now permitted in **machine rooms** or spaces and **control rooms** or spaces where they are not longer than 1.8 m (6 ft), are flame-retardant, are protected from physical damage, and are part of the elevator equipment or driving machine.

5. Verify ampacity of branch-circuit and feeder conductors.

The minimum ampacity of conductors varies according to the equipment supplied, but the conductor sizes are generally based on nameplate values for the controller, transformer, or motor generator set. Conductors for single motors are based on the rules in Article 430, which includes specific factors for **intermittent-duty** motors in Table 430.22(E) of the *NEC*. Where continuous-duty motors are used in intermittent-duty applications, the conductor ampacity must be 140 percent of the nameplate current ratings. Elevator motors that have time ratings may have conductors rated less than the nameplate current ratings. Conductors must be increased in size for any other loads such as the **control system**, and

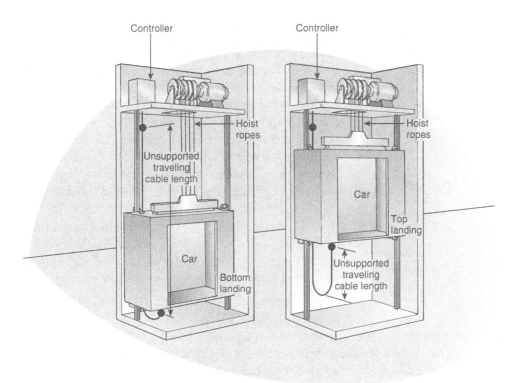

FIGURE 7-29   Unsupported lengths of elevator traveling cable in shaft. (Source: *National Electrical Code®* *Handbook*, NFPA, Quincy, MA, 2011, Exhibit 620.1.)

conductors feeding multiple items must be based on the sum of the nameplate ratings plus any other loads. Where multiple elevators are supplied by a single feeder, demand factors may be applied [620.14].

6. Verify proper wiring methods.

Except for traveling cables, flexible cords of listed cord-and-plug–connected equipment, and some cables for Class 2 circuits, most wiring is required to be installed in rigid metal conduit (RMC), intermediate metal conduit (IMC), electrical metallic tubing (EMT), rigid nonmetallic conduit (PVC or RTRC), or wireways or be in Type MC (metal-clad cable), Type MI (mineral-insulated, metal-sheathed cable), or Type AC (armored cable). Other flexible cords, cables, and raceways are permitted in limited lengths, up to 6 ft (1.8 m) and for specific purposes. The specific requirements of 620.21 for the type of equipment being inspected or installed should be consulted, as the requirements and permitted wiring methods vary by type of equipment (e.g., elevators, escalators, or platform lifts) and by location (e.g., hoistways, cars, and machine rooms). For example, the length limitation on flexible conduits does not apply to conduits run between risers and limit switches, interlocks, operating buttons, and the like in elevator hoistways.

7. Verify the required branch circuits for car lighting, receptacle(s), ventilation, and air conditioning.

A separate branch circuit is required for elevator car lighting. This circuit may also supply receptacles mounted on the car, auxiliary lighting on the car, and ventilation on the car, but the lighting portion of the circuit may not be GFCI protected. Each car requires a separate branch circuit, and the branch-circuit overcurrent device must be located in the machine room or other machine or **control space**. Similarly, another circuit is required for heating and air-conditioning equipment for each car. The branch-circuit overcurrent devices need not necessarily be supplied from a panelboard, although for multiple elevators a panelboard may be more convenient than a group of separate circuit breakers or fusible switches. However, the disconnecting means must be capable of being individually locked in the open (off) position. (See Items 13 and 14 in this checklist.) For a single elevator, a lockable two- or three-pole fusible switch can be considered as the final overcurrent

FIGURE 7-30 Separate disconnecting means and overcurrent protection for elevator power and for car lighting, heating, and air conditioning. (Source: *National Electrical Code® Handbook*, NFPA, Quincy, MA, 2011, Exhibit 620.2.)

device where the two branch circuits required under this item originate (see **FIGURE 7-30** and Items 13 and 14).

8. Verify the required branch circuit for machine-room lighting and receptacle.

Another separate dedicated circuit (in addition to those discussed in Item 7) must be provided for machine room/space and control room/space lighting and receptacles. Although the receptacle(s) must be GFCI protected, the lighting must be connected ahead of such GFCI protection so that the lighting remains on in case the GFCI trips. This effectively mandates the use of GFCI receptacles (rather than GFCI circuit breakers) in most applications unless additional branch circuits are provided. The branch-circuit overcurrent device for this circuit is not required to be located in the machine or control space or room.

9. Verify the required branch circuit for hoistway pit lighting and receptacle.

Each elevator hoistway pit must be supplied by a separate circuit for lighting and receptacles that are required in the pits. Like the machine room, the lighting must be connected ahead of any GFCI protection, but the receptacles are required to be GFCI protected. This branch circuit is not required to originate in the control or machine space or room.

10. Check for required receptacle and light switch provisions in machine rooms and pits.
The switches for lighting in the machine or control room or space must be located at the point of personnel entry to the room. Switches for lighting in hoistway pits must be readily accessible from the pit access door. At least one duplex receptacle must be installed in each machine or control room or space or hoistway pit.

11. Verify that only elevator-associated wiring is installed in hoistways and machine rooms.
The spaces used for the installation of elevator and escalator equipment are dedicated to the elevator or escalator installations, and no other wiring that is not directly associated with the elevator or escalator installation is permitted in such spaces. In addition, elevator hoistways usually may not be used for feeders that are associated with the elevator unless there are no splices in the hoistway and the feeder is exempted by **special permission** or unless the driving machines supplied by the feeder are located in the hoistway.

12. Check elevator machine disconnecting means for proper type, operation, and location.
Disconnects for elevators, escalators, and similar equipment must interrupt all ungrounded conductors at the same time, be listed enclosed and externally operable fused motor circuit switches or circuit breakers, and be furnished with a permanent feature that makes them capable of being locked in the open position as specified in 110.25. The means for adding a lock to secure the device in the open position must be part of the installed equipment and not be of the portable type. The disconnects must be located within sight of the motor controller or, for escalators or moving walks, located in the same space as the controller. (General-use switches or snap switches may be used as disconnects for platform lifts that are 2 hp or less, and stairway chairlifts may use cord-and-plug connections as disconnecting means subject to additional restrictions.) The disconnect must also be readily accessible to qualified persons. Such disconnects are generally located behind locked doors, but those locations are still considered to be **accessible** to qualified persons [see 110.26(G)].

In general, the disconnect may not be operable from any location other than the machine or control room, but where sprinklers are installed in the hoistway or in machine or control spaces, the disconnect may be opened automatically prior to the application of water. This is usually accomplished by installing heat detectors near the sprinkler heads. The heat detectors are set to operate a shunt-trip breaker or similar device at a shaft temperature slightly below the operating temperature of the sprinkler heads. The power may be restored only by manual operation of the disconnect. Disconnects for elevator machines may not disconnect power to car, machine room, or pit lighting or receptacles or car air conditioning or heating. Disconnects must be clearly labeled as required for disconnects generally [see 110.22]. In addition, the disconnects must be marked with the location of the overcurrent device on the supply side of the disconnect.

13. Verify disconnecting means for car lighting, receptacles, and ventilation.

Disconnects for circuits supplying car lighting, receptacles, and ventilation must be of the same types as those permitted for the driving machinery. They must be enclosed, externally operable, fusible motor circuit switches or circuit breakers furnished with permanent (nonportable) features that make them capable of being locked in the open position as required by 110.25, located in the machine or control space or in the same space as the disconnect for the main power supply to the machinery, and clearly marked to indicate the car served and the location of the supply-side overcurrent device. They must be separate from the disconnects for driving machines. However, where the car loads (lighting, receptacle[s], and a ventilation motor that is not over 2 hp) are supplied by an **individual branch circuit**, the disconnect may be a general-use switch or general-use snap switch meeting the requirements of 430.109(C). The switch must still be furnished with a permanent means for locking it in the off position.

14. Verify disconnecting means for car heating and air conditioning.

Disconnects for circuits supplying car heating and air conditioning must be of the same types as those permitted for the driving machinery and the car lighting. They must be enclosed, externally operable, fusible motor circuit switches or circuit breakers furnished with nonportable features that make them capable of being locked in the open position, located in the machine or control space or in the same space as the disconnect for the main power supply to the machinery, and clearly marked to indicate the car served and the location of the supply-side overcurrent device. This could be the same multiple circuit disconnect as that for car lighting and receptacles, but must be separate from the disconnect for driving machines.

15. Check overcurrent protection for proper rating and coordination.

Overcurrent protection for conductors and equipment is required to be provided according to the rules for similar types of motor-operated equipment. Control circuit protection is to be provided in accordance with the requirements of Article 725 as covered in the Remote-Control, Signaling, and Fire Alarm Circuits and Optical Fiber Cables chapter of this manual. Motor overload protection and branch-circuit and feeder protection is required to be provided in accordance with Article 430. Motor and motor circuit protection is covered in this chapter. Where there is more than one driving machine disconnect, the overcurrent protection in each disconnect must be selectively coordinated with the feeder overcurrent protection so that a fault at the point shown as Fault X_1 will operate the overcurrent device in the local disconnect without operating the device in the main distribution switchboard or panelboard **(FIGURE 7-31)**. Selective coordination is required to maximize the reliability of the remaining machines if a fault in one machine causes its overcurrent device to operate. This type of coordination is not easily or reliably demonstrated in

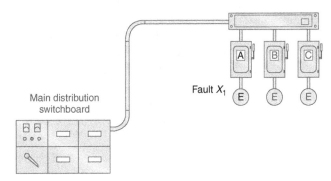

FIGURE 7-31 Overcurrent protection coordination for feeder supplying several elevators. (Reprinted with permission from *National Electrical Code*® *Handbook*, Exhibit 620.7, Copyright © 2008, National Fire Protection Association, Quincy, MA 02169. This reprinted material is not the complete and official position of the NFPA or the referenced subject, which is represented only by the standard in its entirety.)

the field, but must be based on engineering design. Software and manual studies can assist in the selection of proper device types and ratings. The designer should be able to provide evidence of this coordination study, which should demonstrate that for anticipated levels of fault currents, the downstream devices (in the disconnects) have been selected for and will be capable of clearing the fault without operating the upstream device.

16. Verify that only permitted equipment is located in machine rooms.
With a few exceptions, all the driving machines, motor-generator sets, motor controllers, and disconnecting means must be located in rooms or spaces that are dedicated and reserved for such equipment and are secured so that only authorized (usually qualified) persons have access. A dedicated room is usually, but not always, provided. ANSI 17.1 provides requirements for other methods of defining and enclosing a space. For example, sometimes a chain-link fence-type enclosure is used to define and separate the equipment space from a larger space. As is noted in Item 11 of this checklist, only the machine equipment, the associated wiring, and other necessary equipment such as luminaires, sprinklers, and heating, ventilation, and air-conditioning (HVAC) equipment for the space is permitted in the space.

17. Verify GFCI receptacles on car tops and pits and GFCI protection for receptacles in machine rooms.
As is noted in Items 8, 9, and 10 of this checklist, the receptacles required in pits and hoistways, on car tops, and in machine spaces and escalator and moving walk wellways are required to be GFCI type. The receptacles in machine rooms and **machinery spaces** are also required to be GFCI protected, but the protection could be in the form of a branch-circuit GFCI circuit breaker or an upstream receptacle rather than in the specific receptacle, although this method may require additional circuits for lighting. The requirement for a GFCI receptacle means that the user will have immediate and local control of the testing and resetting of the GFCI. This local control is especially important in confined spaces or spaces to which access is somewhat difficult such as elevator pits or escalator wellways but is not as important in a machine room. GFCI protection is not required for a single (not duplex) receptacle that supplies a permanently installed sump pump.

18. Check operation of elevator machine disconnecting means where emergency or standby power is provided.
In most cases where elevators are connected to emergency or standby systems, as in high-rise buildings, the transfer from normal to emergency sources takes place ahead of the elevator machinery disconnecting means, and this tends to satisfy the requirements of 620.91. However, if other transfer schemes are used, the machinery disconnect must be equipped with a means to also disconnect any other power sources when the machinery disconnect is opened. Since some elevator motors will produce regenerative power, a means to absorb this power must be provided. Often, the other building loads that are connected to the alternative power source will be able to absorb the regenerative power. Many (perhaps most) elevators do not regenerate power, so this is not always an issue, but where it is an issue, the designer should be able to provide evidence that the installation complies with the *Code*.

Checklist 7-6: Electric-Vehicle-Charging Equipment

1. Verify that all associated equipment, materials, devices, and fittings are listed.
All electrical equipment used in installations of electric-vehicle-charging systems is required to be listed. Note that some incidental items that are routinely used in electrical installations are not available as listed. Examples include screws, nails, and some types of conduit straps. The intent of the rule is that all equipment that has an electrical function, especially the charging equipment, be tested to meet an applicable product standard that will help to ensure the safety of the installation.

2. Check the suitability of the electric vehicle coupler.

The **coupler** that connects the charger to an **electric vehicle** must be polarized. The coupler must be noninterchangeable with devices of other electrical systems, and grounding couplers must not be interchangeable with non-grounding couplers. Couplers must be designed to minimize the possibility of unintended disconnection. Couplers include a connector that makes electrical connections for the purposes of charging and information exchange. Other types of cord-and-plug assemblies are usually fed from only one end, so the primary restriction on such designs is that the male blades cannot remain energized when exposed (i.e., when not plugged in). Since a coupler can be energized from either the charger or the battery in the vehicle, couplers must be designed to guard against contact on both ends. The best way to ensure compliance with these requirements is to verify listing of the coupler. **FIGURE 7-32** shows a vehicle charger with the connector and **power-supply cord** attached.

3. Verify that coupler and cable safeguards have been provided.

Except for portable charging equipment that is designed to be connected to ordinary 15- or 20-ampere, 125-volt receptacles, chargers must be connected by permanent wiring methods and interlocked so that they are de-energized when disconnected from a vehicle or when the cable or connector is exposed to strain that could otherwise damage the cable or connector. Interlocks are not required for dc supplies less than 50 volts. Types of cables and cords for power supply and output are restricted to specific types, minimum ratings, and maximum lengths. Again, verification of listing and installation according to the instructions should cover most of this issue.

FIGURE 7-32   Electric-vehicle-charging equipment and vehicle supply cord. (Source: *National Electrical Code® Handbook*, NFPA, Quincy, MA, 2011, Exhibit 625.1.)

4. Verify the rating of branch-circuit and feeder overcurrent devices.

Electric vehicle chargers are treated as continuous loads for the purposes of Article 625. The branch-circuit overcurrent device is therefore required to be rated at no less than 125 percent of the maximum load of the charging equipment. Article 625 does not specify an ampacity rating for the charging equipment branch circuit, but aligns with the regular rules of Articles 210 and 240 that would require the branch-circuit conductors to be sized at 125 percent of the continuous load, and where noncontinuous loads are supplied by the same branch circuit, they must be included at 100 percent. Automatic load management equipment can be used to reduce the total load at any given time on feeders and/or the service. This can reduce the overall feeder and service load calculation, resulting in smaller distribution systems and equipment at facilities where multiple **electric vehicle supply equipment** units have been installed.

5. Verify that a personnel protection system has been provided.

Charging equipment must be furnished with **personnel protection systems** similar to the protection provided by double insulation or by a GFCI device. This protection is intended to protect the user from electric shock. However, vehicle chargers may come in sizes and ratings that are not compatible with ordinary GFCI devices, so the protection system is required to be furnished as part of the vehicle charging/supply system rather than in the branch circuit or receptacle. The methods that may be used to provide shock protection are highly variable, so markings on the equipment, including listing and labeling, should be used to verify that personnel protection has been provided. Where the charging equipment is cord-and-plug connected, the protection must be included in the attachment cord or plug within 12 inches (300 mm) of the attachment plug.

6. Check the location of disconnecting means rated over 60 amperes or 150 volts to ground.

Equipment rated over 60 amperes or over 150 volts to ground must have a disconnect that is readily accessible and installed with nonportable means to make the disconnect capable of being locked in the open position in accordance with 110.25. There is no requirement that the disconnect be in sight of the charging equipment, although locating the disconnect in sight may be more convenient for the user.

7. Verify that backfeed prevention provisions have been provided.
Since the batteries in a vehicle will remain energized when power to a charger is interrupted, the charger must be designed to disconnect the battery from the premises wiring system on a power failure. Since this is a required design feature, listing and labeling of the charger should ensure compliance with this requirement. In some cases, the vehicle may be designed and intended to supply power to an optional standby system. Transfer means that complies with Article 702 is required in these cases to prevent backfeed from the vehicle to the normal source. Similarly, where used for parallel power production, equivalent safeguards must be provided through compliance with Article 705. Bidirectional power feed is allowed if the equipment is listed for that use.

8. Check the location and height of indoor and outdoor charging equipment.
The most obvious location requirement for either indoor or outdoor charging equipment is that it must be located where it will permit direct connection to electric vehicles. Section 625.17 limits the length of the cord to 25 ft (7.6 m) when the cord is simply coiled up for storage. Longer lengths are permitted with appropriate listed cable management systems such as cord reels. Generally, if the charger is located so that a vehicle can be parked adjacent to the charger, the shorter cord will be long enough. The other location requirement is for height. In indoor locations, the minimum height of the charger is 18 in. (450 mm) above the floor. In outdoor locations, the minimum height is 24 in. (600 mm) above grade. The maximum height will be limited in practice by cord and connector lengths, but is otherwise not specified.

9. Verify the necessity for and amount of ventilation for indoor charging locations.
Section 625.15 requires that charging equipment be marked by the manufacturer with either "ventilation required" or "ventilation not required." The marking must be clearly visible after installation. Ventilation is an issue only in indoor charging locations. Equipment located outdoors is considered to have sufficient ventilation. The industry trend is toward providing batteries and charging systems that do not require ventilation. This can be done by using nonvented batteries or by designing the battery and charging system so that only small amounts of hydrogen are released. Other battery installations as covered in Article 480 are required to be designed so that gases will be diffused and ventilated in a manner that will "prevent the accumulation of an explosive mixture."

In essence, Article 625 imposes the same requirement as Article 480, but the system can be so designed that ventilation is not needed (**FIGURE 7-33**). If the design of the batteries or the charging system is such that ventilation is required, three options are provided for determining the ventilation rate that is required:

    A. Tables 625.52(B)(1) and (2) of the *NEC* give ventilation requirements in cubic meters or cubic feet per minute based on the voltage and current ratings of the equipment.

    B. If the tables do not include the equipment ratings, formulas are provided in 625.52(B)(2) to calculate the ventilation rate. Separate formulas are provided for single- and 3-phase systems and for rates in cubic feet per minute or cubic meters per minute.

    C. Mechanical engineering designs may be used to determine ventilation rates as part of a total building ventilation system.

A fourth option requires interlocking of the ventilation system with the charging equipment so that the ventilation system remains energized throughout the entire charging operation, but this option does not provide a ventilation rate.

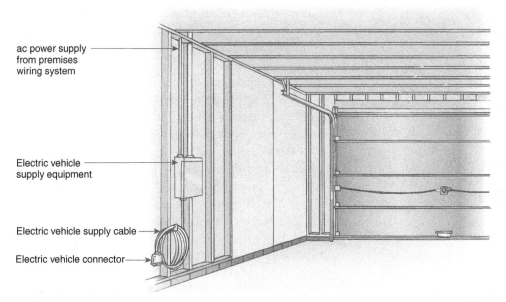

ac power supply
from premises
wiring system

Electric vehicle
supply equipment

Electric vehicle supply cable

Electric vehicle connector

FIGURE 7-33 Mechanical ventilation of garage not required where nonvented batteries are charged. (Source: *National Electrical Code®* *Handbook*, NFPA Quincy, MA, 2008, Exhibit 625.2.)

Actual ventilation rates may vary from design rates depending on many variables, and measurement of air flow requires special equipment, so an electrical inspector can usually verify only that the system has been designed under one of the provisions permitted. The designer of the system should be able to provide evidence or certification that the system design meets the appropriate values for the method used.

Checklist 7-7: Signs and Outline Lighting

1. Check each commercial building and occupancy for sign outlets on dedicated circuits.
Each commercial building and each commercial occupancy is required to have at least one outlet provided for sign or outline lighting use. The outlet(s) for each occupancy must be supplied by a circuit that is dedicated to the outlet(s). Separate occupancies require separate circuits. The outlets are required to be accessible at each entrance to a tenant space where access is provided to pedestrians. These outlets are not required at service entrances. For example, in a shopping mall, at least one outlet would be required at each regular entrance to the mall and another outlet and circuit would be required at each entrance to an individual retail tenant within the mall, but no sign outlet is required at the delivery area. Receptacle outlets are also required to be located within 18 in. (450 mm) of the top of a show window. At least one receptacle outlet is required for every 12 linear feet (3.7 m) of show window or fraction thereof.

The dedicated circuits for sign and outline lighting are required to be rated at least 20 amperes. Many stores will require more than one branch circuit or even a feeder supplying multiple branch circuits, depending on the actual sign or outline lighting load. The requirement for at least one outlet and circuit is a minimum to accommodate loads that are likely to be needed for commercial occupancies. In some commercial occupancies, such as a strip mall or shopping mall, the actual spaces that will be occupied by any individual tenant or the actual locations of all entrances are not known at the beginning of construction or even when the first tenants occupy spaces, so this inspection item may not be applicable until the spaces are assigned.

2. Check signs, section signs, and outline lighting for listing and marking.
All electric signs, **section signs**, and outline lighting systems are required to be listed, whether they are fixed, mobile, or portable. The primary exception is field-installed

FIGURE 7-34   Example of a large section sign. (Courtesy of Kieffer & Co. [http://www.kieffersigns.com].)

FIGURE 7-35   Example of a section sign made up of separate letters.

skeleton tubing (neon) that is not required to be listed as a whole, but is required to meet the requirements of the *NEC,* which in turn requires many of the components to be listed, but not the custom-made **neon tubing** itself. The other exception requires **approval** by special permission, in writing. See Items 1 through 4 in the Field-Installed Skeleton Tubing (Neon) and Wiring checklist. Outline lighting is also not required to be listed as a system where it is assembled from listed luminaires and wired using ordinary methods of Chapter 3 of the *NEC* that are appropriate for the location and occupancy. Outline lighting often is just a series of luminaires used for the purpose of outline lighting, such as the fluorescent luminaires often used under an awning that is used to identify and advertise a commercial establishment. Where section signs are used, the sign will consist of separate listed and labeled components with markings that indicate that installation instructions are required. **FIGURE 7-34** is an example of a large section sign that must be assembled in sections simply because it is so large. Other section signs are sometimes assembled from sections because they are intended to consist of separate components, often just separate letters with or without a logo like the sign shown in **FIGURE 7-35**, which has a label on the end for the overall sign and labels for the sections on the individual letters. In either case, the instructions for assembly and field wiring between signs must be furnished and followed. Signs must also be marked with input voltage and current ratings to facilitate the selection and verification of supply circuits. The markings must be visible after installation of the sign and must be permanent and durable.

3. Check branch circuits supplying signs for permissible ratings.

Branch circuits for signs are limited to 20-ampere ratings for signs that use incandescent or fluorescent lighting sources. This requirement parallels the limitations on the uses of branch circuits of various ratings in 210.21 and 210.23, where larger circuit ratings are limited to lighting applications with heavy-duty lampholders. Fluorescent and incandescent lamps usually are not designed for use with heavy-duty lampholders. (Heavy-duty lampholders are mostly used with high-intensity discharge [HID] lamp sources such as metal halide and high-pressure sodium.) Circuits rated up to 30 amperes may be used for signs that incorporate neon tubing. For the purposes of load calculations, sign loads are treated as continuous loads.

Portable and mobile signs are often connected to ordinary branch circuits and are sometimes used only for short periods, so they are not always in place for an inspection. However, 600.10 requires such signs to be protected by factory-installed GFCI protection that is either integral to the sign or installed in the power supply cord within 12 in. (300 mm) of the attachment cord.

4. Verify that an external disconnect is provided and within sight of each part of a sign or equipped with locking means.

Each sign or outline lighting system, whether supplied by one or more branch circuits or feeders, must be furnished with a disconnecting means that opens all ungrounded conductors of the sign (and no other load) and is external to the sign and located where the conductors enter the sign enclosure or pole that supports the sign. Disconnects for multiwire branch circuits supplying signs must open all ungrounded conductors simultaneously. Signs connected by cord and plug may use the cord and plug as a disconnecting means. Exit directional signs within buildings are not required to have individual disconnecting means (a branch-circuit device can be used to disconnect an exit sign). In general, the disconnecting means should be within sight of the sign or outline lighting system. However, it is allowed to be out of sight where it is installed to comply with 110.25, which requires the disconnect to be equipped with a locking means that remains in place with or without a lock installed and that makes the sign disconnect capable of being locked in the open (off) position. Some signs require controllers, either to turn the sign off-and-on on a schedule or based on ambient lighting conditions, or to control individual elements such as time and temperature signs or signs that flash in a particular sequence. Where a controller is used to operate a sign, the disconnecting means may be out of sight of the sign but must be in sight of the controller or in the same enclosure as the controller. This disconnect must be installed to comply with 110.25. Signs may have additional interior disconnects that allow sections of the sign or separate power supplies to be disconnected for servicing the sign. **FIGURE 7-36** illustrates three possible *Code*-compliant ways that disconnects may be supplied for signs.

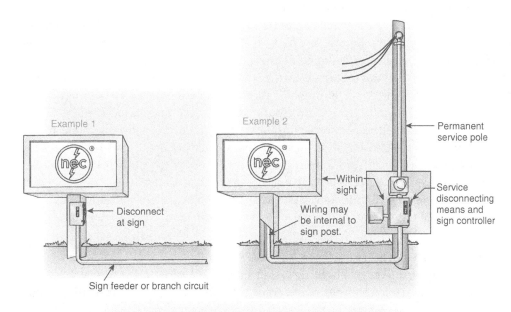

FIGURE 7-36   Three acceptable methods of providing disconnecting means for signs. (Source: *National Electrical Code® Handbook*, NFPA, Quincy, MA, 2008, Exhibit 600.3.)

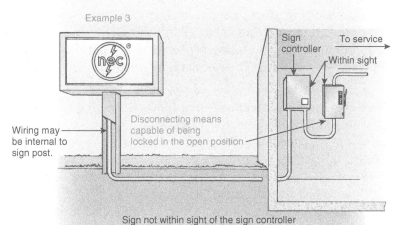

5. Verify that metal parts of signs and outline lighting are bonded together and connected to a properly sized equipment grounding conductor.

Equipment grounding conductors sized on the basis of 250.122 must be provided with the supply to a sign to ground the metal parts of a sign or an outline lighting system. In general, all grounding and bonding requirements must be in accordance with Article 250 of the *NEC* (see the Grounding and Bonding chapter of this manual). There are two significant differences for signs: (1) Except for remote metal parts of a section sign, signs supplied by Class 2 power supplies must have their metal parts connected to an equipment grounding conductor and bonded even if the Class 2 power system is not grounded [600.24(B)] and (2) flexible metal conduit may be used for bonding in longer lengths than permitted in Article 250. Specifically, flexible metal conduit and liquidtight flexible metal conduit used to enclose the secondary circuits from power supplies to neon tubing may be used as a bonding means where the total length of all sections in the path to the power supply does not exceed 100 ft (30 m). However, small metal parts up to 2 in. (51 mm) in any dimension that are part of a neon lighting installation and that are spaced at least ¾ in. (19 mm) from the tubing are not required to be bonded. When a listed nonmetallic conduit is used to contain secondary circuits, a separate bonding means must be provided in the form of a conductor not smaller than 14 AWG that is spaced at least 1½ in. (38 mm) from the conduit for systems operating at up to 100 Hertz (Hz), and at least 1¾ in. (45 mm) for systems operating at higher frequencies.

These rules are intended to help equalize the stresses imposed on the high-voltage Type GTO cable (gas tube sign cable). When the cable is installed in a grounded circular metal raceway, the electrostatic flux lines around the conductor are somewhat equalized, especially if the cable is not too much smaller than the interior of the raceway. But when the same cable is installed in a nonmetallic raceway, the lines of flux will be unevenly distributed toward the side very near a grounded conductor or surface. Unequal distribution results in greater stresses imposed on the insulation on one side of the cable and may result in premature cable insulation failure. Keeping the conductor in a nonmetallic raceway spaced the specified distance from grounded surfaces or objects helps to reduce this problem. Because GTO cable is usually not shielded, surface lettering on GTO cable could interfere with even distribution of stresses across the dielectric (insulation); thus labeling is on an attached tag, on the coil or reel, or on the smallest unit container, and not on the cable itself.

6. Verify that live parts are enclosed.

Live parts of signs and outline lighting systems must be enclosed. For the purposes of this rule, lamps and neon tubing are not considered to be live parts that require enclosures. Transformers and power supplies with an integral enclosure that enclose live parts within the overall enclosure or in primary and secondary splice enclosures do not require another enclosure. Enclosures that are not metal must be listed. Enclosures must be at least 0.020 in. (0.51 mm) thick if made of sheet copper or aluminum and 0.016 in. (0.41 mm) thick if made of sheet steel. (These dimensions are equivalent to 24-gauge aluminum or copper and 27-gauge steel.)

7. Check locations of signs for protection from damage, spacing from combustible materials, and provisions for drainage if in wet locations.

All signs must be protected from damage by vehicles. One method of meeting this requirement is locating the sign at least 14 ft (4.3 m) above areas where vehicles have access. Neon tubing that is not part of a dry-location portable sign must be protected from damage if it is readily accessible to pedestrians. Signs and outline lighting must be located where it will not cause adjacent combustible materials to be subjected to temperatures over 194°F (90°C). For incandescent and HID lamps and lampholders, a minimum distance of 2 in. (50 mm) from combustible materials is required. When signs are installed in wet locations, they must either be listed as a watertight type or be weatherproof and equipped with drain holes. The drain holes must be from ¼ in. to ½ in. (6 mm to 13 mm) in size and be located in each low point and where they will not be obstructed.

8. Check location and verify accessibility of power supplies, ballasts, and transformers.
Ballasts, transformers, Class 2 power sources, and electronic power supplies for signs and outline lighting must be of the self-contained type or otherwise be enclosed and securely fastened in place. They must be located where they are accessible and as near as practicable (as near as can be accomplished in actual practice) to the lamps or tubing they supply. They must have working space not less than 3 ft (900 mm) in width, depth, and height. They are permitted in attics, soffits, and suspended ceilings, but suspended ceiling grids cannot be used for their support, and where they are located in attics or soffits, an access door and walkway must be provided. The access door cannot be less than 36 in. by 22½ in. (900 mm by 562.5 mm), and the walkway may not be less than 12 in. (300 mm) wide. The walkway must be located in a passageway that is not smaller than 3 ft (900 mm) high and 2 ft (600 mm) wide. Power supplies, ballasts, and transformers may be located immediately behind access doors or panels without walkspaces or passageways if the required working space is provided at the access point. If the attic or soffit space does require a passageway walkway, it must also have a lighting outlet at or near the equipment controlled by a switch located at the point of entry. In effect, these requirements for sign power supplies and the like in these spaces are similar to those for similar spaces that contain mechanical equipment that requires servicing [210.70(C)].

9. Check markings and listings of power supplies and ballasts.
Ballasts must be thermally protected and listed. Class 2 power supplies must be listed both as Class 2 power supplies and for use with electric signs and outline lighting. Transformers and electronic power supplies must also be listed as well as being **identified** for the use. Other requirements of 600.23 that establish maximum voltages (15 kV between conductors and 7.5 kV to ground) and current ratings (300 milliamperes) and requirements for secondary-circuit ground-fault protection and markings are covered by the listing and labeling of the equipment. However, the prohibited connections of secondary circuit outputs in parallel (for greater currents) or series (for greater voltages) are specifically related to field practice, and the inspector should verify that such connections are not made.

10. Check secondary wiring for light-emitting diode (LED) signs.
Wiring methods must comply with any sign manufacturer's instructions in addition to the ordinary rules of *NEC* Chapter 3 and, for Class 2 circuits, the requirements of Part III of Article 725. Class 2 cables must be listed as such, may not be smaller than 22 AWG, and, where installed in wet locations, must be identified for that use. The conductors must be protected from physical damage and connected with listed insulating devices. Listed boxes are required for Class 2 conductor splices if the splices are made in a wall.

Checklist 7-8: Field-Installed Skeleton Tubing (Neon) and Wiring

1. Check wiring methods of neon secondary circuits of 1000 volts or less.
Neon installations up to 1000 volts are subject to some special requirements in addition to other requirements for signs and outline lighting. Ordinary wiring methods of Chapter 3 of the *NEC* may be used if they are suitable for the conditions, including the applied voltage. Conductors may be as small as 18 AWG if they are insulated and listed, but they must be protected from physical damage, and they are still subject to the raceway fill limitations of Table 1 in Chapter 9 of the *NEC*. All conductors passing through openings in metal must be protected by bushings.

2. Check wiring methods of neon secondary circuits over 1000 volts.

Installations of field-installed neon tubing over 1000 volts are subject to many requirements that are in addition to or more stringent than the requirements for general wiring methods in Chapter 3 of the *NEC*. Raceways may contain only one conductor of a secondary circuit and may not be smaller than size ½ (metric designator 16). The use of size ⅜ (metric designator 12) flexible metal conduit is not permitted for this application. Wiring methods are generally limited to RMC, IMC, liquidtight flexible nonmetallic conduit (LFNC), flexible metal conduit (FMC), liquidtight flexible metal conduit (LFMC), EMT, and other metal enclosures. Wiring may also be run on insulators in metal raceways or in other equipment that is listed for neon secondary circuits over 1000 volts. Where nonmetallic wiring methods are used, the conduit must be installed to maintain specific spacings from grounded or bonded parts (including metal building parts) **(FIGURE 7-37)**. The minimum spacing is no less than 1½ in. (38 mm) for systems up to 100 Hz and 1¾ in. (44 mm) for higher frequencies. These are the same clearances required and discussed in Item 5 in the Signs and Outline Lighting checklist. Metal parts of the building may not be used as a secondary return conductor or as an equipment grounding conductor. See Item 5 in the Signs and Outline Lighting checklist for grounding and bonding requirements.

3. Check neon secondary circuits over 1000 volts for appropriate installation techniques for conductors.

Many special installation details apply to neon secondary conductors that operate over 1000 volts. A special cable type is used, Type GTO (gas tube sign and ignition cable), which is typically an unshielded cable rated at 5, 10, or 15 kV. These cables are identified as Types GTO-5, GTO-10, or GTO-15, respectively, available in sizes 18 to 10 AWG. The conductor in the cable may not be smaller than 18 AWG. The conductors must be installed so that they are not subject to physical damage. Unless installed in metal conduit or tubing, GTO cables must be spaced at least 1½ in. (38 mm) from objects other than tubing or insulators and from other GTO cables and other conductors including supply conductors. In order to avoid damage due to stresses or tracking on the insulation, sharp bends in the cable must be avoided, and the insulation must remain on the conductor for at least 2½ in. (65 mm) from the point where it exits a metal raceway to the point that the conductor is bared for connection to an electrode (tubing end). The length of a secondary conductor from the power source to the first neon tube electrode is limited to 20 ft (6 m) where installed in metal conduit or tubing and to 50 ft (15 m) where installed in nonmetallic conduit. Other lengths of secondary conductors should be as short as can be accomplished. Splices must be accessible and made in enclosures that are listed and rated for over 1000 volts. Insulators and bushings must also be listed for neon secondary circuits over 1000 volts.

4. Verify that neon tubing has adequate support, spacing, and protection.

Currents in neon tubing are directly related to the size and length of the tubing. The design of the tubing must prevent continuous overloading beyond the rating of the

FIGURE 7-37 Some requirements for wiring methods and installation of secondary circuit conductors.

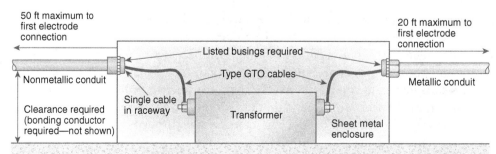

FIGURE 7-38   Three examples of neon tubing supports.

power source. As noted in the Signs and Outline Lighting checklist, neon tubing may not be installed where subject to physical damage and must be guarded from contact by unqualified persons where it is subject to such contact. Listed tube supports must be used to support neon tubing. **FIGURE 7-38** shows three different types of tube supports. Tubing is tied to the one on the left using small wire while the tubing is snapped in place with the ones on the right. Support spacings are not specified in the *NEC* except that a spacing from the nearest surface of at least ¼ in. (6 mm) is required and a support must be provided within 6 in. (152 mm) of the electrode connection (see Item 5 in the Signs and Outline Lighting checklist).

5.   Check electrode connections for locations and suitability.

Electrodes are the conductive points embedded into the ends of neon tubes. Electrode connections are the points where the secondary circuit conductors connect to the neon tubing. The actual connections may be made by an electrode receptacle or by twisting wires together or by another connection device. Electrode receptacles include a spring-loaded contact that is connected to the secondary circuit conductor and makes contact with the neon by pushing the neon tubing end into the receptacle. Such receptacles must be listed. In **FIGURE 7-39**, the spring can be seen in the glass receptacle in the top; the bottom receptacle has an adapter for connection to a metal raceway.

Where twisted wires or other connection methods are used, the enclosure for the connection, called an electrode enclosure, must also be listed. The enclosures must be listed

FIGURE 7-39   Two examples of listed electrode receptacles.

FIGURE 7-40   Listed electrode enclosure suitable for dry and damp locations.

for the location, whether it is dry, damp, or wet, but any listed enclosure may be used in dry locations. **FIGURE 7-40** is an example of a polymeric electrode enclosure that is listed for dry and damp locations. Since electrode terminals may not be located where accessible to unqualified persons, the requirement for a listed enclosure or receptacle helps to meet this requirement. Neon secondary conductors must be supported within 6 in. (150 mm) from an electrode connection. This support may be provided by a tubing support or by a bushing or receptacle. Listed bushings must be used where electrodes penetrate an enclosure if receptacles are not used. Where a bushing or receptacle penetrates a building in a wet location such as on the outside of the building, a listed cap must be used or the opening between neon tubing and a bushing must be sealed.

KEY TERMS

The *NEC* defines the key terms pertaining to commercial and industrial installations as follows (all definitions are from Article 100 unless otherwise noted):

Accessible (as applied to equipment) Admitting close approach; not guarded by locked doors, elevation, or other effective means.

Accessible, Readily (Readily Accessible) Capable of being reached quickly for operation, renewal, or inspections, without requiring those to whom ready access is requisite to use tools, to climb over or remove obstacles, or to resort to portable ladders, and so forth. [*Readily accessible and accessible are separate terms. Electrical equipment does not necessarily have to be accessible to be readily accessible. A piece of equipment may be readily accessible and be behind a locked door, for example. Ready access is concerned with access "for those to whom ready access is requisite." Section 110.26(F) states that "Electrical equipment rooms or enclosures housing electrical apparatus that are controlled by lock(s) shall be considered accessible to qualified persons."*]

Approved Acceptable to the authority having jurisdiction. [*The abbreviation AHJ is often used for "authority having jurisdiction." Everything installed under the NEC is required to be approved. Article 110 provides criteria for an AHJ to determine what should be approved. See also the definition of* Authority Having Jurisdiction.]

Attachment Plug (Plug Cap) (Plug) A device that, by insertion in a receptacle, establishes a connection between the conductors of the attached flexible cord and the conductors connected permanently to the receptacle.

Automatic Performing a function without the necessity of human intervention.

Branch Circuit The circuit conductors between the final overcurrent device protecting the circuit and the outlet(s).

Branch Circuit, Individual A branch circuit that supplies only one utilization equipment. [*Essentially, this definition also implies only one outlet, or in the case of a receptacle outlet, only one receptacle, not a duplex receptacle. The term* individual branch circuit *is not interchangeable with* dedicated branch circuit. *In some cases, such as in Article 600, a circuit is required to "supply no other load." In this case, the circuit is dedicated to a sign, but it could involve more than one outlet or more than one piece of utilization equipment or section that makes up the overall sign or outline lighting.*]

Branch-Circuit Selection Current The value in amperes to be used instead of the rated-load current in determining the ratings of motor branch-circuit conductors, disconnecting means, controllers, and branch-circuit short-circuit and ground-fault protective devices wherever the running overload protective device permits a sustained current greater than the specified percentage of the rated-load current. The value of branch-circuit selection current will always be equal to or greater than the marked rated-load current. [*This definition is taken from Article 440 and applies only to hermetic refrigerant motor-compressors. This value, when provided, or the* Rated Load Current, *whichever is larger, is used in lieu of a full-load current value for this type of equipment, since hermetic refrigerant motor-compressors do not have the horsepower ratings upon which the table values are based.*]

Circuit Breaker A device designed to open and close a circuit by nonautomatic means and to open the circuit automatically on a predetermined overcurrent without damage to itself when properly applied within its rating.

> *Informational Note:* The automatic opening means can be integral, direct acting with the circuit breaker, or remote from the circuit breaker.

> **Instantaneous trip (as applied to circuit breakers)** A qualifying term indicating that no delay is purposely introduced in the tripping action of the circuit breaker.

Inverse time (as applied to circuit breakers)   A qualifying term indicating that there is purposely introduced a delay in the tripping action of the circuit breaker, which delay decreases as the magnitude of the current increases.

Control Circuit   The circuit of a control apparatus or system that carries the electric signals directing the performance of the controller but does not carry the main power current.

Controller   A device or group of devices that serves to govern, in some predetermined manner, the electric power delivered to the apparatus to which it is connected. [*Where used as motor controllers, Article 430 expands on this definition as follows: "a controller is any switch or device that is normally used to start and stop a motor by making and breaking the motor current." (See 430.2.) This is intended to clarify that control devices such as pushbuttons and selector switches are not controllers. Combination controllers, also commonly called* combination starters, *are devices that include disconnecting means, controller, and overload protection in a single unit. Most combination controllers use a fusible switch or circuit breaker as the disconnecting means, so branch-circuit short-circuit and ground-fault protection is also included. Control power transformers, control fuses, and control devices such as pushbuttons and pilot lights are often included as well. Listed self-protected combination controllers are a special category of combination controller that incorporate all of the features mentioned into a single device that may use a single set of contacts for overcurrent, disconnection, and controller functions. Typical examples of controllers are motor starters and lighting contactors, or for small motors, snap switches or manual controllers.*]

Disconnecting Means   A device, or group of devices, or other means by which the conductors of a circuit can be disconnected from their source of supply.

Duty, Continuous   Operation at a substantially constant load for an indefinitely long time. [*This and the following definitions of intermittent, periodic, short-time, and varying duty are used in Article 430 to determine the size of conductors. Motors are assumed to be continuous-duty loads unless the nature of the load is specifically known to be otherwise.*]

Duty, Intermittent   Operation for alternate intervals of (1) load and no load; or (2) load and rest; or (3) load, no load, and rest.

Duty, Periodic   Intermittent operation in which the load conditions are regularly recurrent.

Duty, Short-Time   Operation at a substantially constant load for a short and definite specified time.

Duty, Varying   Operation at loads, and for intervals of time, both of which may be subject to wide variation.

Electric-Vehicle-Charging Systems   [*The definitions that follow are taken from Article 625 and are specific to electric-vehicle-charging systems.*]

Electric Vehicle   An automotive-type vehicle for on-road use, such as passenger automobiles, buses, trucks, vans, neighborhood electric vehicles, electric motorcycles, and the like, primarily powered by an electric motor that draws current from a rechargeable storage battery, fuel cell, photovoltaic array, or other source of electric current. Plug-in hybrid electric vehicles (PHEVs) are considered electric vehicles. For the purpose of this article, off-road, self-propelled electric vehicles, such as industrial trucks, hoists, lifts, transports, golf carts, airline ground support equipment, tractors, boats, and the like, are not included.

Electric Vehicle Coupler   A mating electric vehicle inlet and electric vehicle connector set.

Electric Vehicle Supply Equipment   The conductors, including the ungrounded, grounded, and equipment grounding conductors and the electric vehicle connectors, attachment plugs, and all other fittings, devices, power outlets, or apparatus installed specifically for the purpose of transferring energy between the premises wiring and the electric vehicle.

Informational Note No. 1: For further information, see 625.48 for interactive systems.

Informational Note No. 2: Within this article, the terms *electric vehicle supply equipment* and *electric vehicle charging system equipment* are considered to be equivalent.

Elevators [*The following definitions are specific to elevators, escalators, dumbwaiters, moving walks, and the like and are found in Article 620.*]

Control Room (for Elevator, Dumbwaiter) An enclosed control space outside the hoistway, intended for full bodily entry, that contains the elevator motor controller. The room could also contain electrical and/or mechanical equipment used directly in connection with the elevator or dumbwaiter but not the electric driving machine or the hydraulic machine.

Control Space (for Elevator, Dumbwaiter) A space inside or outside the hoistway, intended to be accessed with or without full bodily entry, that contains the elevator motor controller. This space could also contain electrical and/or mechanical equipment used directly in connection with the elevator or dumbwaiter but not the electric driving machine or the hydraulic machine.

Control System The overall system governing the starting, stopping, direction of motion, acceleration, speed, and retardation of the moving member.

Controller, Motor The operative units of the control system comprised of the starter device(s) and power conversion equipment used to drive an electric motor, or the pumping unit used to power hydraulic control equipment.

Machine Room (for Elevator, Dumbwaiter) An enclosed machinery space outside the hoistway, intended for full bodily entry, that contains the electrical driving machine or the hydraulic machine. The room could also contain electrical and/or mechanical equipment used directly in connection with the elevator or dumbwaiter.

Machinery Space (for Elevator, Dumbwaiter) A space inside or outside the hoistway, intended to be accessed with or without full bodily entry, that contains elevator or dumbwaiter mechanical equipment, and could also contain electrical equipment used directly in connection with the elevator or dumbwaiter. This space could also contain the electric driving machine or the hydraulic machine.

Enclosure The case or housing of apparatus, or the fence or walls surrounding an installation to prevent personnel from accidentally contacting energized parts, or to protect the equipment from physical damage.

Informational Note: See Table 110.28 (of the *NEC*) for examples of enclosure types.
[*Some enclosures are also designed and intended to contain the electrical events such as arcing or sparking that may occur within the enclosure.*]

Energized Electrically connected to, or is, a source of voltage.

Equipment A general term including material, fittings, devices, appliances, luminaires, apparatus, machinery, and the like used as a part of, or in connection with, an electrical installation.

Guarded Covered, shielded, fenced, enclosed, or otherwise protected by means of suitable covers, casings, barriers, rails, screens, mats, or platforms to remove the likelihood of approach or contact by persons or objects to a point of danger.

Hermetic Refrigerant Motor-Compressor A combination consisting of a compressor and motor, both of which are enclosed in the same housing, with no external shaft or shaft seals, the motor operating in the refrigerant. [*Equipment that consists of or contains hermetic refrigerant motor-compressors is the subject of and reason for Article 440. A significant characteristic of hermetic refrigerant motor-compressors is that they have no external shaft and therefore have no horsepower or equivalent ratings on which to base some of the rules of Article 430. Branch-circuit selection current or rated-load current is used for sizing circuit components for equipment under Article 440.*]

Hoistway   Any shaftway, hatchway, well hole, or other vertical opening or space in which an elevator or dumbwaiter is designed to operate.

Identified (as applied to equipment)   Recognizable as suitable for the specific purpose, function, use, environment, application, and so forth, where described in a particular *Code* requirement.

> *Informational Note:* Some examples of ways to determine suitability of equipment for a specific purpose, environment, or application include investigations by a qualified testing laboratory (listing and labeling), an inspection agency, or other organizations concerned with product evaluation.

In Sight From (Within Sight From, Within Sight)   Where the *Code* specifies that one equipment shall be "in sight from," "within sight from," or "within sight of," and so forth, another equipment, the specified equipment is to be visible and not more than 15 m (50 ft) distant from the other.

Leakage-Current Detector-Interrupter (LCDI)   A device provided in a power supply cord or cord set that senses leakage current flowing between or from the cord conductors and interrupts the circuit at a predetermined level of leakage current. [*This type of protective device or an AFCI (arc-fault circuit interrupter) is required in the factory-supplied cord for room air conditioners of the type often used in commercial occupancies such as hotels, nursing homes, and the like, as well as in dwelling units.*] [*from Article 440*]

Live Parts   Energized conductive components.

Motor Control Center (MCC)   An assembly of one or more enclosed sections having a common power bus and principally containing motor control units. [*A motor control center is a specific type of equipment assembly, designed and tested to comply with UL 845. It is not just a centralized point of motor control.*]

Outline Lighting   An arrangement of incandescent lamps, electric-discharge lighting, or other electrically powered light sources to outline or call attention to certain features such as the shape of a building or the decoration of a window.

Overcurrent   Any current in excess of the rated current of equipment or the ampacity of a conductor. It may result from overload, short circuit, or ground fault.

> *Informational Note:* A current in excess of rating may be accommodated by certain equipment and conductors for a given set of conditions. Therefore the rules for overcurrent protection are specific for particular situations.

[*The rules for overcurrent protection in motor circuits are good examples of the statement "rules for overcurrent protection are specific for particular situations." The branch-circuit overcurrent device in a motor or air-conditioning circuit commonly has a rating higher than the ampacity of the conductors, partly because overloads are controlled by other devices.*]

Overload   Operation of equipment in excess of normal, full-load rating, or of a conductor in excess of rated ampacity that, when it persists for a sufficient length of time, would cause damage or dangerous overheating. A fault, such as a short circuit or ground fault, is not an overload.

Personnel Protection System   A system of personnel protection devices and constructional features that when used together provide protection against electric shock of personnel. [*The purpose of a Personnel Protection System is to provide protection roughly equivalent to the GFCI protection required in many other applications.*]

Power-Supply Cord   An assembly consisting of an attachment plug and length of flexible cord that connects the electric vehicle supply equipment (EVSE) to a receptacle.

Qualified Person   One who has skills and knowledge related to the construction and operation of the electrical equipment and installations and has received safety training to recognize and avoid the hazards involved.

> *Informational Note:* Refer to NFPA 70E, *Standard for Electrical Safety in the Workplace*, 2012, for electrical safety training requirements.

[This revised definition is much more specific than the definition of qualified person contained in the NEC prior to the 2002 edition. Skills, knowledge, and training are now clear requirements as well as the intended outcome of the safety training. The previous definition required only that a person be "familiar" with the construction, operation, and hazards.]

Rated-Load Current *[Definition from Article 440]* The rated-load current for a hermetic refrigerant motor-compressor is the current resulting when the motor-compressor is operated at the rated load, rated voltage, and rated frequency of the equipment it serves. *[This definition is taken from Article 440 and applies only to hermetic refrigerant motor-compressors. This value or the branch-circuit selection current, whichever is larger, is used in lieu of a full-load current value for this type of equipment, since hermetic refrigerant motor-compressors do not have the horsepower ratings upon which the table values are based.]*

Signs *[The following definitions are specific to signs and outline lighting, including all installations using neon tubing, and are found in Article 600.]*

    Neon Tubing Electric-discharge luminous tubing, including cold cathode luminous tubing, that is manufactured into shapes to illuminate signs, form letters, parts of letters, skeleton tubing, outline lighting, other decorative elements, or art forms, and filled with various inert gases.

    Section Sign A sign or outline lighting system, shipped as subassemblies, that requires field-installed wiring between the subassemblies to complete the overall sign. The subassemblies are either physically joined to form a single sign unit or are installed as separate remote parts of an overall sign.

    Skeleton Tubing Neon tubing that is itself the sign or outline lighting and is not attached to an enclosure or sign body.

Special Permission The written consent of the AHJ.

Switch, General-Use A switch intended for use in general distribution and branch circuits. It is rated in amperes, and it is capable of interrupting its rated current at its rated voltage.

Switch, General-Use Snap A form of general-use switch constructed so that it can be installed in device boxes or on box covers, or otherwise used in conjunction with wiring systems recognized by the *Code.*

Switch, Motor-Circuit A switch rated in horsepower that is capable of interrupting the maximum operating overload current of a motor of the same horsepower rating as the switch at the rated voltage. *[Many small general-use snap switches also have horsepower ratings to permit their use as motor-circuit switches.]*

Thermally Protected (as applied to motors) The words *Thermally Protected* appearing on the nameplate of a motor or motor-compressor indicate that the motor is provided with a thermal protector.

Transformer The word *transformer* is intended to mean an individual transformer, single- or polyphase, identified by a single nameplate, unless otherwise indicated in this [450] article. *[This definition is taken from Article 450 and, as such, applies only within the scope of Article 450. This is a valuable and important definition because although a 3-phase transformer is technically three transformers in a single enclosure, it may be treated as a single transformer for the purposes of applying overcurrent protection and for applying rules that are based on transformer size. With regard to overcurrent protection, this definition is further clarified to mean "a transformer or polyphase bank of two or more single-phase transformers operating as a unit."]*

KEY QUESTIONS

1. What are the horsepower ratings of motors, which ones are high-efficiency types, and what are the design letters?
2. What are the nameplate FLA ratings of motors?
3. Are there disconnects installed at the controller and at the motor locations?
4. What type of overcurrent device is used for short-circuit and ground-fault protection?
5. What are the minimum circuit ampacities and maximum overcurrent device ratings marked on air-conditioning and refrigeration unit nameplates?
6. Do the nameplates on hermetic refrigerant motor-compressor equipment specify a certain type of overcurrent device?
7. What are the impedances, kVA ratings, insulation class, and primary and secondary current and voltage ratings marked on transformer nameplates?
8. If ventilation clearances are marked on transformers, what are they?
9. Are any transformers liquid-filled, and if so, what type of liquid is used?
10. What are the current ratings of capacitors?
11. What means of discharge is provided on capacitors?
12. What are the liquid capacities of capacitors?
13. Where will elevators or escalators be located?
14. What is the nameplate information for the elevator or escalator equipment?
15. Where will electric-vehicle-charging equipment be located and what types will be used?
16. Where will signs or outline lighting be located and what are the electrical ratings?
17. What type of signs or outline lighting will be used?

CHECKLISTS

Checklist 7-1: Motors				
✔	Item	Inspection Activity	*NEC* Reference	Comments
	1.	Verify that ampacities and sizing of components other than overload devices are based on table values rather than on nameplate values.	430.6, 430.122, Tables 430.247 through 430.250	
	2.	Verify that conductor ampacities for individual motors are at least 125 percent of table FLC.	430.22(A), 430.122	
	3.	Check conductors supplying multiple motors or motors and other loads for ampacities equal to at least the sum of other motors or loads plus 125 percent of largest motor.	430.24, 220.14(C), 220.50	
	4.	Verify that motor overload protection does not exceed permitted values.	430.31 through 430.44, 430.126	
	5.	Verify that short-circuit and ground-fault protection for motor branch circuits does not exceed permitted values.	430.51 through 430.58 430.130 430.131	
	6.	Verify that short-circuit and ground-fault protection for motor feeders does not exceed permitted values.	430.61 through 430.63	
	7.	Check motor control circuits for proper overcurrent protection.	430.71 through 430.74	
	8.	Verify that motor controllers are provided for motors and that they are of the proper type and have adequate ratings, including short-circuit current ratings.	430.8 and 430.81 through 430.90	

✔	Item	Inspection Activity	*NEC* Reference	Comments
		Checklist 7-1: Motors		
	9.	Check MCCs for proper ratings, protection, workspace, and dedicated space.	110.26, 430.92 through 430.98	
	10.	Verify that motor disconnects are of the proper type and rating.	430.109, 430.110	
	11.	Verify that controller disconnects are in sight of controllers, are readily accessible, and have adequate workspace.	110.26, 430.102(A), 430.107, 404.8	
	12.	Verify that motor disconnects are in sight of motors, are readily accessible, and have adequate workspace.	110.26, 430.102(B), 430.107	

✔	Item	Inspection Activity	*NEC* Reference	Comments
		Checklist 7-2: Air-Conditioning and Refrigerating Equipment		
	1.	Identify equipment subject to Article 440—Equipment Containing Hermetic Refrigerant Motor-Compressor(s).	440.1, 440.2	
	2.	Identify the applicable nameplate information for the equipment.	440.4	
	3.	Verify that branch-circuit conductor sizes are adequate on the basis of the applicable nameplate information.	440.31 through 440.35	
	4.	Verify that conductors supplying several units are adequately sized.	430.24, 430.25	
	5.	Verify that branch-circuit overload protection is provided and properly sized.	440.51 through 440.55	
	6.	Verify that branch-circuit short-circuit and ground-fault protection is provided and properly sized.	440.21, 440.22	
	7.	Verify that feeder short-circuit and ground-fault protection is provided and properly sized where applicable.	430.61 through 430.63	
	8.	Verify that controllers have adequate ratings, including short-circuit current ratings, where they are not part of listed multimotor or combination-load equipment.	440.4(B), 440.41	
	9.	Verify that disconnecting means have ratings adequate for the equipment.	440.12, 440.13	
	10.	Verify that the disconnecting means are within sight and readily accessible from the equipment and that working spaces are adequate.	110.26, 440.14	
	11.	Verify that conductors, receptacles, cords, and overcurrent devices for room air conditioners are properly sized and that LCDI devices or AFCI protection is provided for cords.	440.60 through 440.65	
	12.	Check for receptacles and adequate lighting for servicing of mechanical equipment.	210.8, 210.63, 210.70	

CHECKLISTS

Checklist 7-3: Transformers

✔	Item	Inspection Activity	*NEC* Reference	Comments
	1.	Identify transformers that are covered by Article 450.	450.1, 450.2	
	2.	Verify that overcurrent protection for transformers over 1000 volts is provided and properly sized.	Table 450.3(A)	
	3.	Verify that overcurrent protection for transformers 1000 volts or less is provided and properly sized.	Table 450.3(B)	
	4.	Verify that overcurrent protection is provided for transformer primary conductors.	240.4, 240.21(B), 240.100	
	5.	Verify that overcurrent protection is provided for transformer secondary conductors.	240.4, 240.21(C), 240.100	
	6.	Check transformer installations for adequate ventilation and spacing from walls and obstructions.	450.9, 450.10(A)	
	7.	Check transformers for ready access or proper installation in the open or in hollow spaces.	450.13	
	8.	Verify that transformers are supplied with a disconnecting means.	450.14	
	9.	Check indoor dry-type transformers for separation from combustibles or, based on ratings, installation in fire-resistant rooms or vaults.	450.21	
	10.	Check outdoor dry-type transformers for weatherproof enclosures.	450.22	
	11.	Verify that liquid-insulated transformers are installed in accordance with the requirements for the location and type of insulating liquid.	450.23 through 450.28	
	12.	Check transformer vaults for adequate construction, access, ventilation, and drainage and for foreign systems in vaults.	450.41 through 450.48	

Checklist 7-4: Capacitors

✔	Item	Inspection Activity	*NEC* Reference	Comments
	1.	Check capacitors for proper enclosures and guards.	460.2	
	2.	Verify that conductors are properly sized on the basis of the current rating of the capacitor(s).	460.8(A)	
	3.	Verify that capacitors other than those connected to the load side of motor overload devices have disconnects and proper overcurrent protection.	460.8(B) and (C)	

✔	Item	Inspection Activity	*NEC* Reference	Comments
		Checklist 7-4: Capacitors		
	4.	Verify that overload device ratings have been corrected where capacitors are connected to the load side of motor overload devices.	460.9	
	5.	Check capacitors over 1000 volts for proper switching, overcurrent protection, identification, and grounding.	460.24 through 460.27	
	6.	Verify that a proper means for discharge has been provided for capacitors.	460.6, 460.28	

✔	Item	Inspection Activity	*NEC* Reference	Comments
		Checklist 7-5: Elevators, Dumbwaiters, Escalators, Platform Lifts, and Stairway Chairlifts		
	1.	Verify voltage limitations and presence of warning labels where voltage exceeds 600 volts.	620.3	
	2.	Verify that all live parts are enclosed.	620.4	
	3.	Verify required working clearances around elevator electrical equipment.	620.5	
	4.	Check conductors for proper insulation type and minimum size.	620.11, 620.12	
	5.	Verify ampacity of branch-circuit and feeder conductors.	620.13	
	6.	Verify proper wiring methods.	620.21	
	7.	Verify the required branch circuits for car lighting, receptacle(s), ventilation, and air conditioning.	620.22, 620.85	
	8.	Verify the required branch circuit for machine-room lighting and receptacle.	620.23, 620.85	
	9.	Verify the required branch circuit for hoistway pit lighting and receptacle.	620.24 , 620.85	
	10.	Check for required receptacle and light switch provisions in machine rooms and pits.	620.23, 620.24	
	11.	Verify that only elevator-associated wiring is installed in hoistways and machine rooms.	620.37	
	12.	Check elevator machine disconnecting means for proper type, operation, and location.	620.51	

CHECKLISTS

(continues)

Checklist 7-5: Elevators, Dumbwaiters, Escalators, Platform Lifts, and Stairway Chairlifts

✔	Item	Inspection Activity	*NEC* Reference	Comments
	13.	Verify disconnecting means for car lighting, receptacles, and ventilation.	620.53	
	14.	Verify disconnecting means for car heating and air conditioning.	620.54	
	15.	Check overcurrent protection for proper rating and coordination.	620.61, 620.62	
	16.	Verify that only permitted equipment is located in machine rooms.	620.71	
	17.	Verify GFCI receptacles on car tops and pits and GFCI protection for receptacles in machine rooms.	620.85	
	18.	Check operation of elevator machine disconnecting means where emergency or standby power is provided.	620.91	

Checklist 7-6: Electric-Vehicle-Charging Equipment

✔	Item	Inspection Activity	*NEC* Reference	Comments
	1.	Verify that all associated equipment, materials, devices, and fittings are listed.	625.5	
	2.	Check the suitability of the electric vehicle coupler.	625.10	
	3.	Verify that coupler and cable safeguards have been provided.	625.18, 625.19	
	4.	Verify the rating of branch-circuit and feeder overcurrent devices.	625.40, 625.41	
	5.	Verify that a personnel protection system has been provided.	625.22	
	6.	Check the location of disconnecting means rated over 60 amperes or 150 volts to ground.	625.42	
	7.	Verify that backfeed prevention provisions have been provided.	625.46, 625.48	
	8.	Check the location and height of indoor and outdoor charging equipment.	625.50	
	9.	Verify the necessity for and amount of ventilation for indoor charging locations.	625.15, 625.52	

Checklist 7-7: Signs and Outline Lighting

✔	Item	Inspection Activity	*NEC* Reference	Comments
	1.	Check each commercial building and occupancy for sign outlets on dedicated circuits.	600.5, 210.62	
	2.	Check signs, section signs, and outline lighting for listing and marking.	600.3, 600.4	
	3.	Check branch circuits supplying signs for permissible ratings.	600.5(B), 600.10	
	4.	Verify that an external disconnect is provided and within sight of each part of a sign or equipped with locking means.	600.6	
	5.	Verify that metal parts of signs and outline lighting are bonded together and connected to a properly sized equipment grounding conductor.	600.7	
	6.	Verify that live parts are enclosed.	600.8	
	7.	Check locations of signs for protection from damage, spacing from combustible materials, and provisions for drainage if in wet locations.	600.9	
	8.	Check location and verify accessibility of power supplies, ballasts, and transformers.	600.21	
	9.	Check markings and listings of power supplies and ballasts.	600.22, 600.23, 600.24	
	10.	Check secondary wiring for LED signs.	600.33	

Checklist 7-8: Field-Installed Skeleton Tubing (Neon) and Wiring

✔	Item	Inspection Activity	*NEC* Reference	Comments
	1.	Check wiring methods of neon secondary circuits of 1000 volts or less.	600.31	
	2.	Check wiring methods of neon secondary circuits over 1000 volts.	600.32(A)	
	3.	Check neon secondary circuits over 1000 volts for appropriate installation techniques for conductors.	600.32(B) through (K)	
	4.	Verify that neon tubing has adequate support, spacing, and protection.	600.41	
	5.	Check electrode connections for locations and suitability.	600.42	

CHECKLISTS

CHAPTER 8

Hazardous Locations

OVERVIEW

Hazardous (classified) locations are areas where flammable or explosive atmospheres exist or may develop due to the presence of flammable gases, flammable liquid-produced vapors, or combustible liquid-produced vapors; **combustible dusts**; or ignitable quantities of fibers/flyings. The objective of the *National Electrical Code*® (*NEC*®) rules on hazardous locations is to minimize the likelihood that such atmospheres may be ignited by the normal or abnormal operation of the electrical system. This objective may be met by reducing or eliminating the hazardous atmosphere, by locating electrical equipment outside of the hazardous area, or by implementing various protective techniques that are designed to keep electrical equipment from becoming an ignition source.

The terms *hazardous* and *classified* are used interchangeably in relation to areas or locations, *hazardous* being the more common term, which was used exclusively in the *NEC* until the 1975 edition, and *classified* being the more technically accurate term. *Hazardous* areas are *classified* according to the type of flammable or combustible material and the degree of risk that is present.

The *NEC* is intended to define and categorize the hazardous areas, describe protective techniques, and provide installation requirements for electrical wiring and equipment. The *NEC* is generally not intended to define the boundaries or determine the classification of specific areas. Other codes and standards, including a number of National Fire Protection Association (NFPA) standards, are used to classify areas. The *NEC* does provide classification information and specific requirements for five commonly encountered occupancies and uses that involve hazardous locations: (1) commercial garages; (2) aircraft hangars; (3) motor fuel-dispensing facilities; (4) bulk storage facilities; and (5) spray application, dipping, and coating operations. Determining whether an area needs to be electrically classified and, if so, determining the extent of the classified area is generally not the sole purview of an electrical inspector. Those more familiar with the particular process and the materials involved, such as a facility owner, safety manager, electrical engineer, or industrial engineer are vitally important in determining the need for area classification. Fire marshals and similar safety personnel are also stakeholders in the area classification process. Industry standards, including many NFPA documents, provide essential information that can be used in making the proper area classification determination. NFPA 497-2012, *Recommended Practice for the Classification of Flammable Liquids, Gases, or Vapors and of Hazardous (Classified) Locations for Electrical Installations in Chemical Process Areas*, and 499-2012, *Recommended Practice for the Classification of Combustible Dusts and of Hazardous (Classified) Locations for Electrical Installations in Chemical Process Areas* contain recommendations for area classification and provide practical and prudent approaches for assessing processes and materials in respect to whether electrical ignition sources pose an explosion hazard.

SPECIFIC FACTORS

One primary purpose of the *NEC* is to reduce the fire hazards that are related to electrical wiring and equipment. Enclosures and raceways are required for most electrical equipment and wiring to contain any arcs and sparks that may be produced by either normal or abnormal conditions in the electrical system. Hazardous locations present a special problem because the atmosphere itself may be explosive or the electrical equipment may become blanketed by easily ignitable materials. The presence of hazardous materials, not the electrical system itself, creates a classified area. The best practice in classified areas is to reduce the amount of electrical equipment in the area. However, electrical wiring and equipment are often needed in such areas. Since operation of electrical equipment represents a possible ignition source, special measures are needed to reduce the likelihood that the electrical equipment could ignite the atmosphere or the flammable or combustible materials.

Hazardous locations are divided into three basic types: **Class I**, **Class II**, and **Class III**. Class I locations are created by the presence of ignitable quantities of flammable liquid-produced vapors or combustible liquid-produced vapors, or gases. Class II locations are created by combustible dusts that may become suspended in air. Class III locations are created by ignitable fibers/flyings. Each class is subdivided into two divisions: **Division 1 and Division 2**. Division 1 represents a hazardous condition that is normally or frequently present. Division 2 represents a hazardous condition that is present only infrequently or only in the event of accidents or equipment or system failures. Equipment that may be used in Class I and Class II areas must also be suitable for the specific materials that create the classification. Specific materials are divided into groups. Groups A, B, C, and D are Class I materials, and Groups E, F, and G are Class II materials. In most cases, the equipment used must be suitable for the class, division, and group. (**TABLE 8-1** indicates considerations that can be used in the decision-making process as to whether an area should be classified and, if it is, whether it is a Division 1 or Division 2 location.)

Article 500 describes many different protection methods that may be used to reduce the risk of having an electrical system in a hazardous area. These protection methods vary somewhat according to the types of materials that are present. The protection methods include explosionproof equipment for Class I areas, dust-ignitionproof equipment for Class II areas, and **dusttight** equipment for Class II, Division 2, and Class III areas. In addition, **purged and pressurized** equipment and enclosures, **intrinsically safe equipment**, **nonincendive circuits** and equipment, oil immersion, and hermetically sealed equipment may be used in certain classifications.

All of these protection methods provide a high degree of safety if they are properly selected, installed, and maintained. Because proper selection and use are so important, designers, installers, and inspectors of electrical systems in hazardous areas must exercise special care with the installation and maintenance of electrical equipment in classified areas. The design of electrical systems operating in hazardous or classified areas is often well documented in the engineering plans. These drawings, details, and other information are valuable and are generally available to the inspector to aid in establishing the area classification, the boundaries of the classified areas, and the acceptable method(s) of protection for the electrical system. In fact, 500.4 in the *NEC* requires documentation of area classification and requires the documentation to be available to the inspector, and others.

This manual concentrates on the Division classification scheme. **Zone classification** is an alternative scheme that also provides alternative protection methods. This scheme and the protection methods are covered by Articles 505 and 506 in the *NEC*. Article 504: **Intrinsically Safe Systems**, covers a protection method that is used under both the division and zone classifications.

TABLE 8-1   Hazardous (Classified) Locations

Classification		Description	Example
Class I	Division 1	Ignitable concentrations of flammable gases, flammable liquid-produced vapors, or combustible liquid-produced vapors are present: • Under normal operating conditions • During repair or maintenance operations • Where breakdown or faulty operation of process equipment might also cause simultaneous failure of electrical equipment	 Gasoline station (dispensers)
	Division 2	• Flammable gases, flammable liquid-produced vapors, or combustible liquid-produced vapors are normally confined within closed containers or systems. • Ignitable concentrations of gases or vapors are normally prevented by positive mechanical ventilation. • Areas adjacent to Division 1 locations into which ignitable concentrations of gases or vapors might occasionally flow.	 Bulk storage plant or process piping
Class II	Division 1	• Explosive or ignitable mixtures of combustible dusts are present under normal operating conditions. • Explosive or ignitable mixtures of combustible dusts are present where breakdown or faulty operations of process equipment might also cause simultaneous failure of electrical equipment. • Electrically conductive dusts are present.	 Grain elevator
	Division 2	• Explosive or ignitable mixtures of combustible dusts are not normally present and accumulations of dust are normally insufficient to interfere with normal equipment operation. • Dusts may be in suspension in air under infrequent, abnormal handling or processing conditions, and dust accumulations may be sufficient to interfere with electrical equipment heat dissipation or be ignited by abnormal operation or failure of electrical equipment.	Area outside/near grain elevator openings, or near a coal-fired boiler
Class III	Division 1	• Ignitable fibers or materials producing combustible flyings are handled, manufactured, or used in these areas.	Textile mill or some woodworking areas
	Division 2	• Ignitable fibers/flyings are stored or handled in these areas.	Storage of cotton, wool, or similar materials (bales)

KEY QUESTIONS

1. What are the class, the division (or zone), and, for Class I and Class II areas, the group classification of the hazardous areas?

Since most equipment used in hazardous locations must be **identified** for class, division, and group, the answer to this question provides the basis for approval of equipment used in classified areas. Also, in order to answer the question about the group, the specific materials upon which the classification is based must be known. In addition, the wiring methods that are permitted in any particular classified area depend on the class and division of the area. The answer to this question should be provided by the documentation that is required where hazardous locations exist. Where classification of Class I areas is done by zone rather than division, the protection methods may be somewhat different, but the requirement for documentation of the area classification is the same, and equipment will still have to be identified for the use. In many cases protection methods that are intended specifically for use in division-classified areas may also be used in zone-classified areas, but the reverse is not generally true.

2. Where are the boundaries of the classified areas?

Besides defining the areas where hazardous wiring methods and equipment are required to be used, the boundaries establish the locations and types of some of the required seals, especially in Class I areas. This information should always be part of the required documentation. Such a drawing may be obtained in a standard or may have to be developed by the person responsible for area classification (**FIGURE 8-1**).

3. What wiring methods are used?

The wiring methods that may be used in a classified location are determined according to the class and division (or zone) of the area. The wiring methods used must be included in the wiring methods permitted for the specific classified area. The wiring methods that are intended to be used can be determined from plans or by inspection. Although protection methods may be different for zone-classified areas than they are for division-classified areas, the permitted wiring methods are mostly the same.

4. What types of electrical equipment are located in the hazardous location?

Articles 501 through 506 each contain specific rules that apply to various types of equipment. Some types of equipment may not be available for specific group classifications, and heat-producing equipment must be marked with temperature ratings or other information to indicate its suitability for a specific area. The equipment types can be identified on plans or by inspection. The suitability of equipment and proper installation can be verified only by inspection of the equipment.

5. Is the use or the occupancy containing classified areas one of those specifically covered by the *NEC*?

The five articles in the *NEC* that cover specific uses or occupancies, Articles 511 and 513 through 516, contain some specific rules that may permit or require different methods

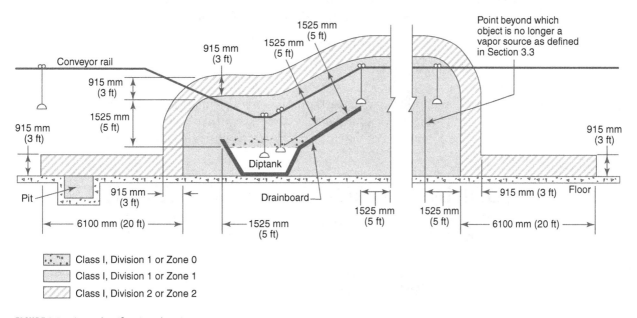

FIGURE 8-1   Area classification drawing.

and equipment than would be permitted or required in other occupancies of the same class and division, and provide more detailed rules about equipment outside classified areas in those occupancies.

PLANNING FROM START TO FINISH

Installations that include hazardous areas vary widely in their extent and in the types of equipment installed. The extent of the areas may vary from a room or a small area to a major portion of an industrial plant. Some installations involve underground wiring, while others do not. If properly and ideally planned, an installation that includes a hazardous location will not have electrical equipment in the hazardous area. For example, a paint spray booth may be designed without any electrical equipment inside the booth or within the classified areas outside the booth. In such a case, the hazardous location inspection may involve only verification that all equipment is outside the classified area boundaries and that the ventilation interlocks, if used, are functional and effective. Similarly, enclosures, which may include entire rooms or buildings in some cases, may be made nonhazardous by purging and pressurization, in which case the inspection should be focused on verifying that the design and use of the purging and pressurization equipment are in accordance with NFPA 496-2013, *Standard for Purged and Pressurized Enclosures for Electrical Equipment.*

In most cases, the inspections of wiring and equipment in classified areas follow a pattern similar to that for other inspections. If underground wiring is involved, the installation will require a separate inspection to verify the suitability of the wiring methods, the adequacy of cover, and encasement in concrete, where required. Underground inspections focus primarily on wiring methods. At least one additional inspection will be required to check above-ground wiring methods, equipment installations, grounding and bonding, location and completion of seals, and other aspects of the installation.

The inspector must be thoroughly familiar with the classification of the areas, especially the class, division or zone, and group classifications. In most cases, the inspection and approval of equipment will involve comparing the area classification with the listing and labeling of the equipment. Appropriate wiring methods are easier to identify, but the suitability of wiring methods still depends on the class and division or zone classification of the areas. Listing and labeling will also identify the equipment that is factory sealed so that required locations for seals can be determined more accurately.

Above-ground inspections can be divided into four main areas:
1. Wiring methods
2. Equipment, including fixtures and wiring devices
3. Grounding and bonding
4. Sealing

This grouping addresses the methods used for distributing power, the equipment used to control and utilize power, the pathways used to equalize potentials and carry fault currents, and the completion of the enclosure system at equipment and, for Class I, at area boundaries.

WORKING THROUGH THE CHECKLISTS

Note: The checklists and the following discussions about Class I and Class II areas are based primarily on classification by division rather than by zone, but the basic considerations are the same either way. Refer to Article 505 of the *NEC* for equipment and protection methods that may be used under the Class I zone classification system and to Article 506 for areas classified due to dust or fibers/flyings.

Checklist 8-1: Class I Locations

1. Confirm the classification of areas, including class, division or zone, and group.

All area classification is required to be documented. The extent of classified areas, the classification of the areas, and the group designation of the flammable liquids, gases, or vapors should be shown on plans or other documents. The extent of the classified areas, as well as the class, division, and group, is necessary to determine the acceptability of equipment and wiring methods. Gasoline-dispensing areas are a common occupancy in which Class I, Division 1, and Division 2 locations exist **(FIGURE 8-2)**.

2. Verify the suitability of the wiring methods being used.

In Class I areas, wiring methods are limited primarily to threaded rigid metal conduit (RMC), threaded intermediate metal conduit (IMC), or Type MI (mineral-insulated, metal-sheathed) cable. Other methods permitted in specific cases include Type PVC (polyvinyl chloride conduit) or Type RTRC (reinforced thermosetting resin conduit) underground encased in concrete, and, in industrial installations, Type MC-HL (metal-clad-hazardous locations) or Type ITC-HL (instrumentation tray cable-hazardous locations) cables with a continuous (noninterlocked) sheath and an overall jacket. Other methods and some additional cable types are permitted in cable trays in Class I, Division 2 areas. All cable types (MI, MC-HL, and ITC-HL) must be terminated in fittings that are listed for the application.

Boxes and other enclosures in **Class I, Division 1** areas are required to be **explosionproof**. **FIGURE 8-3** shows two types of explosionproof junction box construction. The marking (identification) on the boxes aids the inspector in determining suitability for a specific location.

Fittings and connections to enclosures are required to be threaded. Enclosures in **Class I, Division 2** areas are required to be explosionproof only when they contain overcurrent devices, relays, or similar equipment with make-or-break contacts that are not hermetically sealed, oil immersed, or factory sealed.

A termination fitting for Type MI cable is shown in **FIGURE 8-4**. The construction of MI cable prevents the passage of gases and vapors through the cable core. It is one of a limited number of wiring methods permitted in Class I, Division 1 locations, and the only one that is inherently sealed and fire-rated.

In zone-classified areas, only **intrinsically safe circuits** and wiring are permitted in Zone 0. Beyond that, the protection methods may vary somewhat, but wiring methods are essentially the same in Zone 1 as in Division 1 and essentially the same in Zone 2 as in Division 2.

3. Verify that seals are located as required, properly installed, and sealed.

Explosionproof seals in Class I areas are intended to complete the explosionproof enclosure where conduits or cables enter boxes or other enclosures where arcing or sparking may occur. Seals are also used at boundaries of hazardous locations to minimize the passage of gas or vapor between areas and to prevent an internal explosion in a Division 1 area from propagating into other areas. Seals used at boundaries between Class I, Division 2 areas and unclassified areas are not required to be

FIGURE 8-2   Gasoline dispensing is a typical Class I location.

FIGURE 8-3A   Explosionproof junction box with screw-type cover.

FIGURE 8-3B   Explosionproof junction box with bolted flanged cover.

FIGURE 8-4 Type MI cable and termination fitting. (Courtesy of Tyco Thermal Controls.)

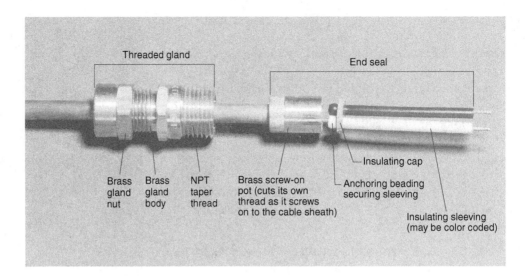

FIGURE 8-5 Explosionproof sealing fitting with automatic drain plug. (Courtesy of Appleton Electric LLC, Emerson Industrial Automation.)

explosionproof, but are required to be identified for minimizing passage of gases and vapors. (Because fittings identified only for minimizing passage of gases and vapors may be difficult or impossible to find and since this requirement is fulfilled by explosionproof seals that are readily available, explosionproof seals are usually used.) **FIGURE 8-5** illustrates an explosionproof seal fitting designed for conduit. This seal fitting is intended for vertical installations only. The diagonal plug is for both installation of the sealing fiber and compound and for inspection of the completed seal. This example includes a drain port and fitting as well. The instructions provided with the seal contain specific information that is critical to the proper installation.

Seals are also used where 2-in. (51-mm) or larger conduits enter enclosures to limit the volume of enclosure-conduit assemblies and reduce the effects of pressure piling. Pressure piling may cause an increase in explosion pressure as an internal explosion or flame front moves through a conduit system.

Some equipment, such as luminaires, switches, and receptacles, are available in factory-sealed assemblies. Factory-sealed equipment is marked for ease of identification in the field. Factory sealing eliminates the need for seals at equipment enclosures but does not change the requirements for seals at area boundaries. **FIGURE 8-6** illustrates the required locations of seals in a Class I, Division 1 location. These seals complete the explosionproof enclosure, prevent pressure piling, and minimize the passage of gases and vapors from the hazardous to the nonhazardous location. The implication in the illustration is that the "light fixture" or luminaire is factory sealed if the luminaire produces high temperatures, but the seals at the pushbutton station or the switch could also be eliminated if these devices were marked as factory-sealed.

4. Check materials used for flexible connections, such as explosionproof flex and flexible cords, for suitability.

Flexible connections are permitted in Class I, Division 1 areas only where "necessary." They should not be used just for convenience, such as to make a connection easier or to avoid bending a conduit. Generally, flexible connections in Class I, Division 1 areas are required to be made with explosionproof flexible fittings (**FIGURE 8-7**). However, extra-hard usage cord terminated with fittings that are listed for the location may be used for connections to portable lighting or utilization equipment including portable assemblies of devices such as receptacles and switches that are listed for the location but are not

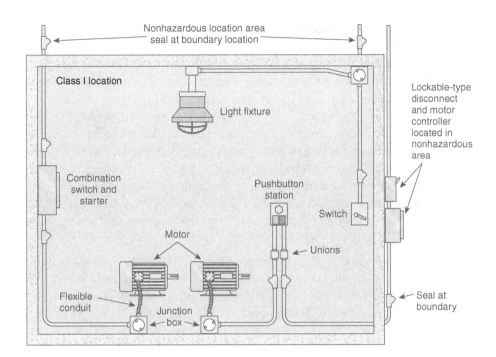

FIGURE 8-6   Required seals in a Class I, Division 1 location. (Source: *National Electrical Code® Handbook*, NFPA, Quincy, MA, 2011, Exhibit 501.8.)

utilization equipment. Although not portable equipment as defined, many submersible pumps and some mixers may also be connected by flexible cord. Section 501.140 covers the specific applications where flexible cord is permitted in Class I, Division 1 and 2 areas. The cords are required to be unspliced, sealed at boxes or other enclosures that are explosionproof, and terminated with fittings listed for the location. Tray cable type TC-ER-HL can be used to make flexible connections to equipment rated 600 volts or less in Division 1 locations, but only within an industrial setting that has qualified persons maintaining the electrical system. Some other wiring methods, such as flexible metal conduit (FMC), liquidtight flexible metal conduit (LFMC), or nonmetallic conduit, may be used in Class I, Division 2 areas.

5. Check flexible cord connectors and receptacles for suitability.

In general, portable cords are permitted in Class I locations only for portable equipment or for equipment connections where a high degree of movement is needed. Receptacles and attachment plugs used in Class I locations must be grounding type and must be identified for the location. Such receptacles and connectors are usually available as products that are listed for the location. Some limited alternatives are permitted for process control instruments. The receptacle and attachment plug assembly shown in **FIGURE 8-8** is designed so that a connection or disconnection cannot be made while the receptacle is energized. The assembly is suitable for use in a Class I, Division 1 or 2 location.

6. Verify that equipment temperature markings are not greater than the ignition temperature of the gases or vapors involved.

Equipment nameplates are furnished with identification numbers that indicate the maximum surface temperature

FIGURE 8-7   Flexible explosionproof fitting. (Courtesy of Cooper Crouse-Hinds.)

FIGURE 8-8   Explosionproof receptacle and cord cap for use in Class I, Division 1 locations. (Courtesy of Appleton Electric LLC, Emerson Industrial Automation.)

of the equipment. This rating must not exceed the ignition temperature of the gases or vapors for which the location is classified. *NEC* Table 500.8(C) provides a key to the temperature codes marked on equipment and the corresponding maximum surface temperatures. Table 4.4.2 in NFPA 497-2012, *Recommended Practice for the Classification of Flammable Liquids, Gases, or Vapors and of Hazardous (Classified) Locations for Electrical Installations in Chemical Process Areas,* provides ignition temperatures (auto-ignition temperatures, or AIT, in °C) for selected common materials. The nameplate on the hazardous location luminaire shown in **FIGURE 8-9** provides the necessary information relative to its suitability for use in a specific class, division, and group. Often, equipment like this is suitable for multiple classified locations and groups. Equipment must also be marked with the special range of ambient temperatures in which the equipment is to be used. Ambient temperature markings are not required for equipment that operates in the range between −25°C and 40°C (−13°F to 104°F), which is not considered "special."

7. Check equipment such as motors, transformers, overcurrent devices, switches and controllers, luminaires, heaters, and appliances for proper ratings and enclosures.
Equipment used in Class I, Division 1 locations must be provided with enclosures identified or otherwise **approved** for the location, including the division and group. Generally, for Class I, Division 1, this means explosionproof enclosures, but some alternatives, such as positive pressure ventilation, are allowed for specific equipment such as motors. General-purpose equipment and enclosures are often allowed in Division 2 areas if the equipment does not include make-or-break contacts that would be likely ignition sources. The *Code* requirements for specific types of equipment should be consulted, as the variations are too numerous to be summarized here.

8. Check for adequate grounding and bonding paths to the disconnecting means of the building or separately derived system.
Bonding methods and requirements in hazardous locations are similar to the requirements at services. Essentially, the methods required at services other than a direct connection to a grounded service conductor, such as threaded connections and bonding jumpers, are extended continuously to the classified area from the service grounding point, the grounding point of a separately derived system, or the grounding point at a separate building. In this case, *grounding point* means the location where the grounded conductor is connected to a grounding electrode. "Service-grade" bonding methods as outlined in 250.92 of the *Code* must be used in conduit runs and around conduit connections even where supplementary equipment grounding conductors are used. Where LFMC is allowed, it must be supplied with internal or external bonding jumpers of the wire type. The jumpers may be eliminated for short runs (6 ft [1.8 m] or less) of listed LFMC with grounding-type fittings if the contained circuits are protected by 10-ampere or smaller overcurrent devices, and they are used for control or similar purposes and are not power utilization circuits.

9. Check for multiwire branch circuits in the classified area.
Multiwire branch circuits may not be used in Class I areas unless the disconnecting device(s) in the circuit, including the branch-circuit overcurrent device, disconnect all ungrounded conductors of the circuit at the same time (**FIGURE 8-10**). This rule was specifically stated in Article 501 prior to the 2014 edition, but because simultaneous disconnection is now required for all multiwire branch circuits in 210.4(B), the rule is no longer included in Article 501. The reason for the rule

FIGURE 8-10 Multiwire branch circuits require simultaneous disconnection.

in hazardous locations is to address (1) the potential shock hazard that may be present in a neutral conductor of a multiwire circuit that is carrying current if all of the ungrounded conductors are not disconnected and (2) the concern that in a classified area, a current-carrying neutral conductor could produce a spark and ignite the surrounding atmosphere.

Checklist 8-2: Class II Locations

1. Confirm the classification of areas, including class, division, and group.
All area classification is required to be documented. The extent of classified areas, the classification of the areas, and the group designation of the combustible dusts present should be shown on plans or other documents. The boundary locations of the classified areas, as well as the class, division, and group, are necessary to determine the acceptability of equipment and wiring methods. Areas in which an ignitable quantity of combustible dust exists under normal conditions require specialized equipment such as the completely enclosed, pipe-**ventilated** motor shown in **FIGURE 8-11**.

2. Verify the suitability of the wiring methods being used.
In **Class II, Division 1** areas, wiring methods are limited primarily to threaded RMC, threaded IMC, or Type MI cable. Type MC (metal-clad) cable with a continuous (noninterlocked) sheath and an overall jacket, listed for use in Class II, Division 1 areas is permitted in some industrial installations. Other wiring methods, including electrical metallic tubing (EMT) and dust-tight wireways, as well as additional cable types in cable trays are permitted in Class I, Division 2 areas.

Boxes and other enclosures in Class II, Division 1 areas are required to be dusttight, with threaded connections. Enclosures used in areas where Group E combustible dusts are present and enclosures used for splice, tap, and junction points must be identified and approved for the Class II area. Enclosures, boxes, and fittings used in **Class II, Division 2** areas must minimize the entrance of dust. Boxes and enclosures are permitted to contain splices or terminal connections in a Class II, Division 1 location, provided that the boxes and enclosures are **dust-ignitionproof** (**FIGURE 8-12**). Such an enclosure prevents the entrance of dust and will contain any arcs or sparks that originate in the enclosure without resulting in exterior temperatures that could ignite the dust. Since there will be no significant amount of dust in the enclosure, they are not intended to contain or control an explosion inside the enclosure like explosionproof equipment is designed to do; however, minimizing the amount of dust within an equipment enclosure is important to reduce the adverse impact that it may have on equipment performance.

FIGURE 8-11   Pipe-ventilated motor suitable for Class II locations. (Courtesy of General Electric Company.)

3. Verify that seals are located where required, properly installed, and sealed, unless raceway arrangements preclude the requirement for seals.
Seals in Class II areas are not required to be explosionproof. Sealing may be accomplished by the use of electrical sealing putty, caulking, or similar products. Seals in Class II areas are intended to prevent the entry of dust into enclosures that are required to be dust-ignitionproof. Since dust will be excluded from properly installed systems in Class II areas, seals are not required at the boundaries of the Class II areas; nor are they required at every enclosure. Seals may be eliminated if raceways and raceway entries are arranged so that dust cannot enter an enclosure through a conduit.

FIGURE 8-12   Dusttight junction boxes with threaded hubs. (Courtesy of Appleton Electric LLC, Emerson Industrial Automation.)

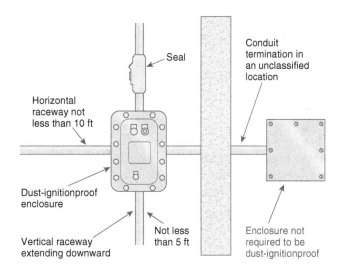

FIGURE 8-13 Permitted methods of preventing dust entrance via raceway into dust-ignitionproof enclosures. (Source: *National Electrical Code® Handbook*, NFPA, Quincy, MA, 2011, Exhibit 502.2.)

FIGURE 8-13 provides four methods of raceway entry into a dust-ignitionproof enclosure. Note that where a horizontal raceway extends 10 ft (3.0 m) or more from the enclosure, a seal is not required. The same concept is applied for a vertical entry into the bottom of the enclosure, but the required distance is reduced to 5 ft (1.5 m). A seal is not required in the conduit leaving the area, as the dust will be kept out of the dust-ignition-proof enclosure and the conduit within the classified area and there is no significant amount of dust in the unclassified area. Other equivalent arrangements, such as a conduit that extends a few feet horizontally from an enclosure and then 5 ft (1.5 m) downward may also be used in lieu of a seal (**FIGURE 8-14**). A seal is required to stop dust from entering the enclosure from a conduit entering the top of the enclosure.

4. Verify that equipment temperature markings are not greater than the ignition temperature of the dusts involved.

Equipment nameplates are furnished with identification numbers that indicate the maximum surface temperature of the equipment. This rating must not exceed the ignition temperature of the dusts for which the location is classified. If the dusts are organic dusts that may dehydrate or carbonize, the temperature is limited to a maximum of 165°C (329°F). *NEC* Table 500.8(C) provides a key to the temperature codes marked on equipment and the corresponding maximum surface temperatures. Table 4.5.2 in NFPA 499-2013, *Recommended Practice for the Classification of Combustible Dusts and of Hazardous (Classified) Locations for Electrical Installations in Chemical Process Areas,* provides ignition temperatures for selected common materials. Because dusts can be ignited either in a cloud or in a layer, sometimes at different temperatures, the lower of the two ignition temperatures is used.

5. Check materials used for flexible connections, including flexible cords, for suitability.

Wiring methods used to provide flexibility in Class II areas are required to be dusttight. Liquidtight flexible conduits, both metal and nonmetallic, may be used with approved fittings. Extra-hard usage cord types may also be used with listed dusttight fittings. The requirements for flexible connections in Class II, Division 2 areas are the same as those for Class II, Division 1 areas and are similar to the requirements for Class I areas. However, the uses of flexible cord are not as restricted in Class II areas as they are in Class I areas. Essentially, where flexible connections are necessary, flexible cord may be used, subject to the restrictions of Article 502 in addition to the usual restrictions found in 400.7 and 400.8. Although flexible cord may sometimes be used for wiring of luminaires, it may not be used for support of luminaires, including pendant luminaires, in Class II areas. Instructions for the use and termination of flexible cords with specific types of equipment are essentially the same for Class II areas in 502.140 as for Class I areas in 501.140.

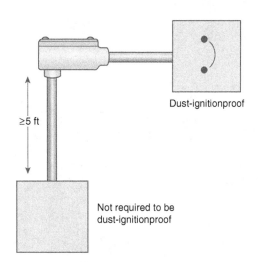

FIGURE 8-14 Equivalent arrangement of preventing dust entrance via raceway into dust-ignitionproof enclosures. (Reprinted with permission from Hazardous Materials Seminar, Slide 173, page 70, Copyright © 2008, National Fire Protection Association, Quincy, MA 02169. This reprinted material is not the complete and official position of the NFPA or the referenced subject, which is represented only by the standard in its entirety.)

6. Check flexible cord connectors and receptacles for suitability.

Where receptacles are needed, grounding-type receptacles and attachment plugs must be used in Class II, Division 1 locations and must be identified for the location. Such receptacles and connectors are usually available as products that are listed for the location. In Class II, Division 2 locations, grounding-type receptacles and attachment plugs are required and must be designed so that contacts that make or break the supply circuit are enclosed when the connection is made or broken. Flexible cords in Class II locations must be terminated with listed fittings or connectors. In Division 1 locations, a cord connector must be installed with a listed seal, but in Division 2 locations, the termination must only be a listed dusttight connector.

7. Check equipment such as motors, transformers, overcurrent devices, switches and controllers, luminaires, heaters, and appliances for proper ratings and enclosures.

Because of the thermal blanketing effects of accumulated dusts, surface temperatures are of particular concern in Class II areas. Appliances, motors, transformers, and other equipment must be excluded from Class II areas if the surface temperatures cannot be kept within required limits. Otherwise, most heat-producing equipment must be approved for Class II areas. Where metal dusts are present, the equipment must also be approved for the specific materials. As noted in the definitions of Class II areas, areas classified because of Group E (electrically conductive) dusts are always Division 1 only. Often, equipment such as luminaires for use in Class II areas is designed to prevent the accumulation of dust, sometimes by using smooth sloped surfaces on the top of the equipment. The temperature rating for Class II, Division 1 equipment must be based on the maximum amount of dust that can accumulate on the equipment. Equipment listed for Class I locations is not suitable for use in Class II locations unless it is identified for both locations.

8. Check for adequate grounding and bonding paths to the disconnecting means of the building or separately derived system.

Bonding methods and requirements in hazardous locations are similar to the requirements at services. Essentially, the methods required at services other than a direct connection to a grounded service conductor, such as threaded connections and bonding jumpers, are extended continuously to the classified area from the service grounding point, the grounding point of a separately derived system, or the grounding point at a separate building. (In this case, *grounding point* means the location where the grounded conductor is connected to a grounding electrode.) "Service-grade" bonding methods as outlined in 250.92 must be used in conduit runs and around conduit connections even where supplementary equipment grounding conductors are used. Where LFMC is allowed, it must be supplied with internal or external bonding jumpers of the wire type. The jumpers may be eliminated for short runs (6 ft [1.8 m] or less) of listed LFMC with grounding-type fittings if the contained circuits are protected by 10-ampere or smaller overcurrent devices and they are used for control or similar purposes and are not power utilization circuits.

9. Check for multiwire branch circuits in the classified area.

Multiwire branch circuits may not be used in Class II areas unless the disconnecting device(s) in the circuit, including the branch-circuit overcurrent device, disconnects all ungrounded conductors of the circuit at the same time. This requirement is no longer explicitly stated in Article 502 because this requirement is the same as the one that applies to Class I areas and, as of the 2008 *NEC*, is required for all multiwire branch circuits. The reason for the rule in hazardous locations is to address (1) the potential shock hazard that may be present in a neutral conductor of a multiwire circuit that is carrying current if all of the ungrounded conductors are not disconnected and (2) the concern that in a classified area, a current-carrying neutral conductor could produce a spark and ignite the surrounding atmosphere.

Checklist 8-3: Class III Locations

1. Confirm the classification of areas, including class and division.

All area classification is required to be documented. The extent of classified areas, the classification of the areas, and the types of fibers or flyings present should be shown on plans or other documents. The boundary locations of the classified areas, as well as the class and division, are necessary to determine the acceptability of equipment and wiring methods within the classified areas.

2. Verify the suitability of the wiring methods being used.

The wiring methods permitted in Class III locations are similar to those permitted in Class II locations. Class II wiring methods are intended to exclude dust; therefore, they will also prevent the entrance of fibers and flyings, which are larger than dusts. Permitted wiring methods in Class III areas include RMC, IMC, PVC, RTRC, EMT, dusttight wireways, Type MI cable, and Type MC cable. Other cable methods are permitted for limited energy and instrumentation circuits and cable trays. Where cable trays are used, spaces at least as great as the larger cable are required between cables. The wiring methods are the same for both Division 1 and Division 2 areas. Boxes and fittings are required to be dusttight.

3. Check materials used for flexible connections for suitability.

Liquidtight flexible conduits, both metal and nonmetallic, and extra-hard usage cords, as well as jacketed interlocked-armor Type MC cable are permitted wiring methods where flexibility is needed in Class III locations. Conduits and cord require listed dusttight fittings. The permitted uses of cords and the restrictions on their use in Class III areas are similar but less specific as for Class II areas, and are also limited by the uses permitted and prohibited in Article 400.

4. Check flexible cord connectors and receptacles for suitability.

Flexible cords are required to be terminated with listed dusttight fittings. Where receptacles are needed, grounding-type receptacles and attachment plugs must be used in Class III locations. Such devices must be designed to keep entry or accumulation of fibers to a minimum. They must also contain the products of any arcing. In some cases, if the accumulation of fibers can be limited to moderate amounts, the authority having jurisdiction (AHJ) may permit the use of general-use receptacles. Fiber accumulations may be limited through appropriate housekeeping and collection procedures.

5. Verify that equipment operating temperatures are acceptable for the conditions.

Luminaires and other types of heat-producing utilization equipment have specific temperature limitations in Class III areas. For equipment such as luminaires that are not subject to overloading, the surface temperatures are limited to 165°C (329°F) under operating conditions. For equipment such as motors and transformers, which are subject to overloading, surface temperatures are limited to 120°C (248°F). The operating conditions must include the equipment being covered with the maximum amount of dust (simulating fibers/flyings) that can accumulate on the equipment.

6. Check equipment such as motors, transformers, overcurrent devices, switches and controllers, luminaires, heaters, and appliances for proper ratings and enclosures.

Equipment in Class III areas, including luminaires, is generally required to be enclosed in a manner that will exclude dust, fibers, and flyings and prevent the release of sparks, burning material, or hot metal. Transformers and capacitors are required to be suitable for Class II, Division 2 locations. Generally, motors and generators are required to be totally enclosed and either nonventilated, pipe-ventilated, or fan-cooled. Heaters are required to be specifically approved for Class III locations. Luminaires are not required to be identified or listed for Class III areas but must not operate with surface temperatures above those mentioned in Item 5 above and must be marked with maximum lamp wattages that will limit surface temperatures to not over 165°C (329°F) in normal use.

7. Check for adequate grounding and bonding paths to the disconnecting means of the building or separately derived system.

Bonding methods and requirements in hazardous locations are similar to the requirements at services. Essentially, the methods required at services, other than direct connection to a grounded service conductor, such as threaded connections and bonding jumpers, are extended continuously to the classified area from the service grounding point, the grounding point of a separately derived system, or the grounding point at a separate building. (In this case, *grounding point* means the location where the grounded conductor is connected to a grounding electrode.) "Service-grade" bonding methods as outlined in 250.92 must be used in conduit runs and around conduit connections even where supplementary equipment grounding conductors are used. Where LFMC is allowed, it must be supplied with internal or external bonding jumpers of the wire type. The jumpers may be eliminated for short runs (6 ft [1.8 m] or less) of listed LFMC with grounding-type fittings if the contained circuits are protected by 10-ampere or smaller overcurrent devices, and they are used for control or similar purposes and are not power utilization circuits.

8. Check installations of cranes, hoists, and battery chargers for appropriate location and installation.

Cranes and hoists operating in or over Class III areas must have power supplies that are isolated from all other systems and detectors that will recognize the occurrence of a ground fault and automatically de-energize contact conductors and produce audible and visible alarms as long as power is supplied to contact conductors and a ground fault remains. In addition, special contact conductors and current collectors must be used with the traveling power supply to guard against accidental contact, reduce sparking, and contain sparks and hot particles. Battery chargers must be located in separate rooms. The rooms must have noncombustible interiors, be well ventilated, and prevent the entrance of ignitable quantities of fibers/flyings.

Checklist 8-4: Commercial Garages

1. Confirm the applicability of Article 511.

See the discussion under the definition of **Garage** in the Key Terms section of this chapter.

2. Identify the extent and division of the Class I areas.

The classified areas in a typical commercial garage are described in Article 511. Section 511.3 provides detailed descriptions of areas that are classified as well as areas that are not classified. Subsection (A) deals with parking garages that are usually unclassified as long as they are used only for parking and storage. Garages with dispensing of motor fuels are covered by both Article 511 for the repair areas and Article 514 for the dispensing areas according to 511.3(B). Subsection (C) covers garage areas where there is no dispensing but where transfer of fuels such as gasoline, natural gas, hydrogen, or liquefied petroleum gas (LPG) may take place (**major repair garages**) and where classification of floor, ceiling, and pit areas depends on the amount of ventilation provided. Ventilation is also the basis of area classification for floor, ceiling, and pit areas of **minor repair garages**, all of which can be unclassified if ventilation rates are as specified in subsection D. Without adequate ventilation, minor repair garages are still mostly unclassified except around below-grade or subfloor pits or work areas. Finally, 511.3(E) allows some adjacent areas such as stock and parts rooms to be unclassified and clarifies that classification will not be based on alcohol-based windshield washer fluid. **FIGURE 8-15** illustrates some of these provisions for a major repair garage. In a minor repair garage without ventilation, the Division 2 area above the floor would only extend 3 ft (900 mm) from the edges of the pit in the illustration. Most flammable liquids in garages are Group D materials, but other flammable materials may also be present, and where areas are classified, equipment must be approved for the specific liquids and gases present.

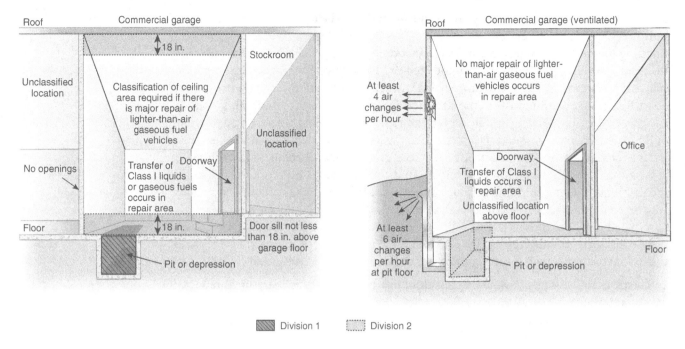

Division 1   Division 2

FIGURE 8-15   Class I locations within a commercial garage where fuels are transferred.

3. Verify that suitable wiring methods are used within Class I areas.

Wiring methods are required to meet the same requirements as in other Class I areas. Wiring in the Class I areas in a commercial garage must meet the requirements of Article 501 for the class and division. See the discussion of wiring methods in Item 2 of the Class I Locations checklist. Where there is inadequate ventilation for lighter-than-air fuels and the area within 18 in. (457 mm) of the ceiling is classified, the areas below are not classified except, depending on ventilation rates, for some areas near the floor level. Within commercial garages the areas below the floor, where the wiring method is in earth or concrete, are not classified, as there is insufficient air in these areas to form an ignitable mixture.

4. Verify that seals are located, installed, and sealed as required in 501.15.

Seals are required to be located according to the same rules as for other Class I locations considering both horizontal and vertical boundaries. See the discussion of seals in Item 3 of the Class I Locations checklist. Where conduits leave a classified area and extend below grade, the vertical boundary may be considered to be crossed below grade and the rules of 501.15(A)(4), Exception No. 2, will apply, which requires a seal as the first fitting after the conduit emerges from below grade.

5. Verify that suitable wiring methods and equipment are used where installed above Class I areas.

All metal raceways, most nonmetallic raceways, and a number of cable types are permitted for wiring above the Class I areas in commercial garages. Type NM cable is not permitted by Article 334 to be used in a commercial garage that has hazardous locations, whereas Type AC is not similarly restricted. Hard-usage flexible cords may be used for pendants if they are otherwise suitable for the application and area. Equipment other than receptacles and lighting equipment that is less than 12 ft (3.7 m) above the floor level and that may produce arcs, sparks, or particles of hot metal, or other equipment having make-and-break or sliding contacts, must be totally enclosed or constructed to contain any sparks or particles of hot metal. The same enclosure and construction requirements apply to fixed lighting equipment that is in locations where it is subject to contact with vehicles or other physical damage. However, fixed lighting must be located at least 12 ft (3.7 m) above such locations or it must be totally enclosed or otherwise constructed to prevent escape of sparks or hot metal particles. Fixed lighting cannot be in the classified area at the ceiling (if there is one due to the presence of lighter-than-air gases) unless the luminaires are identified for the application and the classification of the area.

6. Verify that receptacles have ground-fault circuit interrupter (GFCI) protection where required.

GFCI protection must be provided for convenience-type receptacles installed where electrical hand tools, portable lamps, or diagnostic equipment are to be used. This rule requires GFCI protection on all 125-volt, 15- and 20-ampere receptacles located in the areas where the equipment mentioned is to be used, not only on those receptacles specifically intended for such equipment. Generally, these receptacles should be located above any classified area at the floor unless the receptacles are of the explosionproof type. If portable lighting is used, the portable lighting must be of mostly nonconductive material, and if the portable lighting will be used in the classified areas, it must be suitable for Class I, Division 1 locations. Portable lighting that is not explosionproof can be used if it is supported in such a way that it cannot be used in the classified areas near or below the floor. For example, cord reels can be arranged in a manner that will keep the lighting from being pulled into any classified area at the floor. Adequate ventilation or the fact that major repairs do not occur in a specific location allows the floor area (other than near a pit) to be considered unclassified.

7. Verify that battery chargers are not located in classified areas.

Battery chargers, where used, must be located outside of the classified areas in a commercial garage. Where there are classified areas above the garage floor, the battery chargers are permitted to be located above the classified area(s).

8. Verify that connectors for electric-vehicle-charging equipment are not located in Class I locations.

Connectors and the plug connections for electric vehicles may not be located in the classified areas of a garage. Cords must be arranged so that they do not lie on the floor. They must be installed and managed in such a way that they will remain at least 6 in. (150 mm) above the floor level.

Checklist 8-5: Aircraft Hangars

1. Confirm the applicability of Article 513.

See the definition and discussion of **Aircraft Hangar** in the Key Terms section of this chapter.

2. Identify the extent and division of the Class I areas.

The classified areas in an aircraft hangar are as described in Article 513 (**FIGURE 8-16**). The Class I, Division 2 areas include the area of a hangar up to 18 in. (450 mm) above the floor and the areas within 5 ft (1.5 m) of aircraft fuel tanks or engines up to 5 ft (1.5 m) above the wings or engine enclosures. Pits or depressions below floor level, such as sumps and service pits, are Class I, Division 1 up to the floor level. Most flammable liquids in aircraft hangars are in Group D, but other flammable materials also may be present, and equipment must be approved for the specific liquids and gases present. Some hangars house aircraft that use only combustible liquids and no flammable fuels, but the areas are still classified if the combustible liquids are handled above their flashpoints. Also, alternative fuels are sometimes used, and other types of aircraft may occasionally be serviced. The classified areas for a painting hangar are much larger than those described above, but unlike similar paint booths covered by Article 516, the classified area is not the entire inside of the **aircraft painting hangar**.

FIGURE 8-16 Area classification in aircraft hangars. (Source: *National Electrical Code® Handbook*, NFPA, Quincy, MA, 2011, Exhibit 513.1.)

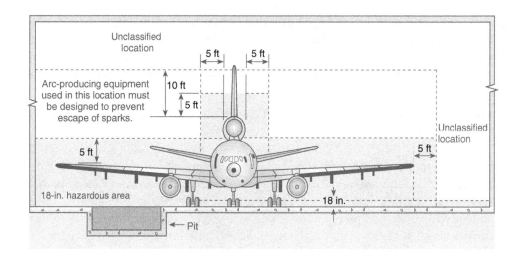

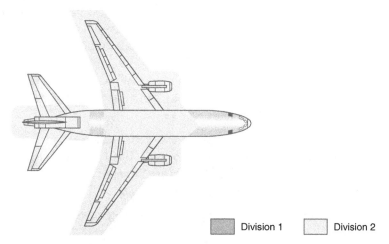

Division 1   Division 2

3. Verify that suitable wiring methods are used within and below Class I areas.

Wiring in and below the Class I areas in a hangar, including wiring below grade, must meet the requirements of Article 501 for the Class and Division (or Article 505 where classification is by Zone). See the discussion of wiring methods in Item 2 of the Class I Locations checklist. The area below the floor in a hangar is treated as if it were Class I, Division 1, even where the raceway descends below the floor or rises above it outside the classified area.

4. Verify that suitable wiring methods and equipment are used in unclassified areas of the hangar.

Wiring that is inside hangars but not in Class I areas must be in metal raceways or cable Types MI, MC, or TC (power and control tray cable) unless the area is cut off from classified areas and ventilated to prevent the accumulation of a flammable atmosphere, in which case any wiring method from the Wiring Methods and Devices chapter of this manual that is otherwise suitable is permitted. Flexible cords may be used in the unclassified areas of a hangar, but they must contain grounding conductors and must be hard-usage type where used for pendants, or extra-hard usage type where used for **portable equipment**. Equipment such as luminaires that are located within 10 ft (3.0 m) above engine enclosures or wings of aircraft must be made to contain any sparks or particles of hot metal so that the sparks or particles cannot fall into the classified area below. Portable lighting and other portable equipment must be approved for the locations where they are or may be used. Portable equipment other than lighting must be suitable for Class I, Division 2 or Zone 2 locations.

5. Verify that equipment used on stanchions, rostrums, and docks is appropriate.

Electrical wiring and equipment that is mounted on stanchions, rostrums, or docks must meet the requirements for Class I, Division 2 locations if the wiring or equipment is likely to be located in Class I locations or if the wiring or equipment is located up to 18 in. (450 mm) above the floor. Such wiring or equipment that is not in the classified area and more than 18 in. (450 mm) above the floor must meet the requirements described in Item 4 for wiring in unclassified areas. Locking-type receptacles and attachment plugs are required where receptacles are used on stanchions, rostrums, or docks.

6. Verify that seals are located, installed, and sealed in accordance with 501.15.

Seals are required to be located according to the same rules as for other Class I locations, considering both horizontal and vertical boundaries. Where conduits leave a classified area and extend below grade, the vertical boundary may be considered to be crossed below grade and the rules of 501.15(A)(4), Exception No. 2, will apply. See the discussion of seals in Item 3 of the Class I Locations checklist.

7. Verify that battery chargers are not located within the Class I locations.

Battery chargers may not be located in the classified areas of a hangar. Battery chargers should be located in separate buildings or areas that are isolated from the classified areas of a hangar and ventilated. Batteries installed in aircraft may not be charged while any part of the aircraft is inside a hangar.

8. Verify that external power sources and mobile equipment are properly located and installed.

External power sources used to energize aircraft may not be used in Class I locations, and all such wiring and equipment must be at least 18 in. (450 mm) above the floor in a hangar. The wiring and electrical parts of mobile servicing equipment also must be located at least 18 in. (450 mm) above the floor unless it is suitable for use in Class I, Division 2 areas. Cords and connectors must meet the same requirements as cords for other portable equipment, as described in Item 4 of this manual.

9. Verify that receptacles have GFCI protection where required.

GFCI protection must be provided for convenience-type receptacles installed where electrical hand tools, portable lighting, or diagnostic equipment are to be used. This rule requires GFCI protection on all 125-volt, 15- and 20-ampere receptacles located in the areas where the equipment mentioned is to be used, not only on those receptacles specifically intended for such equipment. Generally, these receptacles should be located outside the classified areas unless the receptacles and enclosures are of the explosionproof type.

Checklist 8-6: Motor Fuel-Dispensing Facilities

1. Confirm the applicability of Article 514.

See the definition of Motor Fuel-Dispensing Facility in the Key Terms section.

2. Identify the extent and division of the Class I areas.

The extent of Class I, Division 1 and Division 2 areas is described in detail in Article 514 and in NFPA 30A, *Code for Motor Fuel Dispensing Facilities and Repair Garages*, 2012 edition. Commonly used motor fuels such as gasoline, propane, and natural gas are Group D materials. Dispensers only for diesel fuel do not create classified areas unless the liquid is likely to be handled or stored above the flashpoint of diesel (about 130–140°F [54.4–60°C], depending on the grade). Where diesel and gasoline are both dispensed from the same equipment, the area will be classified because of the gasoline. Dispensers for compressed natural gas (CNG), LPG, or liquefied natural gas (LNG) are required to be separated a minimum of 5 ft (1.5 m) from dispensers for Class I flammable liquids, such as gasoline. The areas that are classified around dispensers for LPG are essentially the same as for gasoline. However, the classified areas around CNG or LNG dispensers are quite different because the vapors from these materials are lighter than air, so the extended classified areas near the ground are eliminated for these products. Since some areas in dispensing facilities may be used as commercial garages, and Article 511 refers to Article 514 for classification of such areas, Table 514.3(B)(1) of the *NEC* includes the descriptions of some classified areas in commercial garages.

3. Verify that suitable wiring methods and equipment are used within and below Class I areas.

Wiring methods are required to meet the same requirements as other Class I areas, except that PVC or RTRC raceways with an equipment grounding conductor may be used underground without concrete encasement as long as it is under at least 2 ft (600 mm) of cover. Cover includes dirt, as well as paving materials. Extensions from the 2-ft (600-mm) depth that emerge above grade must be threaded RMC or IMC. Wiring below the Class I areas of a dispensing facility must be sealed within 10 ft (3.0 m) of where it emerges from below grade, even though the underground area is considered unclassified, in order to minimize the passage of gases and vapors in the event that spilled fuel infiltrates the underground conduit. Underground wiring is limited to RMC, IMC, and raceway Types PVC and RTRC.

4. Verify that suitable wiring methods and equipment are used above classified areas.

The requirements for wiring and equipment above classified areas in **gasoline-dispensing and service stations** are the same as for commercial garages, in fact Article 514 refers to Article 511 for these requirements (see Item 5 of the Commercial Garages checklist). All metal raceways, most nonmetallic raceways, and a number of cable types are permitted for wiring above the Class I areas in gasoline-dispensing and service stations. Certain cable types, such as Type NM, are excluded. Hard-usage flexible cords may be used for pendants if they are otherwise suitable. Equipment other than receptacles and lighting equipment that is less than 12 ft (3.7 m) above the floor level and that may produce arcs, sparks, or particles of hot metal, or other equipment having make-and-break or sliding contacts, must be totally enclosed or constructed to contain any sparks or particles of hot metal. The same enclosure and construction requirements apply to fixed lighting equipment that is in locations where it is subject to contact with vehicles or other sources of physical damage. Generally, fixed lighting must be located at least 12 ft (3.7 m) above such locations. Although most modern gasoline-dispensing stations no longer include service areas, the service area requirements apply to the canopy areas above dispensing units (**FIGURE 8-17**).

5. Verify that seals are located, installed, and sealed at dispensers and in accordance with 501.15.

FIGURE 8-17 Motor fuel-dispensing station with carwash and canopy but without service areas.

In general, seals are required to be located according to the same rules as for other Class I locations. Seals are required at dispensers; however, the areas below ground are not classified areas, so the seals shown in the unclassified areas in **FIGURE 8-18** and **FIGURE 8-19** are not boundary seals in the usual sense. Nevertheless, they are specifically required to be installed wherever the raceway emerges from grade [514.8]. Thus, the raceway supplying the vacuum in the foreground of Figure 8-17 would require a seal at the vacuum unless the underground raceway is routed around all of the classified areas around the dispensers.

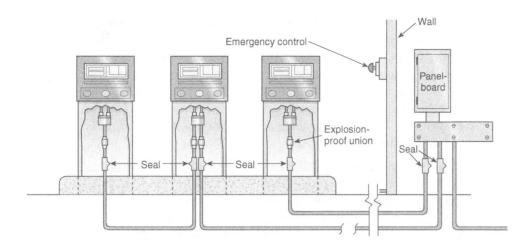

FIGURE 8-18   Seals required at gasoline-dispensing devices. (Source: *National Electrical Code® Handbook*, NFPA, Quincy, MA, 2011, Exhibit 514.4.)

The seals shown in Figures 8-18 and 8-19 that emerge away from the dispensers are in unclassified areas and not at boundaries, but they are at the first accessible location after the raceway passes under a classified area. Therefore, they are not necessarily required to be explosionproof, but they are required to be identified as suitable for minimizing the passage of gas and vapor through the seal. In the absence of this specific limited identification, explosionproof seals would have to be used. This is consistent with the risk in this case. The conduit is sealed at the dispenser, which also qualifies as a boundary seal between the Class I, Division 1 area in the base of the dispenser and the unclassified area below ground. However, dispensing areas are subject to spills in the Division 2 areas surrounding the dispensers. Spilled liquid may result in liquid or vapor migration into the underground conduit, so a seal against vapor migration at the point of emergence in an unclassified area is needed. (See the discussion of seals in Item 3 of the Class I Locations checklist.)

6. Verify that circuit and emergency disconnecting means are provided and that they disconnect all circuit conductors, including any grounded conductors.
A clearly identified, readily accessible disconnecting means must be provided for each circuit that goes to or through a gasoline dispenser. The disconnect must simultaneously disconnect all circuit conductors, including a grounded conductor. This is applicable not only to power conductors, but all communications, data, video, and remote control

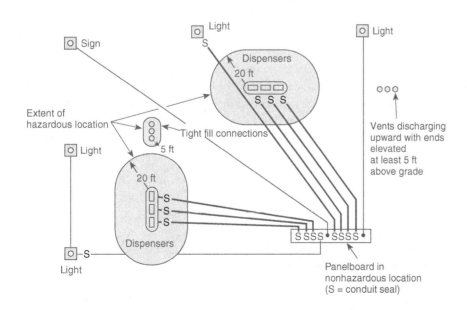

FIGURE 8-19   Seals required in raceways terminating in and passing through Class I locations. (Source: *National Electrical Code® Handbook*, NFPA, Quincy, MA, 2011, Exhibit 514.5.)

circuit conductors must also be disconnected. Although overcurrent devices are not normally permitted in grounded conductors, they are allowed when all conductors are disconnected at the same time. Special circuit breakers that have provisions for switching a grounded circuit conductor are available and have often been used for this purpose. Individual fuses or single-pole circuit breakers with handle ties do not satisfy the requirements of this rule. Multipole switches or contactors could be used for this purpose. Special equipment is available for disconnecting the many circuits for data and communications that are now common in dispensers.

7. Verify that means to disconnect all voltage sources, including feedback voltages, are provided.

Emergency controls and controls for maintenance and servicing are required for dispensers. The emergency controls are required to disconnect all circuit conductors simultaneously, as described earlier. The controls for maintenance and servicing must also disconnect all communications, data, and video circuits and all sources of power, including feedback from other equipment. Feedback is a concern because of the multiple voltage sources that exist at modern dispensers. Many modern dispensers control multiple pumps and include pay-at-pump services, communications, and other remote controls. Disconnects for emergency use are required to be within 100 ft (30 m) of dispensers at attended self-service stations and between 20 ft (6 m) and 100 ft (30 m) of dispensers at unattended self-service stations. When self-service stations are unattended, additional emergency shutoff controls are required on each group of dispensers or on the outdoor equipment that controls the dispensers, and the emergency controls must be manually reset in an approved manner. Emergency disconnects are required to shut off all power to all dispensers in a dispensing station. The disconnecting means for maintenance and servicing can be, but are not required to be, at the dispensers, so an emergency disconnect may also be used for servicing if it removes all sources of voltage including feedback voltages. The disconnecting means for servicing and maintenance must have provisions so that they can be locked in the open position in accordance with 110.25.

Checklist 8-7: Bulk Storage Plants

1. Confirm the applicability of Article 515.

See the definition of **Bulk Storage Plant** in the Key Terms section of this chapter. Also see the description of the scope of Article 515 in Section 515.1.

2. Identify the extent and division of the Class I areas.

The Class I, Division 1 and Division 2 areas of bulk storage plants are described in detail in Article 515 and in NFPA 30-2012, *Flammable and Combustible Liquids Code.* The group classifications must be determined from the specific materials that are stored, but Group D liquids are the most common. Combustible liquids such as diesel fuel and fuel oil usually do not create classified areas because they are usually not transferred above their **flash points. FIGURE 8-20** shows an example of the areas classified in and around a bulk storage tank like those found in large tank farms and petroleum transportation terminals. The areas shown in the drawing are also described in Table 515.3 in the *NEC.* Bulk storage plants and terminals usually include many different areas and types of areas that require classification, including some that are covered in other articles or standards.

3. Verify that suitable wiring methods and equipment are used within and below Class I areas.

Wiring methods are required to meet the same requirements as Class I areas in general, except that, like dispensing areas covered by Article 514, a PVC or RTRC raceway that includes an equipment grounding conductor may be used underground (below Class I areas) without concrete encasement as long as it is under at least 2 ft (600 mm) of cover.

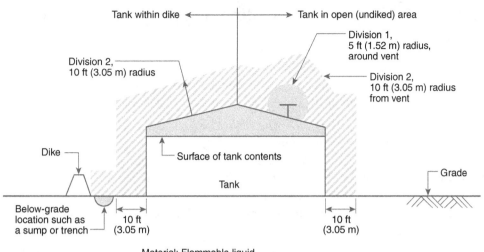

FIGURE 8-20   Area classification at a flammable liquid storage tank located outdoors above grade.

Wiring in the Class I areas in a bulk storage plant must meet the requirements of Article 501 for the class and division. In bulk storage plants, the wiring below Class I locations is treated as if it is within a Class I, Division 1 or Zone 1 area. See the discussion of wiring methods in Item 2 of the Class I Locations checklist.

4. Verify that suitable wiring methods and equipment are used above classified areas.

All metal raceways, Schedule 80 PVC, Type RTRC-XW (reinforced thermosetting resin conduit—extra heavy wall), and Type MI, MC, and TC cables may be used above Class I areas in bulk storage plants. Except for the PLTC, PLTC-ER, ITC, and ITC-ER cables that are permitted for remote control, signaling, and instrumentation circuits, other wiring methods are not permitted in these areas. Fixed equipment, including fixed lighting, must be totally enclosed or constructed to contain any arcs, sparks, or particles of hot metal that may be produced by the equipment. A specific minimum height is not specified as it is for commercial garages and motor fuel dispensing. Portable equipment and lamps and their flexible cords must be approved for the class and division of the locations below their areas of use.

5. Verify that seals are located, installed, and sealed in accordance with 501.15.

Seals are required to be located according to the same rules as for other Class I locations and must consider both horizontal and vertical boundaries. Wiring buried below Class I areas is considered to be within the Class I, Division 1 or Zone 1 area. (Although the area below ground does not meet the definition of a classified area due to the lack of air to form an ignitable mixture, in this case it is treated as a classified area, primarily for selection of wiring methods and placement of seals.) See the discussion of seals in Item 3 of the Class I Locations checklist.

6. Verify that dispensing areas comply with Article 514.

Article 514 applies to portions of bulk storage plants where **volatile flammable liquids** or liquefied flammable gases are dispensed. Classification of areas where transfer to drums, containers, tank vehicles, and tank cars takes place is covered by Article 515.

Checklist 8-8: Spray Application, Dipping, and Coating Processes

1. Confirm the applicability of Article 516.
See the definition of <u>Spray Application, Dipping, and Coating Processes</u> and the definitions of <u>Spray Area</u>, <u>Spray Booth</u>, and <u>Spray Room</u> in the Key Terms section.

2. Identify the extent and division of the Class I and Class II areas.
The Class I and Class II, Division 1 and 2 areas related to spray applications and dipping and coating processes are extensively described and illustrated in Article 516. Further information and a more detailed description of the classified areas can be found in NFPA 33, *Standard for Spray Application Using Flammable or Combustible Materials*, 2011 edition, and NFPA 34, *Standard for Dipping and Coating Processes Using Flammable or Combustible Liquids*, 2011 edition.

3. Identify the extent of any areas that are unclassified due to interlocks and ventilation.
See the definition of <u>Interlocked</u> in the Key Terms section. Interlocks can be used with spray booths to reduce the dimensions of the Division 2 area adjacent to the open front of the spray booth. In some cases, ventilation can reduce the extent or classification of other hazardous areas. The inspector should investigate the reliability and integrity of such interlocks and ventilation systems. For example, an interlock from the starter of a fan only proves the position of the contactor and does not prove that the fan is running, so the interlock should be based on proof of airflow, and the method used to prove airflow should not be compromised by accumulations of spray residue on the device or sensor used.

4. Verify that suitable wiring methods and equipment are used within the Class I areas.
Wiring methods in the Class I areas of spray, dipping, and coating operations are required to meet the same requirements as other Class I areas as long as the area is subject to vapor only (not accumulations of residue). Wiring in such Class I areas must meet the requirements of Article 501 for the class and division. See the discussion of wiring methods in Item 2 of the Class I Locations checklist.

Additional restrictions apply to spray areas where there are both vapors and residues. Where hazardous quantities of residues (from overspray, for example) may collect, Class I wiring and junction boxes and fittings that contain no splices, taps, or terminals are permitted, but other types of equipment are prohibited unless the equipment is specifically listed for locations where hazardous deposits and residues may accumulate.

5. Verify that luminaires are suitably listed for the classified areas or are located behind fixed panels and outside the classified areas.

Luminaires that are located within a classified area must be approved for the class, division, and group of the materials that create the classified area and must comply with the requirements for other types of equipment discussed in Item 4 above. Luminaires within the spray area are likely to be subject to the buildup of spray and residue, so often, luminaires for spray, dipping, and coating operations can be located outside of the classified area and still illuminate the work area (**FIGURE 8-21**). For example, luminaires are often located behind transparent or translucent panels that are incorporated in the wall or barrier forming the boundary of the classified area. Thus, ordinary luminaires can be used on the other side of the barrier because they are outside of

FIGURE 8-21   Spray booth with luminaires mounted outside booth in unclassified area.

the classified area. Some restrictions apply to this use of panels. The panels must be suitable to form the boundary of the classified area, be designed or protected to make breakage unlikely, and be arranged so that any residue will not be heated by the light source to temperatures that might cause ignition of the residue. In addition, the luminaires must be secured in place and suitable for their location. Article 516 provides detailed requirements for the materials used for panels for luminaires or observation panels including the types of glass and the methods of framing and attachment.

6. Verify that suitable wiring methods and equipment are used above and below classified areas.

Permitted wiring methods for fixed wiring above Class I and II locations are primarily metal raceways, rigid nonmetallic conduit, electrical nonmetallic tubing, and Type MI, TC, or MC cable. Equipment such as luminaires, control equipment, and appliances installed above Class I or II locations or above areas where freshly finished products are handled must be totally enclosed or otherwise designed and constructed to contain any sparks or particles of hot metal that may be produced by the equipment.

7. Verify that seals are located, installed, and sealed in accordance with 501.15.

Seals are required to be located according to the same rules as for other Class I locations. See the discussion of seals in Item 3 of the Class I Locations checklist. Where Class II areas are involved, sealing should be in accordance with Article 502.

8. Check electrostatic spraying equipment for proper installation.

The requirements for electrostatic spraying equipment are grouped under two general categories: fixed electrostatic equipment and electrostatic hand-spraying equipment. Rules that are common to both types of equipment include the following:

- All spray equipment must be listed.
- Most power and control equipment must be located outside the classified area or be identified for the area.
- Electrically conductive objects in the spraying area must be grounded, except for those parts that are required by the process to be at a higher voltage.
- Objects being coated must be supported in a manner that will ensure that the objects are grounded.
- Signs are required to indicate the need for grounding in the areas. A number of other rules apply only to fixed equipment or only to hand-spraying equipment. For example, fixed equipment is required to have automatic controls and interlocks with ventilation systems. Persons in the hand-spraying areas are required to be grounded to prevent the buildup of electrostatic charges. Consult the specific *NEC* requirements for additional detailed information.

9. Check powder coating equipment for proper installation.

Powder coating is a specialized type of **electrostatic coating** that uses combustible dry powders. Many of the rules for other types of electrostatic coating also apply to powder coating. A primary distinction is that because powder coating areas are Class II rather than Class I, equipment must be suitable for Class II, Division 1 locations. Consult the specific *NEC* requirements and Article 502 for more information.

KEY TERMS

The *NEC* defines the key terms pertaining to hazardous locations as follows (all definitions are from Article 100 unless otherwise noted):

Aircraft Hangar [*Definition derived from 513.1*] Buildings or structures, in any part of which aircraft containing Class I (flammable) liquids or Class II (combustible) liquids whose temperatures are above their flash points are housed or stored and in which aircraft might undergo service, repairs, or alterations. It shall not apply to locations used exclusively for aircraft that have never contained fuel or unfueled aircraft.

> *Informational Note No. 1:* For definitions of aircraft hangar and unfueled aircraft, see NFPA 409-2011, *Standard on Aircraft Hangars.*

> *Informational Note No. 2:* For further information on fuel classification see NFPA 30-2012, *Flammable and Combustible Liquids Code.*

[*Various classification systems are occasionally a source of some confusion. Do not confuse the Class I and Class II flammable liquids referenced in the foregoing definition with Class I and Class II classified locations. The NEC also refers to Class 1, Class 2, and Class 3 circuits, as covered in Article 725. These are easier to keep track of because of the use of Arabic rather than Roman numerals, but they still sound the same. See also the definition of* Volatile Flammable Liquid *later in this section.*]

Aircraft Painting Hangar [*Definition from 513.2*] An aircraft hangar constructed for the express purpose of spray/coating/dipping applications and provided with dedicated ventilation supply and exhaust. [*These applications are much like spray rooms as defined and covered by Article 516. However, the classification of the area is different. Under Article 516, the entire area inside an enclosed spray area is classified, but under Article 513, the classified areas are somewhat smaller than the entire interior of the hangar.*]

Approved Acceptable to the authority having jurisdiction (AHJ). [*The relationship between listing and approval of equipment for use in hazardous locations is interesting. Everything covered by the NEC is required to be approved. Most equipment contemplated by Articles 500 through 516 is also required to be "identified." Listing is not often mentioned, although it is one means of identification. Few inspectors will approve explosionproof or dust-ignitionproof equipment without listing, as few inspectors have the ability or facilities to test equipment to the applicable standards. Also, many of the wiring methods and materials that are allowed in hazardous locations, such as rigid metal conduit, are required to be listed under their respective articles. Alternatives to listing and labeling can be found in the description of* Identified *in 500.8(A)(1). See the definition of* Identified *in this section.*]

Bulk Storage Plant [*Definition from 515.1 Scope*] That property or portion of a property where liquids are received by tank vessel, pipelines, tank car, or tank vehicle and are stored or blended in bulk for the purpose of distributing such liquids by tank vessel, pipeline, tank car, tank vehicle, portable tank, or container. [*NFPA 30, Flammable and Combustible Liquids Code, 2012 edition, provides additional information about bulk storage plants. Note that the size of the containers is not specified, but on the basis of the vehicles that are used to transport materials to and from the storage facility, storage and mixing will not usually be in drums or barrels. Storage and handling will be in "bulk," but the materials may be shipped in "containers" such as 55-gallon (208-liter) drums. See Table 8-1 for an example of the type of facility covered by Article 515.*]

Class I, Division 1 Location [*Definition from 500.5(B)(1)*] A Class I, Division 1 location is a location:

In which ignitable concentrations of flammable gases, flammable liquid-produced vapors, or combustible liquid-produced vapors can exist under normal operating conditions, or

In which ignitable concentrations of such flammable gases, flammable liquid-produced vapors, or combustible liquids above their flash points may exist frequently because of repair or maintenance operations or because of leakage, or

In which breakdown or faulty operation of equipment or processes might release ignitable concentrations of flammable gases, flammable liquid-produced vapors, or combustible liquid-produced vapors and might also cause simultaneous failure of electrical equipment in such a way as to directly cause the electrical equipment to become a source of ignition.

Informational Note No. 1: This classification usually includes the following locations:

(1) Where volatile flammable liquids or liquefied flammable gases are transferred from one container to another
(2) Interiors of spray booths and areas in the vicinity of spraying and painting operations where volatile flammable solvents are used
(3) Locations containing open tanks or vats of volatile flammable liquids
(4) Drying rooms or compartments for the evaporation of flammable solvents
(5) Locations containing fat- and oil-extraction equipment using volatile flammable solvents
(6) Portions of cleaning and dyeing plants where flammable liquids are used
(7) Gas generator rooms and other portions of gas manufacturing plants where flammable gas may escape
(8) Inadequately ventilated pump rooms for flammable gas or for volatile flammable liquids
(9) The interiors of refrigerators and freezers in which volatile flammable materials are stored in open, lightly stoppered, or easily ruptured containers
(10) All other locations where ignitable concentrations of flammable vapors or gases are likely to occur in the course of normal operations

Informational Note No. 2: In some Division 1 locations, ignitable concentrations of flammable gases or vapors may be present continuously or for long periods of time. Examples include the following:

(1) The inside of inadequately vented enclosures containing instruments normally venting flammable gases or vapors to the interior of the enclosure
(2) The inside of vented tanks containing volatile flammable liquids
(3) The area between the inner and outer roof sections of a floating roof tank containing volatile flammable fluids
(4) Inadequately ventilated areas within spraying or coating operations using volatile flammable fluids
(5) The interior of an exhaust duct that is used to vent ignitable concentrations of gases or vapors

Experience has demonstrated the prudence of avoiding the installation of instrumentation or other electric equipment in these particular areas altogether or where it cannot be avoided because it is essential to the process and other locations are not feasible [see 500.5(A), Informational Note], using electric equipment or instrumentation approved for the specific application or consisting of intrinsically safe systems as described in Article 504.

Class I, Division 2 Location [*Definition from 500.5(B)(2)*] A Class I, Division 2 location is a location:

In which volatile flammable gases, flammable liquid-produced vapors, or combustible liquid-produced vapors are handled, processed, or used but in which the liquids, vapors, or gases will normally be confined within closed containers or closed systems from which they can escape only in case of accidental rupture or breakdown of such containers or systems, or in case of abnormal operation of equipment; or

In which ignitable concentrations of flammable gases, flammable liquid-produced vapors, or combustible liquid-produced vapors are normally prevented by positive mechanical ventilation and that might become hazardous through failure or abnormal operation of the ventilating equipment; or

That is adjacent to a Class I, Division 1 location, and to which ignitable concentrations of flammable gases, flammable liquid-produced vapors, or combustible liquid-produced vapors above their flash points might occasionally be communicated unless such communication is prevented by adequate positive-pressure ventilation from a source of clean air and effective safeguards against ventilation failure are provided.

Informational Note No. 1: This classification usually includes locations where volatile flammable liquids or flammable gases or vapors are used but that, in the judgment of the AHJ, would become hazardous only in case of an accident or of some unusual operating condition. The quantity of flammable material that might escape in case of accident, the adequacy of ventilating equipment, the total area involved, and the record of the industry or business with respect to explosions or fires are all factors that merit consideration in determining the classification and extent of each location.

Informational Note No. 2: Piping without valves, checks, meters, and similar devices would not ordinarily introduce a hazardous condition even though used for flammable liquids or gases. Depending on factors such as the quantity and size of the containers and ventilation, locations used for the storage of flammable liquids or liquefied or compressed gases in sealed containers may be considered either hazardous (classified) or unclassified locations. See NFPA 30-2012, *Flammable and Combustible Liquids Code,* and NFPA 58-2014, *Liquefied Petroleum Gas Code.*

[*See the definition of* Flash Point. *Definitions for combustible liquid and flammable liquid are contained in NFPA 30-2012,* Flammable and Combustible Liquids Code, *as well as in NFPA 497-2012,* Recommended Practice for the Classification of Flammable Liquids, Gases, or Vapors and of Hazardous (Classified) Locations for Electrical Installations in Chemical Process Areas.]

Class II, Division 1 Location [*Definition from 500.5(C)(1)*] A Class II, Division 1 location is a location:

In which combustible dust is in the air under normal operating conditions in quantities sufficient to produce explosive or ignitable mixtures; or

Where mechanical failure or abnormal operation of machinery or equipment might cause such explosive or ignitable mixtures to be produced and might also provide a source of ignition through simultaneous failure of electric equipment, operation of protection devices, or from other causes; or

In which Group E combustible dusts may be present in quantities sufficient to be hazardous.

Informational Note: Dusts containing magnesium or aluminum are particularly hazardous, and the use of extreme precaution will be necessary to avoid ignition and explosion.

Class II, Division 2 Location [*Definition from 500.5(C)(2)*] A Class II, Division 2 location is a location:

In which combustible dust due to abnormal operations may be present in the air in quantities sufficient to produce explosive or ignitible mixtures; or

Where combustible dust accumulations are present but are normally insufficient to interfere with the normal operation of electrical equipment or other apparatus, but could as a result of infrequent malfunctioning of handling or processing equipment become suspended in the air; or

In which combustible dust accumulations on, in, or in the vicinity of the electrical equipment could be sufficient to interfere with the safe dissipation of heat from electrical equipment, or could be ignitible by abnormal operation of failure of electrical equipment.

Informational Note No. 1: The quantity of combustible dust that may be present and the adequacy of dust removal systems are factors that merit consideration in determining the classification and may result in an unclassified area.

Informational Note No. 2: Where products such as seed are handled in a manner that produces low quantities of dust, the amount of dust deposited may not warrant classification.

Class III Locations [*Definition from 500.5(D)*] Class III locations are those that are hazardous because of the presence of easily ignitable fibers or where materials producing combustible flyings are handled, manufactured, or used, but in which such fibers/flyings are not likely to be in suspension in the air in quantities sufficient to produce ignitable mixtures. Class III locations shall include those specified in 500.5(D)(1) and (D)(2).

    (D)(1) Class III, Division 1: A Class III, Division 1 location is a location in which easily ignitable fibers/flyings are handled, manufactured, or used.

    Informational Note No. 1: Such locations usually include some parts of rayon, cotton, and other textile mills; combustible fiber manufacturing and processing plants; cotton gins and cottonseed mills; flax-processing plants; clothing manufacturing plants; woodworking plants; and establishments and industries involving similar hazardous processes or conditions.

    Informational Note No. 2: Easily ignitable fibers/flyings include rayon, cotton (including cotton linters and cotton waste), sisal or henequen, istle, jute, hemp, tow, cocoa fiber, oakum, baled waste kapok, Spanish moss, excelsior, and other materials of similar nature.

[Section 503.5 offers the following explanation of the requirements for equipment in Class III areas: "Equipment installed in Class III locations shall be able to function at full rating without developing surface temperatures high enough to cause excessive dehydration or gradual carbonization of accumulated fibers/flyings. Organic material that is carbonized or excessively dry is highly susceptible to spontaneous ignition. The maximum surface temperatures under operating conditions shall not exceed 165°C (329°F) for equipment that is not subject to overloading, and 120°C (248°F) for equipment (such as motors or power transformers) that may be overloaded." In a Class III, Division 1 location, the operating temperature shall be the temperature of the equipment when blanketed with the maximum amount of dust (simulating fibers/flyings) that can accumulate on the equipment.]

    (D)(2) Class III, Division 2: A Class III, Division 2 location is a location in which easily ignitable fibers/flyings are stored or handled other than in the process of manufacture.

Combustible Dust [*Definition from 500.2*] Dust particles that are 500 microns or smaller (material passing a U.S. No. 35 Standard Sieve as defined in ASTM E 11-09, *Standard Specification for Wire Cloth and Sieves for Testing Purposes*) and present a fire or explosion hazard when dispersed and ignited in air.

Division 1 and Division 2 [*Division 1 and Division 2 are distinguished in the definitions of Class I and Class II areas. For any given type of vapor or dust source that may produce a Class I or Class II hazardous area, the Division number indicates the degree of likelihood that a hazardous condition will actually exist or develop in the area. Division 1 indicates a greater likelihood than Division 2.*]

Dust-Ignitionproof [*Definition from 500.2*] Equipment enclosed in a manner that excludes dusts and does not permit arcs, sparks, or heat otherwise generated or liberated inside of the enclosure to cause ignition of exterior accumulations or atmospheric suspensions of a specified dust on or in the vicinity of the enclosure.

    Informational Note: For further information on dust-ignitionproof enclosures, see Type 9 enclosure in ANSI/NEMA 250-2008, *Enclosures for Electrical Equipment*, and ANSI/UL 1203-2009, *Explosionproof and Dust-Ignitionproof Electrical Equipment for Hazardous (Classified) Locations.*

[Equipment installed in Class II locations must be able to function at full rating without developing surface temperatures high enough to cause excessive dehydration or gradual carbonization of any organic dust deposits that may occur. The surface temperature and its effect on dust accumulations is a concern because dust that is carbonized or excessively dry is highly susceptible to spontaneous ignition. Section 502.5 says "Explosionproof equipment and wiring shall not be required and shall not be acceptable in Class II locations unless identified for such locations." In other words, dust-ignitionproof and explosionproof requirements are quite different, and although much equipment is listed for both uses, equipment that is suitable for either Class I or Class II areas is not necessarily suitable for

the other type of classification. An assumption in the design of explosionproof equipment is that vapors and gases will enter the equipment and there will be internal explosions that must be controlled. By definition, dust-ignitionproof equipment is designed to exclude the dust so internal explosions will not occur, but surface temperatures are critical, and must include considerations of the effects of accumulated dust blankets.]

Dusttight   Constructed so that dust will not enter the enclosing case under specified test conditions. [*This term is also defined in 500.2. Some wiring methods, fittings, and boxes for Class II, Division 2 areas and Class III areas are required only to be dusttight and are not necessarily required to be specifically listed for the area.*]

Electrostatic Coating Equipment   [*Definition derived from 516.10(A) and (B)*] Equipment using electrostatically charged elements for the atomization, charging, and/or precipitation of hazardous materials for coatings on articles or for other similar purposes. [*In this coating method, the "spray" usually consists of electrically charged particles that are applied to an object that has an electrical charge of the opposite polarity. Thus, the coating particles are attracted to the object to be coated. This method is used in applying "powder coated" finishes.*]

Explosionproof   Equipment enclosed in a case that is capable of withstanding an explosion of a specified gas or vapor that may occur within it and of preventing the ignition of a specified gas or vapor surrounding the enclosure by sparks, flashes, or explosion of the gas or vapor within and that operates at such an external temperature that a surrounding flammable atmosphere will not be ignited thereby.

Informational Note: For further information, see ANSI/UL 1203-2009, *Explosionproof and Dust-Ignitionproof Electrical Equipment for Use in Hazardous (Classified) Locations.*

[*Explosionproof equipment is not expected to exclude a flammable atmosphere. Rather, it is assumed that the atmosphere will enter the conduit system and the explosionproof enclosures by the normal "breathing" caused by atmospheric and temperature-induced pressure changes. Arcs and sparks produced by normal operation of the enclosed equipment will then ignite the contained gases or vapors. The function of the conduit system and the explosionproof enclosures is to withstand the explosion pressures and control (not contain) the release of the flames and hot gases in a manner that prevents the ignition of the surrounding atmosphere.*]

Flash Point   [*Definition from NFPA 497-2008, Recommended Practice for the Classification of Flammable Liquids, Gases, or Vapors and of Hazardous (Classified) Locations for Electrical Installations in Chemical Process Areas, 3.3.7*] The minimum temperature at which a liquid gives off vapor in sufficient concentration to form an ignitible mixture with air near the surface of the liquid, as specified by test. [*The 2008 NEC clarified that combustible liquids—that is, liquids with flash points above 100°F (37.8°C)—may be the basis of area classification if they are handled above their flash points. Generally, flammable and combustible liquids are not easily ignited until they give off vapors that form an ignitible mixture with air. NFPA 497 provides some information about the characteristics of common liquids and gases. For example, gasoline has a flash point of 51°F (11°C) while kerosene and No. 1 fuel oil have flash points of 162°F (72°C), so they are much less likely to be handled above their flash points than gasoline, and therefore are less likely to be the basis for a classified area. Flash points for other flammable and combustible liquids are provided with Material Safety Data Sheets (MSDS).*]

Garage   A building or portion of a building in which one or more self-propelled vehicles carrying volatile flammable liquid for fuel or power are kept for use, sale, storage, rental, repair, exhibition, or demonstration purposes.

Informational Note: For commercial garages, repair, and storage, see Article 511.

[*See the definition of Volatile Flammable Liquid. Article 511 describes its scope, Commercial Garages, in 511.1 as follows: "These occupancies shall include locations used for service and repair operations in connection with self-propelled vehicles (including, but not limited to, passenger automobiles, buses, trucks, and tractors) in which volatile flammable liquids are used for fuel or power." Clarification of this scope and the extent of classified areas in garages is found in Article 511. "Where fuel-dispensing units (other than liquid petroleum gas, which is prohibited) are located within buildings, the requirements of Article 514 shall govern" [511.4(B)(1)]. "Parking garages used for parking or storage shall be permitted to be unclassified" [511.3(A)]. NFPA 88A-2011, Standard for Parking Structures, and NFPA 30A-2012, Code for Motor Fuel-Dispensing Facilities and Repair Garages, provide further information on these occupancies.*]

Garage, Major Repair [*Definition from 511.2*] A building or portions of a building where major repairs, such as engine overhauls, painting, body and fender work, and repairs that require draining of the motor vehicle fuel tank are performed on motor vehicles, including associated floor space used for offices, parking, or showrooms. [*This term is used to determine where the area within 18 in. (457 mm) of the ceiling is required to be classified due to lighter-than-air gaseous fuels, but also affects the floor and pit areas that are classified. See* Garages, Minor Repair *below.*]

Garage, Minor Repair [*Definition from 511.2*] A building or portions of a building used for lubrication, inspection, and minor automotive maintenance work, such as engine tune-ups, replacement of parts, fluid changes (e.g., oil, antifreeze, transmission fluid, brake fluid, air conditioning refrigerants), brake system repairs, tire rotation, and similar routine maintenance work, including associated floor space used for offices, parking, or showrooms. [*Many areas in major repair garage that are classified are unclassified in minor repair garages because the "major repairs" that might involve draining a fuel tank or handling flammable liquids are not performed in minor repair garages. Note that the liquids mentioned in this definition are not flammable liquids; some are combustible liquids, but they are not likely to be handled above their flash points.*]

Gasoline-Dispensing and Service Station [*See* Motor Fuel-Dispensing Facility.]

Identified (as applied to equipment) Recognizable as suitable for the specific purpose, function, use, environment, application, and so forth, where described in a particular *Code* requirement.

> *Informational Note:* Some examples of ways to determine suitability of equipment for a specific purpose, environment, or application include investigations by a qualified testing laboratory (listing and labeling), an inspection agency, or other organizations concerned with product evaluation.

[*Identified equipment is described in more detail in Articles 500 and 505. Section 500.8(B) (1) states that "Equipment shall be identified not only for the class of location but also for the explosive, combustible, or ignitable properties of the specific gas, vapor, dust, or fibers/ flyings that will be present. In addition, Class I equipment shall not have any exposed surface that operates at a temperature in excess of the ignition temperature of the specific gas or vapor. Class II equipment shall not have an external temperature higher than that specified in 500.8(D)(2). Class III equipment shall not exceed the maximum surface temperatures specified in 503.5."*

Section 500.8(A) says "Suitability of identified equipment shall be determined by one of the following:

(1) Equipment listing or labeling

(2) Evidence of equipment evaluation from a qualified testing laboratory or inspection agency concerned with product evaluation

(3) Evidence acceptable to the authority having jurisdiction such as a manufacturer's self-evaluation or an owner's engineering judgment."

Similar language appears in 505.9(A) for equipment identified for zone use. Identified *has a more specific meaning and is described in more detail for hazardous areas than it is for the rest of the NEC.*]

Interlocked, Interlocks [*Definition derived from Article 516*] As the term relates to spray booths, *interlocked* means "that the spray application equipment cannot be operated unless the exhaust ventilation system is operating and functioning properly and spray application is automatically stopped if the exhaust ventilation system fails" (516.3(C)(2)). [*In general,* interlocks *refer to some form of electrical or mechanical circuitry or devices that make the operation of some equipment or circuit dependent on a specific state of some other equipment or circuit. For example, conveyor systems are often interlocked so that an upstream conveyor cannot operate unless a downstream conveyor is operating. This prevents the pileup of materials when some component fails. Similar electrical and/or mechanical interlocks are used between the forward and reverse contactors of a reversing motor starter to prevent short circuits. Electrical circuit interlocks between spray and ventilation equipment usually consist of a differential pressure switch or some similar device that verifies air movement and closes a switch. The spray equipment is connected through the switch to allow the spray equipment to operate only when air is moving, and failure of the ventilation system will disable the spray equipment. Using a contact on a ventilation fan starter does not verify operation of the fan itself due to the possibility of broken drive belts or similar failures, but such connections are still called* electrical interlocks. *Article 516 establishes a higher standard than is normally implied by electrical interlocks.*]

Intrinsically Safe Apparatus [*Definition from 504.2*] Apparatus in which all the circuits are intrinsically safe.

Intrinsically Safe Circuit [*Definition from 504.2*] A circuit in which any spark or thermal effect is incapable of causing ignition of a mixture of flammable or combustible material in air under prescribed test conditions.

> *Informational Note:* Test conditions are described in ANSI/UL 913-2006, *Standard for Safety, Intrinsically Safe Apparatus and Associated Apparatus for Use in Class I, II, and III, Division 1, Hazardous (Classified) Locations.*

Intrinsically Safe System [*Definition from 504.2*] An assembly of interconnected intrinsically safe apparatus, associated apparatus, and interconnecting cables in that those parts of the system that may be used in hazardous (classified) locations are intrinsically safe circuits.

> *Informational Note:* An intrinsically safe system may include more than one intrinsically safe circuit.

Mobile Equipment (as applied to aircraft hangars) [*Definition from 513.2*] Equipment with electric components suitable to be moved only with mechanical aids or that are provided with wheels for movements by person(s) or powered devices. [*See also* Portable Equipment.]

Motor Fuel-Dispensing Facility [*Definition from 514.2*] That portion of a property where motor fuels are stored and dispensed from fixed equipment into the fuel tanks of motor vehicles or marine craft or into approved containers, including all equipment used in connection therewith.

> *Informational Note:* Refer to Articles 510 and 511 with respect to electric wiring and equipment for other areas used as lubritoriums, service rooms, repair rooms, offices, salesrooms, compressor rooms, and similar locations.

Nonincendive Circuit [*Definition from 500.2*] A circuit, other than field wiring, in which any arc or thermal effect produced under intended operating conditions of the equipment is not capable, under specified test conditions, of igniting the flammable gas-, vapor-, or dust-air mixture.

> *Informational Note:* For test conditions, see ANSI/ISA-12.12.01-2012, *Nonincendive Electrical Equipment for Use in Class I and II, Division 2, and Class III, Divisions 1 and 2 Hazardous (Classified) Locations.*

Portable Equipment (as applied to aircraft hangars) [*Definition from 513.2*] Equipment with electric components suitable to be moved by a single person without mechanical aids.

Purged and Pressurized [*Definition from 500.2*] The process of (1) purging, supplying an enclosure with a protective gas at a sufficient flow and positive pressure to reduce the concentration of any flammable gas or vapor initially present to an acceptable level; and (2) pressurization, supplying an enclosure with a protective gas with or without continuous flow at sufficient pressure to prevent the entrance of a flammable gas or vapor, a combustible dust, or an ignitible fiber.

> *Informational Note:* For further information, see ANSI/NFPA 496-2013, *Purged and Pressurized Enclosures for Electrical Equipment.*

Spray Application, Dipping, and Coating Processes [*Definition derived from 516.1*] Spray application, dipping, and coating process locations are those involving the regular or frequent application of flammable liquids, combustible liquids, and combustible powders by spray operations and the application of flammable liquids, or combustible liquids at temperatures above their flashpoint, by dipping, coating, or other means.

> *Informational Note:* For further information regarding safeguards for these processes, such as fire protection, posting of warning signs, and maintenance, see NFPA 33-2011, *Standard for Spray Application Using Flammable and Combustible Materials,* and NFPA 34-2011, *Standard for Dipping and Coating Processes Using Flammable or Combustible Liquids.* For additional information regarding ventilation, see NFPA 91-2010, *Standard for Exhaust Systems for Air Conveying of Vapors, Gases, Mists, and Noncombustible Particulate Solids.*

Spray Area [*As used in Article 516, Definition from 516.2*] Any fully enclosed, partly enclosed, or unenclosed area in which ignitible quantities of flammable or combustible vapors, mists, residues, dusts, or deposits are present due to the operation of spray processes, including (1) any area in the direct path of a spray application process; (2) the interior of a spray booth or spray room or limited finishing workstation, as herein defined; (3) the interior of any exhaust plenum, eliminator section, or scrubber section; (4) the interior of any exhaust duct or exhaust stack leading from a spray application process; (5) the interior of any air recirculation filter house or enclosure, including secondary recirculation particulate filters; (6) any solvent concentrator (pollution abatement) unit or solvent recovery (distillation) unit. The following are not considered to be a part of the spray area: (1) fresh air make-up units; (2) air supply ducts and air supply plenums; (3) recirculation air supply ducts downstream of secondary filters; (4) exhaust ducts from solvent concentrator (pollution abatement) units.

> *Informational Note:* Unenclosed spray areas are locations outside of buildings or are localized operations within a larger room or space. Such are normally provided with some local vapor extraction/ventilation system. In automated operations, the area limits are the maximum area in the direct path of spray operations. In manual operations, the area limits are the maximum area of spray when aimed at 90 degrees to the application surface.

[*This definition was extensively revised in the 2014* NEC, *and an informational note was added to include the many areas not previously considered such as the interior of some ducts or other areas.*]

Spray Booth [*As used in Article 516, Definition from 516.2*] A power-ventilated enclosure for a spray application operation or process that confines and limits the escape of the material being sprayed, including vapors, mists, dusts, and residues that are produced by the spraying operation and conducts or directs these materials to an exhaust system.

> *Informational Note:* A spray booth is an enclosure or insert within a larger room used for spray/coating/dipping applications. A spray booth may be fully enclosed or have open front or face and may include a separate conveyor entrance and exit. The spray booth is provided with a dedicated ventilation exhaust but may draw supply air from the larger room or have a dedicated air supply.

Spray Room [*As used in Article 516, Definition from 516.2*] A power-ventilated fully enclosed room used exclusively for open spraying of flammable or combustible materials. A spray room is a purposefully enclosed room built for spray/coating/dipping applications provided with dedicated ventilation supply and exhaust. Normally, the room is configured to house the item to be painted, providing reasonable access around the item/process. Depending on the size of the item being painted, such rooms may actually be the entire building or the major portion thereof.

Ventilated Provided with a means to permit circulation of air sufficient to remove an excess of heat, fumes, or vapors. [Adequate Ventilation *is defined in NFPA 497-2012,* Recommended Practice for the Classification of Flammable Liquids, Gases, or Vapors and of Hazardous (Classified) Locations for Electrical Installations in Chemical Process Areas, *as "A ventilation rate that affords either 6 air changes per hour, or 1 cfm per square foot of floor area, or other similar criteria that prevent the accumulation of significant quantities of vapor-air mixture from exceeding 25 percent of the lower flammable limit." The criteria in this definition may be modified for some specific application. For example, in commercial garages [see 511.3(C)(1)(a) and (D)(1)(a)], a rate of four air changes per hour is considered adequate for declassifying the areas immediately above the floor and adjacent to classified areas. Other ventilation rates or criteria are also used in Article 625 for electric vehicle recharging systems.*]

Volatile Flammable Liquid A flammable liquid having a flashpoint below 38°C (100°F), or a flammable liquid whose temperature is above its flashpoint, or a Class II combustible liquid having a vapor pressure not exceeding 40 psi (276 kPa) at 38°C (100°F) whose temperature is above its flashpoint.

Zone Classification [*Definitions from 505.5(B) and 506.5(B)*]

Class I, Zone 0: A Class I, Zone 0 location is a location

(1) In which ignitable concentrations of flammable gases or vapors are present continuously, or

(2) In which ignitable concentrations of flammable gases or vapors are present for long periods of time.

Informational Note No. 1: As a guide in determining when flammable gases or vapors are present continuously or for long periods of time, refer to ANSI/American Petroleum Institute (API) RP 505-1997, *Recommended Practice for Classification of Locations for Electrical Installations of Petroleum Facilities Classified as Class I, Zone 0, Zone 1, or Zone 2;* ISA 12.24.01-1998, *Recommended Practice for Classification of Locations for Electrical Installations Classified as Class I, Zone 0, Zone 1, or Zone 2;* ANSI/ISA-TR 12.24.01-1998 (IEC 60079-10-Mod), *Recommended Practice for Classification of Locations for Electrical Installations Classified as Class I, Zone 0, Zone 1, or Zone 2;* IEC 60079-10-1995, *Electrical Apparatus for Explosive Gas Atmospheres, Classifications of Hazardous Areas;* and *Area Classification Code for Petroleum Installations, Model Code, Part 15,* Institute of Petroleum.

Informational Note No. 2: This classification includes locations inside vented tanks or vessels that contain volatile flammable liquids; inside inadequately vented spraying or coating enclosures where volatile flammable solvents are used; between the inner and outer roof sections of a floating roof tank containing volatile flammable liquids; inside open vessels, tanks, and pits containing volatile flammable liquids; the interior of an exhaust duct that is used to vent ignitable concentrations of gases or vapors; and inside inadequately ventilated enclosures that contain normally venting instruments utilizing or analyzing flammable fluids and venting to the inside of the enclosures.

Informational Note No. 3: It is not good practice to install electrical equipment in Zone 0 locations except when the equipment is essential to the process or when other locations are not feasible [see 505.5(A) Informational Note No. 2]. If it is necessary to install electrical systems in a Zone 0 location, it is good practice to install intrinsically safe systems as described by Article 504.

Class I, Zone 1: A Class I, Zone 1 location is a location

(1) In which ignitable concentrations of flammable gases or vapors are likely to exist under normal operating conditions; or

(2) In which ignitable concentrations of flammable gases or vapors may exist frequently because of repair or maintenance operations or because of leakage; or

(3) In which equipment is operated or processes are carried on of such a nature that equipment breakdown or faulty operations could result in the release of ignitable concentrations of flammable gases or vapors and also cause simultaneous failure of electrical equipment in a mode to cause the electrical equipment to become a source of ignition; or

(4) That is adjacent to a Class I, Zone 0 location from which ignitable concentrations of vapors could be communicated, unless communication is prevented by adequate positive-pressure ventilation from a source of clean air and effective safeguards against ventilation failure are provided.

Informational Note No. 1: Normal operation is considered the situation when plant equipment is operating within its design parameters. Minor releases of flammable material may be part of normal operations. Minor releases include the releases from mechanical packings on pumps. Failures that involve repair or shutdown (such as the breakdown of pump seals and flange gaskets, and spillage caused by accidents) are not considered normal operation.

Informational Note No. 2: This classification usually includes locations where volatile flammable liquids or liquefied flammable gases are transferred from one container to another; areas in the vicinity of spraying and painting operations where flammable solvents are used; adequately ventilated drying rooms or compartments for evaporation of flammable solvents; adequately ventilated locations containing fat and oil extraction equipment using volatile flammable solvents; portions of cleaning and dyeing plants where volatile flammable liquids are used; adequately ventilated gas generator rooms and other portions of gas manufacturing plants where flammable gas may escape; inadequately ventilated pump rooms for flammable gas or for volatile flammable liquids; the interiors of refrigerators and freezers in which volatile flammable materials are stored in the open, lightly stoppered, or in easily ruptured containers; and other locations where ignitable concentrations of flammable vapors or gases are likely to occur in the course of normal operation but not classified Zone 0.

Class I, Zone 2: A Class I, Zone 2 location is a location

(1) In which ignitable concentrations of flammable gases or vapors are not likely to occur in normal operation and, if they do occur, will exist only for a short period; or

(2) In which volatile flammable liquids, flammable gases, or flammable vapors are handled, processed, or used but in which the liquids, gases, or vapors normally are confined within closed containers of closed systems from which they can escape only as a result of accidental rupture or breakdown of the containers or system or as a result of the abnormal operation of the equipment with which the liquids or gases are handled, processed, or used; or

(3) In which ignitable concentrations of flammable gases or vapors normally are prevented by positive mechanical ventilation but that may become hazardous as a result of failure or abnormal operation of the ventilation equipment; or

(4) That is adjacent to a Class I, Zone 1 location, from which ignitable concentrations of flammable gases or vapors could be communicated, unless such communication is prevented by adequate positive-pressure ventilation from a source of clean air, and effective safeguards against ventilation failure are provided.

Informational Note: The Zone 2 classification usually includes locations where volatile flammable liquids or flammable gases or vapors are used but that would become hazardous only in case of an accident or of some unusual operating condition.

[*The zone classification system is an alternative to the division classification system for Class I areas. Like Divisions 1 and 2, Zones 0, 1, and 2 represent decreasing degrees of risk or decreasing degrees of likelihood that a hazardous condition will develop in a given area. Zones 0 and 1 are roughly equivalent to Division 1, and Zone 2 is roughly equivalent to Division 2. Although many wiring methods are common to both classification systems, the protection methods that may be used differ. Protection methods permitted under the Division classification system are listed in 500.7. Zone classification protection methods are listed in 505.8. Equipment marking systems also differ as described in 505.9(C). Equipment is required to be specifically listed for the application. Many types of equipment may provide markings for both systems. Classification, engineering, design, selection of equipment and wiring methods, installation, and inspection under this classification scheme must be done by persons experienced and skilled in these methods.]*

Zone 20, Zone 21, and Zone 22 Locations [*From 506.5(B)*] Zone 20, Zone 21, and Zone 22 locations are those in which combustible dust or ignitible fibers/flyings are or may be present in the air or in layers, in quantities sufficient to produce explosive or ignitible mixtures. Zone 20, Zone 21, and Zone 22 locations shall include those specified in 506.5(B)(1), (B)(2), and (B)(3).

Informational Note: Through the exercise of ingenuity in the layout of electrical installations for hazardous (classified) locations, it is frequently possible to locate much of the equipment in a reduced level of classification and, thus, to reduce the amount of special equipment required.

Zone 20: A Zone 20 location is a location in which (a) ignitible concentrations of combustible dust or ignitible fibers/flyings are present continuously and (b) ignitible concentrations of combustible dust or ignitible fibers/flyings are present for long periods of time.

Informational Note No. 1: As a guide to classification of Zone 20 locations, refer to ANSI/ISA-61241-10 (12.10.05)-2004, Electrical Apparatus for Use in Zone 20, Zone 21, and Zone 22 Hazardous (Classified) Locations—Classification of Zone 20, Zone 21, and Zone 22 Hazardous (Classified) Locations.

Informational Note No. 2: Zone 20 classification includes locations inside dust containment systems; hoppers, silos, etc., cyclones and filters, dust transport systems, except some parts of belt and chain conveyors, etc.; blenders, mills, dryers, bagging equipment, etc.

Zone 21: A Zone 21 location is a location:

(1) In which ignitible concentrations of combustible dust or ignitible fibers/flyings are likely to exist occasionally under normal operating conditions; or

(2) In which ignitible concentrations of combustible dust or ignitible fibers/flyings may exist frequently because of repair or maintenance operations or because of leakage; or

(3) In which equipment is operated or processes are carried on, of such a nature that equipment breakdown or faulty operations could result in the release of ignitible concentrations of combustible dust or ignitible fibers/flyings and also cause simultaneous failure of electrical equipment in a mode to cause the electrical equipment to become a source of ignition; or

(4) That is adjacent to a Zone 20 location from which ignitible concentrations of dust or ignitible fibers/flyings could be communicated, unless communication is prevented by adequate positive pressure ventilation from a source of clean air and effective safeguards against ventilation failure are provided.

Informational Note No. 1: As a guide to classification of Zone 21 locations, refer to ANSI/ISA-61241-10 (12.10.05)-2004, Electrical Apparatus for Use In Zone 20, Zone 21, and Zone 22 Hazardous (Classified) Locations—Classification of Zone 20, Zone 21, and Zone 22 Hazardous (Classified) Locations.

Informational Note No. 2: This classification usually includes locations outside dust containment and in the immediate vicinity of access doors subject to frequent removal or opening for operation purposes when internal combustible mixtures are present; locations outside dust containment in the proximity of filling and emptying points, feed belts, sampling points, truck dump stations, belt dump over points, etc., where no measures are used to prevent the formation of combustible mixtures; locations outside dust containment where dust accumulates and where due to process operations the dust layer is likely to be disturbed and form combustible mixtures; locations inside dust containment where explosive dust clouds are likely to occur (but neither continuously, nor for long periods, nor frequently) as, for example, silos (if filled and/or emptied only occasionally) and the dirty side of filters if large self-cleaning intervals are occurring.

Zone 22: A Zone 22 location is a location:

  (1) In which ignitible concentrations of combustible dust or ignitible fibers/flyings are not likely to occur in normal operation and, if they do occur, will only persist for a short period; or

  (2) In which combustible dust or fibers/flyings are handled, processed, or used but in which the dust or fibers/flyings are normally confined within closed containers of closed systems from which they can escape only as a result of the abnormal operation of the equipment with which the dust or fibers/flyings are handled, processed, or used; or

  (3) That is adjacent to a Zone 21 location, from which ignitible concentrations of dust or fibers/flyings could be communicated, unless such communication is prevented by adequate positive pressure ventilation from a source of clean air and effective safeguards against ventilation failure are provided.

Informational Note No. 1: As a guide to classification of Zone 22 locations, refer to ANSI/ISA-61241-10 (12.10.05)-2004, *Electrical Apparatus for Use in Zone 20, Zone 21, and Zone 22 Hazardous (Classified) Locations—Classification of Zone 20, Zone 21, and Zone 22 Hazardous (Classified) Locations.*

Informational Note No. 2: Zone 22 locations usually include outlets from bag filter vents, because in the event of a malfunction there can be emission of combustible mixtures; locations near equipment that has to be opened at infrequent intervals or equipment that from experience can easily form leaks where, due to pressure above atmospheric, dust will blow out; pneumatic equipment, flexible connections that can become damaged, etc.; storage locations for bags containing dusty product, since failure of bags can occur during handling, causing dust leakage; and locations where controllable dust layers are formed that are likely to be raised into explosive dust-air mixtures. Only if the layer is removed by cleaning before hazardous dust-air mixtures can be formed is the area designated unclassified.

Informational Note No. 3: Locations that normally are classified as Zone 21 can fall into Zone 22 when measures are used to prevent the formation of explosive dust-air mixtures. Such measures include exhaust ventilation. The measures should be used in the vicinity of (bag) filling and emptying points, feed belts, sampling points, truck dump stations, belt dump over points, etc.

[*Zones 20, 21, and 22 include both Class II and Class III areas. These areas are further defined in 506.5 and follow Zones 0, 1, and 2 in degree of risk or likelihood of exposure. However, this classification scheme was new in the 2005 NEC, and classification experience for these zones in the United States is not readily available yet. Like Divisions 1 and 2, Zones 20, 21, and 22 represent decreasing degrees of risk or decreasing degrees of likelihood that a hazardous condition will develop in a given area. Zones 20 and 21 are roughly equivalent to Division 1, and Zone 22 is roughly equivalent to Division 2. However, in this classification method, combustible dusts are grouped together with fibers/flyings and are not broken out into two separate classes, and groups are not defined for combustible dusts. Like the Class I zones, this scheme offers alternative protection techniques. These alternative protection methods are detailed in 506.8, and use and identification of equipment are covered by 506.9. Certain division equipment may be used in Zone-classified areas if it has additional markings as specified in 506.9(C)(1). Since this classification scheme and adequate details for its use are very new in the United States, they are not covered further in this manual. Classification, engineering, design, selection of equipment and wiring methods, installation, and inspection under this classification scheme must be done by persons experienced and skilled in these methods.*]

KEY QUESTIONS

1. What are the class, the division (or zone), and, for Class I and Class II areas, the group classification of the hazardous areas?
2. Where are the boundaries of the classified areas?
3. What wiring methods are used?
4. What types of electrical equipment are located in the hazardous location?
5. Is the use or the occupancy containing classified areas one of those specifically covered by the *NEC*?

CHECKLISTS

✔	Item	Inspection Activity	*NEC* Reference	Comments
		Checklist 8-1: Class I Locations		
	1.	Confirm the classification of areas, including class, division or zone, and group.	500.4(A), 500.5, 500.6	
	2.	Verify the suitability of the wiring methods being used.	501.10	
	3.	Verify that seals are located as required, properly installed, and sealed.	501.15	
	4.	Check materials used for flexible connections, such as explosionproof flex and flexible cords, for suitability.	501.10(A)(2), 501.10(B)(2), 501.140	
	5.	Check flexible cord connectors and receptacles for suitability.	501.140, 501.145	
	6.	Verify that equipment temperature markings are not greater than the ignition temperature of the gases or vapors involved.	500.8(B) and (C)	
	7.	Check equipment such as motors, transformers, overcurrent devices, switches and controllers, luminaires, heaters, and appliances for proper ratings and enclosures.	500.8, 501.100 through 501.135	
	8.	Check for adequate grounding and bonding paths to the disconnecting means of the building or separately derived system.	501.30, 250.92, 250.100	
	9.	Check multiwire branch circuits in the classified area for simultaneous disconnection of all ungrounded conductors.	210.4(B)	

✔	Item	Inspection Activity	*NEC* Reference	Comments
		Checklist 8-2: Class II Locations		
	1.	Confirm the classification of areas, including class, division, and group.	500.4(A), 500.5, 500.6	
	2.	Verify the suitability of the wiring methods being used.	502.10	

Checklist 8-2: Class II Locations

✔	Item	Inspection Activity	*NEC* Reference	Comments
	3.	Verify that seals are located where required, properly installed, and sealed, unless raceway arrangements preclude the requirement for seals.	502.15	
	4.	Verify that equipment temperature markings are not greater than the ignition temperature of the dusts involved.	500.8(B), 500.8(C)(3), 500.8(D)(2), 502.5	
	5.	Check materials used for flexible connections, including flexible cords, for suitability.	502.10(A)(2) and (B)(2), 502.130(A) (3) and (B)(4), 502.140	
	6.	Check flexible cord connectors and receptacles for suitability.	502.140, 502.145	
	7.	Check equipment such as motors, transformers, overcurrent devices, switches and controllers, luminaires, heaters, and appliances for proper ratings and enclosures.	502.5, 502.100 through 502.135	
	8.	Check for adequate grounding and bonding paths to the disconnecting means of the building or separately derived system.	250.92, 250.100, 502.30	
	9.	Check multiwire branch circuits in the classified area for simultaneous disconnection of all ungrounded conductors.	210.4(B)	

Checklist 8-3: Class III Locations

✔	Item	Inspection Activity	*NEC* Reference	Comments
	1.	Confirm the classification of areas, including class and division.	500.4(A), 500.5, 500.6	
	2.	Verify the suitability of the wiring methods being used.	503.10	
	3.	Check materials used for flexible connections for suitability.	503.10(A)(2), 503.10(B)	
	4.	Check flexible cord connectors and receptacles for suitability.	503.140, 503.145	
	5.	Verify that equipment operating temperatures are acceptable for the conditions.	503.5	
	6.	Check equipment such as motors, transformers, overcurrent devices, switches and controllers, luminaires, heaters, and appliances for proper ratings and enclosures.	500.8(B) and (C), 503.5, 503.100 through 503.135	

(*continues*)

CHECKLISTS

Checklist 8-3: Class III Locations

✔	Item	Inspection Activity	*NEC* Reference	Comments
	7.	Check for adequate grounding and bonding paths to the disconnecting means of the building or separately derived system.	250.92, 250.100, 503.30	
	8.	Check installations of cranes, hoists, and battery chargers for appropriate location and installation.	503.155, 503.160	

Checklist 8-4: Commercial Garages

✔	Item	Inspection Activity	*NEC* Reference	Comments
	1.	Confirm the applicability of Article 511.	511.1, 511.3	
	2.	Identify the extent and division of the Class I areas.	511.3	
	3.	Verify that suitable wiring methods are used within Class I areas.	511.4	
	4.	Verify that seals are located, installed, and sealed as required in 501.15.	511.9	
	5.	Verify that suitable wiring methods and equipment are used where installed above Class I areas.	511.7	
	6.	Verify that receptacles have GFCI protection where required.	511.12, 511.4(B)(2)	
	7.	Verify that battery chargers are not located in classified areas.	511.10(A)	
	8.	Verify that connectors for electric-vehicle-charging equipment are not located in Class I locations.	511.10(B)	

Checklist 8-5: Aircraft Hangars

✔	Item	Inspection Activity	*NEC* Reference	Comments
	1.	Confirm the applicability of Article 513.	513.1	
	2.	Identify the extent and division of the Class I areas.	513.3	

Checklist 8-5: Aircraft Hangars

✔	Item	Inspection Activity	*NEC* Reference	Comments
	3.	Verify that suitable wiring methods are used within and below Class I areas.	513.4, 513.8	
	4.	Verify that suitable wiring methods and equipment are used in unclassified areas of the hangar.	513.7, 513.10	
	5.	Verify that equipment used on stanchions, rostrums, and docks is appropriate.	513.4(B), 513.7(E)	
	6.	Verify that seals are located, installed, and sealed in accordance with 501.15.	513.9	
	7.	Verify that battery chargers are not located within the Class I locations.	513.10(A) and (B)	
	8.	Verify that external power sources and mobile equipment are properly located and installed.	513.10(C), (D), and (E)	
	9.	Verify that receptacles have GFCI protection where required.	513.12	

Checklist 8-6: Motor Fuel-Dispensing Facilities

✔	Item	Inspection Activity	*NEC* Reference	Comments
	1.	Confirm the applicability of Article 514.	514.1, 514.2	
	2.	Identify the extent and division of the Class I areas.	514.3, Table 514.3(B)(1) and (2)	
	3.	Verify that suitable wiring methods and equipment are used within and below Class I areas.	514.4, 514.8	
	4.	Verify that suitable wiring methods and equipment are used above classified areas.	511.7, 514.7	
	5.	Verify that seals are located, installed, and sealed at dispensers and in accordance with 501.15.	514.8, 514.9	
	6.	Verify that circuit and emergency disconnecting means are provided and that they disconnect all circuit conductors, including any grounded conductors.	514.11, 514.13	
	7.	Verify that means to disconnect all voltage sources, including feedback voltages, are provided.	514.11, 514.13	

CHECKLISTS

Checklist 8-7: Bulk Storage Plants

✔	Item	Inspection Activity	*NEC* Reference	Comments
	1.	Confirm the applicability of Article 515.	515.1	
	2.	Identify the extent and division of the Class I areas.	515.3, Table 515.3	
	3.	Verify that suitable wiring methods and equipment are used within and below Class I areas.	515.4, 515.8, 515.9	
	4.	Verify that suitable wiring methods and equipment are used above classified areas.	515.7	
	5.	Verify that seals are located, installed, and sealed in accordance with 501.15.	515.9	
	6.	Verify that dispensing areas comply with Article 514.	515.10	

Checklist 8-8: Spray Application, Dipping, and Coating Processes

✔	Item	Inspection Activity	*NEC* Reference	Comments
	1.	Confirm the applicability of Article 516.	516.1	
	2.	Identify the extent and division of the Class I and Class II areas.	516.3	
	3.	Identify the extent of any areas that are unclassified due to interlocks and ventilation.	516.3(C)(2)(a), 516.3(F)	
	4.	Verify that suitable wiring methods and equipment are used within the Class I areas.	516.4	
	5.	Verify that luminaires are suitably listed for the classified areas or are located behind fixed panels and outside the classified areas.	516.4(B) and (C)	
	6.	Verify that suitable wiring methods and equipment are used above and below classified areas.	516.7	
	7.	Verify that seals are located, installed, and sealed in accordance with 501.15.	516.4(A), 516.7(A)	
	8.	Check electrostatic spraying equipment for proper installation.	516.4(E), 516.10(A) and (B)	
	9.	Check powder coating equipment for proper installation.	516.10(C)	

Special Occupancies

OVERVIEW

This chapter covers the following seven selected occupancy types for which special requirements are provided in Chapter 5 of the *National Electrical Code*® (*NEC*®):

- Health care facilities
- Assembly occupancies
- Theaters and audience areas of motion picture studios
- Carnivals, circuses, fairs, and similar events
- Agricultural buildings
- Recreational vehicle (RV) parks
- Marinas and boatyards

In some cases, as in hospitals, the typical requirements of the *Code* are modified in very significant ways, whereas in others, such as places of assembly, the modifications are primarily restrictions on permitted wiring methods. In these occupancies, the general rules are changed to address any unique hazards or considerations that are inherent in that type of occupancy.

Not all the special occupancies covered in Chapter 5 of the *NEC* are included here (see the Hazardous Locations chapter of this manual for those that commonly include hazardous areas and the Dwelling Units and Mobile/Manufactured Home Sites chapter of this manual for sites for mobile and manufactured homes).

The rules for mobile homes and park trailers are not covered here; nor are the interior wiring requirements for RVs. Because the interior wiring of mobile homes, park trailers, and RVs is usually inspected at the factory, most electrical inspectors are concerned primarily with the locations where these products are used. The issues and rules for mobile home parks and the locations where park trailers are used are similar to the issues covered in the section on RV parks in this chapter.

SPECIFIC FACTORS

The various special occupancies covered by the checklists in this chapter each have characteristics that require some modification of the general rules of Chapters 1 through 4 of the *NEC*. Many of the occupancies covered here include some special or slightly increased hazard. Therefore, the permitted wiring methods, grounding requirements, load calculations, or other aspects of the electrical design are modified in various ways to address the particular hazards. For example, because of the use of electrical diagnostic and treatment equipment and the invasive nature of many medical procedures, patients in health care areas are more vulnerable to electric shock than they are elsewhere. Patients in certain areas, such as operating rooms and critical care areas, are also very reliant on the power supply to their area, especially if they are dependent on life-support equipment. Therefore, the permitted wiring methods are restricted, "redundant" grounding is mandated, special types of wiring devices must be used, and more than one power source is required in many patient care areas. Because the risks are greater in hospitals, more restrictions apply, and more rules are modified for hospitals than are for areas such as medical and dental offices, where there usually are no critical care areas, and where the primary special requirement is redundant grounding.

Increased or different hazards drive the modification of rules in other special occupancies as well. For example, because large numbers of people may gather in assembly occupancies or theaters, the fire resistance of structures and wiring methods in these occupancies merit special consideration. Theaters and audience areas of motion picture studios combine a large occupant load with the specialized electrical equipment required to produce plays and television shows. From an electrical inspector's standpoint, circuses and fairs are often just massive temporary power installations, where the temporary and portable nature of equipment and the numbers of people add another element of risk.

Agricultural buildings present a special problem because animals may be inclined to chew on, rub against, or otherwise damage electrical wiring within their reach and because:

- The environment tends to be wet or corrosive or both.
- Animals are more exposed to contact with electrical equipment in containment areas than they are elsewhere.
- Some animals are more sensitive to electrical currents than are people.

Similarly, the wet and perhaps corrosive environment in a marina or boatyard requires special attention to electrical installations. In many cases the requirements of the *Code* are not changed as much as they are repeated and emphasized where a particular aspect of the *Code* needs such emphasis.

RV parks, marinas, and boatyards share some common issues that are not concerns in many other occupancies. In both cases, the equipment supplied is highly mobile and likely to be moved. In both cases, the mobile equipment is connected by cord and plug and the equipment is not inherently well connected to earth as an electrical reference. In the case of a marina or boatyard, wave action and tides make the powered equipment likely to move, even when it is docked. These are significant concerns covered by *NEC* Articles 551 and 555.

KEY QUESTIONS

1. What wiring methods are being used?
The answer to this question is vital to all of the special occupancies covered in this chapter. Restrictions on permitted wiring methods are common modifications to the general rules of the *NEC* pertaining to an occupancy. An inspector should verify that the wiring method used in any occupancy is suitable for that occupancy. Special occupancies place restrictions on wiring methods and may disallow a wiring method that might well be suitable for the same type of building with a different use. In places such as hospitals, some of the wiring methods allowed in general-use areas such as offices and lobbies would not be permitted in the areas used for patient care. Similarly, in places such as fairs, carnivals, and marinas, portable cords are used for feeders and for other purposes that might not be permissible in other occupancies.

2. What and where is the service or power source?
The answer to this question provides an important common point for all of the special occupancies. It also provides a common point for inspections of the other aspects of an installation that are not covered by special rules from Chapter 5 of the *NEC*. For example, most inspections of grounding are logically started at the power source, which is usually the system grounding point.

3. Is temporary wiring or portable equipment being used?
Temporary wiring is the basis for special rules in many special occupancies, especially carnivals, fairs, theaters, and the like. Portable equipment is an issue in all of the special

occupancies. In the case of RVs and boats, the entire load is portable and interchangeable. Similarly, much of the equipment in a fair, carnival, circus, or theater production may be portable, including the power distribution equipment.

4. For health care facilities, what are the essential electrical system requirements?

This and a closely related question—namely, what type of health care facility is it?— establish the context that determines most of the requirements for essential electrical systems. Some facilities (hospitals and some nursing homes and **ambulatory care centers**) are required to have separated critical and life safety branches, while others (medical and dental offices and ambulatory care centers without critical care areas) are not.

5. For health care facilities, what type of alternative power supply is provided?

This question and its answer are closely related to the previous question and answer. Essential power requirements in health care facilities vary from a large system backed up by generators and supplied through multiple transfer switches, to a few luminaires with battery packs (unit equipment). Any system requiring a prime mover or any **alternate power source** other than unit equipment (luminaires with battery packs) should be laid out on drawings to show the components of the system along with sizing, overcurrent device coordination, and load information.

6. For health care facilities, which areas are general care spaces, critical care spaces, and wet procedure locations?

The locations identified by this question are required to have special grounding, specific minimum numbers of receptacles, alternative power sources, or perhaps ground-fault circuit interrupter (GFCI) protection or isolated power systems. The precise requirements depend on the class of care and the type of facility.

7. For theaters and audience areas, are dressing rooms provided?

Switches and pilot lights for certain receptacles are required for dressing rooms. Thus, identifying those areas is necessary to complete the inspection of theaters.

8. For RV parks, marinas, and boatyards, how many sites or slips are supplied with power?

RV parks, marinas, and boatyards are not required to be supplied with power for all individual sites, but if they are, specific circuit ratings and receptacle types are required, and the load calculations are dependent on the number of sites or slips supplied with power. Certain receptacle types and power supplies must be supplied in specified minimum quantities in RV parks, the quantity being a percentage of the total sites supplied with power. The receptacle types (ratings) in a boatyard are determined by the owner or designer of the boatyard, but the total quantity is still used in the load calculation.

9. For marinas and boatyards, does the marina or boatyard include fuel-dispensing facilities?

Dispensing of gasoline or other fuels may create hazardous areas that are covered by the Hazardous Locations chapter of this manual and *NEC* Article 514, *Motor Fuel-Dispensing Facilities.*

10. For marinas and boatyards, what level or reference is being used to define the electrical datum plane?

The **electrical datum plane** defines the minimum height for much of the electrical equipment used in marinas and boatyards and may also be used to determine the extent of some hazardous (classified) areas.

PLANNING FROM START TO FINISH

Although all of the special occupancies covered by this chapter have specific requirements for circuits and equipment, they also have many aspects in common with other installations. For example, each has a service or power source. The system grounding still takes place at the service or structure disconnecting means or other power source. The usual requirements for working space, conductor arrangements, conductor ampacities, support of wiring methods, placement of boxes, and equipment grounding still apply to these special occupancies. For this reason, the stages of construction and the timing of inspections is not significantly different for special occupancies. The checklists from other chapters of this manual also apply to special occupancies; Chapter 5 in the *NEC* only modifies the rules of *NEC* Chapters 1 through 4. Any requirements that are not modified or not specifically addressed in *NEC* Chapter 5 still apply to the special occupancy. Thus, the rules (and checklists) for services, grounding, and wiring methods still apply, except as modified in *NEC* Chapter 5.

The first inspections will still be inspections of underground or embedded work. Facilities such as hospitals, agricultural buildings, and RV parks will probably include some underground or embedded work. Some parts of agricultural buildings must have an equipotential plane embedded in the floor slab. RV parks usually distribute much of the power to the sites through the use of underground wiring methods. Other occupancies may involve only aboveground rough and finish inspections. Some occupancies, such as a carnival or fair, where almost everything is aboveground and little is concealed unless it is part of a permanent installation, may require only a single inspection.

In any case, the service or power source is a good place to start. This location provides a good place to check grounding, bonding, and feeder or branch-circuit arrangements and conductor sizes. From here, the checklists from this chapter can be used to check for compliance with the specific requirements of the occupancy. The checklists from the other chapters should be used at the same time to check for compliance with the general requirements that apply to all installations. For example, service equipment or panelboards represent the beginning of the required special equipment grounding for patient care areas in health care facilities. Similar special requirements for separate insulated (or covered) copper equipment grounding conductors are also required in agricultural buildings, RV parks, and marinas and boatyards.

In health care facilities, such as hospitals and nursing homes with critical care areas, an essential electrical system with an on-site power supply is a usual requirement. This additional power source presents another logical place to start the inspection of the essential electrical power system and all its related wiring methods, wiring separations, transfer switches, essential system branches, and so on.

In each occupancy, following the power distribution through feeders and branch circuits brings the inspector to the locations where other significant aspects of special occupancies can be inspected. At a circus or fair, this path may include the flexible cords and overhead wiring, the placement and protection of disconnects, the protection of lamps and cords, and the GFCI protection of receptacles. In a health care facility, the feeders will lead to the **reference grounding point** at the essential and normal system panelboards, through the redundant grounding requirements, and on to the patient care areas where numbers and types of receptacles, GFCI protection, patient equipment grounding points, and isolated power systems, if any, may be inspected. The same approach may be used with other special occupancies, although the nature of the items that must be checked will vary.

WORKING THROUGH THE CHECKLISTS

Checklist 9-1: Health Care Facilities

1. Review definitions and determine the proper classification of the areas to be inspected.
See the definition of **Health Care Facilities** in the Key Terms section of this chapter for a description of the various categories of health care facilities. Some of the types of health care facilities are further defined in Article 517. The health care facilities covered by Article 517 all have patient care areas or patient bed locations. *NEC* requirements, especially requirements for essential electrical power systems, vary according to the type of facility.

The rules for hospitals, for example, are much more detailed and demanding than are those for clinics or medical or dental offices, and therefore the precise classification of the facility under consideration is needed. A frequent misunderstanding is that Article 517 is just about hospitals, while in reality it covers any health care facilities that provide services to human beings. It therefore covers a much larger scope, including medical, dental, and psychiatric care clinics and offices. This understanding of Article 517 also means that an ordinary office space may require modifications other than just different signs and furniture for use as a medical office.

2. Verify that insulated copper conductors in metal raceways or equivalent cables are used to provide equipment grounding for branch circuits in patient care spaces and that receptacles with an insulated grounding terminal (Type IG) are not used.

"Patient Care Areas" were changed in 500.2 to "Patient Care Spaces," but the change was not made everywhere in Article 517. Where the term was not changed in the 2014 *NEC*, "area" should be considered the same as "space." Receptacles, boxes containing receptacles, and other fixed electrical equipment that operates over 100 volts in patient care areas are required to be grounded by both insulated copper conductors and metal raceways or certain cable sheaths. The wiring method for grounding of receptacles and fixed electrical equipment serving patient care spaces must itself qualify as equipment grounding conductor per Section 250.118, in addition to the requirements found in Section 517.13(B). Where cable wiring methods are used, the cables must be listed Types MI, MC, or AC that include a metallic sheath that is suitable for grounding and also include an insulated copper grounding conductor. Certain types of MC cables have an outer sheath that is listed and identified as an equipment grounding conductor without having to be used in combination with an internal equipment ground wire, but an internal ground wire is required for patient care areas. Because the sheath of Type AC provides a ground path, it is not required to have a separate insulated grounding conductor for many applications, but is required to have an internal ground wire for wiring in patient care spaces. This requirement for two separate grounding paths is often referred to as *redundant grounding*. Normally, either of the two grounding paths would be adequate to comply with Article 250 in most other areas or occupancies covered by the *Code*. Because of the requirement for two equipment grounding paths from virtually all electrical equipment in patient care spaces, isolated ground receptacles are prohibited in these areas in new installations.

3. Check for bonding between normal and essential branch-circuit panelboards serving any single patient care space.

Where patient care spaces include inpatient sleeping beds, the receptacles at the patient bed location are required to be supplied by at least two branch circuits. At least one circuit must be supplied by the normal power system (except where the receptacles are supplied from two critical branch circuits connected to separate transfer switches), and at least one circuit must be supplied by the essential power system. The equipment grounding terminal buses of the two or more normal or essential system panelboards that supply these circuits must be bonded together with an insulated copper conductor not smaller than 10 American Wire Gauge (AWG). This bonding helps to minimize any potential difference between various conductive parts of electrical equipment in the vicinity of the patient. A patient equipment grounding point may be established in the vicinity of a critical care patient bed. The patient equipment grounding point is not typically used in newer hospitals, except in critical care spaces with isolated power systems such as operating rooms. The local grounding points are optional, but they help to minimize potential differences in the patient care space by making bonding paths very short. The grounding and bonding requirements that are unique to patient care spaces are illustrated in **FIGURE 9-1**. Note that these rules apply to branch circuits only.

Panelboards

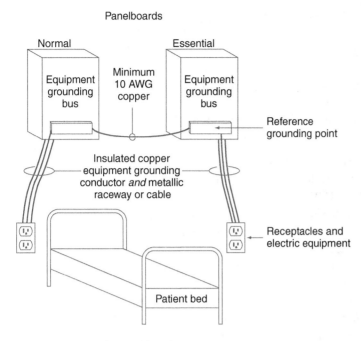

FIGURE 9-1   Grounding and bonding in patient care spaces.

4. Verify that each general care space patient bed location has at least two branch circuits, one from the normal system and one from the critical branch system.

As noted in Item 3, at least two circuits must supply receptacles at a patient bed location. The two circuits must be supplied by normal and essential power systems, with at least one circuit supplied by a normal power system and one by a critical branch power system, or both circuits can be supplied by two critical branch circuits connected to panelboards supplied by two separate transfer switches. This requirement reduces the likelihood that all receptacle power in the vicinity of a patient bed could fail at the same time because each system "backs up" the other. If more than one normal power circuit is used, all such branch circuits must be supplied by the same panelboard, unless the additional normal power circuits supply only special-purpose outlets or receptacles. More than one essential power system is allowed to serve a patient bed location.

Critical branch circuits for general care spaces are not required to be dedicated to one bed location. In order to enhance the independence and reliability of power to the circuits supplying patient bed locations, these circuits are not permitted to be part of a multiwire branch circuit.

Patient bed locations in clinics, in medical and dental offices, and in outpatient facilities are generally not required to have two circuits from separate systems. Beds in such health care areas are not usually considered the sleeping beds referred to in the definition of **Patient Bed Location**. Rooms with patient sleeping beds in psychiatric, substance abuse, and rehabilitation hospitals and sleeping rooms in **nursing homes** and limited-care facilities are not covered if they are used exclusively for sleeping rooms. In some facilities, the receptacles may not be needed, and in others, such as psychiatric or substance abuse facilities, the receptacles may present an unnecessary hazard to patients. The rule requiring multiple circuits from multiple sources, including critical branch power sources, is intended to provide a high degree of reliability in areas where life support or other critical equipment is used. Some outpatient surgical facilities have critical care spaces. In these areas, the requirements to ensure reliable power are applicable.

5. Verify that each general care space patient bed location is provided with a minimum of eight receptacles.

A minimum of eight receptacles must be provided at a patient bed location in a general care space. In most cases, this requires a minimum of four duplex receptacles, because of the requirement for a minimum of two circuits to supply the space. Eight single receptacles or some other combination of single, duplex, or quadruplex receptacles could be used to meet this requirement. Again, rooms with patient beds in psychiatric, substance abuse, and rehabilitation hospitals and sleeping rooms in nursing homes and limited-care facilities are not covered if they are used exclusively for sleeping rooms. No receptacles are required in psychiatric security rooms.

In hospitals, the required receptacles are often furnished as a part of a preassembled headwall unit that also includes communications, nurse call, and medical gases. Those receptacles that are supplied from the critical branch or their cover plates must be clearly identified to distinguish them from receptacles that are supplied from only the normal power source.

6. Check for tamper-resistant receptacles or covers in the pediatric locations of general care spaces.

Where receptacles of any rating are installed in general care spaces that are designated for pediatric patients, either the receptacles or the covers used with the receptacles must be listed as tamper-resistant. For the purposes of this rule, pediatric general care spaces include bathrooms, playrooms, activity rooms, and other rooms in pediatric wards. The *NEC* specifies a wiring device or cover assembly listed as tamper-resistant for these receptacles. A physical barrier that reduces the likelihood of a child coming into contact with live parts is required by the product standard for these devices and covers (**FIGURE 9-2**). Although these devices are sometimes referred to as "tamperproof," manufacturers do not label receptacles that way because nothing is truly tamperproof; therefore, the term that is used is "tamper-resistant." These are the same types of receptacles required in dwelling units, where the option for a tamper-resistant cover is not recognized. However, in this case the receptacle is also required to be hospital-grade (see Item 11 of this checklist), which is not a requirement for dwelling

FIGURE 9-2 Hospital-grade tamper-resistant receptacle for additional shock protection in pediatric spaces. (Courtesy of Pass & Seymour/Legrand.)

units. Currently, tamper-resistant receptacles provide the only option to meet this requirement, as there are no listed tamper-resistant covers available in the marketplace.

7. Verify that each critical care space patient bed location has at least two branch circuits, one from the normal system and one from the critical branch.

Except as covered in Item 8, this rule is essentially the same as the rule for general care spaces as explained in Item 4 of this checklist. Receptacles on the critical branch system are required to be identified as such, and in critical care spaces they must also be marked to indicate the panel and circuit number from which they are supplied. The marking is required to be a distinctive color or marking (traditionally red) on either the receptacle or the cover plate. Using both colored receptacles and plates is also acceptable. The *NEC* does not say where the circuit identification is to be, but usually the panel and circuit numbers are engraved into the cover, or labels are attached to the cover. In order to enhance the independence and reliability of power to the circuits supplying patient bed locations in critical care spaces, these circuits are not permitted to be part of a multiwire branch circuit.

8. Verify that each critical care space patient bed location has at least one receptacle supplied by a critical branch circuit dedicated to that bed location.

At least one dedicated branch circuit from the critical branch is required to supply one or more receptacles at a patient bed location in a critical care space. Because such a circuit is permitted to supply more than one outlet or receptacle, it is not an *individual branch circuit* as defined in Article 100, but it may not supply any other loads or any other areas. **FIGURE 9-3** shows a typical arrangement for branch circuits supplying receptacles at two patient bed locations in a critical care space. If the two critical branch panels are supplied by different transfer switches, it is not necessary to supply each bed from the normal panelboard, if each bed is supplied by both critical branch panelboards. The illustration shows two circuits from the normal source for each patient bed location, and that exceeds the minimum requirement, as is quite common.

9. Verify that each critical care space patient bed location has at least 14 receptacles.

A minimum of 14 receptacles must be provided at a patient bed location in a critical care space. In most cases, seven or more duplex receptacles are used to meet this requirement

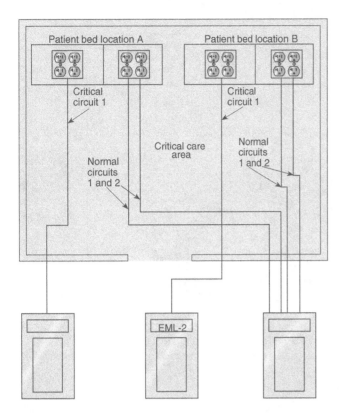

FIGURE 9-3 Examples of normal and critical branch circuits supplying patient bed locations in a critical care space. (Source: *National Electrical Code® Handbook*, NFPA, Quincy, MA, 2011, Exhibit 517.5.)

because of the related requirement for a minimum of two branch circuits to supply the area receptacles. Fourteen or more single receptacles or some other combination of single, duplex, and quadruplex receptacles could be used to meet this requirement as implied by Figure 9-3, which does not show all the required receptacles but does illustrate the number of circuits required.

In hospitals, the required receptacles are often furnished as a part of a preassembled headwall unit that also includes communications, nurse call, and medical gases, but the supply to the headwall units is part of the electrical installation and inspection.

10. Verify that at least one receptacle in a critical care space bed location is connected to a branch circuit from a separate normal or critical branch source.
Of the 14 required receptacles in a critical care space patient bed location, at least one must be on a normal system branch circuit or on a critical branch circuit supplied by a different transfer switch than that which supplies other receptacles at the same location. Although this rule is very similar to that for general care spaces, it more specifically requires that receptacles served by separate sources be available at the bed location. Highly reliable power must be available where life support or other critical equipment is used or invasive medical procedures are performed. This rule, combined with the rule discussed in Item 8, ensures such reliability. Note that in Figure 9-3, some receptacles are supplied from a critical branch panelboard, and some receptacles are supplied from a normal panelboard at each bed location. This example does not show all required receptacles but represents a common practice. However, only one receptacle is actually required from the normal branch or second critical branch source.

11. Verify that a minimum of 36 receptacles are installed in each operating room.
This requirement, first appearing in the 2014 *NEC*, is extracted from NFPA 99. Many of today's surgical procedures involve extensive use of electrically powered medical equipment. The 36 receptacle requirement can be met through the use of any combination of single, duplex, and quadruplex receptacles. At least 12 of these receptacles are to be supplied from one or more normal system branch circuits or from a branch circuit supplied by a critical branch that is connected to a different transfer switch than other critical branch receptacle circuit supply devices in the same operating room.

12. Verify that all patient bed location receptacles are hospital-grade and that all critical branch receptacles are identified.
All new or replacement receptacles at patient bed locations must be listed as hospital-grade. Hospital-grade receptacles are made to be more durable than ordinary receptacles. They have impact-resistant faces and bodies, and the contact surfaces are designed to provide higher levels of pull-out resistance than do ordinary receptacles. Hospital-grade receptacles are identified by a green dot on the face of the receptacle. (The dot can be seen on the receptacle in Figure 9-2.) All of the patient bed receptacles in general and critical care spaces are required to be hospital-grade. **FIGURE 9-4** illustrates the supply-circuit and quantity requirements for patient bed location receptacles. Receptacles on any branch of the critical branch are required to be marked by a distinctively colored cover or receptacle

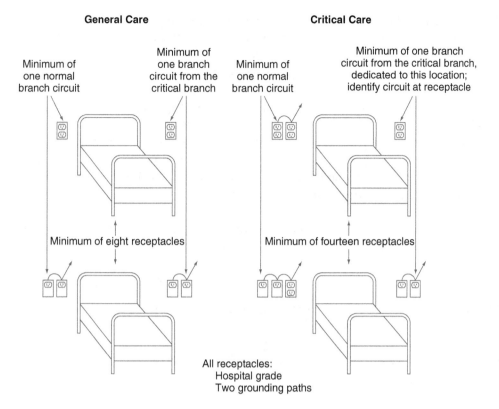

General Care

Minimum of one normal branch circuit

Minimum of one branch circuit from the critical branch

Minimum of eight receptacles

Critical Care

Minimum of one normal branch circuit

Minimum of one branch circuit from the critical branch, dedicated to this location; identify circuit at receptacle

Minimum of fourteen receptacles

All receptacles:
Hospital grade
Two grounding paths

FIGURE 9-4   Patient bed location receptacles and circuits.

(or both). Neither the *NEC* nor NFPA 99, *Standard for Health Care Facilities*, specifies a color, but NFPA 99 requires that if color is used for identification of critical branch receptacles, the color be the same throughout the facility. Red is the traditional marking. As is noted in Item 7 of this checklist, critical branch receptacles in critical care bed locations must also be labeled with the panel and circuit number that supplies them.

13. Check wet procedure locations for protection by GFCI devices or, where interruption cannot be tolerated, use of isolated power systems.

See the definition of **Wet Procedure Location** in the Key Terms section. The term *wet procedure location* has a different meaning in relation to health care facilities than *wet location* in the rest of the *NEC*. Receptacles and fixed equipment in wet locations must be GFCI protected or supplied through an isolated power system. Because virtually all electrical equipment is either fixed or connected by cord and plug to a receptacle, all such equipment in wet locations is required to be either GFCI protected or served by an isolated power system. According to Section 6.3.2.2.8.1 of NFPA 99, *Standard for Health Care Facilities*, 2012 edition, "Wet procedure locations shall be provided with special protection against electric shock." Section 6.3.2.2.8.2 proceeds to say, "This special protection shall be provided as follows:

"(1) Power distribution system that inherently limits the possible ground-fault current due to a first fault to a low value, without interrupting the power supply

"(2) Power distribution system in which the power supply is interrupted if the ground-fault current does, in fact, exceed the trip value of a Class A GFCI."

The first option describes an isolated power system. So, for example, where interruptions can usually be tolerated, such as in hydrotherapy areas, GFCIs can be used. However, interruptions generally cannot be tolerated in places such as operating rooms. Some operating rooms are wet procedure locations, depending on the types of procedures performed, and where an operating room is a wet location, isolated power systems are installed instead of GFCI protection where power interruption is unacceptable. Only in this circumstance does a risk assessment under NFPA 99 require use of an isolated power system. They are permitted to be used elsewhere in a health care facility where continuity of power is desirable.

See the definition of **Isolated Power System** in the Key Terms section. Isolated power systems provide for some additional reliability in a system, because a single ground fault can occur without necessarily operating an overcurrent device or creating a shock hazard. This feature is described in 517.20(A)(1) as follows: "Power distribution system that inherently limits the possible ground-fault current due to a first fault to a low value, without interrupting the power supply." Isolated power systems are provided with **line isolation monitors** to warn when insulation leakage and such faults do occur, but power is not automatically interrupted on the occurrence of a ground fault.

GFCI protection is not required for receptacles in critical care spaces where the toilet and basin are within the patient room rather than in a separate bathroom. This may or may not be considered a wet location. A matching exception is also provided in 210.8(B).

For Hospitals and Ambulatory Care Facilities with Critical Care Spaces

14. Review the essential electrical systems, and verify that critical, life safety, and equipment branches are provided.

See the definitions of **Essential Electrical System**, **Critical Branch**, **Equipment Branch**, and **Transfer Switch** in the Key Terms section. The critical life safety and equipment branches are subdivisions of the essential electrical system. Details of the essential electrical system should be provided on electrical design plans. Examples of typical essential system layouts can be found as Informational Notes in Part III of Article 517. Additional information is available in NFPA 99, *Standard for Health Care Facilities*. The number and arrangement of transfer switches depend on the size (load) of the system supplied. For loads of not more than 150 kilovolt-amperes (kVA), a single automatic transfer switch may be used. Larger loads may require multiple transfer switches and may require connection of the equipment branch to be delayed until after connection of the critical and **life safety branches** is complete. Both branches require automatic connection to an alternative power source on failure of the normal source.

Certain equipment system loads may be connected to the alternative power source by a manual transfer switch, and the equipment branch is generally required to be connected after the critical and life safety branches. **FIGURE 9-5A** illustrates the required separation between branches of a hospital essential electrical system with the usual minimum requirements for separation of systems and transfer switches. **FIGURE 9-5B** illustrates a possible arrangement for a much larger hospital system.

Note that a single feeder from the generator is permitted to supply the entire essential electrical system. The number of transfer switches on the individual branches is not mandated, but is a function of reliability, design, and load considerations. **FIGURE 9-6** shows the minimum requirement for an essential electrical system arrangement permitted for hospitals with small electrical loads (150 kVA or less) where a single transfer switch is acceptable. In this case, the requirement for a time lag between the connection of the critical and life safety branches and the connection of the equipment branch to the critical branch source is waived, but the alternate source must be adequate for all the anticipated loads.

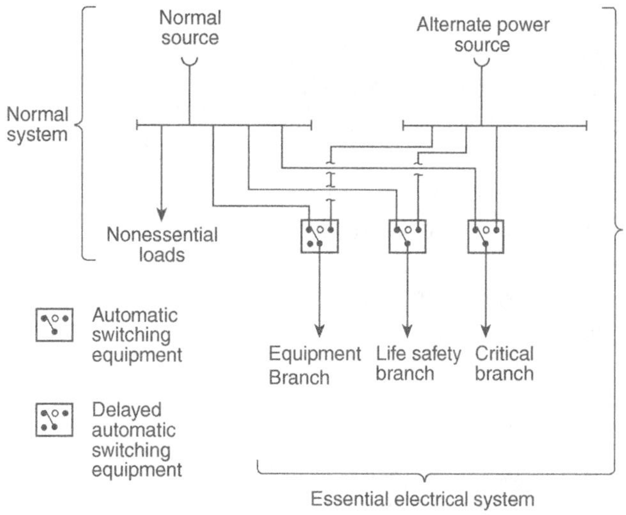

FIGURE 9-5A Small electrical system configuration in hospital. (Source: *National Electrical Code® Handbook*, NFPA, Quincy, MA, 2011, Informational Note Figure 517.30, No. 1.)

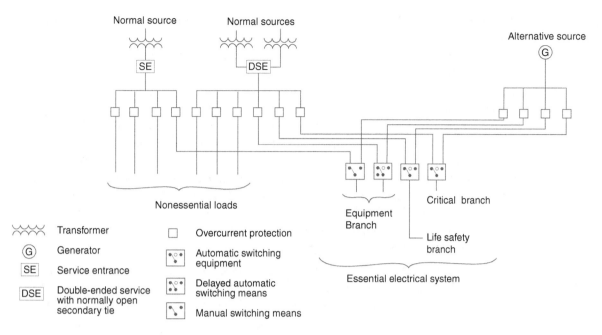

Transformer

Generator

SE Service entrance

DSE Double-ended service with normally open secondary tie

Overcurrent protection

Automatic switching equipment

Delayed automatic switching means

Manual switching means

FIGURE 9-5B Large electrical system configuration in hospital.

15. Review load calculations for the essential system and verify that capacity of power sources and feeders is adequate.

Load calculations should be shown on electrical drawings or otherwise furnished by the electrical designer. Feeders are required to be sized in accordance with Articles 215 and 220. Generators may be sized on the basis of one or more of the following: prudent demand factors and historical data, connected load, or Article 220 feeder calculations. In this respect, Article 517 provides more design latitude in applying demand factors than would be permitted under Articles 700 or 701 for emergency or legally required standby systems.

16. Verify that the wiring of the life safety and critical branches is independent of, and separated from, other wiring and equipment.

Generally, emergency system wiring as covered in Article 700 may not share the same raceway, box, or cabinet with the wiring of any other system. In Article 517, separation of essential system wiring (life safety or critical branch) applies even to other essential system wiring from the same source, including wiring of the equipment branch, the normal system, or circuits from the other branches of the essential system, even when they supply the same patient care location. Essential system wiring may occupy the same enclosure with other circuits in transfer switches or exit or emergency lighting fixtures or boxes. Two or more circuits of the same essential branch (life safety or critical) may also share an enclosure or raceway, but circuits of the life safety branch may not be mixed with circuits of the critical branch. Equipment branch wiring may occupy the same raceways, cabinets, or boxes with other circuits that are not essential system circuits. The requirements for separation are intended to increase the reliability of the life safety and critical branches.

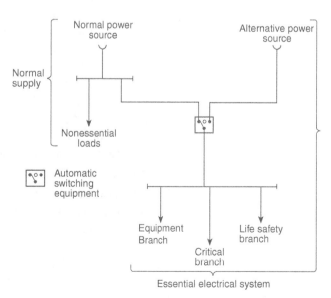

FIGURE 9-6 Small electrical system arrangement single transfer switch permitted. (Source: *National Electrical Code® Handbook*, NFPA, Quincy, MA, 2011, Informational Note Figure 517.30, No. 2.)

17. Verify that mechanical protection for essential system wiring is provided by nonflexible metallic raceways or Type MI cable.

Nonflexible metal raceways such as rigid metal conduit (RMC), intermediate metal conduit (IMC), Schedule 80 polyvinyl chloride conduit (PVC), or electrical metallic tubing (EMT) must be used for essential system wiring unless Type MI cable is used. Where the circuits supply patient care spaces, the circuit must also include an insulated copper equipment grounding conductor. Schedule 80 PVC without encasement or Schedule 40 PVC encased in 2 in. (51 mm) of concrete may be used for essential system circuits that do not serve patient care spaces. (The nonmetallic conduit does not provide the required redundant ground path.) Certain limited uses of flexible cords and flexible metal raceways or cables are permitted for flexibility or in listed prefabricated equipment and for "fishing" new circuits into wall and ceiling spaces that are not otherwise accessible. The wiring of secondary circuits of Class 2 or Class 3 limited power systems used for communications or signaling, such as nurse call systems, does not require raceways for protection.

18. Verify that only those loads that are intended for connection to the life safety branch are supplied by the life safety branch.

The following types of loads may be connected to the life safety branch of an essential system:

- Illumination of means of egress
- Exit signs and exit directional signs
- Fire alarms, alarms required for nonflammable medical gas piping systems, and alarms or alerting systems for mechanical, control, or other accessories required for the operation of life safety systems
- Hospital communications systems used for issuing emergency instructions
- **Task illumination** battery chargers for emergency **battery-powered lighting units** and **selected receptacles** at the generator set or essential transfer switch locations
- Generator set accessories
- Automatic doors in egress corridors
- Lighting, control, communications, and signal systems for elevator cabs

The life safety branch must supply these listed loads where they exist. No other loads may be connected to the life safety branch.

These permitted loads are all related to direct safety issues for patients and other personnel and to emergency evacuation of buildings. The critical branch is intended to accommodate other types of loads that are essential to patient care and life support. The loads to be applied to the critical branch are listed in 517.33. See NFPA 101-2012, *Life Safety Code,* for more information on life safety equipment and the use of the life safety branch.

Heating, ventilation, and air-conditioning (HVAC) controls are permitted to be installed on the life safety branch if required for effective operation of life safety systems. This equipment includes dampers, mechanical controls, and other accessories needed for life safety. Section 517.33 specifies the equipment that is permitted and required to be connected to the critical branch. For example, the receptacle from the critical branch that is required in patient bed locations is connected to the critical branch.

19. Verify that two independent sources are provided for the essential electrical system and that the alternative source is suitable.

To ensure a high level of reliability, the essential electrical system is required to be provided with two separate power sources. These sources are usually a utility source and an on-site generator. Either the utility or the generator may be used as the normal power source, with the other acting as the alternate source. Where there is more than one generator and no utility source, one or more generators may be used as the normal power source, with others used as the alternate source. When judging the suitability of the alternate source, locations and exposure to natural hazards such as storms, earthquakes, and floods must be considered. Possible interruptions of the normal system due to damage to or failure of premises wiring or equipment must also be considered. Separation of feeders from the normal and alternate sources may also be necessary to prevent simultaneous disruption of both sources.

Transfer of the life safety branch from the normal source to the alternate source must be accomplished automatically within 10 seconds of a failure of the normal source. Connection of the critical branch loads to an alternate source may be connected after transfer of the life safety branch. Some equipment must be automatically connected after an appropriate delay, and other critical branch loads, such as some elevators or some heating equipment, may be automatically or manually transferred. As noted in Item 13 of this checklist, equipment system loads may be transferred by delayed automatic connections or by a manual transfer switch, but the connection must be delayed until after life safety and critical branch loads are connected, except in those instances in which the essential system load (150 kVA or less) permits a single transfer switch.

20. Determine that the overcurrent protective devices of the life safety branch are selectively coordinated and that the overcurrent devices of the critical branch are coordinated to 0.01 seconds.

Determination of compliance with these requirements cannot be achieved through visual inspection; nor is it practical or even feasible to require actual overcurrent protective device operation to demonstrate compliance. However, compliance with these rules results in an extremely important operational condition that increases the electrical system reliability. Localizing faults and overloads in an electrical system is required in order to minimize large-scale power outages that could put patients and hospital occupants in peril. Engineering documentation is the surest method for an authority having jurisdiction (AHJ) to determine compliance with these requirements. In fact, the 2014 *NEC* requires that the means of achieving selective coordination be chosen by a licensed professional engineer or other person similarly qualified and "engaged in the design, installation, or maintenance of electrical systems." Documentation of the chosen method(s) is required to be provided to those responsible for designing, installing, inspecting, maintaining, or operating the electrical system.

Noteworthy is the difference in level of system coordination required for the *life safety branch* versus that required for the *critical branch*. Because 517.26 specifies that the life safety branch meet the requirements of Article 700, selective coordination, as defined in Article 100, must be a feature of operation of all overcurrent protective devices used to protect branch circuits and feeders that comprise the life safety branch. This requirement means that localization of the faulted circuit must occur under any condition of overcurrent, whether it is a short circuit, ground fault, or overload (see the definition of *Coordination, Selective* in the Emergency and Standby Systems and Fire Pumps chapter of this manual). In contrast, the overcurrent devices of the critical branch are required to be coordinated to only those fault conditions that exceed .01 seconds (6 cycles). Coordination at this level cannot be relied upon to localize electrical system problems under all overcurrent conditions. However, the AHJ's responsibility is to verify that the requisite levels of overcurrent device coordination have been incorporated into the selection of protective devices, and the most effective way of verifying this is to require the system coordination documentation from the system designer. This documentation is likely to include coordination information from the manufacturer of the overcurrent protective devices used in the life safety and critical branches

For Nursing Homes and Limited-Care Facilities Where Life-Support
Equipment or General Anesthesia Is Used

21. Review the essential electrical system and verify that life safety and critical branches are provided.

The requirements for an essential electrical system in a nursing home or a **limited-care facility** are very similar to those for a hospital. One significant difference is that there is no requirement for a separate equipment branch in nursing homes or limited-care facilities. Instead, some loads that would be supplied by the equipment branch in a hospital may be supplied by the critical branch in a nursing home or limited-care facility (see Item 13 of this checklist).

22. Review load calculations for the essential system and verify that capacity of power sources and feeders is adequate.

Essential electrical system power sources for nursing homes or limited-care facilities are required to be adequate to supply the load that will be connected at one time. This is a simpler rule than the rule for sizing such systems in hospitals, where the types of essential loads are often larger and more diverse. Article 517 does not specify the length of time that the alternate source must supply the essential system, but Articles 700 and 701 do require that batteries last at least 1½ hours and that fuel supplies for generators be adequate for at least 2 hours of operation. Longer times may be required by other standards or authorities.

23. Verify that the life safety branch wiring is independent and separated from other wiring and equipment. (Separation of the critical branch is not required.)

Generally, life safety branch wiring may not share the same raceway, box, or cabinet with the wiring of any other system, including wiring of the critical branch and the normal branch. Life safety branch wiring may occupy the same raceway or enclosure as other circuits in transfer switches or exit or emergency lighting fixtures or boxes. Two or more circuits of the life safety branch may share an enclosure. Critical branch wiring may occupy the same raceways, cabinets, or boxes as other circuits, except for circuits of the life safety branch. Note that this rule is considerably less restrictive than are the circuit separation rules for hospitals. In this respect, the treatment of critical branch wiring is more like the legally required standby systems of Article 701. Life safety branch wiring separations are similar to those for emergency systems covered by Article 700.

24. Verify that only those loads intended for connection to the life safety branch are supplied by the life safety branch.

The following types of loads may be connected to the life safety branch of an essential electrical system:

- Illumination of means of egress
- Exit signs and exit directional signs including such lighting in dining and recreation areas
- Fire alarms and alarms required for nonflammable medical gas piping systems
- Communications systems used for issuing emergency instructions
- Selected outlets for illumination and receptacles at the generator set location
- The lighting, control, communications, and signal systems for elevator cabs

The life safety branch must supply these loads. No other loads may be connected to the life safety branch. The permitted loads are all related to direct safety issues for patients and other personnel and emergency evacuation of buildings. The critical branch is intended to accommodate other types of loads that are essential to continuing patient care and life support. See NFPA 101 for more information on life safety equipment and the use of the life safety branch. Note that the limitations on the use of the life safety branch are almost identical to the rules for hospitals, except that lighting of dining and recreational room exit ways are specifically mentioned and automatic doors and emergency battery sets for illumination at generator sets are not specified in the rule for nursing homes and limited-care facilities.

Those portions of facilities that admit patients who must be sustained by life-support equipment must meet the requirements for hospital essential electrical systems found in 517.30 through 517.35.

25. Verify that two independent sources are provided for the essential electrical system and that the alternative source is suitable.

Unless the normal power source is an on-site generator, the alternative source must be a generator driven by a prime mover. A utility source may be the alternative source if the normal source is a generator. This rule and the previous rules apply only to those nursing homes and limited-care facilities where life-support equipment or general anesthesia is used for surgical treatment (see the heading above Item 21 of this checklist). Other types of nursing homes and limited-care facilities may use a battery system or self-contained battery-powered equipment.

For Other Health Care Facilities, Including Clinics, Medical and Dental Offices, and Ambulatory Health Care Facilities Without Critical Care Spaces

26. Verify that an essential electrical system is provided where required.

An essential electrical system is required where task illumination is related to the safety of life and is necessary for the safe cessation of procedures in progress or where inhalation anesthetics are administered to patients. The essential electrical system is required to provide power to equipment that is required for life safety. If life-support equipment is used or critical care spaces are present, the essential electrical system must comply with the requirements for hospitals.

27. Verify that an alternative power source is provided that is adequate and designed specifically for the purpose.

The alternative power source must be designed specifically for the purpose and is required to be a generator or a battery system, or the equipment is required to have integral batteries. If a hospital-grade system or a generator is not required, the requirements of Article 700 will govern the battery installation. Transfer from normal to alternative power source(s) must be automatic and must take place within 10 seconds of the loss of normal power. An alternative source consisting of batteries must be capable of supplying the essential loads for 1½ hours after a failure of the normal branch. Because most medical offices do not include critical care spaces or areas where general anesthesia or life-support equipment is used, emergency lighting and other essential power are often provided entirely by unit equipment (**FIGURE 9-7**). Where generator systems are used, they must comply with the requirements for essential power systems in hospitals.

Note: Checklist Items 28 through 38 could apply to any type of health care facility.

28. Determine whether hazardous (classified) anesthetizing locations exist in the facility.

Few, if any, modern **anesthetizing locations** still use **flammable anesthetics**. Ethyl ether, which is a Group C flammable liquid, was the first widely used anesthetic, but it is no longer commonly used. Most modern operating rooms and anesthetizing locations are termed *other-than-hazardous (classified) locations* in Article 517. The

FIGURE 9-7 Self-contained, battery-supplied unit equipment meets requirement for emergency lighting in most medical and dental offices. (Courtesy of Dual-Lite, Inc.)

hospital administration should be able to confirm whether the anesthetizing areas are classified or unclassified.

29. Verify that appropriate wiring methods and equipment are used in and above hazardous (classified) anesthetizing locations.

See the Hazardous Locations chapter of this manual for a discussion of the wiring methods appropriate for Class I areas. Where flammable anesthetics are used, the area up to 5 ft (1.5 m) above the floor is a Class I, Division 1 area. The remainder of the area is considered "above a hazardous (classified) location" where 517.61(B) applies.

30. Verify that power circuits in flammable anesthetizing locations are isolated from other power distribution systems.

Power circuits in flammable **anesthetizing locations** are required to be isolated from other circuits. One method of providing isolation is through a listed isolated power system. See the definition of Isolated Power System in the Key Terms section. Because isolated power systems are required in operating rooms that are also wet procedure locations and because isolated power systems are often used where not specifically required by the *Code,* the presence of an isolated power system should not be taken as an indication of a classified area. Ordinary grounded power systems are permitted for some circuits and equipment in anesthetizing locations, but they must be isolated from other distribution systems.

31. Verify that one or more battery-powered lighting units are provided in anesthetizing locations.

Even though most anesthetizing locations are required to be served by essential electrical systems, at least one battery-powered lighting unit that meets the requirements of Article 700 ("unit equipment") is required in an anesthetizing location. Such lighting units transfer battery-powered illumination almost instantly when normal power fails, whereas the essential system may require up to 10 seconds to come on line. The battery-powered units also provide an additional level of redundancy to the lighting supplied from the essential electrical system that serves an anesthetizing location. These lighting units are permitted to be connected to the critical branch lighting circuit for the area and connected ahead of any local switches. This will allow the battery-powered equipment to come on whenever the power fails, whether the area lighting is on or not.

32. Verify that supply circuits to X-ray equipment are adequate and supplied through appropriate wiring methods and connections.

See Item 35 for sizing of branch-circuit conductors and overcurrent devices. Generally, equipment connections to supply circuits should be made by a wiring method that meets the general requirements of the *NEC*, although the grounding requirements and the restrictions on wiring methods that apply to patient care spaces must also be observed (see Items 2 and 17 in this checklist). Fixed or stationary X-ray equipment that is supplied by branch circuits rated not over 30 amperes may be connected by cord and plug if hard-service cable or cord is used. **Portable, mobile,** or **transportable X-ray equipment** can also be connected by cord and plug under the normal permitted uses of flexible cords covered by Article 400. Receptacles for such moveable equipment or circuits up to 60 amperes are not required to be on individual branch circuits. Article 490 covers equipment operating on supply circuits of over 1000 volts.

33. Check location, capacity, and type of disconnecting means for X-ray equipment.

The **disconnecting means** must be installed in a location that is readily accessible from the X-ray control. In addition, the disconnect must have a rating at least equal to the rating required for the branch-circuit overcurrent device—that is, the larger of 50 percent of the **momentary current rating** or 100 percent of the **long-time current rating**. A grounding-type attachment plug and receptacle may be used as the disconnecting means for portable X-ray equipment operating at 120 volts and not over 30 amperes.

34. Verify that supply circuits to X-ray equipment meet minimum ampacity and overcurrent rating requirements.

Branch-circuit sizing requirements depend on the use of the X-ray equipment. Conductor ampacity and overcurrent device ratings for diagnostic equipment are required to be equal to the greater of 50 percent of the momentary rating or 100 percent of the long-time rating of the equipment. Conductors and overcurrent devices for therapeutic equipment must be sized at 100 percent of the equipment current rating. Special demand factors cover feeder conductors supplying more than one branch circuit for X-ray equipment. The equipment manufacturer usually specifies conductor sizes and overcurrent protective device sizes. The specified sizes may be larger than the minimum *NEC* requirement because of voltage drop or other considerations. Of course, the manufacturer's instructions must be followed in sizing conductors and other circuit elements.

35. Verify that enclosures for high-voltage parts and noncurrent-carrying metal parts of X-ray equipment are grounded.

Noncurrent-carrying metal parts must be connected to an equipment grounding conductor in accordance with the grounding requirements for patient care spaces as covered in Item 2 of this checklist. In addition to meeting the requirements for patient care spaces, high-voltage parts (over 1000 volts) must be within grounded enclosures and insulated from the enclosures. Connections between high-voltage parts must be made with high-voltage shielded cables.

36. Verify that low-voltage systems in patient care spaces have insulation and isolation equivalent to power distribution systems.

Communications, signaling, data-gathering, fire alarm, and other systems that are commonly less than 120 volts are required to be insulated and isolated. The isolation and insulation are required to be equivalent to that required for the power distribution system. To comply with this requirement, many systems are kept out of reach of patients. Other systems, such as nurse call systems, may have nonelectrical interfaces where they are intended to be within reach of patients; otherwise the low-voltage wiring must be installed using ordinary wiring or wiring with insulation equivalent to that for other wiring (600-volt insulation) rather than some lower-rated limited-energy cables. This is not a requirement that such systems be installed in conduit except where specified in Chapter 7 or 8 of the *NEC* for the type of circuit being considered. Nor are such circuits required to be provided with redundant grounding. Note that according to 517.30(C)(3)(5), even limited-energy system wiring on the essential electrical system is permitted to be installed using limited-energy wiring methods without raceway protection.

37. Check isolated power systems (where installed) for proper installation, features, and conductor identification.

See the definitions of Isolated Power System and **Hazard Current** in the Key Terms section. The requirements for isolated power systems are covered by Part VII of Article 517. Some of those requirements can be summarized as follows: Isolated power systems must have disconnecting means that provide for the simultaneous disconnection of each isolated circuit conductor. Circuits supplying primaries of isolated power systems are limited to 600 volts and must have overcurrent protection. Isolated power circuits are also limited to 600 volts between conductors. Because the objective of an isolated power system is complete isolation from other systems and circuits, the system must be isolated from ground, from the supply circuit, and from conductors of other systems. Isolation from the supply circuit is usually provided by **isolation transformers**, but motor generator sets or batteries may also provide the required isolation. Isolated power systems are required to be equipped with a line isolation monitor that measures and displays hazard current and conspicuous green and red lamps and sounds an alarm when hazard current reaches 5 milliamperes. An example of a line isolation monitor is shown in **FIGURE 9-8**. An ammeter

FIGURE 9-8 Components of an isolated power system. (This material and associated copyrights are proprietary to, and used with the permission of, Schneider Electric.)

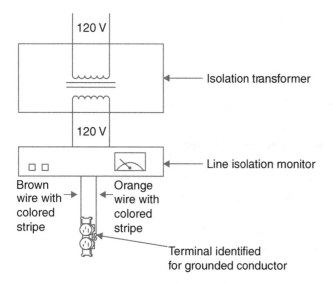

FIGURE 9-9 Wiring devices with isolated power system conductors. (Reprinted with permission from *National Electrical Code® Changes*, page 330, as modified, Copyright © 1999, National Fire Protection Association, Quincy, MA 02169. This reprinted material is not the complete and official position of the NFPA or the referenced subject, which is represented only by the standard in its entirety.)

and indicating lights provide personnel with a constant status report relative to the operation of the isolated power system. Additional remote annunciators may be installed for monitoring in more than one location.

A green lamp indicates that the unit is properly isolated from ground, and a red lamp indicates excessive hazard current. Generally, an isolated power system is permitted to serve only one operating room and its associated anesthetic induction room, although exceptions are provided for specific cases. Isolated circuit conductors are required to be identified by orange and brown colors. For a 3-phase system, the third conductor is identified as yellow. The orange, brown, and, where used, yellow conductors must be further identified by a colored stripe that runs along the entire length of the conductors. The stripe must be distinctive and not white, green, or gray. Where the isolated system supplies convenience-type receptacles (single-phase, 125 volts, 15 and 20 amperes), the orange conductor is connected to the terminal identified for a grounded conductor. **FIGURE 9-9** shows a 125-volt receptacle connected to an isolated power system. Inspectors should be aware that plug-in receptacle polarity testers will not work properly with receptacles on isolated power systems.

38. Verify second-level ground-fault protection of equipment (GFPE) where applicable in hospitals and other health care facilities.

Where GFPE is required for the service or feeder disconnecting means of a health care facility, GFPE equipment is required in the next level of feeders or branch circuits downstream. The required selectivity of the second level isolates a ground fault on a feeder or branch circuit to that circuit and does not cause the main supply overcurrent protective device to open. This requirement applies only to those health care facilities or buildings containing or supplying essential services or utilities to health care facilities where the health care facilities include critical care spaces or utilize electrically powered life-support equipment. For example, it applies not only to hospitals, but also to multiple-occupancy buildings that include some occupancy that uses **electrical life-support equipment** or has critical care spaces. The inspector cannot be reasonably expected to test the device for selectivity, but some evidence of engineering to achieve the required 100 percent selectivity should be provided by the designer.

Checklist 9-2: Assembly Occupancies

1. Determine the applicability of Article 518.

See the definition of **Assembly Occupancies** in the Key Terms section. Article 520 applies to assembly occupancies used as audience areas for motion pictures or stage productions. **FIGURE 9-10** shows a building in which areas have been designated as assembly occupancies in accordance with Article 518. In this example, the double lines with dark shading represent the walls that are required to be fire-rated construction by the local building code. The wiring in these walls must meet the requirements of Article 518 if they are part of or within the assembly occupancies. The walls represented by single lines represent walls that are not required to be fire-rated construction. Any wiring method from Chapter 3 of the *NEC*, installed in accordance with its respective requirements, can be used in these areas unless they are within assembly occupancies. For example, the walls defining the serving corridors are not fire-rated but are within the assembly occupancies in Figure 9-10, while office area walls are outside the assembly area.

2. Verify that wiring methods are suitable for the occupancy and fire rating of the area(s).

The wiring methods permitted in assembly occupancies are metal raceways, including flexible metal raceways, and cables with metallic sheaths, specifically Type MI and MC cables and Type AC cable. The wiring method itself must qualify as an equipment grounding conductor or contain a separate insulated equipment grounding conductor. Nonmetallic raceways may be used, but only when encased in at least 2 in. (50 mm) of concrete. The wiring for audio systems, communications circuits, Class 2 and 3 circuits, and fire alarm circuits may be installed in accordance with their respective articles, Articles 640, 800, 725, and 760, which allow appropriately rated cables without raceways in most cases.

The preceding general rules are modified based on the fire rating of the building. Where the applicable building code permits an assembly occupancy to be nonrated (not fire-resistive) construction, nonmetallic-sheathed (Type NM) and Type AC cables can be used along with electrical nonmetallic tubing (ENT) and PVC or reinforced thermosetting resin conduit (RTRC). The general rules are also modified for certain occupancies where ENT, PVC, and RTRC are concealed by material with a 15-minute **finish rating**. The conduit or tubing must remain behind the rated material. The occupancies where this is permitted are listed in 518.4(C). **FIGURE 9-11** illustrates how a finish rating for a wall, ceiling, or floor is

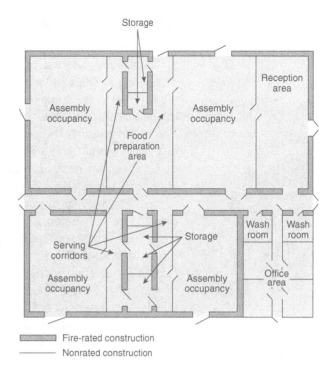

FIGURE 9-10   For Purposes of the *NEC*, a place of assembly is based on 100 or more persons. (Source: *National Electrical Code®* *Handbook*, NFPA, Quincy, MA, 2011, Exhibit 518.2.)

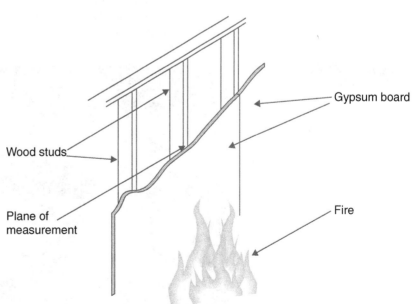

FIGURE 9-11   Finish rating is based on temperature measured inside of partition. (Source: *National Electrical Code® Changes*, NFPA, Quincy, MA, 1999, page 333.)

determined. The term *finish rating* is established for assemblies containing combustible (wood) supports.

3. Check temporary wiring for compliance with Article 590, except for GFCI requirements.
Temporary wiring is often used in large assembly rooms, such as exhibition halls where display booths are set up (**FIGURE 9-12**). Temporary wiring is also frequently needed in many other meeting and convention areas. Such temporary wiring must comply with Article 590, which modifies the general rules of the *Code* primarily with respect to wiring methods. An exception to 518.3(B) allows hard-usage or extra-hard-usage flexible cords and cables in cable trays to be used for temporary wiring if the cable trays are designated for temporary wiring use and equipped with signs and if servicing will be only by qualified persons. If protected from contact by the general public, hard- or extra-hard-usage cords may also be laid on floors. GFCI protection for receptacles is not required for temporary wiring in assembly occupancies where the receptacles are supplying power for exhibits or display booths at trade shows and similar functions, unless one of the usual requirements for GFCI protection applies, such as one of the requirements in 210.8(B) for areas outdoors or near sinks. Where GFCI protection is supplied through cord-and-plug connections, the GFCI protection must be listed as portable GFCI protection or provide equivalent protection. In addition to the usual functions of a GFCI, a portable GFCI will open the circuit on a loss of either the grounded or ungrounded conductor.

4. Check portable distribution equipment for adequate ratings and supply from listed power outlets.
Portable switchboards and portable power distribution equipment must be supplied from properly rated listed power outlets. See the definition of **Power Outlet** in the Key Terms section. The power outlets must have voltage and ampere ratings adequate for the load. Power outlets must include connections for equipment grounding conductors. Neutral conductors supplying dimmer systems may or may not be considered current carrying, depending on the type of dimmer. See Item 8 of the following checklist, Theaters and Audience Areas of Motion Picture Studios.

5. Check portable distribution equipment for overcurrent protection and isolation from the general public.
Overcurrent devices must protect power outlets serving portable distribution equipment. The overcurrent devices and power outlets cannot be located where they are accessible to the general public.

FIGURE 9-12 Exhibition hall with display booths. (Photographed by Tony Greco, FDIC 2008. Used with permission of PennWell Corporation.)

Checklist 9-3: Theaters and Audience Areas of Motion Picture Studios

1. Review definitions and determine the applicability of Article 520.

See the definition of <u>Theaters, Audience Areas of Motion Picture and Television Studios, and Similar Areas</u> in the Key Terms section. Where this definition overlaps with the definition of *Assembly Occupancies*, Article 520 applies according to 518.2(C). Many terms have special meanings in *Theaters*. Some aspects of *NEC* Articles 540 (*Motion Picture Projection Rooms*) and 640 (*Audio Signal Processing, Amplification, and Reproduction Equipment*) may also apply.

2. Verify that wiring methods are suitable for the occupancy and fire rating of the area(s).

The requirements for wiring methods under Article 520 are similar to those in Article 518 (*Assembly Occupancies*). The wiring methods permitted in theaters and audience areas of motion picture studios are metal raceways, including flexible metal raceways, and Type MI and Type MC cables. Type AC cable may be used if it is a type that includes a separate insulated equipment grounding conductor. Nonmetallic raceways may be used, but only when encased in at least 2 in. (50 mm) of concrete. The wiring for audio systems, communications circuits, Class 2 and 3 circuits, and fire alarm circuits may be installed in accordance with their respective articles—Articles 640, 800, 725, and 760.

The preceding general rules are modified based on the fire rating of the building. Where the applicable building code permits theaters or portions of theaters that are not fire-rated construction, Type NM cable, Type AC cable (without a separate insulated grounding conductor), ENT, and rigid nonmetallic conduit can be used. Unlike assembly occupancies, finish ratings are not a factor in selecting theater wiring methods.

3. Check for compliance with raceway fill requirements.

The raceway fill provisions in Chapter 9 of the *NEC* apply to installations in theaters. Wireways and auxiliary gutters are limited to the 20 percent fill limitation that applies elsewhere when wireways or auxiliary gutters are used. Wireways and auxiliary gutters are permitted to contain more than 30 current-carrying conductors without having to apply ampacity adjustment factors as would be required for general applications where the number of current-carrying conductors exceeds 30.

4. Check portable equipment used outdoors for supervision by qualified personnel and isolation from the general public.

Live parts are required to be enclosed and guarded from contact. <u>Portable equipment</u> for stage and studio lighting may be used outdoors, but the equipment must be isolated from the general public and must be supervised while it is energized. Portable stage and studio lighting may be used outdoors temporarily even if it is not identified for outdoor use, but it must be where it is isolated from the general public and supervised by qualified persons.

5. Check fixed stage switchboards for suitability.

Fixed <u>stage switchboards</u> must be listed. Stage switchboards containing dimmers must comply with the specific requirements for the type of dimmer installed. Four general types of stage switchboards are permitted: manual, remote controlled, intermediate, and constant power.

A switchboard may be a combination of types. Manual switchboards contain switches that are operated by handles with mechanical linkages. The switches in remote-controlled switchboards are electrically controlled from a local or remote-control panel. Intermediate switchboards, sometimes called patch panels, are used for interconnections. Constant power switchboards contain only overcurrent devices without control elements.

6. Check stage switchboard feeders for type and capacity.

Stage switchboard feeders may be one of three types: a single feeder, a multiple feeder to an intermediate stage switchboard, or a separate feeder to a single primary stage switchboard. Generally, the supply capacity of a stage switchboard feeder is calculated according to Article 220. However, the supply capacity may be calculated on the basis of the maximum load the switchboard is intended to control rather than on the entire connected load. This provision is true if the supply feeder is protected at its ampacity and the operation of exit and egress lights is not affected by the opening of the feeder overcurrent device.

Neutral conductors in 3-phase four-wire supplies to dimmer systems are considered as current carrying for the purpose of derating, or more specifically for the application of adjustment factors, if the dimmers are solid-state phase-control type, or if the conductor supplies a combination of solid-state phase-control and sine wave dimmers. Neutrals that supply only solid-state sine wave dimmers are not considered current carrying. See the definitions of **Solid-State Phase-Control Dimmer** and **Solid-State Sine Wave Dimmer** in the Key Terms section.

7. Check fixed stage equipment other than switchboards for suitability and compliance with specific requirements for the equipment type.

Part III of Article 520 covers fixed **stage equipment** other than switchboards. Part III includes requirements for **stage lighting hoists**, circuit loads, conductor insulation and ampacities, **footlights**, **border lights** and **proscenium** sidelights, **drop boxes**, **connector strips**, cord and cable uses and ampacities, receptacles, outlet enclosures, backstage lamps, curtain machines, and smoke ventilator controls.

8. Check portable switchboards on stage for proper supply, overcurrent protection, construction, and feeders.

Part IV of Article 520 covers portable switchboards on stage (**FIGURE 9-13**). Part IV includes installation and construction requirements for portable switchboards as well as rules for supplies, overcurrent protection, branch circuits, and feeders. These rules are too extensive to be summarized meaningfully here. Requirements for neutral conductors are specified and are similar to the requirements discussed in Item 6 of this checklist, except that in some cases, the neutral is required to be larger than the ungrounded circuit conductors.

9. Check portable stage equipment other than switchboards for appropriate construction, conductors, protection, and listings (where required).

Part V of Article 520 covers many types of portable stage equipment. The specific rules that apply to a type of equipment should be consulted for permissible uses, required connections, and listing requirements. Equipment covered in Part V includes arc lamp fixtures, **portable power distribution units**, bracket fixture wiring, portable strips, festoons, special effects, multipole branch-circuit cable connectors, conductors for portables including flexible cords for luminaires, and adapters. Many of these types of equipment are highly specialized and are not used in other occupancies.

10. Verify that pendant lampholders are not used and that exposed incandescent lamps less than 8 ft (2.5 m) above the floor have guards in dressing rooms.

Pendant lampholders are not permitted in dressing rooms. Incandescent lamps that are exposed in dressing rooms and less than 8 ft (2.5 m) above the floor must be equipped with guards. The guards must be riveted or locked in place and have an open end to facilitate lamp exchange.

11. Verify that dressing rooms are equipped with switches and pilot lights for lights and receptacles adjacent to mirrors.

All the lights in a dressing room and receptacles located adjacent to mirrors and above dressing table counters must be provided with wall switches. The switches must be in the dressing room, and switches that control receptacles in the areas mentioned must be connected to a pilot light located outside the dressing room. The pilot light is to be lit when the lights and receptacles are left energized. Receptacles near mirrors and above dressing counters are often used with appliances such as hair dryers, which create a hazard when left on and unattended. Receptacles that are not located adjacent to mirrors or above counters, such as those for refrigerators or com-

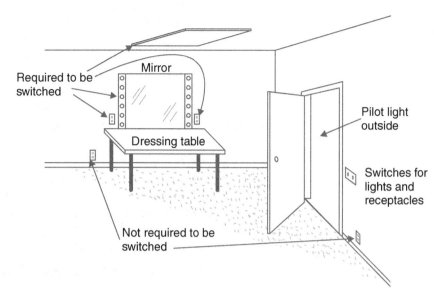

FIGURE 9-14   Switched receptacles and lighting outlets in dressing rooms. (Source: *National Electrical Code® Changes*, NFPA, Quincy, MA, 1999, page 337.)

puters, need not be switched. The requirements for the control of lighting fixtures and of receptacles located in close proximity to the dressing table are shown in **FIGURE 9-14**.

12. Verify that all metal raceways, metal-sheathed cables, and metal frames of equipment are grounded.

Grounding requirements in theater areas are essentially the same as for other electrical equipment. All metal equipment frames and enclosures for electrical equipment such as lighting fixtures are required to be connected to an equipment grounding conductor in accordance with Article 250.

Checklist 9-4: Carnivals, Circuses, Fairs, and Similar Events

Note: This checklist and Article 525 are not meant to apply to the permanent installations at fairgrounds, parks, and other locations where carnivals, circuses, and the like are held. Permanent facilities are covered by other requirements of the *NEC*. This checklist is concerned primarily with the temporary wiring and mobile equipment associated with carnivals, fairs, circuses, and so on. Control systems for permanent amusement attractions are covered by Article 522 and are not covered in detail in this manual.

Remember that the nonpermanent parts of these installations are essentially temporary wiring that is used repeatedly and that electrical cables and equipment for traveling shows, circuses, and carnivals endure significant amounts of abuse in setup and teardown in addition to abuse in normal operation.

1. Verify that mechanical protection is provided for equipment subject to damage.

Electrical wiring methods and equipment that are located on **portable structures** and subject to physical damage must be provided with mechanical protection from such damage. The *NEC* does not specify how protection is to be provided, so the form of protection must be based on a judgment of the nature of the risk of damage. Guards, barriers, additional enclosures, burial, or certain wiring methods may provide the needed protection.

2. Verify that transformers, generators, or services used as power sources meet the requirements of their respective articles and Article 250 and that rides or attractions are bonded to the same grounding electrode system when supplied from different power sources and separated by less than 12 ft (3.7 m).

Transformers, generators, and services are required to meet the usual requirements of Articles 450, 445, and 230, respectively, and be properly protected against overcurrent. Service equipment must be securely fastened to a solid backing, and if the equipment is not weatherproof, it must be protected from the weather. In addition, service equipment must be guarded from access by unauthorized persons or be lockable. Where portable structures are supplied by multiple sources, and the structures are within 12 ft (3.7 m) of each other, the power supply sources must be bonded together by connecting the grounded conductor (and equipment grounding conductors) of each supply source to the same grounding electrode system. This bonding conductor is to be sized using *NEC* Table 250.122 based on the rating of the largest overcurrent device supplying the portable structures, but may not be smaller than 6 AWG.

3. Check overhead conductors for adequate clearances from the ground and from rides and attractions.

Clearances from the ground for overhead conductors must meet the same standards as other outside feeder or branch-circuit conductors. The required clearances are found in Article 225. Overhead conductors other than those supplying a portable structure are required to be at least 15 ft (4.5 m) in all directions from other portable structures (**FIGURE 9-15**). This rule applies regardless of voltage. However, conductors over 600 volts may not pass directly over a ride or attraction or directly over an area within 15 ft (4.5 m) of a ride or attraction. Of course, conductors over 600 volts are more likely to be part of permanent installations, so this rule is usually more about where a portable structure is placed relative to overhead lines than it is about placement of overhead lines. It should be considered in the placement of overhead lines if the intended use of the area includes portable structures.

4. Check wiring methods, especially flexible cords, for suitability and protection from damage.

In general, wiring methods, including flexible cords, should be chosen and installed according to the rules of Chapters 1 through 4 of the *NEC*. A significant difference is that extra-hard-usage flexible cords may be used for permanent wiring on portable rides and attractions. Such cords must be protected from (not subject to) physical damage. Hard-usage flexible cords may be used for other purposes if they are not subject to physical damage. Otherwise flexible cords must be suitable for extra hard usage. Flexible cords used outdoors must be sunlight-resistant and suitable for wet locations. Flexible cords may not be spliced between boxes or fittings. Flexible cord connectors must be guarded from contact by the public and must be listed for wet locations if laid on the ground. The cords or cables may not be placed where they are a tripping hazard. Nonconductive mats may be used to protect flexible cords or cables that are laid on the ground, provided the mats do not create a greater tripping hazard. Because of the temporary nature of installations for circuses and the like, cables are permitted to be directly

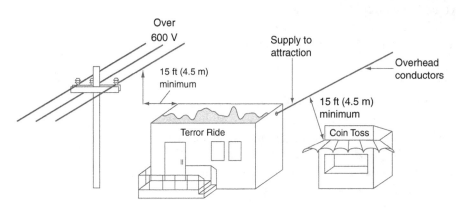

FIGURE 9-15   Overhead conductor clearances at carnivals, fairs, and similar events.

buried for physical protection and to reduce tripping hazards. The burial depths do not have to comply with 300.5. Generally, the wiring for one ride or attraction may not be supported by the structure of another ride or attraction unless specifically designed for that purpose.

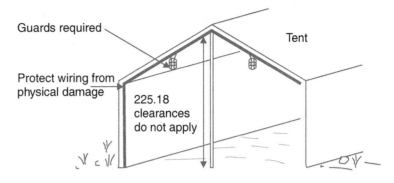

FIGURE 9-16   Protection of exposed lamps with tents. (Source: *National Electrical Code® Changes*, NFPA, Quincy, MA, 1999, page 339.)

5. Check wiring inside tents and concessions for mechanical protection for wiring and guards on temporary lamps.
Temporary lamps used for general lighting in tents and concessions must be protected from accidental breakage. Temporary wiring in such locations must be securely installed and protected where exposed to physical damage. The wiring inside tents is not required to comply with the overhead clearances from Article 225. However, wiring and luminaires installed in tents must be secured in place and protected from mechanical damage (**FIGURE 9-16**).

6. Verify that suitable boxes are provided at all outlets and at connection points, junction points, and switch points.
The requirements for boxes are essentially the same as those for other occupancies, as outlined in Section 300.15. Boxes or fittings are required at all outlets and at all connection, junction, and switch points. The boxes or fittings used must be suitable for the location and conditions, such as exposure to weather [314.15].

7. Check the construction and suitability of portable distribution and termination boxes.
Portable power and termination boxes are available as preassembled listed products. However, Article 525 provides construction requirements for this type of equipment. The boxes must be designed so that no live parts are exposed other than when they are exposed for examination, adjustment, servicing, or maintenance by qualified persons. Boxes used outdoors must be weatherproof and designed so the bottoms of the enclosures are at least 6 in. (150 mm) above the ground. Internal busbars must be protected by overcurrent devices rated not more than the ampacity of the busbars. **FIGURE 9-17** shows a type of preassembled portable power distribution equipment that may be used for temporary wiring and may also include GFCI protection. Except for receptacles supplying motor loads, overcurrent protection matching the ratings of any receptacles must be installed in the box. Receptacles supplying motors may be protected as permitted in Article 430.

8. Verify that overcurrent protection is provided in accordance with Article 240.
Overcurrent protection for conductors and equipment must be provided according to the usual rules of Article 240, as these rules are not modified by Article 525. Installations covered by Article 525 must comply with provisions for ratings of overcurrent devices and locations of overcurrent devices in the circuits and on the premises.

9. Verify that general-use receptacles are provided with GFCI protection.
All 125-volt, single-phase, 15- and 20-ampere non-locking-type receptacle outlets used for assembly and

FIGURE 9-17   Portable power outlet for supply of equipment at construction or other location where temporary power is used. (Courtesy of Hubbell Wiring Device-Kellems.)

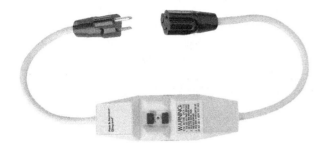

FIGURE 9-18   Portable GFCI device for protection of personnel equipment supplied by temporary wiring. (Courtesy of Pass & Seymour/Legrand.)

disassembly of equipment or readily accessible to the public must be GFCI protected. These rules for GFCI protection are similar to temporary power requirements as covered in Article 590, except that 590.6(A) also applies GFCI requirements to 30-ampere, 125-volt receptacles. Also, equipment that is readily accessible to the public and supplied by 15- or 20-ampere, 120-volt branch circuits must have GFCI protection. The GFCI protection may be provided by a listed cord set that includes GFCI protection (**FIGURE 9-18**). GFCI protection is not required for locking-type receptacles that are used only to facilitate quick connection and disconnection of electrical equipment or for receptacles and equipment that are not accessible from grade level. Egress lighting may not be connected to the load side of GFCI receptacles.

10. Verify that grounding continuity is assured each time portable equipment is connected.

Each time portable equipment is connected, the continuity of the grounding conductor system must be verified. The grounding requirements themselves are not different from the requirements for other equipment or occupancies, but grounding continuity must be verified on each use. Article 525 does not mandate a written program, but some proof of a procedure to satisfy this requirement should be available.

11. Verify that metal frames, enclosures, and equipment are bonded.

The requirements for grounding and bonding in circuses, carnivals, and the like are covered by Article 250. As required by Article 250 and repeated in 525.30, metal raceways, metal cable sheaths, metal enclosures for electrical equipment, and metal frames and parts of rides and other structures must be bonded. Like "other metal piping" as covered in Article 250, the equipment grounding conductor that supplies the equipment likely to energize a conductive part is permitted to serve as the bonding means. This helps to reduce shock hazards and provides fault current pathways so that equipment will not become and remain energized.

12. Verify that equipment grounding conductors of the proper size and type are supplied to all equipment requiring grounding.

Equipment grounding conductors must comply with Article 250 with regard to type and size and isolation from grounded conductors. Sizing is based on the rating of the overcurrent device supplying the circuit or equipment in question. The continuity of all portions of the grounding conductor circuits used to prevent shock hazards must be verified.

13. Check each portable structure for a disconnect in the form of a fusible switch or circuit breaker located within sight and within 6 ft (1.8 m) of the operator's station.

Disconnecting means must be furnished for each ride and concession and be located in sight and within 6 ft (1.8 m) of the **operator's** station. The disconnect must be readily accessible to the operator at all times. The disconnect may be remotely controlled by a shunt-trip device operated from the operator's location. The disconnect must be lockable if accessible to unqualified persons.

14. Verify that any attractions using contained volumes of water meet the applicable requirements of Article 680.

Attractions utilizing pools, fountains, and similar installations with contained volumes of water, such as log rides and children's boat rides, must comply with the applicable requirements of Article 680 (see the Swimming Pools and Related Installations chapter of this manual).

Checklist 9-5: Agricultural Buildings

1. Determine the applicability of Article 547.

Article 547 applies only to certain agricultural buildings and certain parts of agricultural buildings. Article 547 applies to buildings where excessive dust and dust with water may accumulate. It also applies where a corrosive atmosphere exists in agricultural buildings.

The dust and water areas include all confinement areas for fish, livestock, and poultry. Some buildings used to house animals in zoos or aviaries may also be considered agricultural buildings. Corrosive areas are those where corrosive vapors may be produced by poultry and animal excrement, where corrosive particles may combine with water, where periodic washing with water and cleansing agents causes the area to be damp and wet, or where similar wet or corrosive conditions exist. Where none of these conditions exists, buildings, although agricultural in nature, are not covered by Article 547, and the other applicable rules of the *NEC* must be followed.

2. Check wiring methods for suitability for the occupancy and conditions and for protection from physical damage.

Wiring methods must be suitable for the wet and corrosive conditions described in Item 1. Underground feeder and branch-circuit cable (UF), corrosion-resistant nonmetallic-sheathed cable (NMC), jacketed MC cable, and copper service entrance (SE) cable are specifically mentioned as suitable. Rigid nonmetallic conduit and liquidtight flexible non-metallic conduit (LFNC) are also specifically mentioned. In dust and water areas, wiring methods suitable for Class II areas are permitted. However, some metallic wiring methods may not be suitable in corrosive locations. (Stainless steel generally is suitable, and aluminum conduit is suitable for certain conditions, but nonmetallic wiring methods are often preferred due to their resistance to a wide range of corrosive agents.)

Normally, boxes and wiring methods in a wet location or a location subject to hose-down are required to be mounted with an airspace from surfaces of ¼ in. (6 mm) to allow for drainage. However, that rule does not apply to nonmetallic enclosures and raceways in agricultural buildings (or elsewhere in indoor wet locations; see 300.6(D), Exception). Dusttight flexible connectors, liquidtight flexible metal conduit (LFMC) or LFNC, or flexible cord listed and identified for hard usage may be used for flexible connections. Connectors and fittings must be listed and identified for the purpose. Boxes must be suitable for the conditions. Type 4 and 4X enclosure ratings are examples of suitable ratings for boxes, depending on the degree of exposure to corrosion. All wiring methods and equipment must be protected where subject to physical damage. Physical damage caused by housed livestock should be considered. Many horses, for example, will attempt to chew on almost anything within reach. The likelihood of such damage occurring can be reduced by proper placement and selection of wiring and equipment.

3. Verify that any equipment grounding conductors installed underground are insulated or covered.

Equipment grounding conductors installed underground must be either insulated or covered to help minimize stray currents in livestock areas.

4. Check switches, circuit breakers, controllers, and the like for enclosures suitable for the conditions.

Fuses and devices with contacts, such as switches, pushbuttons, relays, controllers, and circuit breakers, must be housed in suitable enclosures. In areas with excessive dust and dust with water, enclosures must be dusttight and weatherproof. (See the definitions of these terms and the difference in the term *Watertight* in the Key Terms section of the Wiring Methods and Devices chapter of this manual.) Enclosure Types 4 and 4X are common examples of such enclosures, but external operators, if any, must also be operable under the conditions created by the dust and water. Type 4X enclosures are also suitable for corrosive atmospheres. See *NEC* Table 110.28 for the characteristics and intended uses of numbered enclosure types.

5. Verify that luminaires are installed to minimize the entry of dust and water and that fixtures exposed to damage are supplied with guards.

Luminaires must be selected and installed to minimize the entrance of dust and other foreign matter, including moisture and corrosive material. Where exposed to physical damage, luminaires must be protected by guards, and where exposed to water, including condensation, they must be listed as suitable for use in wet locations. All luminaires are required by Article 410 to be listed.

6. Verify that the arrangement of service equipment, distribution equipment, overcurrent protection, and grounding complies with requirements.

Two basic arrangements of services and disconnecting means are outlined and permitted for groups of agricultural buildings. Both of these configurations are arranged around a **distribution point**. The configurations vary according to whether the overcurrent protection and disconnecting means are located at the distribution point or at the building(s) or structure(s). Figure 9-19 illustrates the two arrangements recognized for power distribution at a multibuilding agricultural complex. Most farms are arranged in multibuilding complexes.

The first method (**FIGURE 9-19A**) is permitted only with overhead distribution from the **site-isolating device**. The disconnecting means and overcurrent protection are located at the building(s) with the disconnecting means and overcurrent protection at the load end of the supply (service or feeder) conductors. Another disconnecting means without overcurrent protection, called a site-isolating device, is pole mounted and installed at the distribution point for the supply of one or more additional buildings. Service-type grounding takes place at the distribution point in the site-isolating device, even though the service overcurrent protection is not located at the distribution point. Additional electrodes are established at each building or structure. A site-isolating device must be permanently marked as a site-isolating device. The marking must be on or immediately adjacent to the operating handle.

Grounding at the buildings follows a certain protocol. A separate equipment grounding conductor sized equivalent to the largest supply conductor for each building or structure must be run from the distribution point to each building disconnect. The equipment grounding conductor is bonded to the grounded circuit conductor at the distribution point as previously mentioned. The equipment grounding conductor for each building or structure is connected to a grounding electrode system at each building disconnecting means. The grounded circuit conductor is not connected to the grounding electrode(s) or disconnects at the buildings. The special equipment grounding conductor-sizing rule is intended to help reduce stray voltages between the equipment grounding conductor and the grounded conductor.

The second method (**FIGURE 9-19B**) may be used for either overhead or underground supplies. The disconnecting means and overcurrent protection for each set of branch circuit or feeder conductors are located at the distribution point. There is no site-isolating device as such, although there may be a single main disconnecting means if there

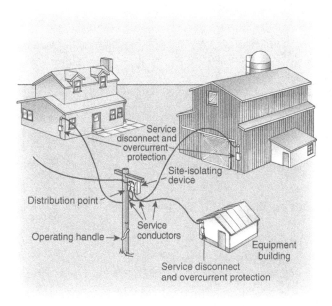

FIGURE 9-19A Centrally located site-isolation device supplying service conductors run to each structure. (Source: *National Electrical Code® Handbook*, NFPA, Quincy, MA, 2011, Exhibit 547.1.)

are more than six devices for supplying individual structures. Additional disconnecting means are required at the buildings in accordance with Article 225, Parts I and II, and grounding at the separate buildings must comply with Article 250, specifically 250.32. Grounding at separate buildings is essentially the same as described in the first option in Figure 9-19A, but the equipment grounding conductor in the second option (Figure 9-19B) is not required to be the same size as the other feeder conductors. However, any equipment grounding conductor run underground must be insulated or covered copper.

The method described in Figure 9-19B is essentially the normal method of multibuilding distribution under the *Code*, as evidenced by the references to Articles 225 and 250. The method described in Figure 9-19A is a special method for agricultural buildings only. Although the site-isolating device used in Figure 9-19A is required to include a means for connecting the grounded conductor to an electrode, it is not required to be listed as suitable for use as service equipment. Nonfused switches or pole-top switches are often used for this purpose. When there is more than one distribution point, each must be marked with a permanent plaque or directory that indicates the location of the other distribution points.

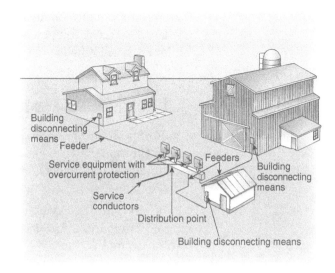

FIGURE 9-19B   Centrally located service equipment supplying multiple structures. (Source: *National Electrical Code® Handbook*, NFPA, Quincy, MA, 2011, Exhibit 547.2.)

7. Verify that an equipotential plane has been installed in concrete floors of livestock containment areas and bonded to electrodes and conductive elements.

See the definition of **Equipotential Plane** in the Key Terms section. An equipotential plane is established by installing wire mesh or a conductive grid of rebar or other conductive elements in the concrete floor of a livestock confinement area. The mesh or grid is then connected to the building grounding electrode system with a solid copper conductor not smaller than 8 AWG. Bonding all other conductive elements in the area of the equipotential plane to the equipotential plane is required to eliminate (or at least minimize) voltage differences between metal parts that may be contacted by personnel or livestock.

As defined, equipotential planes are not required in dirt confinement areas but are required in concrete floor confinement areas indoors or outdoors where any metallic equipment containing insulated conductors (likely to become energized) is accessible to animals. For more information on equipotential planes, see American Society of Agricultural and Biological Engineers (ASABE) EP473.2-2001, *Equipotential Planes in Animal Containment Areas.*

8. Verify that GFCI protection has been provided where required.

All convenience-type receptacles installed in the area of an equipotential plane must have GFCI protection (**FIGURE 9-20**). GFCI protection is also required for convenience-type (125-volt, 15- or 20-ampere)

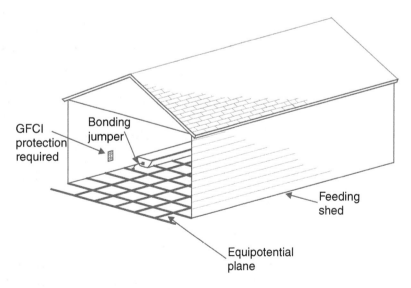

FIGURE 9-20   GFCI protection of receptacles in areas protected by equipotential plane. (Source: *National Electrical Code® Changes*, NFPA, Quincy, MA, 1999, page 349, as modified.)

receptacles installed outdoors, in damp or wet locations, or in dirt confinement areas where the electric equipment supplied is accessible to animals. The exceptions that allowed the elimination of GFCI protection in certain cases were removed in the 2011 *NEC*.

Checklist 9-6: Recreational Vehicle Parks

1. Review definitions and determine the applicability of Article 551 and Part VI.
The definitions of terms applicable to **recreational vehicle (RV)** parks appear in the Key Terms section. This checklist is not meant to cover the RVs themselves. It is meant to cover the **RV sites** and distribution systems in **RV parks** only.

2. Verify that all sites with power have 20-ampere, 125-volt receptacles, at least 20 percent have 50-ampere, 125/250-volt receptacles, and at least 70 percent have 30-ampere, 125-volt receptacles.
Part VII of Article 551 specifies the receptacle ratings and configurations that must be used in an RV park. It also specifies the distribution of certain ratings of receptacles. The percentages listed are minimum percentages of sites that must have specific receptacle types where power is provided. Higher percentages are permitted, and some sites may have both 50- and 30-ampere receptacles. All sites with power must have at least one of the listed types. If a site is provided with a 50-ampere receptacle, it must also have a 30-ampere receptacle. This section does not require all sites in a campground to be provided with electric power, only that all sites with electric power have a 20-ampere, 125-volt, GFCI-protected receptacle. Any 15- or 20-ampere, 125-volt receptacle installed must be GFCI protected. These rules do not require power at all RV parks or sites; they govern only parks and sites that do have electric power. The percentages also do not apply to sites dedicated to tents even if receptacles are provided to the tent sites. The *Code* percentages are based on the percentage of each RV type manufactured by the RV industry. There are more larger-ampacity RVs manufactured in recent years; therefore the requirement for 50-ampere power sources increased from 10 percent to 20 percent of all new sites installed in an RV park.

The standardization of receptacle configurations and distribution of those configurations ensure that RVs will be compatible with the power distribution at various RV parks and that a reasonable number of the higher-powered receptacles will be available for larger RVs. This uniformity will reduce the likelihood of tampering or bypassing of devices to adapt vehicles to power sources. **TABLE 9-1** summarizes the requirements for the use of various receptacle configurations at RV sites with electric power.

3. Verify that the voltages of distribution systems are appropriate for the sites supplied.
Sites with 50-ampere receptacles must be supplied from a nominal 120/240-volt, single-phase, three-wire system or from a 208Y/120-volt 3-phase, four-wire system. Receptacles rated 20 or 30 amperes, single-phase may be supplied by any system that supplies power at 120 volts, such as 120/240-volt or 208/120-volt, single- or 3-phase systems. Since the 2008 edition, the *NEC* also permits the use of either 208- or 240-volt systems to supply line-to-line loads.

4. Review load calculations, and check sizing or ratings of transformers, panelboards, and feeders.
Part VI of Article 551 includes load values for each type of site based on the rating of the receptacles installed at the site. Part VI also includes a table of demand factors based on the number of sites served by a feeder or service. These loads and demand factors can also be used to size transformers and panelboards. Loads for park amenities such as swimming pools or buildings for stores, restaurants, recreation, or service should be calculated according to Article 220. These loads may be added to the calculated load for the RV sites if such

| TABLE 9-1 | Receptable Configurations at RV Sites |

Receptacle Configurations	Requirements at Sites with Power
20-ampere, 125-volt grounding	Required at every site
30-ampere, 125-volt grounding	Required at a minimum of 70% of sites, excluding tent sites, and at sites with 50-ampere receptacles
50-ampere, 125/250-volt grounding	Required at a minimum of 20% of sites, excluding tent sites

loads are all supplied by a single service. The designer of the site distribution system should furnish calculations for review. An example of a load calculation for a 100-site RV park in which all of the sites are equipped with electric power and all are supplied by one service or feeder is shown in **BOX 9-1**.

Box 9-1   Load Calculation for a 100-Site RV Park

Load calculation for a 100-site RV park, in which all of the sites are equipped with electric power and all are supplied by one service or feeder.

Given:

An RV park has 100 sites. Twenty sites are dedicated to tents. Sixteen sites are equipped with both 50-ampere, 125/250-volt receptacles and 30-ampere 125/250-volt receptacles. Sixty sites are equipped with 30-ampere, 125-volt receptacles. All sites have 20-ampere, 125-volt receptacles.

Find:

Calculated load for service or feeder serving all 100 sites. Determine whether receptacles are properly supplied based on percentage of sites.

Minimum Number of Each Type of Receptacle:

Total sites	=	100
Less tent sites	=	<u>20</u>
		80 (number of sites to which percentages apply)

20% × 80 = 16 sites required to have 50-A receptacles

70% × 80 = 56 sites required to have 30-A receptacles

Therefore, planned receptacle distribution complies.

Calculated Load:

Sixteen 50-A sites—16 × 9600 VA	=	153,600 VA
Sixty 30-A sites—60 × 3600 VA	=	216,000 VA
Four 20-A sites—4 × 2400 VA	=	9,600 VA
Twenty tent sites—20 × 600 VA	=	<u>12,000 VA</u>
Total load before demand factor	=	391,200 VA
Demand factor for 100 sites	=	41%
Total Calculated Load:		
0.41 × 391,200	=	160,392 VA, 668 A, 120/240 V

5. Verify that grounded feeder conductors have the same ampacity as ungrounded conductors.

The grounded feeder conductors may not be reduced in size from the size of the ungrounded conductors. Most loads in RVs are 120 volts, and because reduced loading of a grounded conductor due to 240-volt loads cannot be reliably predicted, grounded conductors are required to be sized for the full load on the ungrounded conductors. Therefore, the calculated load used to size the ungrounded conductors also determines the size of the grounded conductors.

6. Verify that separate equipment grounding conductors extending from a service or secondary distribution system are run to equipment requiring grounding.

Equipment requiring grounding is determined according to Article 250 and repeated or summarized in 551.76(A). All such equipment must be grounded by direct connection (bonding) to other grounded equipment or by a continuous equipment grounding conductor run with the circuit conductors from the service equipment or from the transformer of a secondary distribution system. Grounded conductors may not be used for grounding of equipment. Equipment grounding conductors must be sized, installed, and connected in accordance with Article 250. Equipment grounding conductors may be spliced by listed means. Although Article 250 would allow metal conduits to be used for equipment grounding, Article 551 requires a separate equipment grounding conductor sized in accordance with 250.122 regardless of the wiring method. Direct-buried cables or conductors are commonly used as the feeder and branch-circuit wiring method in RV parks.

RV site equipment (pedestals) is treated as the interface with the RV park electrical supply system rather than a separate "structure" for the purposes of applying the grounding requirements specified in 250.32. Therefore a separate grounding electrode system is not required to be installed at each RV site provided with electric power. Installing an auxiliary grounding electrode and connecting it to the equipment grounding conductor at RV site equipment is not prohibited by the *NEC*; however, connecting the grounded conductor to a grounding electrode(s) at the RV site equipment is expressly prohibited by 551.76(D).

7. Check RV site supply equipment for proper location relative to the vehicle parking stand.

Electrical supply equipment for RV sites must be located according to the dimensions shown in **FIGURE 9-21**. This arrangement reflects the standardization in the RV industry in regard to locating the RV power-supply cord.

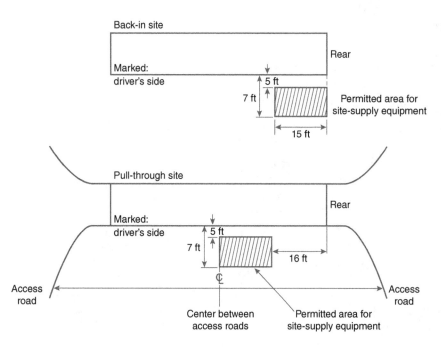

Location. Where provided on back-in sites, the recreational vehicle site electrical supply equipment shall be located on the left (road) side of the parked vehicle, on a line that is 5 ft to 7 ft (1.5 m to 2.1 m) from the left edge (driver's side of the parked RV) of the stand and shall be located at any point on this line from the rear of the stand to 15 ft (4.5 m) forward of the rear of the stand.

For pull-through sites, the electrical supply equipment shall be permitted to be located at any point along the line that is 5 ft to 7 ft (1.5 m to 2.1 m) from the left edge (driver's side of the parked RV) from 16 ft (4.9 m) forward of the rear of the stand to the center point between the two roads that gives access to and egress from the pull-through sites.

The left edge (driver's side of the parked RV) of the stand shall be marked.

FIGURE 9-21   Dimensions for back-in and pull-through RV sites supplied with electric power.

Where provided on back-in sites, the RV site electrical supply equipment shall be located on the left (road) side of the parked vehicle, on a line that is 5 to 7 ft (1.5 to 2.1 m) from the left edge (driver's side of the parked RV) of the <u>RV stand</u> and shall be located at any point on this line from the rear of the stand to 15 ft (4.6 m) forward of the rear of the stand. For pull-through sites, the electrical supply equipment shall be permitted to be located at any point along the line that is 5 to 7 ft (1.5 to 2.1 m) from the left edge (driver's side of the parked RV) from 16 ft (4.9 m) forward of the rear of the stand to the center point between the two roads that gives access to and egress from the pull-through sites. Additionally, the left edge (driver's side of the parked RV) of the stand must be marked.

8. Check site supply equipment and disconnecting means for proper access, mounting height, working space, and marking.

Each site must have a means to disconnect power to the RV. Access to the disconnect must be through an unobstructed entrance at least 2 ft (600 mm) wide and 6½ ft (2.0 m) high. (This does not mean that there will necessarily be an entrance, as the equipment may be located outdoors.) The disconnect must be located not less than 2 ft (600 mm) and not more than 6½ ft (2.0 m) above the ground. Workspace must meet the general requirements of Section 110.26(A). If a 125/250-volt receptacle is supplied at a site, a sign with the following notice must be provided on the equipment next to the receptacle: "Turn disconnecting switch or circuit breaker off before inserting or removing plug. Plug must be fully inserted or removed." **FIGURE 9-22** is an example of <u>RV site supply equipment</u>. A 125/250-volt, 50-ampere receptacle and a 125-volt, 20-ampere receptacle are provided by this particular site supply configuration. The change in 2014 requires this example to also include a 30-ampere, 125-volt receptacle, but previously existing sites are likely to have equipment like that shown in the figure. Note that the circuit breakers provide the required disconnecting means and that the 125-volt receptacle is GFCI protected.

FIGURE 9-22   RV site supply equipment. (This material and associated copyrights are proprietary to, and used with the permission of, Schneider Electric.)

9. Check overhead conductors for vertical and horizontal clearances from finished grade in areas subject to vehicle movement.

Overhead conductors must be not less than 18 ft (5.5 m) above grade in areas where vehicles are moved and must be not less than 3 ft (900 mm) horizontally from such areas. The overhead clearances for outside feeders and branch circuits as described in Article 225 apply in other areas. If there are any overhead conductors that are over 1000 volts, the clearances specified in Part III of Article 225 should be applied.

10. Check underground conductors for appropriate ratings, insulation, identification, and, where emerging from the ground, proper physical protection.

All underground conductors must be suitable for the conditions—that is, suitable for wet locations. In addition, conductors that are direct buried must be insulated and identified for direct burial. Underground conductors must be continuous from equipment to equipment. Any splices or taps must be in approved junction boxes or be made with listed materials. Where direct-buried conductors are used, they must be protected where they emerge from a trench. The protection may be RMC, IMC, EMT with supplementary corrosion protection, PVC, RTRC, high-density polyethylene (HDPE), nonmetallic underground conduit with conductors (NUCC), LFNC, LFMC, or other approved raceways or enclosures. The protection must extend down to at least 18 in. (450 mm) below finished grade. If the protective raceway is subject to physical damage, the raceway must be RMC, IMC, RTRC-XW, or Schedule 80 PVC. These rules are substantially the same as the rules found in Article 300.

Checklist 9-7: Marinas and Boatyards

1. Determine the applicability of Article 555.

Marinas and boatyards are not defined in the *NEC*, but the scope of Article 555 provides some assistance. This article covers the installation of wiring and equipment in the areas comprising fixed or floating piers, wharves, docks, and other areas in marinas, boatyards, boat basins, boathouses, and similar occupancies that are used, or intended for use, for the purpose of repair, berthing, launching, storage, or fueling of small craft and the moorage of floating buildings. It does not apply to similar private, noncommercial facilities associated with single-family dwellings. Requirements for GFCI protection of 15- and 20-ampere, 125-volt, single-phase receptacles for boathouses at one-family dwellings are contained in 210.8(A)(8). Boat hoists at one-family dwellings are required to be GFCI protected where supplied by 120-volt, 15- or 20-ampere branch circuits. The requirement found in 210.8(C) applies to any "outlets not exceeding 240 volts," so it applies to both cord-and-plug–connected equipment and to hard-wired equipment. Article 555 supplements other requirements of the *Code,* so other applicable code rules may also apply to marinas and boatyards and to docks at one-family dwellings.

2. Verify that shore power receptacles are of an appropriate grounding type and have appropriate ampere ratings.

Receptacles used to supply shore power to boats are required to be of a single grounding and locking type if rated 50 amperes or less. The receptacles must be at least 30-ampere rated. The typical configurations of permitted receptacles are shown in **FIGURE 9-23**. Receptacles rated 60 or 100 amperes must be pin and sleeve type as shown in **FIGURE 9-24**. Receptacles must be located at or above the electrical datum plane on a fixed pier and at least 12 in. (305 mm) above the deck of a floating pier. Strain relief must be provided for plugs and receptacles.

3. Verify that general-use receptacles not used for shore power are GFCI protected.

General-use receptacles rated 15 and 20 amperes, single-phase, 125 volts are required to be protected by GFCIs if located outdoors, in boathouses, and in buildings used for storage, maintenance, or repair where portable electrical hand tools, electrical diagnostic equipment, or portable lighting equipment will be used. Receptacles in other areas are required to have GFCI protection in accordance with 210.8(B) (in bathrooms, on rooftops, outdoors, near sinks, and in kitchens).

4. Verify that disconnecting means are provided within sight of shore power connections to isolate each boat from its supply circuit.

Each boat with a shore power connection must be provided with a means of cutoff of the electrical supply to the boat. The disconnect must be in the supply circuit ahead of and not more than 30 in. (762 mm) from the receptacle it controls. It must be a circuit breaker, or a switch, or both. **FIGURE 9-25** shows an assembly specifically designed for the harsh environment encountered at marinas. The assembly provides protection against excessive moisture for the shore power connections and also is equipped with integral disconnecting means.

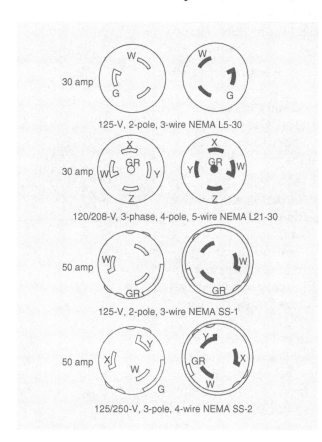

30 amp — 125-V, 2-pole, 3-wire NEMA L5-30

30 amp — 120/208-V, 3-phase, 4-pole, 5-wire NEMA L21-30

50 amp — 125-V, 2-pole, 3-wire NEMA SS-1

50 amp — 125/250-V, 3-pole, 4-wire NEMA SS-2

FIGURE 9-23   Locking and grounding receptacles and cord caps rated 30 to 50 amperes. (Source: *National Electrical Code® Handbook*, NFPA, Quincy, MA, 2011, Exhibit 551.46(C).)

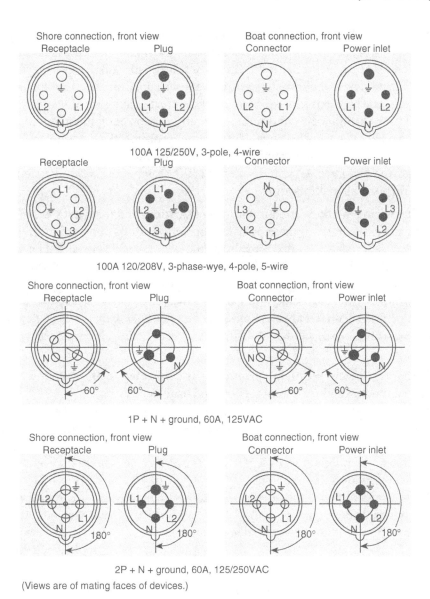

Shore connection, front view
Receptacle   Plug

Boat connection, front view
Connector   Power inlet

100A 125/250V, 3-pole, 4-wire

Receptacle   Plug   Connector   Power inlet

100A 120/208V, 3-phase-wye, 4-pole, 5-wire

Shore connection, front view
Receptacle   Plug

Boat connection, front view
Connector   Power inlet

1P + N + ground, 60A, 125VAC

Shore connection, front view
Receptacle   Plug

Boat connection, front view
Connector   Power inlet

2P + N + ground, 60A, 125/250VAC

(Views are of mating faces of devices.)

FIGURE 9-24 Pin and sleeve receptacles and inlets rated 60 or 100 amperes. (Source: *National Electrical Code® Handbook*, NFPA, Quincy, MA, 2011, Exhibit 555.2.)

5. Verify that receptacles used for shore power are supplied by individual branch circuits with voltage and current ratings corresponding to the receptacles.

A wide variety of grounding and locking receptacle configurations are available as shown in Figures 9-23 and 9-24. Each configuration is intended to be used with a specific voltage system and maximum load current. Supplied equipment may overheat or otherwise malfunction if it is not supplied with the proper voltage. Therefore, receptacles must be selected to correspond to the voltage and current ratings of the branch circuit supplying the receptacle. An individual branch circuit is required for each single shore power receptacle. This makes the circuit for each shore power receptacle independent of all others.

6. Review feeder and service calculations for compliance with requirements.

Article 555 provides load values for feeders and services supplying shore power circuits. The load calculation is based on the sum of the ratings of the shore power receptacles. The sum of the ratings is multiplied by a demand factor to get the calculated load. The demand factors are based on the total number of

FIGURE 9-25 Power outlet assembly for docks, wharves, piers, and other marina locations. (Courtesy of Hubbell Wiring Device-Kellems.)

shore power receptacles. Where more than one receptacle is supplied at the shore power connection for a single boat, only the receptacle with the higher demand load is required to be included in the calculation. Modifications of this calculation method may be made where individual kilowatt-hour meters are used at each slip. Calculated loads should be increased in areas where extreme temperatures may create higher loading on circuits due to increased use of heating, air-conditioning, or refrigeration equipment, but this information is provided in an Informational Note, so the recommendation is not enforceable. The application of demand factors to feeder neutrals provided in 220.61 are not applicable when the demand factors in Article 555 are used. The designer of the site distribution system should furnish calculations for review.

7. Verify that wiring methods are suitable for wet locations and that portable power cables, where used, are extra-hard-usage type listed for wet locations and sunlight resistance.
Portable power cables must be rated for at least 167°F (75°C), 600 volts, listed for wet locations and sunlight resistance. Such cables are permitted only for permanent wiring on the underside of piers and where flexibility is needed. Types G and W portable power cables additionally rated for wet locations, sunlight resistance, and resistance to gasoline, ozone, abrasion, acids, and chemicals are examples of acceptable wiring methods on floating piers. Portable power cables are in a different listing category than flexible cords; see *NEC* Table 400.4.

8. Verify that equipment requiring grounding is connected to an insulated equipment grounding conductor included with feeders and branch circuits.
The types of equipment required to be grounded at a marina or boatyard are the same types required to be grounded by Part VI of Article 250. However, grounding methods are somewhat restricted in marinas and boatyards. Equipment grounding conductors must meet the following requirements: Except where Type MI cable is used, they must be insulated with a continuous outer finish that is green or green with one or more yellow stripes. (Type MI cable is identified at terminations.) They must be sized in accordance with Article 250 but may not be smaller than 12 AWG. Where part of a branch circuit, they must terminate on a grounding terminal in a panelboard or at the main service. Where part of a feeder, they must run from the service equipment to the supplied panelboard.

9. Check wiring over or under navigable water for suitability and clearances.
The AHJ in this case is likely to be a harbormaster or perhaps the Coast Guard. Minimum clearances from navigable water are not provided by the *NEC*. The *NEC* simply requires that overhead wiring be installed so that possible contact with masts and other parts of boats is avoided. Warning signs may be required in accordance with NFPA 303-2011, *Fire Protection Standard for Marinas and Boatyards.*

10. Check wiring in motor fuel-dispensing stations for compliance with Article 514.
Gasoline stations that are part of a marina or boatyard must meet the requirements of Article 514 in addition to the requirements of Article 555. Article 555 provides additional details for coordinating the area classification of Article 514 with the specific construction type of the docks, piers, or wharfs encountered in marinas and boatyards. See the Hazardous Locations chapter of this manual for additional information. Further information is also available in NFPA 30A-2012, *Motor Fuel Dispensing Facilities and Repair Garages,* and NFPA 303-2011, *Fire Protection Standard for Marinas and Boatyards.* Power wiring supplying equipment on a wharf, pier, or dock must be installed on the opposite side of the dock from any piping that contains motor fuels.

11. Verify that service equipment for floating docks or marinas is located adjacent to, but not on or in, the floating structure.
For additional information on services, see *NEC* Article 230 and the Services, Feeders, and Branch Circuits chapter of this manual.

12. Verify that ground-fault protection is provided for the main overcurrent device on the supply to the marina.

Ground-fault protection or GFCI protection of individual feeders or branch circuits does not meet this requirement. A GFCI could be applied to meet the requirement but may not be usable in the marine environment due to the very low currents (4–6 mA) they are designed to sense and interrupt. A Class B or Class C GFCI will trip at a higher level (15 to 20 mA). This requirement calls for protection at not over 100 mA. Under current listing standards, protection at this level is most likely to be achieved with equipment listed as "Ground-Fault Sensing and Relaying Equipment." At 100 mA, this may not provide ground-fault protection for personnel, but the rule does not call for that level of protection. However, individual GFCI protection on each feeder and branch circuit is permitted as an alternative to the 100-mA protection on the main supply and may be preferred because it will more closely isolate the problem when a ground fault is detected.

KEY TERMS

The *NEC* defines the key terms pertaining to the special occupancies covered in this chapter as follows. Most of these definitions come from the *NEC* article on the specific occupancy. Those definitions are applicable to that article only. The abbreviations following the key terms relate to the occupancy type to which the definition applies, as follows:

AG—Agricultural buildings
AS—Assembly occupancies
CF—Circuses, fairs, carnivals, and similar events
HC—Health care facilities
MB—Marinas and boatyards
RV—Recreational vehicle parks
TH—Theaters and audience areas of motion picture studios

Alternate Power Source (HC) [*Definition from 517.2*] One or more generator sets, or battery systems where permitted, intended to provide power during the interruption of the normal electrical services; or the public utility electrical service intended to provide power during interruption of service normally provided by the generating facilities on the premises.

Ambulatory Health Care Occupancy (HC) [*Definition from 517.2*] A building or portion thereof used to provide services or treatment simultaneously to four or more patients that provides, on an outpatient basis, one or more of the following:

(1) Treatment for patients that renders the patients incapable of taking action for self-preservation under emergency conditions without assistance of others.

(2) Anesthesia that renders the patients incapable of taking action for self-preservation under emergency conditions without the assistance of others.

(3) Emergency or urgent care for patients who, due to the nature of their injury or illness, are incapable of taking action for self-preservation under emergency conditions without the assistance of others.

Anesthetizing Location (HC) [*Definition from 517.2*] Any area of a facility that has been designated to be used for the administration of any flammable or nonflammable inhalation anesthetic agent in the course of examination or treatment, including the use of such agents for relative analgesia.

Assembly Occupancies (AS) [*Definition derived from 518.2(A), (B), and (C)*] The assembly occupancies covered by Article 518 are buildings or portions of buildings or structures designed or intended for the gathering of 100 or more persons for such purposes as deliberation, worship, entertainment, eating, drinking, amusement, awaiting transportation, or similar purposes. Assembly occupancies include, but are not limited to, armories, courtrooms, assembly halls, dance halls, auditoriums, dining and drinking facilities, exhibition halls, gymnasiums, mortuary chapels, multipurpose rooms, museums, bowling lanes, places of awaiting transportation, pool rooms, club rooms, restaurants, conference rooms, and skating rinks. Assembly occupancies do not include those areas specifically covered in Article 520 (Theaters, Audience Areas of Motion Picture and Television Studios, Performance Areas, and Similar Locations, which includes many large stadiums and arenas). Auditoriums within business establishments, mercantile establishments, and other occupancies may also be assembly occupancies if intended for occupancy by 100 or more persons. Occupancy of any room or space for assembly purposes by less than 100 persons in a building of other occupancy, and incidental to such other occupancy, shall be classified as part of the other occupancy and subject to the provisions applicable thereto.

Where any such building structure, or portion thereof, contains a projection booth or stage platform or area for the presentation of theatrical or musical productions, either fixed or portable, the wiring for that area, including associated audience seating areas, and all equipment that is used in the referenced area, and portable equipment and wiring for use in the production that will not be connected to permanently installed wiring, shall comply with Article 520. [*See the definition of* Theaters, Audience Areas of Motion Picture and Television Studios, and Similar Areas.]

> *Informational Note:* For methods of determining population capacity, see local building code or, in its absence, NFPA 101-2012, *Life Safety Code.*

Battery-Powered Lighting Units (HC) [*Definition from 517.2*] Individual unit equipment for backup illumination consisting of the following:

  (1) Rechargeable battery
  (2) Battery-charging means
  (3) Provisions for one or more lamps mounted on the equipment, or with terminals for remote lamps, or both
  (4) Relaying device arranged to energize the lamps automatically upon failure of the supply to the unit equipment

[*This definition is derived from 700.12(F), which describes other performance characteristics of unit equipment.*]

Border Light (TH) [*Definition from 520.2*] A permanently installed overhead strip light.

Critical Branch (HC) [*Definition from 517.2*] A system of feeders and branch circuits supplying power for task illumination, fixed equipment, select receptacles, and select power circuits serving areas and functions related to patient care and that is automatically connected to alternative power sources by one or more transfer switches during interruption of the normal power source.

Disconnecting Means (RV) [*Definition from 551.2*] The necessary equipment usually consisting of a circuit breaker or switch and fuses, and their accessories, located near the point of entrance of supply conductors in a recreational vehicle and intended to constitute the means of cutoff for the supply to that recreational vehicle. [*This special definition for use with Article 551 is specific to RVs because it describes not only the function, but the location with respect to the RV. In this respect, it modifies the definition in Article 100.*]

Distribution Point (AG) [*Definition from 547.2*] An electrical supply point from which service drops, service conductors, feeders, or branch circuits to buildings or structures utilized under single management are supplied.

> *Informational Note No. 1:* Distribution points are also known as the center yard pole, meterpole, or the common distribution point.

> *Informational Note No. 2:* The service point as defined in Article 100 is typically at the distribution point.

[*Article 547 provides special rules for power distribution to groups of agricultural buildings. Under some of these rules, distribution from the supply point or distribution point may be done without providing overcurrent protection at the distribution point, and the individual supplies are still service conductors. Where overcurrent protection is provided at the distribution point, the supplies to individual buildings are feeders. See the definition of* Site-Isolating Device.]

Drop Box (TH) [*Definition from 520.2*] A box containing pendant- or flush-mounted receptacles attached to a multiconductor cable via strain relief or a multipole connector.

Electrical Datum Plane (MB) [*Definition from 555.2*] The electrical datum plane is defined as follows:

  (1) In land areas subject to tidal fluctuation, the electrical datum plane is a horizontal plane 606 mm (2 ft) above the highest tide level for the area occurring under normal circumstances, that is, highest high tide.
  (2) In land areas not subject to tidal fluctuation, the electrical datum plane is a horizontal plane 606 mm (2 ft) above the highest water level for the area occurring under normal circumstances.

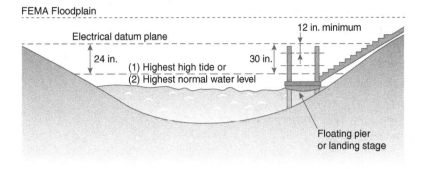

FIGURE 9-26   Establishing the electrical datum plane for the purpose of locating electrical equipment.

(3) The electrical datum plane for floating piers and landing stages that are (a) installed to permit rise and fall response to water level, without lateral movement, and (b) that are so equipped that they can rise to the datum plane established for (1) or (2), is a horizontal plane 762 mm (30 in.) above the water level at the floating pier or landing stage and a minimum of 305 mm (12 in.) above the level of the deck. [*The heights of equipment above water level are measured from the* electrical datum plane *in Article 555. This level may vary for a given body of water depending on whether the measurement applies to a fixed or floating structure. The datum plane considers normal high tides and normal high water levels but does not consider flood levels due to natural or other disasters (FIGURE 9-26). A similar definition is found in Article 682 for* Natural and Artificially Made Bodies of Water, *but Article 682 does not cover bodies of water intended for navigation or boating.*]

Electrical Life-Support Equipment   (HC) [*Definition from 517.2*] Electrically powered equipment whose continuous operation is necessary to maintain a patient's life.

Equipment Branch   (HC) [*Definition from 517.2*] A system of feeders and branch circuits arranged for delayed, automatic, or manual connection to the alternate power source and that serves primarily 3-phase power equipment.

Equipotential Plane   (AG) [*Definition from Article 547*] An area where wire mesh or other conductive elements are embedded in or placed under concrete, bonded to all metal structures and fixed nonelectrical equipment that may become energized, and connected to the electrical grounding system to minimize voltage potentials within the plane and between the planes, the grounded equipment, and the earth. [*For the purposes of this definition in Article 547—that is, where an equipotential plane is required—livestock does not include poultry (see 547.10). A similar definition is found in Article 682 and the idea is also applied to equipotential bonding in Article 680.*]

Essential Electrical System   (HC) [*Definition from 517.2*] A system comprised of alternate sources of power and all connected distribution systems and ancillary equipment, designed to ensure continuity of electrical power to designated areas and functions of a health care facility during disruption of normal power sources and also to minimize disruption within the internal wiring system. [*The essential electrical system in a health care facility includes the life safety branch, the critical branch, and the equipment branch. Essential electrical systems include more types of equipment and circuits than the emergency systems covered by Article 700. The life safety branch of the essential electrical system in a health care facility serves a similar function to the emergency system in other occupancies. The life safety branches of essential electrical systems in hospitals are installed according to the rules of Article 700 as modified by Article 517.*]

Finish Rating   (AS) [*Definition from 518. 4(C), Informational Note*] A finish rating is established for assemblies containing combustible (wood) supports. The *finish rating* is defined as the time at which the wood stud or wood joist reaches an average temperature rise of 121°C (250°F) or an individual temperature rise of 163°C (325°F) as measured on the plane of the wood nearest the fire. A finish rating is not intended to represent a rating for a membrane ceiling.

Flammable Anesthetics   (HC) [*Definition from 517.2*] Gases or vapors, such as fluroxene, cyclopropane, divinyl ether, ethyl chloride, ethyl ether, and ethylene, which may form flammable or explosive mixtures with air, oxygen, or reducing gases such as nitrous oxide. [*The air, oxygen, and nitrous oxide mentioned are not themselves flammable, although they may increase the flammability of other materials. Most anesthetics presently in common use are not flammable.*]

Flammable Anesthetizing Location (HC) [*Definition from 517.2*] Any area of the facility that has been designated to be used for the administration of any flammable inhalation anesthetic agents in the normal course of examination or treatment.

Footlight (TH) [*Definition from 520.2*] A border light installed on or in the stage (**FIGURE 9-27**).

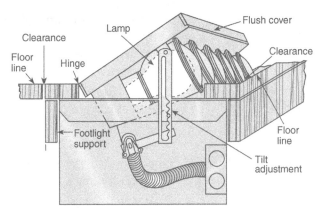

FIGURE 9-27 Disappearing stage footlights. (Source: *National Electrical Code® Handbook*, NFPA, Quincy, MA, 2011, Exhibit 520.5.)

Hazard Current (HC) [*Definition from 517.2*] For a given set of connections in an isolated power system, the total current that would flow through a low impedance if it were connected between either isolated conductor and ground.

Health Care Facilities (HC) [*Definition from 517.2*] Buildings or portions of buildings in which medical, dental, psychiatric, nursing, obstetrical, or surgical care are provided. Health care facilities include, but are not limited to, hospitals, nursing homes, limited-care facilities, clinics, medical and dental offices, and ambulatory care centers, whether permanent or movable. [*This definition was revised in the 2002 NEC to more clearly include psychiatric facilities in the scope of Article 517. Health care facilities such as medical and dental offices are often part of mixed-use buildings such as strip malls and office buildings. The requirements of Article 517 apply to those portions of these buildings that are used for patient care and may affect the choice of wiring methods in those areas. For example, Type NM cable or other nonmetallic wiring methods may be suitable for all parts of such buildings except the patient care areas, where a metallic wiring method that provides an inherent equipment grounding conductor supplemented with an insulated equipment grounding conductor is required. See the definition of* Patient Care Space.]

Hospital (HC) [*Definition from 517.2*] A building or portion thereof used on a 24-hour basis for the medical, psychiatric, obstetrical, or surgical care, of four or more inpatients.

Isolated Power System (HC) [*Definition from 517.2*] A system comprising an isolating transformer or its equivalent, a line isolation monitor, and its ungrounded circuit conductors.

Isolation Transformer (HC) [*Definition from 517.2*] A transformer of the multiple-winding type, with the primary and secondary windings physically separated, which inductively couples its secondary winding(s) to circuit conductors connected to its primary winding(s).

Life Safety Branch (HC) [*Definition from 517.2*] A system of feeders and branch circuits supplying power for lighting, receptacles, and equipment essential for life safety that is automatically connected by one or more transfer switches to alternative power sources during interruption of the normal power source. [*See* Essential Electrical System.]

Limited-Care Facility (HC) [*Definition from 517.2*] A building or portion thereof used on a 24-hour basis for the housing of four or more persons who are incapable of self-preservation because of age, physical limitation due to accident or illness, or limitations such as mental retardation/developmental disability, mental illness, or chemical dependency. [*These facilities are distinguished from hospitals because the persons housed are not "inpatients" undergoing active medical treatments of the types administered in hospitals.*]

Line Isolation Monitor (HC) [*Definition from 517.2*] A test instrument designed to continually check the balanced and unbalanced impedance from each line of an isolated circuit to ground and equipped with a built-in test circuit to exercise the alarm without adding to the leakage current hazard.

Nursing Home (HC) [*Definition from 517.2*] A building or portion of a building used on a 24-hour basis for the housing and nursing care of four or more persons who, because of mental or physical incapacity, might be unable to provide for their own needs and safety without the assistance of another person.

Operator (CF) [*Definition from 525.2*] The individual responsible for starting, stopping, and controlling an amusement ride or supervising a concession.

Patient Bed Location (HC) [*Definition from 517.2*] The location of a patient sleeping bed or the bed or procedure table of a critical care area.

Patient Care Space (HC) [*Definition from 517.2*] Spaces within a health care facility wherein patients are intended to be examined or treated.

> *Basic Care Space.* Space in which failure of equipment or a system is not likely to cause injury to the patients or caregivers but may cause patient discomfort.
> *General Care Space.* Space in which failure of equipment or a system is likely to cause minor injury to patients or caregivers.
> *Critical Care Space.* Space in which failure of equipment or a system is likely to cause major injury or death to patients or caregivers.
> *Support Space.* Space in which failure of equipment or a system is not likely to have a physical impact on patients or caregivers.

Informational Note No. 1: The governing body of the facility designates patient care space in accordance with the type of patient care anticipated and with the definitions of the area classification. Business offices, corridors, lounges, day rooms, dining rooms, or similar areas typically are not classified as patient care space.

Informational Note No. 2: Basic care space is typically a location where basic medical or dental care, treatment, or examinations are performed. Examples include, but are not limited to, examination or treatment rooms in clinics, medical and dental offices, nursing homes, and limited care facilities.

Informational Note No. 3: General care space includes areas such as patient bedrooms, examining rooms, treatment rooms, clinics, and similar areas where the patient may come into contact with electromedical devices or ordinary appliances such as a nurse call system, electric beds, examining lamps, telephones, and entertainment devices.

Informational Note No. 4: Critical care space includes special care units, intensive care units, coronary care units, angiography laboratories, cardiac catheterization laboratories, delivery rooms, operating rooms, and similar areas in which patients are intended to be subjected to invasive procedures and are connected to line-operated, electromedical devices.

Informational Note No. 5: Spaces where a procedure is performed that subjects patients or staff to wet conditions are considered as wet procedure areas. Wet conditions include standing fluids on the floor or drenching of the work area. Routine housekeeping procedures and incidental spillage of liquids do not define wet procedure areas. It is the responsibility of the governing body of the health care facility to designate the wet procedure areas.

Patient Care Vicinity (HC) [*Definition from 517.2*] A space, within a location intended for the examination and treatment of patients, extending 1.8 m (6 ft) beyond the normal location of the patient bed, chair, table, treadmill, or other device that supports the patient during examination and treatment and extending vertically to 2.3 m (7 ft 6 in.) above the floor (**FIGURE 9-28**).

Patient Equipment Grounding Point (HC) [*Definition from 517.2*] A jack or terminal bus that serves as the collection point for redundant grounding of electric appliances serving a patient care vicinity or for grounding other items to eliminate electromagnetic interference problems. [*A patient equipment grounding point is permitted in a patient vicinity as a means of eliminating differences of potential between equipment or between* *equipment and the patient in the patient vicinity. Other methods may be used. See 517.19(C).*]

Portable Equipment (TH) [*Definition from 520.2*] Equipment fed with portable cords or cables intended to be moved from one place to another.

Portable Power Distribution Unit (TH) [*Definition from 520.2*] A power distribution box containing receptacles and overcurrent devices.

Portable Structures (CF) [*Definition from 525.2*] Units designed to be moved including, but not limited to, amusement rides, attractions, concessions, tents, trailers, trucks, and similar units.

Power Outlet (TH, RV) [*Definition from Article 100*] An enclosed assembly that may include receptacles, circuit breakers, fuseholders, fused switches, buses, and watt-hour meter mounting means intended to supply and control power to mobile homes, recreational vehicles, park trailers, or boats or to serve as a means for distributing power required to operate mobile or temporarily installed equipment.

Proscenium (TH) [*Definition from 520.2*] The wall and arch that separates the stage from the auditorium (house).

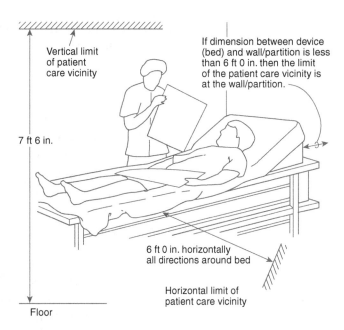

FIGURE 9-28 A patient bed at one kind of patient care vicinity. (Reprinted with permission from *Health Care Facilities Handbook*, Exhibit 3.8, Copyright © 2005, National Fire Protection Association, Quincy, MA, 02169. This reprinted material is not the complete and official position of the NFPA or on the referenced subject, which is represented only by the standard in its entirety.)

Psychiatric Hospital (HC) [*Definition from 517.2*] A building used exclusively for the psychiatric care, on a 24-hour basis, of four or more inpatients.

Recreational Vehicle (RV) [*Definition from 551.2*] A vehicular-type unit primarily designed as temporary living quarters for recreational, camping, or travel use, which either has its own motive power or is mounted on or drawn by another vehicle.

> *Informational Note:* The basic entities are travel trailer, camping trailer, truck camper, and motor home as referenced in NFPA 1192-2011, *Standard on Recreational Vehicles*. See 3.3.46, *Recreational Vehicle*, and A.3.3.46 of NFPA 1192.

Recreational Vehicle Park (RV) [*Definition from 551.2*] A plot of land upon which two or more RV sites are located, established, or maintained for occupancy by recreational vehicles of the general public as temporary living quarters for recreation or vacation purposes.

Recreational Vehicle Site (RV) [*Definition from 551.2*] A plot of ground within a recreational vehicles park intended for the accommodation of a recreational vehicle on a temporary basis or used as a camping unit site.

Recreational Vehicle Site Supply Equipment (RV) [*Definition from 551.2*] The necessary equipment, usually a power outlet, consisting of a circuit breaker or switch and fuse and their accessories, located near the point of entrance of supply conductors to a recreational vehicle site and intended to constitute the disconnecting means for the supply to that site.

Recreational Vehicle Stand (RV) [*Definition from 551.2*] That area of a recreational vehicle site intended for the placement of a recreational vehicle.

Reference Grounding Point (HC) [*Definition from 517.2*] The ground bus of the panelboard or isolated power system panel supplying the patient care area. [*In some areas of health care facilities such as in operating rooms, the reference grounding point is extended to the patient equipment grounding point from which temporary connections are made to a patient and all persons or conductive objects in contact with the patient. This reduces* *any difference of potential between people or items that may contact the patient, especially during invasive procedures.*]

Selected Receptacles (HC) [*Definition from 517.2*] A minimum number of electric receptacles to accommodate appliances ordinarily required for local tasks or likely to be used in patient care emergencies.

Site-Isolating Device (AG) [*Definition from 547.2*] A disconnecting means installed at the distribution point for the purposes of isolation, system maintenance, emergency disconnection, or connection of optional standby systems.

Solid-State Phase-Control Dimmer (AS, TH) [*Definition from 520.2*] A solid-state dimmer where the wave shape of the steady-state current does not follow the wave shape of the applied voltage, such that the wave shape is nonlinear.

Solid-State Sine Wave Dimmer (AS, TH) [*Definition from 520.2*] A solid-state dimmer where the wave shape of the steady-state current follows the wave shape of the applied voltage such that the wave shape is linear. [*Although this term and the term* Solid-State Phase-Control Dimmer *are defined in Article 520, they are also used in Article 518. The practical difference from an* NEC *standpoint is that the neutral conductor is counted as a current-carrying conductor in ampacity calculations for the purposes of applying adjustment factors if the wave shape is nonlinear, but the neutral is not considered to be current-carrying if the wave shape is linear.*]

Stage Equipment (TH) [*Definition from 520.2*] Equipment at any location on the premises integral to the stage production including, but not limited to, equipment for lighting, audio, special effects, rigging, motion control, projection, or video.

Stage Lighting Hoist (TH) [*Definition from 520.2*] A motorized lifting device that contains a mounting position for one or more luminaires, with wiring devices for connection of luminaires to branch circuits, and integral flexible cables to allow the luminaires to travel over the lifting range of the hoist while energized.

Stage Switchboard (TH) [*Definition from 520.2*] A switchboard, panelboard, or rack containing dimmers or relays with associated overcurrent protective devices, or overcurrent protective devices alone, used primarily to feed stage equipment.

Task Illumination (HC) [*Definition from 517.2*] Provision for the minimum lighting required to carry out necessary tasks in the described areas, including safe access to supplies and equipment, and access to exits.

Theaters, Audience Areas of Motion Picture and Television Studios, and Similar Areas (TH) [*Definition derived from 520.1*] All buildings or that part of a building or structure, indoor or outdoor, designed or used for presentation, dramatic, musical, motion picture projection, or similar purposes and to specific audience seating areas within motion picture or television studios.

Transfer Switch (HC) [*Definition from Article 100, see Switch, Transfer*] An automatic or nonautomatic device for transferring one or more load conductor connections from one power source to another.

Wet Procedure Location The area in a patient care space where a procedure is performed that is normally subject to wet conditions while patients are present, including standing fluids on the floor or drenching of the work area, where either such condition is intimate to the patient or staff.

> *Informational Note:* Routine housekeeping procedures and incidental spillage of liquids do not define a wet procedure location.

[*This definition is specific to health care occupancies and differs from the more general definition of* wet location *given in Article 100.*]

X-ray Installations, Long-Time Rating (HC) [*Definition from 517.2*] A rating based on an operating interval of 5 minutes or longer.

X-ray Installations, Mobile (HC) [*Definition from 517.2*] X-ray equipment mounted on a permanent base with wheels, casters, or a combination of both to facilitate moving the equipment while completely assembled.

X-ray Installations, Momentary Rating (HC) [*Definition from 517.2*] A rating based on an operating interval that does not exceed 5 seconds.

X-ray Installations, Portable (HC) [*Definition from 517.2*] X-ray equipment designed to be hand carried.

X-ray Installations, Transportable (HC) [*Definition from 517.2*] X-ray equipment to be conveyed by a vehicle or that is readily disassembled for transport by a vehicle.

KEY QUESTIONS

1. What wiring methods are being used?
2. What and where is the service or power source?
3. Is temporary wiring or portable equipment being used?
4. For health care facilities, what are the essential electrical system requirements?
5. For health care facilities, what type of alternative power supply is provided?
6. For health care facilities, which areas are general care spaces, critical care spaces, and wet procedure locations?
7. For theaters and audience areas, are dressing rooms provided?
8. For RV parks, marinas, and boatyards, how many sites or slips are supplied with power?
9. For marinas and boatyards, does the marina or boatyard include fuel-dispensing facilities?
10. For marinas and boatyards, what level or reference is being used to define the electrical datum plane?

CHECKLISTS

✔	Item	Inspection Activity	*NEC* Reference	Comments
		Checklist 9-1: Health Care Facilities		
	1.	Review definitions and determine the proper classification of the areas to be inspected.	517.1, 517.2	
	2.	Verify that insulated copper conductors in metal raceways or equivalent cables are used to provide equipment grounding for branch circuits in patient care spaces and receptacles with an insulated grounding terminal (Type IG) are not used.	517.13(A) and (B), 517.16	
	3.	Check for bonding between normal and essential branch-circuit panelboards serving any single patient care space.	517.14, 517.18, 527.19(D)	
	4.	Verify that each general care space patient bed location has at least two branch circuits, one from the normal system and one from the critical branch.	517.10(B)(2), 517.18(A)	
	5.	Verify that each general space area patient bed location is provided with a minimum of eight receptacles.	517.18(B), 517.30(E)	
	6.	Check for tamper-resistant receptacles or covers in the pediatric locations of general care spaces.	517.18(B) and (C)	
	7.	Verify that each critical care space patient bed location has at least two branch circuits, one from the normal system and one from the critical branch.	517.19(A), 517.30(E)	
	8.	Verify that each critical care space patient bed location has at least one receptacle supplied by a critical branch circuit dedicated to that bed location.	517.19(A)	
	9.	Verify that each critical care space patient bed location has at least 14 receptacles.	517.19(B)	
	10.	Verify that at least one receptacle in a critical care space bed location is connected to a branch circuit from a separate normal or critical branch source.	517.19(B)	

Checklist 9-1: Health Care Facilities

✔	Item	Inspection Activity	*NEC* Reference	Comments
	11.	Verify that a minimum of 36 receptacles are installed in each operating room.	517.19(C)	
	12.	Verify that all patient bed location receptacles are hospital-grade and that all critical branch receptacles are identified.	517.18(B), 517.19(B)(2), 517.30(E)	
	13.	Check wet procedure locations for protection by GFCI devices or, where interruption cannot be tolerated, use of isolated power systems.	517.20, 517.21	
	*Checklist Items 14 through 20 apply only to hospitals and ambulatory care facilities with critical care areas.*			
	14.	Review the essential electrical systems and verify that critical, life safety, and equipment branches are provided.	517.30(B)(1)	
	15.	Review load calculations for the essential system and verify that capacity of power sources and feeders is adequate.	517.30(D)	
	16.	Verify that the wiring of the life safety and critical branches is independent of, and separated from, other wiring and equipment.	517.30(C)(1)	
	17.	Verify that mechanical protection for critical branch wiring is provided by nonflexible metallic raceways or Type MI cable.	517.30(C)(3)	
	18.	Verify that only those loads that are intended for connection to the life safety branch are supplied by the life safety branch.	517.32	
	19.	Verify that two independent sources are provided for the essential electrical system and that the alternative source is suitable.	517.35	
	20.	Determine that the overcurrent protective devices of the life safety branch are selectively coordinated and that the overcurrent devices of the critical branch are coordinated to .01 seconds.	517.26 (700.28) 517.30(G)	
	*Checklist Items 21 through 25 apply only to nursing homes and limited-care facilities where life-support equipment or general anesthesia is used.*			
	21.	Review the essential electrical system, and verify that life safety and critical branches are provided.	517.41(A)	
	22.	Review load calculations for the essential system, and verify that capacity of power sources and feeders is adequate.	517.41(C)	
	23.	Verify that the life safety branch wiring is independent and separated from other wiring and equipment. (Separation of the critical branch is not required.)	517.41(D)	
	24.	Verify that only those loads intended for connection to the life safety branch are supplied by the life safety branch.	517.42	
	25.	Verify that two independent sources are provided for the essential electrical system and that the alternative source is suitable.	517.44	

(continues)

✔	Item	Inspection Activity	*NEC* Reference	Comments
		Checklist 9-1: Health Care Facilities		

Checklist Items 26 and 27 apply only to other health care facilities, including clinics, medical and dental offices, and ambulatory health care facilities without critical care areas.

✔	Item	Inspection Activity	*NEC* Reference	Comments
	26.	Verify that an essential electrical system is provided where required.	517.45(B) and (C)	
	27.	Verify that an alternative power source is provided that is adequate and designed specifically for the purpose.	517.45(A) and (D)	

Note: Checklist items 28 through 38 could apply to any type of health care facility.

✔	Item	Inspection Activity	*NEC* Reference	Comments
	28.	Determine whether hazardous (classified) anesthetizing locations exist in the facility.	517.60	
	29.	Verify that appropriate wiring methods and equipment are used in and above hazardous (classified) anesthetizing locations.	517.61(A) and (B)	
	30.	Verify that power circuits in flammable anesthetizing locations are isolated from other power distribution systems.	517.63(F)	
	31.	Verify that one or more battery-powered lighting units are provided in anesthetizing locations.	517.63(A)	
	32.	Verify that supply circuits to X-ray equipment are adequate and supplied through appropriate wiring methods and connections.	517.71	
	33.	Check location, capacity, and type of disconnecting means for X-ray equipment.	517.72	
	34.	Verify that supply circuits to X-ray equipment meet minimum ampacity and overcurrent rating requirements.	517.73	
	35.	Verify that enclosures for high-voltage parts and noncurrent-carrying metal parts of X-ray equipment are grounded.	517.78	
	36.	Verify that low-voltage systems in patient care areas have insulation and isolation equivalent to power distribution systems.	517.80	
	37.	Check isolated power systems (where installed) for proper installation, features, and conductor identification.	517.160	
	38.	Verify second-level GFPE where applicable in hospitals and other health care facilities.	517.17	

CHECKLISTS

Checklist 9-2: Assembly Occupancies

✔	Item	Inspection Activity	*NEC* Reference	Comments
	1.	Determine the applicability of Article 518.	518.1, 518.2	
	2.	Verify that wiring methods are suitable for the occupancy and fire rating of the area(s).	518.4	
	3.	Check temporary wiring for compliance with Article 590, except for GFCI requirements.	518.3(B)	
	4.	Check portable distribution equipment for adequate ratings and supply from listed power outlets.	518.5	
	5.	Check portable distribution equipment for overcurrent protection and isolation from the general public.	518.5	

Checklist 9-3: Theaters and Audience Areas of Motion Picture Studios

✔	Item	Inspection Activity	*NEC* Reference	Comments
	1.	Review definitions and determine the applicability of Article 520.	520.1, 520.2, 520.3, 520.4	
	2.	Verify that wiring methods are suitable for the occupancy and fire rating of the area(s).	520.5	
	3.	Check for compliance with raceway fill requirements.	520.6	
	4.	Check portable equipment used outdoors for supervision by qualified personnel and isolation from the general public.	520.7, 520.10	
	5.	Check fixed stage switchboards for suitability.	520.21 through 520.26	
	6.	Check stage switchboard feeders for type and capacity.	520.27	
	7.	Check fixed stage equipment other than switchboards for suitability and compliance with specific requirements for the equipment type.	520.40 through 520.49	
	8.	Check portable switchboards on stage for proper supply, overcurrent protection, construction, and feeders.	520.50 through 520.53	

CHECKLISTS

(continues)

Checklist 9-3: Theaters and Audience Areas of Motion Picture Studios

✔	Item	Inspection Activity	NEC Reference	Comments
	9.	Check portable stage equipment other than switchboards for appropriate construction, conductors, protection, and listings (where required).	520.61 through 520.69	
	10.	Verify that pendant lampholders are not used and that exposed incandescent lamps less than 8 ft (2.4 m) above the floor have guards in dressing rooms.	520.71, 520.72	
	11.	Verify that dressing rooms are equipped with switches and pilot lights for lights and receptacles adjacent to mirrors.	520.73	
	12.	Verify that all metal raceways, metal-sheathed cables, and metal frames of equipment are grounded.	520.81, Article 250	

Checklist 9-4: Carnivals, Circuses, Fairs, and Similar Events

✔	Item	Inspection Activity	NEC Reference	Comments
	1.	Verify that mechanical protection is provided for equipment subject to damage.	525.6	
	2.	Verify that transformers, generators, or services used as power sources meet the requirements of their respective articles and Article 250 and that rides and attractions are bonded to the same grounding electrode system when supplied from different power sources and separated by less than 12 ft (3.7 m).	525.10, 525.11	
	3.	Check overhead conductors for adequate clearances from the ground and from rides and attractions.	525.5	
	4.	Check wiring methods, especially flexible cords, for suitability and protection from damage.	525.20	
	5.	Check wiring inside tents and concessions for mechanical protection for wiring and guards on temporary lamps.	525.21(B)	
	6.	Verify that suitable boxes are provided at all outlets and at connection points, junction points, and switch points.	525.20(H)	
	7.	Check the construction and suitability of portable distribution and termination boxes.	525.22	
	8.	Verify that overcurrent protection is provided in accordance with Article 240.	Article 240	
	9.	Verify that general-use receptacles are provided with GFCI protection.	525.23(A)	

CHECKLISTS

Checklist 9-4: Carnivals, Circuses, Fairs, and Similar Events

✔	Item	Inspection Activity	*NEC* Reference	Comments
	10.	Verify that grounding continuity is assured each time portable equipment is connected.	525.32	
	11.	Verify that metal frames, enclosures, and equipment are bonded.	525.30	
	12.	Verify that equipment grounding conductors of the proper size and type are supplied to all equipment requiring grounding.	525.31	
	13.	Check each portable structure for a disconnect in the form of a fusible switch or circuit breaker located within sight and within 6 ft (1.8 m) of the operator's station.	525.21(A)	
	14.	Verify that any attractions using contained volumes of water meet the applicable requirements of Article 680.	525.3(D)	

Checklist 9-5: Agricultural Buildings

✔	Item	Inspection Activity	*NEC* Reference	Comments
	1.	Determine the applicability of Article 547.	547.1	
	2.	Check wiring methods for suitability for the occupancy and conditions and for protection from physical damage.	547.5	
	3.	Verify that any equipment grounding conductors installed underground are insulated or covered.	547.5(F)	
	4.	Check switches, circuit breakers, controllers, and the like for enclosures suitable for the conditions.	547.5(C), 547.6	
	5.	Verify that luminaires are installed to minimize the entry of dust and water and that fixtures exposed to damage are supplied with guards.	547.8	
	6.	Verify that the arrangement of service equipment, distribution equipment, overcurrent protection, and grounding complies with requirements.	547.9	
	7.	Verify that an equipotential plane has been installed in concrete floors of livestock containment areas and bonded to electrodes and conductive elements.	547.10	
	8.	Verify that GFCI protection has been provided where required.	547.5(G)	

✔	Item	Inspection Activity	*NEC* Reference	Comments
		Checklist 9-6: Recreational Vehicle Parks		
	1.	Review definitions and determine the applicability of Article 551 and Part VI.	551.1, 551.2, 551.4	
	2.	Verify that all sites with power have 20-ampere, 125-volt receptacles, at least 20 percent have 50-ampere, 125/250-volt receptacles, and at least 70 percent have 30-ampere, 125-volt receptacles.	551.71	
	3.	Verify that the voltages of distribution systems are appropriate for the sites supplied.	551.31, 551.40(A)	
	4.	Review load calculations, and check sizing or ratings of transformers, panelboards, and feeders.	551.73	
	5.	Verify that grounded feeder conductors have the same ampacity as ungrounded conductors.	551.73(D)	
	6.	Verify that separate equipment grounding conductors extending from a service or secondary distribution system are run to equipment requiring grounding.	551.76	
	7.	Check RV site supply equipment for proper location relative to the vehicle parking stand.	551.77(A)	
	8.	Check site supply equipment and disconnecting means for proper access, mounting height, working space, and marking.	551.77(B) through (F)	
	9.	Check overhead conductors for vertical and horizontal clearances from the finished grade in areas subject to vehicle movement.	551.79	
	10.	Check underground conductors for appropriate ratings, insulation, identification, and, where emerging from the ground, proper physical protection.	551.80	

Checklist 9-7: Marinas and Boatyards

✔	Item	Inspection Activity	*NEC* Reference	Comments
	1.	Determine the applicability of Article 555.	555.1	
	2.	Verify that shore power receptacles are of an appropriate grounding type and have appropriate ampere ratings.	555.19(A)	
	3.	Verify that general-use receptacles not used for shore power are GFCI protected.	555.19(B)	
	4.	Verify that disconnecting means are provided within sight of shore power connections to isolate each boat from its supply circuit.	555.17	
	5.	Verify that receptacles used for shore power are supplied by individual branch circuits with voltage and current ratings corresponding to the receptacles.	555.19(A)(3)	
	6.	Review feeder and service calculations for compliance with requirements.	555.12	
	7.	Verify that wiring methods are suitable for wet locations and that portable power cables, where used, are extra-hard-usage type listed for wet locations and sunlight resistance.	555.13	
	8.	Verify that equipment requiring grounding is connected to an insulated equipment grounding conductor included with feeders and branch circuits.	555.15	
	9.	Check wiring over or under navigable water for suitability and clearances.	555.13(B)(3)	
	10.	Check wiring in motor fuel-dispensing stations for compliance with Article 514.	555.21	
	11.	Verify that service equipment for floating docks or marinas is located adjacent to, but not on or in, the floating structure.	555.7	
	12.	Verify that ground-fault protection is provided for the main overcurrent device on the supply to the marina.	555.3	

Swimming Pools and Related Installations

OVERVIEW

This chapter covers the installations and equipment within the scope of Article 680 of the *National Electrical Code*® (*NEC*®). The scope of Article 680 is "the construction and installation of electrical wiring for and equipment in or adjacent to all swimming, wading, therapeutic, and decorative pools, fountains, hot tubs, spas, and hydromassage bathtubs, whether permanently installed or storable, and to metallic auxiliary equipment, such as pumps, filters, and similar equipment. The term *body of water* applies to all bodies of water covered in this scope unless otherwise amended." It is evident what these installations and equipment have in common: bodies of water that are frequently associated with electrical equipment and that are likely to be contacted by people and that are constructed specifically for one of the purposes described in the scope of the article.

People being intentionally in contact with or immersed in a body of water that has associated electrical equipment or electrical equipment in the vicinity creates the potential hazards that the requirements of Article 680 are intended to mitigate. Fountains (and decorative pools) are a little different. Most fountains are not intended as wading pools, but as they are often used that way, it is reasonable to expect that persons will contact the contained water of a fountain.

This chapter covers the special rules of Article 680 that are intended to eliminate, or minimize, the risk of electrical installations combined with bodies of water. The general rules for equipment such as branch circuits and feeders and the use and installation of wiring methods are covered in other sections of this manual. This chapter does not cover the "Natural and Artificially Made Bodies of Water" such as fish farm ponds, storm retention ponds, and treatment ponds that are covered by Article 682. This chapter also does not cover the bodies of water associated with marinas and boatyards; these are covered by Article 555 and the Special Occupancies chapter of this manual.

SPECIFIC FACTORS

This chapter covers the rules of Article 680 that are intended to eliminate or minimize the increased shock hazards associated with electrical installations combined with bodies of water. Note, however, that the definition of *Permanently Installed Swimming, Wading, Immersion, and Therapeutic Pools* covers such **pools** even if they are not served by any electrical circuits. Also, note that nothing in the definitions or the scope of Article 680 applies to swimming beaches at oceans, lakes, or ponds, even though a swimming pool next to a beach is covered. Article 680 does not cover the other natural or artificial bodies of water that are included in the scope of Article 682.

The general intent of the *NEC* is to protect persons and property from the hazards of electricity. The risks of electric shock are much greater when a person is in a wet or damp environment. This fact is recognized and addressed by the **ground-fault circuit interrupter (GFCI)** requirements for receptacles in bathrooms, outdoors, and in basements. Risks are greater still when persons are standing or immersed in water, and where electrical equipment such as circulating pumps, heaters, pool covers, and luminaires are in direct or indirect contact with the pool water or located in close proximity to the pool. The voltage limits for contact by persons in the wet environment of a swimming pool are given in the definition of *Low Voltage Contact Limit*, which can be found in the Key Terms section of this chapter or in Section 680.2 of the *Code*. Even where there is no electrical equipment intentionally installed or connected to a pool, voltage gradients may occur in a pool area due to lightning or due to currents resulting from the grounding of utility distribution systems or from other electrical equipment in the general area. For this reason, bonding requirements apply to pools even where there is no electrical equipment directly associated with the pool. However, nearly all swimming pools do involve the use of electrical equipment, and electrical equipment is essential to the operation of other installations covered by Article 680. For example, spas, hot tubs, therapeutic tubs, hydromassage bathtubs, and fountains require electrical pumps for their normal operation.

Because of the increased risk of electrical shock, Article 680 includes many special requirements for bonding, grounding, GFCI protection, special enclosures suitable for the wet and often corrosive environment, and minimum separations of electrical equipment from the covered areas. Many of the rules are intended to provide more reliable grounding, insulation, and isolation and a higher degree of corrosion resistance than might be required in other areas or for other equipment.

KEY QUESTIONS

1. Which parts of Article 680 apply to the installation?

The definitions and the statements at the beginning of each part of Article 680 can be used to answer this question. Some types of portable equipment, such as portable hydromassage units and portable fountains, are not covered by Article 680 and are treated instead as appliances and covered by Article 422. In another case, equipment may not be covered entirely in the part of Article 680 whose title is most like the name of the equipment. This situation can be found where fountains that have water in common with a swimming pool are subject to both the requirements for pools and those for fountains. A large therapeutic pool built as a permanent installation may also be considered a swimming pool for the purposes of applying Article 680. Furthermore, the requirements for a hydromassage bathtub are quite different from the requirements for a spa, although some examples of each may look very similar. Thus, the classification and definition of the equipment or installation in question are critical to the proper application of the *NEC* requirements.

For the most part, the organization of Article 680 precludes the need to cross-reference and apply various parts of the article to some specific installation, unless, as noted, the installation actually does contain more than one type of body of water. Part I applies to all installations under Article 680, along with the part or parts that cover the specific type(s) of body of water. The only exception to this is hydromassage bathtubs, which must comply with Part VII only. Because Part I is applicable to all "pool-type" installations other than hydromassage bathtubs, it uses the generic term *body of water*. Parts II through VII name the specific type of installation or body of water they cover. The requirements for wiring methods in Part II, Permanently Installed Pools, also apply to Part IV, Spas and Hot Tubs.

2. What wiring methods are used?

Most of the installations covered by Article 680 are subject to some restrictions on wiring methods. Many of the restrictions are due to, or closely related to, grounding requirements. Some installations require insulated copper conductors, while others do not. Many installations require brass or other corrosion-resistant conduits, whereas hydromassage bathtubs may be supplied by any wiring method that is otherwise suitable in a bathroom. Wiring methods also must be compatible with other equipment, enclosures, and boxes. In some cases, such as in swimming pool decks, specially listed boxes are required. Wiring methods may be specified on electrical plans, but they must be checked in the field.

3. What electrical equipment is associated with the installation, and where is it located?

Electrical equipment, including luminaires, switches, receptacles, and motors, are subject to restrictions on where they may be located in relation to pools and similar installations. Some equipment is required to be GFCI protected and/or **bonded** based on location; other equipment associated with pools, tubs, and the like may have to be bonded or GFCI protected regardless of its location. Much of the equipment may be shown on plans or shop drawings, but detailed shop drawings usually are not included with plans submitted for review and may be available only at the job site. The final answer to this question can be found only by a site inspection.

4. What type of underwater luminaires are used?

Depending on the type of installation, swimming pool, spa, hot tub, or fountain, the luminaires that may be used are varyingly restricted. Luminaires for fountains, for example, often rely on submersion for proper operation and are frequently used with low-water cutoff devices. Such devices are not as common in swimming pools. However, swimming pools may use wet-, dry-, or no-niche luminaires. The wiring methods and grounding requirements for these three types of luminaires vary somewhat. Luminaire types are often specified on electrical plans, but field inspections are usually necessary to determine the types of luminaires actually being used.

5. What equipment and structures are required to be bonded, and how will these items be connected together in order to establish equipotential bonding in the pool area?

The need for equipotential bonding of the noncurrent-carrying conductive surfaces of electrical equipment and the conductive surfaces of nonelectrical equipment, in order to minimize voltage gradients (differences in potential) in the pool area, is determined by its distance from the body of water and the function of the equipment. Electrical equipment associated with pool recirculation or conditioning of the water is usually required to be bonded. Double-insulated equipment is not required to be bonded, but internal metal parts may have to be grounded. Other metal equipment must be bonded if it is within the distances specified for the type of installation. Distances within which metal parts must be bonded are greatest for swimming pools and are reduced to 5 ft (1.5 m) for most other installations. Some equipment and parts that must be bonded can be identified from electrical plans. A complete list of parts requiring bonding must be based on what is actually installed and can be determined only by a site inspection.

6. What are the voltages of the equipment used in and around the body of water?

The voltage limits for contact by persons in the wet environment of a swimming pool are given in the definition of "Low Voltage Contact Limit," as noted in the Specific Factors section of this chapter. The requirements for some items such as luminaires vary according to the voltage of the equipment.

PLANNING FROM START TO FINISH

The installations that definitely require more than one inspection are divided into two checklist sections. Swimming pools, or other facilities treated as swimming pools, will usually require an inspection to verify proper equipotential bonding and to inspect embedded wiring and embedded parts such as **forming shells**. Many fountains are constructed at least partially of concrete or masonry and will require an inspection of the embedded equipment. Embedded items and bonding details will not be accessible at a later inspection. Otherwise, most of the items that must be inspected can be examined at an intermediate or final inspection. Exactly when any aspect of an installation can be inspected depends on the timing of the project. Packaged and self-contained equipment can usually be inspected all at one time. Something like the suitability of access to hydromassage bathtub equipment can be verified only at the final inspection, after all finishes are in place.

On an initial inspection of a swimming pool or other permanent installation with embedded equipment, a logical place to start can be found by identifying the equipotential bonding grid. While inspecting the bonding methods, most other aspects of the installation will also be encountered. For example, forming shells are items that must be bonded and connected to a wiring method and, usually, to a deck box. In many cases, other aspects of the installation can also be checked on the initial inspection. For example, locations of pool equipment, luminaires, switches, and receptacles may be evident at an initial (rough) inspection, especially when the installations are located indoors.

On later inspections, the electrical equipment related to water conditioning and recirculation is a good place to start. Some GFCI requirements apply to this equipment. Bonding methods should also be evident here and can be followed back to the pool or tub that will contain water. Because most other checklist items are based on the water location and its boundaries, inspections of switches, receptacles, lighting, and GFCI requirements can be completed in the area of the water.

WORKING THROUGH THE CHECKLISTS

PERMANENTLY INSTALLED SWIMMING POOLS— ARTICLE 680, PARTS I AND II

Checklist 10-1: Initial Inspection: Prior to Pouring of Concrete or Burial

1. Review definitions and determine the applicability of Article 680.
Definitions for terms used in Article 680 are provided in the article itself and in the Key Terms section of this chapter. The scope of Article 680 is described as "the construction and installation of electrical wiring for and equipment in or adjacent to all swimming, wading, therapeutic, and decorative pools, fountains, hot tubs, spas, and hydromassage bathtubs, whether permanently installed or storable, and to metallic auxiliary equipment, such as pumps, filters, and similar equipment. The term *body of water* used throughout Part I applies to all bodies of water covered in this scope unless otherwise amended. See also the discussion under "Factors Specific to Swimming Pools and Related Installations" of Article 680. Careful review of the Article 680 definitions of different types of bodies of water and different types of equipment helps to ensure that the correct set of requirements will be applied to the equipment involved.

2. Check overhead conductor clearances for conformance with requirements.
The clearances from grade for conductors around swimming pools are significantly different from the requirements of Article 225 or Article 230. The clearances from pools also take into account the structures related to the pool, such as diving boards, diving platforms, and any observation structures that are within 10 ft (3.0 m) horizontally from the inside edge of a pool. The required clearances are not only vertical dimensions; clearances must be maintained in all directions from the pool and pool structure. Required clearances vary by voltage and type of overhead conductor, but the basic minimum distances for cabled conductors supported by a solidly grounded bare messenger or neutral conductor are 22.5 ft (6.9 m) from water level or edge of water and 14.5 ft (4.4 m) from diving platforms and towers. **FIGURE 10-1** summarizes and illustrates the clearance requirements for conductors installed above or around a swimming pool. The presence of diving boards, platforms, or other structures must be considered in establishing the minimum clearance.

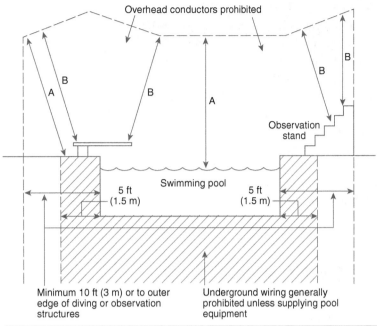

Clearances to overhead conductors				
Dimension	Supply and Service Drops Condition			Overhead conductors of communications, radio and T.V. antenna and CATV systems
	Condition 1	Condition 2	Condition 3	
A	22.5 ft (6.9 m)	25 ft (7.5 m)	27 ft (8.0 m)	10 ft (3 m)
B	14.5 ft (4.4 m)	17 ft (5.2 m)	18 ft (5.5 m)	10 ft (3 m)

Conditions:
1 — Supported on and cabled together with solidly grounded bare messenger or solidly grounded neutral 0–750 V to ground and network-powered broadband communications conductors

2 — 0–15,000 V to ground

3 — Over 15,000–50,000 V to ground

FIGURE 10-1 Overhead and underground conductor clearances.

3. Check underground wiring for suitability, clearances from pool, and minimum cover requirements.

Underground wiring is generally prohibited from passing under or within 5 ft (1.5 m) of a pool (see Figure 10-1) unless the wiring is necessary for pool equipment. Where underground wiring must be installed within 5 ft (1.5 m) of a pool, because of the limited footprint of the site where the pool is installed, the wiring methods are restricted to complete raceway systems of rigid metal conduit (RMC), intermediate metal conduit (IMC), or a nonmetallic raceway ([polyvinyl chloride conduit [PVC], reinforced thermosetting resin conduit [RTRC], electrical nonmetallic tubing [ENT], liquidtight flexible nonmetallic conduit, Type B [LFNC-B]).

Such raceways must be buried at least 6 in. (150 mm) if RMC or IMC is used and at least 18 in. (450 mm) if the raceways are nonmetallic with or without concrete encasement. However, nonmetallic raceways under at least 4 in. (102 mm) of concrete that extends 6 in. (152 mm) beyond the underground raceway may have the minimum cover reduced to 6 in. (150 mm). This allows for connections to pool equipment such as deck boxes or pool covers that are within 5 ft (1.5 m) of the pool. Depending on the location, Article 300 may specify a greater depth. Outside the 5-ft (1.5-m) limit, wiring methods must comply with their respective articles in Chapter 3 of the *NEC* and Article 300. For example, in accordance with 356.12(3), LFNC may not be used in lengths of more than 6 ft (1.8 m) unless it is Type LFNC-B.

4. Check underwater luminaires for locations, wiring methods, and connections to wiring methods.

Underwater luminaires must be designed so that they present no shock hazard during normal use. Relamping is not considered normal use for the purposes of this rule. However, luminaires operating at over 15 volts must be GFCI protected to eliminate shock hazard at any time, including relamping. Luminaires must be located not less than 18 in. (457 mm) below normal water level unless the luminaire is specifically listed for a lesser depth, which may not be less than 4 in. (102 mm) below normal water level. Luminaires facing upward must be guarded from contact by people. Any luminaire that relies on water for cooling must have inherent protection against overheating when not submerged. In order to comply with these requirements, the luminaires, and where used, the GFCIs or transformers must be listed for use with swimming pools, spas, hot tubs, etc.

Underwater luminaires are divided into three types: <u>wet-niche</u>, <u>dry-niche</u>, and <u>no-niche</u>. Wet-niche luminaires are designed so that water can enter the forming shell and the lamp housing can be removed from the forming shell. A cord is used for connection to the lamp housing so that the lamp housing can be removed from the forming shell and raised above water level for relamping. The cord of a wet-niche luminaire must be long enough to allow the luminaire to be relamped from the pool deck or similar location that is above the water level of the pool. Thus, wet-niche luminaires can be relamped without lowering the water level or requiring a person to enter the water for relamping. Dry niche luminaires are designed to exclude water from the forming shell and from the area around the lamp housing. This type of luminaire is often used for indoor swimming pools where access to the rear of the luminaire is provided by a tunnel or crawl space. Relamping can be accomplished from these locations. No-niche luminaires are essentially surface mounted to a bracket and do not require a forming shell. They can be removed from the water for relamping.

Connections to forming shells for wet-niche luminaires must be made with RMC, IMC, LFNC, or rigid nonmetallic conduit (typically PVC). Metal conduit must be brass or other corrosion-resistant metal. Because IMC is steel by definition (see Article 342), rigid conduit made of red brass or PVC conduit has been most commonly used, but RMC is also available in stainless steel. Any of the types of nonmetallic conduit must have an 8 American Wire Gauge (AWG)-insulated copper equipment bonding jumper installed for connection to the forming shell, unless listed low-voltage lighting that does not require grounding is used. The 8 AWG bonding jumper is permitted to be solid or stranded. One of the types of conduit mentioned must extend from the forming shell of a wet-niche luminaire to a listed swimming pool junction box or transformer enclosure (**FIGURE 10-2**). These luminaires are typically equipped with a flexible cord that extends through the conduit to the junction box or transformer enclosure.

5. Check pool-related equipment for appropriate wiring methods.

The wiring for pool-related equipment is restricted to specific wiring methods. The permitted methods vary somewhat according to the type of equipment supplied. All types of pool-related equipment may be wired using RMC, IMC, or rigid nonmetallic conduit. (Aluminum RMC is not permitted for feeders in the pool area where it is subject to corrosion, such as where contact with pool water may occur or underground.) Wiring on or within buildings may be installed in electrical metallic tubing. Electrical nonmetallic tubing may be used for wiring within buildings, except for wiring to motors. Some additional wiring methods are permitted for certain types of equipment. For example, liquidtight flexible metal conduit (LFMC) or LFNC may be used for wiring motors where flexibility is needed, and motors may be wired using Type MC (metal-clad) cable if the cable is listed for the location. A Type MC cable that contains an insulated equipment grounding conductor, is listed for direct burial, and is listed as sunlight-resistant is an example of a cable that is listed for the location and can be used as an underground wiring method to supply a swimming pool pump motor.

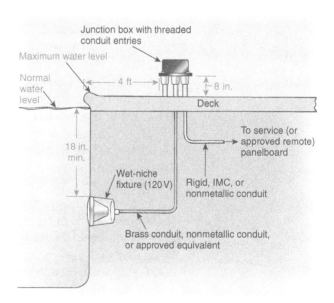

FIGURE 10-2   Wet-niche luminaire installation with junction box supported above pool deck. (Source: *National Electrical Code® Handbook*, NFPA, Quincy, MA, 2011, Exhibit 680.1.)

Panelboards and pool luminaires may be wired using LFNC. Either LFNC or LFMC may be used for wiring to transformers and junction boxes associated with pool luminaires, but where LFMC is used for wiring to pool lighting transformers, it is limited to 6 ft (1.8 m) in any one length or 10 ft (3.0 m) in the total length used unless it is type LFNC-B, which is not limited in length where properly secured and supported.

All permitted wiring methods are required to include an insulated copper equipment grounding conductor, except that insulated aluminum equipment grounding conductors are permitted in the supply circuit to feeder-supplied panelboards (subpanels). According to 680.25(B), equipment grounding conductors are permitted by 680.25(A), Exception, to be uninsulated in existing feeders and in accordance with 680.25(B)(2) for existing feeders run to separate buildings; however, for new feeder installations, a wire-type equipment grounding conductor must be installed with feeder conductors that are run to a separate building or structure to comply with 250.32, and 680.25(B)(2) requires that it be insulated. Aluminum is permitted because the rule does not require copper in feeders as it does for branch circuits. However, 250.120(B) requires aluminum equipment grounding conductors to be insulated if they are subject to corrosive conditions, and the aluminum or copper-clad aluminum equipment grounding conductors may not be terminated within 18 in. (450 mm) of the earth.

TABLE 10-1 provides an overview of the wiring methods that are acceptable for equipment covered in the scope of Article 680. Notice that nonmetallic or brass or other highly corrosion-resistant conduits and enclosures are required in those locations where the conduit is likely to be in contact with pool water.

6. Check junction boxes and enclosures connecting to underwater luminaires for listing and labeling, appropriate size and location, and proper materials.

Special rules apply to junction boxes and enclosures for transformers, GFCIs, and similar devices. Such boxes or enclosures that connect to conduits coming directly from forming shells or no-niche luminaire mounting brackets must be listed as a swimming pool junction box, be equipped with threaded entries or hubs, and be provided with an integral means to ensure electrical continuity between all conduit entries. These boxes must also be provided with no less than one more grounding terminal than the number of conduit entries.

Boxes that are listed as swimming pool junction boxes will comply with the other requirements. The boxes or enclosures must meet the size requirements of Articles 312 or 314, as applicable. Boxes and enclosures must be located so that the inside bottom of the box or enclosure is at least 4 in. (100 mm) above the ground level or pool deck surrounding the pool, or at least 8 in. (200 mm) above the **maximum water level**, whichever is higher. They must also be located at least 4 ft (1.2 m) from the inside wall of the pool unless they are separated from the pool by a solid barrier, such as a wall or a fence. In addition to these requirements, which apply to both junction boxes and other enclosures, boxes and enclosures must be made of copper, brass, suitable plastic, or other approved corrosion-resistant material, and conduit entries to enclosures must be sealed against air circulation. Boxes used with low-voltage lighting (operating at the defined low voltage contact limit or less) may be flush mounted in the pool deck if they are filled with a potting compound to exclude moisture and are located at least 4 ft (1.2 m) from the inside wall of the pool.

7. Check forming shells and wiring methods for underwater audio equipment.

Wiring methods and connections to forming shells of underwater audio equipment (speakers) must meet the same requirements as for wet-niche luminaires, and the wiring methods must extend to a listed junction box or enclosure. (See Items 5 and 6 of this checklist and Table 10-1.) However, where underwater speakers are used, forming shells for the speakers must also be installed in a recess in the pool wall or floor. In addition, each speaker and forming shell must be covered by a grounded screen of brass or other approved corrosion-resistant metal.

TABLE 10-1 ▸ **Wiring Methods for Swimming Pools**

Application	All Wiring Methods Installed per Applicable Chapter 3 Article and Flexible Cords Installed per Article 400	
	Acceptable Wiring Methods Outside Building or Structures	Acceptable Wiring Methods Inside Building or Structures
Branch circuit supplying swimming pool pump motor. All wiring methods must include an insulated equipment grounding conductor [680.21].	IMC, RMC, PVC, RTRC. EMT on buildings. Type MC cable listed for the location. LFMC or LFNC for flexible connections. Flexible cord: Up to 3 ft (1 m) for cord-and-plug connections. Listed, double-insulated pool motors: Any wiring method of Chapter 3 that is suitable for the location.	IMC, RMC, PVC, RTRC, EMT. Type MC cable. LFMC or LFNC for flexible connections. Flexible cord: Up to 3 ft (1 m) for cord-and-plug connections. Listed, double-insulated pool motors: Any wiring method of Chapter 3 that is suitable for the location. One- and two-family dwellings Any Chapter 3 wiring method cable assemblies may use an uninsulated equipment grounding conductor that is within the overall cable jacket.
Branch circuit from overcurrent protection device (OCPD) to deck/junction box supplying wet-niche and no-niche luminaires or to the wiring compartment of dry-niche luminaires installed in swimming pools and site-built spas and hot tubs.	IMC, RMC, PVC, RTRC, LFNC-B. EMT on or within buildings. ENT, MC cable, or AC cable within buildings. LFMC up to 6 ft (1.8 m) in any one length, 10 ft (3.0 m) total cumulative length.	IMC, RMC, PVC, RTRC, LFNC-B. EMT on or within buildings. ENT, MC cable, or AC cable within buildings. LFMC up to 6 ft (1.8 m) in any one length, 10 ft (3.0 m) total cumulative length.
Branch-circuit conductors or cord of underwater luminaires from deck/junction box to forming shell of wet-niche luminaires or mounting bracket of no-niche luminaires installed in swimming pools and site-built spas and hot tubs.	Red brass or equivalent corrosion-resistant RMC. LFNC, PVC, RTRC with 8 AWG solid bonding conductor.	Red brass or equivalent corrosion-resistant RMC. LFNC, PVC, RTRC with 8 AWG solid bonding conductor.
Branch circuit from OCPD to swimming pool and outdoor spa and hot tub equipment other than filter pump motors and underwater luminaires.	Any *NEC* Chapter 3 wiring method suitable for the location. Flexible cord in accordance with Article 400.	Any *NEC* Chapter 3 wiring method suitable for the location. Flexible cord in accordance with Article 400.
Underwater speakers—audio circuit from deck/junction box to speaker forming shell.	Red brass or equivalent corrosion-resistant RMC. LFNC-B, PVC, RTRC with 8 AWG solid bonding conductor.	Red brass or equivalent corrosion-resistant RMC. LFNC-B, PVC, RTRC with 8 AWG solid bonding conductor.
Feeder to panelboard supplying branch circuits for swimming pool equipment.	IMC, RMC, PVC, RTRC, LFNC-B. EMT on or within buildings. LFMC up to 6 to 10 ft (1.8 to 3.0 m) cumulative length. Existing installations: FMC or cable with wire-type equipment grounding conductor within outer sheath.	IMC, RMC, PVC, RTRC, LFNC-B, ENT. EMT on or within buildings. LFMC up to 6 to 10 ft (1.8 to 3.0 m) cumulative length. Existing installations: FMC or cable with wire-type equipment grounding conductor within outer sheath.
Branch circuit from OCPD to listed outdoor self-contained or packaged spa or hot tub with underwater luminaires.	IMC, RMC, PVC, RTRC. EMT on or within buildings. LFNC or LFMC up to 6 ft (1.8 m) for flexible connections. Up to 15 ft (4.6 m) of GFCI-protected flexible cord.	IMC, RMC, PVC, RTRC, ENT. EMT on or within buildings. LFNC-B.
Branch circuit from OCPD to listed outdoor self-contained or packaged spa or hot tub without underwater luminaires.	IMC, RMC, PVC, RTRC, EMT. MC cable listed for the location. EMT installed on or within buildings. LFNC or LFMC for flexible connections to motors. Up to 15 ft (4.5 m) of GFCI-protected flexible cord. For listed double-insulated pump motors: Any Chapter 3 method suitable for the location.	IMC, RMC, PVC, RTRC, EMT, MC. Interior of one-family dwellings: Any *NEC* Chapter 3 wiring method suitable for the location—raceways must have insulated equipment grounding conductor, cables must have wire-type equipment grounding conductor within sheath. For listed double-insulated pump motors: Any Chapter 3 method suitable for the location.
Branch circuit from OCPD to hydromassage bathtubs.	Any *NEC* Chapter 3 wiring method suitable for the location—if applicable (not usually installed outdoors).	Any *NEC* Chapter 3 wiring method suitable for the location. Flexible cord in accordance with Article 400.

8. Verify that metal parts of pools and other nearby electrical equipment and metal parts are connected together to create equipotential bonding in the pool area, using appropriate methods.

Metal parts associated with or in the vicinity of a swimming pool must be bonded together. This includes conductive pool shells, bonding systems for perimeter surfaces, all metallic components of the pool structure, metallic forming shells, and all metal parts of electrical equipment or wiring methods. The pool water must also be in contact with bonded metal parts of the swimming pool, or, if there are no such parts in contact with pool water, a conductive corrosion-resistant surface that provides at least a 9 in² (5800 mm²) contact area with the pool water must be installed. Small, isolated metal parts, such as rope hooks, that are no larger than 4 in. (100 mm) in any dimension and do not penetrate the pool structure by more than 1 in. (25 mm) are excluded from the bonding requirement.

Bonding in this case is not necessarily a grounding requirement and is not intended to provide a path for ground-fault current. No grounding conductor is required to extend from the bonding grid to a panel or other grounding point. However, in most cases, one or more items that must be connected to the equipotential bonding grid will also be connected to an equipment grounding conductor. The bonding at swimming pools applies to both electrical and nonelectrical metallic equipment and is intended to eliminate voltage differences between different parts or areas of a pool.

The equipotential bonding grid for a conductive pool shell may be the steel reinforcing of the concrete pool structure or the wall of a bolted or welded metal pool, or the grid may be constructed with 8 AWG or larger solid copper conductors bonded together at no more than 12-in. (300-mm) intervals. A tolerance of up to 4 in. (100 mm) is permitted in the 12-in. (300-mm) intervals. Therefore, in one place they could be 16 in. (400 mm) apart, and in another place they would be as little as 8 in. (200 mm) apart. This tolerance is necessary to accommodate obstacles such as forming shells or structural items. A means of bonding the perimeter surfaces surrounding a pool must extend under the paved or unpaved surfaces for at least 3 ft (1 m) horizontally from the inside walls of the pool. Where the surface abutting the pool is less than 3 ft (1 m) from the pool to a building or wall that is at least 5 ft (1.5 m) high, the bonding grid must extend only to the building or wall. The field-constructed bonding grid consisting of 8 AWG copper conductors is required to cover the contour of the pool.

However, where the steel reinforcing of a pool does not exist, is encapsulated, or does not extend under the perimeter surfaces, or where the pool shell is nonconductive, alternative methods are permitted for the perimeter surfaces. In such cases, a single 8 AWG solid copper conductor (rather than a grid) may be used if it follows the contour of the pool and is 18 to 24 in. (450 to 600 mm) from the inside walls of the pool and secured 4 to 6 in. (100 to 150 mm) below the subgrade. Connections to the grid or perimeter bonding conductor may be made by exothermic welding (**FIGURE 10-3**) or by stainless steel, copper, brass, or copper alloy pressure connectors or clamps that are labeled as suitable for the purpose, meaning those that meet the requirements of 250.8 for grounding connections. The harsh environment in the vicinity of the pool, due to chemically treated water, causes rapid deterioration of connections if made with devices that are not identified for this application.

Except as noted, most metal parts are required to be bonded together. Metal parts included are metal structural parts and uncoated reinforcing bars; forming shells and mounting brackets; metal parts of fences, awnings, and doors or window frames; metal parts of associated electrical equipment; and metal-sheathed cables and raceways within 5 ft (1.5 m) horizontally or 12 ft (3.7 m) vertically from the pool, the water surface, or any towers, platforms, or other structures that

FIGURE 10-3   Permanent bonding connections to pool structure.

are not separated by permanent barriers. Double-insulated electrical equipment, listed nonmetallic forming shells for low-voltage lighting, and rebar coated with insulating compounds are not required to be bonded. In fact, the internal non-current-carrying metal parts of double-insulated equipment are prohibited from being bonded, even though such parts may be grounded according to 680.21(B). Where double-insulated water pump motors are used, a bonding means must be provided for replacement motors that may not be double-insulated. Where there is no other connection between the equipment grounding system and the bonding grid, this bonding means for replacement motors must be connected to the equipment grounding conductor of the motor circuit, which will indirectly bond the internal parts. The prohibition bars only direct bonding connections, which prevents an installer from compromising the system of double insulation.

Note that some of the equipment that is required to be bonded to the bonding grid is also required to be grounded. Although the bonding requirements apply to all types of metal equipment, the grounding requirement applies only to electrical equipment such as pool motors that could become energized due to insulation failure. However, since some of the parts that are required to be bonded are also required to be grounded, most pool parts are also directly or indirectly connected to the electrical grounding system.

FIGURE 10-4A illustrates acceptable arrangements for bonding electrical and nonelectrical equipment at a swimming pool. Through the required connection to electrical equipment, such as the pump motor and wet-niche luminaire, the bonding conductor is connected to the electrical grounding system through the metal motor housing or metal forming shell for the underwater luminaire.

For pools with nonconductive shells, the example in **FIGURE 10-4B** applies, except that since there would be no connection to the shell itself, there would have to be bonding connections to any metallic elements, such as a ladder, and at least one bonding connection at least 9 in.² (5800 mm²) in area to the pool water must be established in accordance with 680.26(C). This connection could be made with other metal parts that are bonded and in contact with the pool water, such as forming shells and pockets for ladders. The single 8 AWG copper conductor for perimeter surfaces may also be used in lieu of the grid as shown in Figure 10-4B.

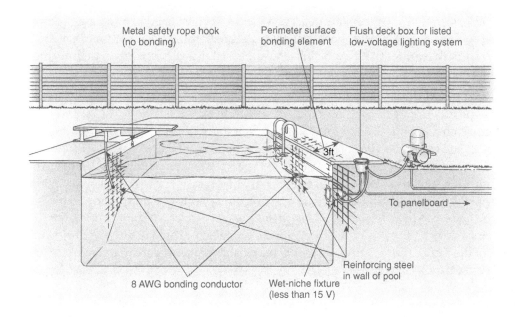

FIGURE 10-4A   Required bonding connections at a poured-concrete pool. (Source: *National Electrical Code® Handbook*, NFPA, Quincy, MA, 2008, Exhibit 680.12)

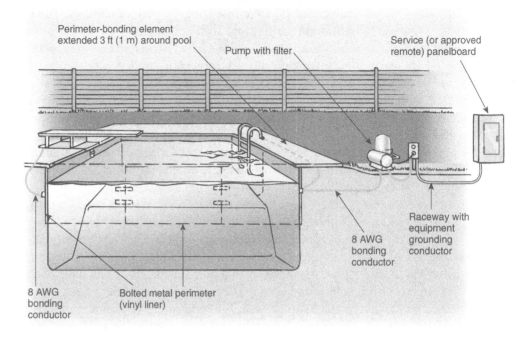

FIGURE 10-4B   Required bonding connections at a pool with steel or aluminum walls.

Checklist 10-2: Intermediate and Final Inspections

1. Check pool equipment for suitability for approval.

All equipment installed in the water, walls, or decks of pools must comply with Article 680 and be suitable for the purpose. Generally, equipment listed or identified for the purpose and used accordingly should be approved. In some cases, the *Code* specifically requires listed equipment, and in some cases the *Code* requires listing for the specific application, such as swimming pool junction boxes or deck boxes. Because most inspectors are not equipped to test equipment for *Code* compliance, listing may be the only reasonable way for an inspector to determine that equipment meets the requirements for approval under the *NEC*. Section 110.3(A) provides a list of issues to be considered in evaluating equipment for approval when equipment is not available as a listed product. For products that are listed, 110.3(B) requires that they be installed and used in accordance with their listing and labeling instructions.

2. Check transformers, power supplies, and GFCIs for identification and suitability for the purpose.

Where a transformer is used to supply underwater luminaires, the transformer and its enclosure are required to be listed for swimming pool and spa use. The power supply or transformer must include either an isolation-type transformer with grounded metal barriers (often referred to as *shields*) between primary and secondary windings or an approved system of double insulation. The transformer secondary must be ungrounded. GFCIs may be furnished in a number of forms, including receptacles, circuit breakers, or self-contained devices. Generally, the inspector should look for listed devices regardless of the form of the device, since listing is the only reasonable way to ensure that the device meets the standard for a Class A device as specified in the definition of a GFCI. See Article 100 and the Key Terms section. GFCIs are not required for luminaires that operate at or below the low voltage contact limit. The installation must assure that there will be no shock hazard during relamping. See the definition of **Low Voltage Contact Limit** in the Key Terms section of this chapter.

3. Verify that conductors on the load side of transformers or GFCIs are separated from conductors not protected by GFCIs.

GFCI-protected conductors (and conductors on the secondary side of isolating transformers or power supplies) may not share spaces with other conductors that are not protected by GFCIs unless the other conductors are grounding conductors. This rule does not apply to conductors in panelboards and the supply conductors of feed-through GFCIs, where some conductors are protected by GFCI circuit breakers and others are not. GFCI-protected conductors are not always (in other parts of the *Code*) required to be separated from other conductors. This rule is specific to swimming pool-type installations and is intended to increase safety in the wet and somewhat corrosive pool environment by physically isolating GFCI-protected circuits from contact with circuits that are not GFCI protected.

4. Verify that general-use receptacles are not located within 6 ft (1.8 m) of pool walls and that all receptacles within 20 ft (6.0 m) of pool walls are GFCI protected.

Receptacles must be at least 6 ft (1.8 m) from the inside walls of a pool. All 125-volt receptacles within 20 ft (6.1 m) of the inside walls must be GFCI protected, regardless of their purpose. This does not include receptacles that are separated from the pool by an effective barrier, such as a wall or door (**FIGURE 10-5**). The measurement is based on the shortest path a cord would follow without piercing a permanent barrier. The distance measurement does not apply to 120-volt through 240-volt receptacles that supply pool pump motors, all of which must be GFCI protected.

5. Verify that any receptacle installed between 6 and 10 ft (1.8 and 3.0 m) of pool walls that is used for pool equipment is a single, grounding-type device and is GFCI protected.

No receptacles may be located less than 6 ft (1.8 m) from the inside walls of a pool. Receptacles for loads directly related to pool circulation and sanitation may be located between 6 and 10 ft (1.8 and 3.0 m) from pool walls. All such receptacles must be GFCI protected and must be single, grounding-type devices. Where pools are permanently installed at dwelling units, at least one 125-volt, 15- or 20-ampere, GFCI-protected receptacle is required to be installed not less than 6 ft (1.8 m) but not more than 20 ft (6.0 m) from the inside walls of the pool as explained in Item 6. For any occupancy type, all 125-volt receptacles within 20 ft (6.0 m) are required to be GFCI protected. In all occupancies, pool pump motors supplied by 15- or 20-ampere, 125- or 240-volt, single-phase branch circuits must be GFCI protected. Measurements are made as in Item 4 of this checklist. **FIGURE 10-6** illustrates the requirements for receptacle outlets located in the vicinity of a swimming pool.

6. Verify that at least one GFCI receptacle on a general-purpose branch circuit is installed between 6 and 20 ft (1.8 and 6.1 m) of pool walls at dwelling units.

At least one general-use receptacle (125-volt, 15- or 20-ampere) on a general-purpose branch circuit is required between 6 and 20 ft (1.8 m and 6.0 m) of the inside walls of a pool. Because the receptacle is within 20 ft (6.0 m) of the pool, it must be GFCI protected. This requirement ensures that a convenient GFCI-protected receptacle is provided, so that users will be less likely to run extension cords to receptacles that are not GFCI protected. Also, because most appliances are supplied with cords that are about 6 ft (1.8 m) or less in length, an appliance that is plugged into the required receptacle will be kept a safe distance from the pool.

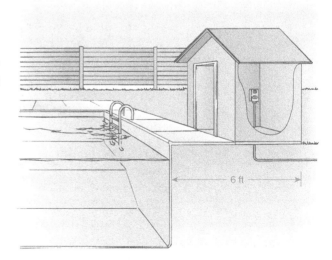

FIGURE 10-5   Receptacle in structure is considered separated from the swimming pool. (Source: *National Electrical Code® Handbook*, NFPA, Quincy, MA, 2008, Exhibit 680.5.)

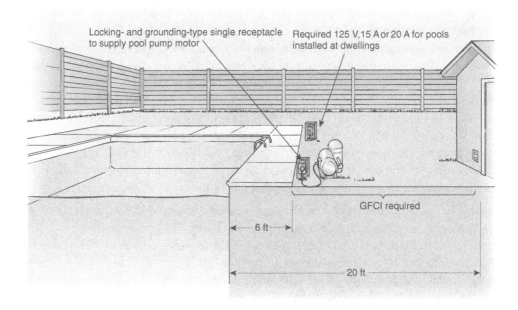

Locking- and grounding-type single receptacle to supply pool pump motor

Required 125 V, 15 A or 20 A for pools installed at dwellings

GFCI required

6 ft

20 ft

7. Verify GFCI protection of outlets for pool circulating pump motors that are supplied by single-phase, 15- or 20-ampere, 120- through 240-volt branch circuits.

Regardless of whether the motor is cord-and-plug connected or direct (hard) wired, swimming pool pump motors supplied from single-phase, 15- or 20-ampere, 120- through 240-volt branch circuits are required to be GFCI protected. The GFCI protection can be provided at the outlet installed at the pump location or can be provided elsewhere in the branch circuit.

8. Verify that luminaires and ceiling fans are located so that required clearances are maintained.

Luminaires and ceiling-suspended fans in new outdoor installations must be 12 ft (3.7 m) or more above the maximum water level within the area that extends 5 ft (1.5 m) horizontally from the pool walls. Luminaires, paddle fans, and lighting outlets that are placed more than 5 ft (1.5 m) and less than 10 ft (3.0 m) horizontally from pool walls must be GFCI protected unless they are more than 5 ft (1.5 m) above the maximum water level and rigidly attached to the structure. Existing luminaires on existing structures within 5 ft (1.5 m) of pool walls are permitted if they are rigidly attached to the structure, are at least 5 ft (1.5 m) above the water level, and are GFCI protected. Luminaires and paddle fans in indoor pool areas may be less than 12 ft (3.7 m) vertically or 5 ft (1.5 m) horizontally from the pool water if they are totally enclosed, GFCI protected, and at least 7½ ft (2.3 m) above the maximum water level. **FIGURE 10-7** shows the areas in which luminaires and ceiling-suspended paddle fans are allowed. The figure also shows the areas in which GFCI protection is required for the branch circuit supplying the luminaire(s).

9. Verify that luminaires and ceiling fans are GFCI protected where GFCI protection is required.

As mentioned in Item 7 of this checklist, GFCI protection is required on luminaires and fans that are within 12 ft (3.7 m) vertically and 5 ft (1.5 m) horizontally from the pool water, including existing installations. Lighting between 5 and 10 ft (1.5 and 3.0 m) horizontally and less than 5 ft (1.5 m) above the water level must also be GFCI protected.

10. Verify that switches are located at least 5 ft (1.5 m) from pool walls or separated from the pool by a permanent barrier.

Switching devices must be separated from the pool by solid and permanent barriers such as fences or walls or be at least 5 ft (1.5 m) horizontally from the pool walls. The solid

Outdoor Pools

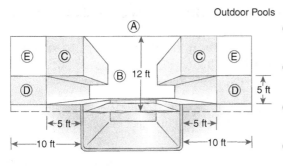

Ⓐ Luminaires, lighting outlets, and ceiling-suspended (paddle) fans permitted above 12 ft.

Ⓑ Luminaires, lighting outlets, and ceiling-suspended (paddle) fans not permitted below 12 ft.

Ⓒ Existing luminaires and lighting outlets permitted in this space if rigidly attached to existing structure (GFCI required).

Ⓓ Luminaires and lighting outlets permitted if protected by a GFCI.

Ⓔ Luminaires and lighting outlets permitted if rigidly attached.

FIGURE 10-7 Clearances for luminaires and paddle fans installed above and adjacent to a swimming pool. (Source: *National Electrical Code®* *Handbook*, NFPA, Quincy, MA, 2014, Exhibit 680.4.)

Indoor Pools

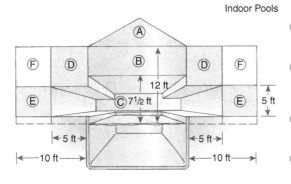

Ⓐ Luminaires, lighting outlets, and ceiling-suspended (paddle) fans permitted above 12 ft.

Ⓑ Totally enclosed luminaires protected by a GFCI and ceiling-suspended (paddle) fans protected by a GFCI permitted above 7 1/2 ft.

Ⓒ Luminaires, lighting outlets, and ceiling-suspended (paddle) fans not permitted below 5 ft.

Ⓓ Existing luminaires and lighting outlets permitted in this space if rigidly attached to existing structure (GFCI required).

Ⓔ Luminaires and lighting outlets permitted if protected by a GFCI.

Ⓕ Luminaires and lighting outlets permitted if rigidly attached.

permanent barrier is intended to prevent people from reaching through a barrier such as a fence to operate the switching device, so the barrier should be tall enough to fulfill this purpose. Other types of switching devices may be used within the 5-ft (1.5-m) boundary only if they are listed for the use. Typically, such switches are nonelectrical or operate below the low voltage contact limit, so that they do not present a shock hazard.

11. Verify that other outlets are at least 10 ft (3.0 m) from the pool.

Outlets for things other than pool equipment, receptacles, luminaires, and paddle fans may not be installed closer than 10 ft (3.0 m) from the inside walls of the pool. These could include outlets for space heaters or other appliances that are not connected by cord and plug, or for remote-control, signaling, fire alarm, and communications circuits. The measurement from the pool wall is the same as for outlets covered in Item 4 of this checklist.

12. Check flexible cords (where used) for compliance with equipment grounding conductor requirements and length limitations.

Flexible cords may be used for easy removal of fixed or stationary equipment for repairs or maintenance. The cords may not be longer than 3 ft (900 mm), must be equipped with a grounding-type plug, and must include a 12 AWG or larger copper equipment grounding conductor that otherwise complies with 250.122. This rule does not apply to wet-niche or no-niche luminaires supplied by flexible cords or cables, which are permitted to use longer cords with smaller equipment grounding conductors (see Item 18 in this checklist and the discussion in Item 4 in the previous checklist, Initial Inspection: Prior to Pouring of Concrete or Burial, which requires a bonding jumper for a forming shell in addition to the equipment grounding conductor for the luminaire).

13. Check underwater luminaires for locations, wiring methods, and connections to wiring methods.

See the discussion in Item 4 in the previous checklist, Initial Inspection: Prior to Pouring of Concrete or Burial. Some aspects of underwater lighting installations cannot be checked until after concrete is poured or some finishes are completed.

14. Check pool-related equipment for appropriate wiring methods.

See the discussion of wiring methods in Item 5 in the previous checklist, Initial Inspection: Prior to Pouring of Concrete or Burial.

15. Check junction boxes and enclosures connecting to underwater luminaires for listing and labeling, appropriate size and location, and proper materials.

See the discussion in Item 6 in the previous checklist, Initial Inspection: Prior to Pouring of Concrete or Burial.

16. Check forming shells and wiring methods for underwater audio equipment.

See the discussion in Item 7 in the previous checklist, Initial Inspection: Prior to Pouring of Concrete or Burial.

17. Verify that metal parts of pools and nearby equipment and metal parts are bonded to an appropriate equipotential bonding grid, using appropriate methods.

See the discussion in Item 8 in the previous checklist, Initial Inspection: Prior to Pouring of Concrete or Burial. In order to verify the required attachments to a bonding grid, most aspects of bonding can and must be inspected prior to burial or pouring of concrete. However, some equipment that is required to be bonded, such as pool motors, stairs, and platforms, may not be complete until the final inspection, so bonding must also be inspected at intermediate and final inspections.

18. Verify that all equipment required to be grounded is grounded by insulated copper equipment grounding conductors and/or bonding jumpers of the proper size.

As noted in Item 5 of this checklist, the grounding requirements for pools also dictate the permitted wiring methods. In each case, even where metallic conduits are used, an insulated copper conductor is required for grounding/bonding electrical equipment. The equipment grounding conductor is permitted to be uninsulated only in cable assemblies inside dwelling units. The size of the equipment grounding conductor is required to be based on the rating of the circuit, which is the rating of the overcurrent device. The usual rules of Article 250 are used in general, but in some cases, such as for cord-and-plug–connected pool motors, a minimum of 12 AWG is required. However, the equipment grounding conductor is permitted to be as small as 16 AWG for a wet-niche or no-niche luminaire that is supplied by a flexible cord or cable. (If a nonmetallic conduit is used to connect a wet-niche forming shell, the forming shell is still required to be bonded and grounded by an 8 AWG or larger insulated copper bonding jumper, and the forming shell is required to be bonded to the bonding grid by an 8 AWG or larger solid copper wire, even though the luminaire itself may be grounded by a 16 AWG copper equipment grounding conductor in a flexible cord.)

Grounding connections to underwater wet-niche luminaires and audio equipment should be checked for the use of a listed potting compound to protect the connections from corrosion caused by contact with pool water. Such grounding connections are often incomplete at the first inspection.

19. Verify that a disconnecting means is provided, accessible, located within sight of pool equipment, and not located within 5 ft (1.5 m) of pool walls.

This disconnecting means is required to be within sight of swimming pool, hot tub, and spa equipment. It is required to disconnect all ungrounded conductors of the circuit simultaneously. This enables the person servicing equipment other than pool lighting to have access to a local disconnecting means (**FIGURE 10-8**). One or more disconnects may be required. Maintenance disconnecting means are not required for pool lighting. The prohibition regarding switch locations within 5 ft (1.5 m) of a pool is a rule that applies generally to switching devices including those used as disconnecting means, but it is specific to swimming pools. A switching device may be closer than 5 ft (1.5 m) from a pool if it is separated from the pool by a permanent barrier and the shortest path to reach the disconnect is at least 5 ft (1.5 m). Section 680.12 applies to all equipment covered by Article 680 except hydromassage bathtubs. Disconnecting means may be provided in some form other than a switching device.

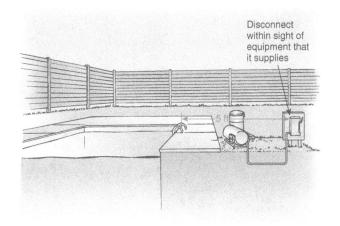

FIGURE 10-8   Disconnect for pool, spa, or hot tub equipment that is within sight of equipment supplied and not less than 5 ft (1.5 m) from the inside walls of the pool, spa, or hot tub. (Source: *National Electrical Code® Handbook*, NFPA, Quincy, MA, 2011, Exhibit 680.2.)

20. Check electric pool heaters for subdivision of heating elements and sizing of branch-circuit conductors.

Like most resistance space-heating equipment covered by Article 424, resistance heating elements for pool heaters must be broken up into groups so that no group presents a load of more than 48 amperes and no group is protected at over 60 amperes. The nameplate loads are treated as continuous loads for sizing branch-circuit conductors and overcurrent devices; that is, the minimum ampacity and rating must be at least 125 percent of the nameplate current rating of the pool heater.

21. Check equipment rooms or pits for adequate drainage.

Electric pool equipment must be installed in locations that have adequate drainage. Such equipment may not be installed where water will accumulate during normal operation or filter maintenance. Abnormal conditions such as flooding are not covered by this rule.

22. Check electrically operated pool covers for proper motor and controller location, motor enclosure, and GFCI protection.

See the definition of **Pool Cover, Electrically Operated** in the Key Terms section of this chapter. Motors, controllers, and wiring for electrically operated pool covers must be located at least 5 ft (1.5 m) from the inside walls of a pool unless separated from the pool by a permanent barrier or an access cover. The motor and controller must be connected to a GFCI-protected branch circuit. Because such motors, and perhaps their controllers, are often located below grade in recesses near a pool end, such motors must be totally enclosed, and the motor, switches, or enclosures may have to be suitable for wet locations. The controller or the control device for operating the pool cover must be located in a place that gives the operator a clear view of the entire pool.

23. Check deck-area heaters for suitability and proper clearances from pool.

Article 680 covers only those heaters that are installed within 20 ft (6.0 m) of the inside walls of a pool. Unit heaters and radiant heaters must be guarded or enclosed, and they must be installed at least 5 ft (1.5 m) horizontally from the inside walls of a pool. Radiant heaters may not be located less than 12 ft (3.7 m) above the pool deck. Radiant-heating cable cannot be used in or below pool decks. The *NEC* does not address radiant heating that operates by circulating water or other fluid through tubing, as this type of nonelectrical heating is outside of the *NEC*'s scope.

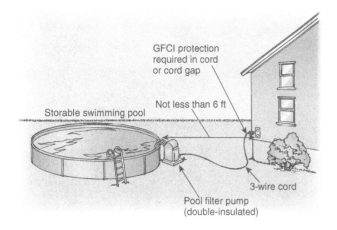

GFCI protection
required in cord
or cord gap

Not less than 6 ft

Storable swimming pool

3-wire cord

Pool filter pump
(double-insulated)

FIGURE 10-9   Storable swimming pool requirements. (Source: *National Electrical Code® Handbook*, NFPA, Quincy, MA, 2008, Exhibit 680.16, as modified.)

Checklist 10-3: Storable Swimming Pools and Storable Spas and Hot Tubs

1. Review definitions and determine the applicability of Part III of Article 680.

See the definition of <u>Storable Swimming, Wading, or Immersion Pool or Storable/Portable Spas and Hot Tubs</u> in the Key Terms section. Although many small portable pools meet the definition of *storable swimming, wading, or immersion pool*, the *NEC* does not provide any requirements for such pools unless the installation involves some electrical equipment. Many storable pools may be used without any electrical equipment, but most spas and hot tubs will include pumps and heaters. Storable pools are defined by their dimensions (**FIGURE 10-9**). The provisions of Part I of Article 680 (680.1 through 680.12) also apply to storable pools. Some of these requirements may not be applicable, such as the restrictions on underground wiring. However, others, such as overhead clearances, limitations on flexible cords, and maintenance disconnecting means likely will apply.

2. Verify that cord-connected pool pumps are double-insulated and that internal metal parts are grounded through a grounding-type attachment plug.

In addition to double insulation, internal, nonaccessible, non-current-carrying metal parts of pool filter pumps must be grounded. The system of double insulation is used in lieu of grounding any external non-current-carrying metal parts, and grounding or bonding connections to these parts are not permitted. Most double-insulated appliances are supplied without grounding conductors. However, because water, especially water laden with chemicals, may come into contact with internal parts of pumps, such parts must be grounded to eliminate any voltage difference. The grounding means must be a conductor that is part of the flexible cord, and it must terminate in a grounding-type attachment plug. A GFCI device must be part of the attachment plug or located in the cord and within 12 in. (300 mm) of the attachment plug.

3. Verify that all electrical equipment associated with the storable pool is provided with GFCI protection.

Storable pools are required to have GFCI protection for all electrical equipment associated with the pool. Also, receptacles rated 125 volts and 15 or 20 amperes that are within 20 ft (6.0 m) of a storable pool (measured along the shortest length of a cord from an appliance to the receptacle) must be GFCI protected. Receptacles separated from the pool by a permanent barrier are not counted if the cord would have to pass through an opening in that barrier in a manner that is prohibited by 400.8(2) and (3). The GFCI protection for a pool pump must be an integral part of the power supply cord as noted in Item 2 in this checklist.

4. Check luminaires for compliance with requirements based on the voltage of the luminaires.

Luminaires installed with storable pools must have no exposed metal parts, must have impact-resistant polymeric lenses and enclosures, and must be listed as an assembly for the purpose. Luminaires operating at the low voltage contact limit or less must include an isolation transformer or power supply and enclosure, and the primary of the transformer may not be over 150 volts. Luminaires may be over the low voltage contact limit but may not operate at over 150 volts. Such luminaires that operate over the low voltage contact limit must be designed to eliminate any shock hazard during normal use and must be protected by a GFCI with open neutral conductor protection that is an integral part of the assembly and is permanently connected to the lamp. Open neutral protection is also employed in portable-type GFCI devices listed in accordance with UL 943.

5. Verify that receptacles are not located within 6 ft (1.83 m) of the pool.
Receptacles may not be located within 6 ft (1.83 m) of the inside edge of a storable pool. That is, a receptacle may not be within 6 ft (1.83 m) measured along a cord from the pool to a receptacle, and the cord may not pass through a floor, ceiling, or wall, or a door, window, or similar opening. Since a storable pool may be dismantled and relocated, but a receptacle is likely to remain where installed, this rule is really about where the storable pool may be located. The rule may, in some cases, dictate where receptacles can be installed if the area available for the pool to be installed is limited.

SPAS AND HOT TUBS

Checklist 10-4: All Installations

1. Review definitions and determine the applicability of Part IV of Article 680.
See the definitions of <u>Spa or Hot Tub</u>, <u>Packaged Spa or Hot Tub Equipment Assembly</u>, and <u>Self-Contained Spa or Hot Tub</u> in the Key Terms section.

2. Check spa and hot tub equipment for suitability for approval.
All equipment installed in the water, walls, or decks of spas and hot tubs must comply with Article 680 and be suitable for the purpose. Generally, equipment that is listed or identified for the purpose should be approved. In some cases, the *Code* specifically requires listed equipment or permits certain installations only with listed equipment. For example, many of the rules that apply to packaged spas and hot tubs apply only to listed units. Often, listing is the only reasonable way for an inspector to determine that equipment meets the requirements for approval under the *NEC*, because most inspectors are not equipped to test equipment for *Code* compliance. However, many spas and hot tubs are assembled from parts, many of which are listed and all of which must comply with the applicable provisions of Article 680. For spas and hot tubs, the applicable provisions are found in Parts I and IV, and, for wiring methods and installation details, in Part II. Section 110.3(A) provides a list of issues to be considered in evaluating equipment for approval if the equipment is not listed.

3. Review the checklist for permanent pools for compliance with the applicable provisions of Parts I and II of Article 680 (modified for indoor installations).
Generally, outdoor spas and hot tubs are required to meet the same provisions as are swimming pools. Some modifications of the rules apply to outdoor spas and hot tubs, and other modifications apply to indoor spas and hot tubs. For example, equipotential bonding of perimeter surfaces is not required for some listed self-contained spas installed outdoors if certain other requirements are met. Also, the metal hoops used with wooden hot tubs are not required to be bonded, and bonding can be accomplished by mounting equipment on a common frame or base. Special rules are provided for flexible connections to outdoor spa and hot tub equipment. For example, flexible cords with GFCI protection are allowed to be up to 15 ft (4.6 m) long to connect some listed packaged units.

Some of the rules for swimming pools are modified by Part IV for indoor spa and hot tub installations. The modified rules cover aspects of receptacle installation, mounting of luminaires and fans, locations of wall switches and other electrical devices, bonding and grounding of metal parts, electric water heaters, and underwater audio equipment. See the next checklist, Indoor Installations Only, as it pertains to indoor spas and hot tubs for more information.

Interior wiring that supplies an outdoor self-contained or packaged spa or hot tub at a single-family dwelling unit is permitted according to essentially the same rules that apply to interior wiring for swimming pools in dwellings. All wiring methods are subject to the restrictions in the article of *NEC* Chapter 3 that covers that wiring method. For example, Type NM (nonmetallic-sheathed) cable cannot be installed in wet locations and is typically restricted to interior wiring.

4. Check outlets supplying a self-contained spa or hot tub or a packaged spa or hot tub assembly for integral or separately provided GFCI protection.

GFCI protection must be provided for an outlet that supplies a spa or hot tub. However, the outlet is not required to be protected if a self-contained or packaged equipment assembly is listed and provided with integral GFCI protection that protects all electrical parts that are included in the assembly. Larger field-assembled spas and hot tubs, those rated 3-phase, over 250 volts, or having heater loads over 50 amperes, are not required to have GFCI protection in the supply, primarily because GFCI devices are not generally available above the ratings noted.

5. Verify compliance with requirements for disconnecting means.

Disconnecting means must comply with the requirements for swimming pools. Therefore, maintenance disconnects must be readily accessible, within sight of the spa or hot tub, and not less than 5 ft (1.5 m) from the spa or hot tub.

6. Verify that spas or hot tubs in other than single-family dwellings are provided with an emergency shutoff or control switch, as required.

In other than single-family dwellings, a means must be provided for stopping the circulation and jet pump motor(s) in an emergency. The emergency shutoff is required only to stop the pump motor(s) and is not required to meet the requirements of the disconnecting means described in Item 5. The emergency shutoff must be located where it is in sight and readily accessible to the users and at least 5 ft (1.5 m) from the spa or hot tub. (The disconnecting means described in Item 5 is required to be readily accessible to the operators, owners, or management, but not necessarily the users, in other than single-family dwellings.) This requirement can be met through the use of a remote-control device that operates the switch or circuit breaker supplying the spa or hot tub.

FIGURE 10-10 shows the permitted location for the emergency control of a hot tub or spa. As noted, the disconnect and emergency shutoff have different functions and are usually different devices, although in some cases, both functions could be combined into one

FIGURE 10-10 Location of the emergency control device for spas and hot tubs. (Source: *National Electrical Code®* *Handbook*, NFPA, Quincy, MA, 2014, Exhibit 680.13.)

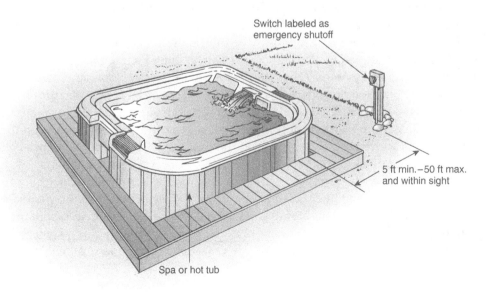

Switch labeled as emergency shutoff

5 ft min.–50 ft max. and within sight

Spa or hot tub

device. For some spas and hot tubs, the required disconnecting means for servicing could be located out of sight of the spa or hot tub location.

7. Check electric pool heaters for subdivision of heating elements and sizing of branch-circuit conductors.

Like most resistance space-heating equipment covered by Article 424, resistance heating elements must be broken up into groups so that no group presents a load of more than 48 amperes and no group is protected at over 60 amperes. The nameplate loads are treated as continuous loads for the purpose of sizing branch-circuit conductors and overcurrent devices; that is, the minimum ampacity and rating must be at least 125 percent of the nameplate current rating of the heater. This is a general requirement for pools in Part I that is not modified for spas and hot tubs.

Checklist 10-5: Indoor Installations Only

1. Verify that a suitable wiring method of Chapter 3 of the *NEC* is used to supply and connect spas and hot tubs, unless cord-and-plug connections are permitted.

The restrictions on wiring methods that apply to swimming pools are not imposed on indoor spas and hot tub installations. Wiring methods must be suitable for the location and conditions of use. Where needed for easy removal or disconnection of equipment for repair or maintenance, flexible cords may be used for connections to listed packaged spa or hot tub equipment rated 20 amperes or less. Equipotential bonding requirements for perimeter surfaces (deck areas) of pools do not apply to listed self-contained spas or hot tubs that are installed above a finished floor.

2. Verify that at least one GFCI-protected receptacle on a general-purpose branch circuit is located between 6 and 10 ft (1.83 and 3.0 m) of the walls of the spa or hot tub.

Receptacles may be within 10 ft (3.0 m) of an indoor hot tub or spa but must be at least 6 ft (1.83 m) away. All 125-volt receptacles rated 30 amperes or less within 10 ft (3.0 m) must be GFCI protected, and one such receptacle rated 15 or 20 amperes is required between 6 and 10 ft (1.83 and 3.0 m) from the hot tub or spa. Measurements are based on the shortest path a cord could take without passing through a permanent barrier. This rule makes a receptacle convenient to the spa or hot tub, but far enough from the spa or hot tub so that most standard appliance cords will not reach the water, which helps to ensure that portable electric equipment brought into the area will be GFCI protected.

3. Verify that any receptacles used to supply power to a spa or hot tub are GFCI protected.

All receptacles that provide power for a spa or hot tub are required to be GFCI protected. This is not a general requirement for all outlets, only for receptacles. For example, GFCI protection is not necessarily required for "hardwired" water heaters. For outlets supplying self-contained or packaged equipment, see Item 4 of the checklist titled Spas and Hot Tubs: All Installations.

4. Verify that luminaires and paddle fans are spaced as required from spa or hot tub walls and above maximum water level and that GFCI protection is provided, as required.

The rules for indoor spas and hot tubs differ from the requirements for indoor swimming pools with regard to luminaires and paddle fans. Luminaires and fans are permitted less than 12 ft (3.7 m) above an indoor spa or hot tub, but if they are less than 12 ft (3.7 m) above an indoor spa or hot tub, they are required to be GFCI protected. Generally, luminaires and ceiling-suspended fans must be at least 7½ ft (2.3 m) above the water level. However, surface or recessed luminaires that have glass or plastic lenses or globes, have enclosures that are nonmetallic or isolated from contact, and have GFCI protection are permitted to be less than 7½ ft (2.3 m) above water level. This will accommodate installations of spas and hot tubs where ceiling heights are only 8 ft (2.4 m) or so above the floor. Recessed luminaires are permitted to have trims with metal parts if the metal parts are electrically isolated.

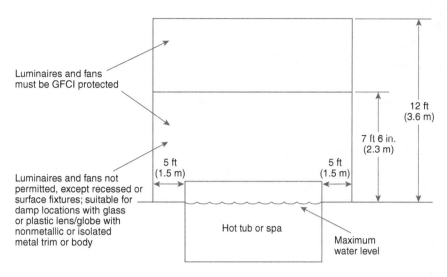

Luminaires and fans must be GFCI protected

Luminaires and fans not permitted, except recessed or surface fixtures; suitable for damp locations with glass or plastic lens/globe with nonmetallic or isolated metal trim or body

12 ft (3.6 m)

7 ft 6 in. (2.3 m)

5 ft (1.5 m)   5 ft (1.5 m)

Hot tub or spa

Maximum water level

FIGURE 10-11   Clearance requirements for luminaires and paddle fans installed above and adjacent to spas and hot tubs.

FIGURE 10-11 illustrates the locations in which luminaires of various types and ceiling fans can be installed relative to the location of an indoor spa or hot tub.

5. Verify that switches are located at least 5 ft (1.5 m) from the inside walls of the spa or hot tub.

Unlike the similar rule for swimming pools, this rule does not contain any provision for switches separated from the spa or tub by permanent barriers, and this rule modifies the requirements from Part II. However, nonelectrical controls, such as pneumatic devices, may be located within reach of a spa or hot tub. This rule is concerned with switches that are part of the premises wiring. Switching devices that are an integral part of listed spa and hot tub units are not covered by this rule.

6. Verify that all parts that are required to be grounded or bonded are grounded or bonded using appropriate methods.

The rules for swimming pools (and outdoor spas and hot tubs) are modified and somewhat less stringent for indoor spas and hot tubs. The distance within which items are required to be bonded to an indoor tub or spa is reduced to 5 ft (1.5 m) in all directions, from the 12 ft (3.7 m) vertical and 5 ft (1.5 m) horizontal rule that applies to swimming pools. The excepted metallic items are not based strictly on dimensions. Small conductive parts that are not likely to become energized such as towel bars, drain fittings, and water jets are not required to be bonded. As with swimming pools, all metal parts of electric equipment and fittings associated with a spa or hot tub must be bonded unless they are part of a listed self-contained spa or hot tub. However, methods of bonding and grounding also are modified. For example, a number of items of electrical equipment may be bonded by mounting them to a common metal frame or base with direct metal-to-metal connections. Other bonding methods include interconnections of threaded piping that is not specifically required to be brass or similarly corrosion-resistant and 8 AWG or larger solid copper bonding jumpers. The requirements for equipment grounding conductors that are part of or installed with a wiring method are not modified by 680.43, so the requirements of Part II apply. Perimeter surface bonding, like that required for a swimming pool, is not required for listed self-contained spas and hot tubs installed indoors above a finished floor.

7. Check underwater audio equipment for compliance with Part II.

See the discussion in Item 7 of the checklist titled Initial Inspection: Prior to Pouring of Concrete or Burial. This discussion is based on Part II, which covers swimming pools and related installations in general. Part II includes restrictions on wiring methods for underwater audio equipment that also apply to spas and hot tubs.

Checklist 10-6: Fountains

1. Review definitions and determine the applicability of Part V of Article 680.

See the definitions of **Fountain** and **Permanently Installed Decorative Fountains and Reflection Pools** in the Key Terms section of this chapter. Self-contained portable fountains are not covered by Article 680. A fountain that has water common to a swimming pool is considered part of a swimming pool and must comply with the requirements for swimming pools (Part II) in addition to the requirements of Part V. Like other bodies of water covered by Article 680, fountains must also comply with Part I of Article 680.

2. Check fountain equipment for suitability for approval.

All equipment installed in the water, walls, or other surfaces of a fountain must comply with Article 680 and be suitable for the purpose. Some items are specifically required to be listed either by Article 680 or by other articles of the *Code*. Generally, equipment listed or identified for the purpose and used accordingly should be approved. Because most inspectors are not equipped to test equipment for code compliance, often listing is the only reasonable way for an inspector to determine that equipment meets the requirements for approval under the *NEC*. Section 110.3(A) provides a list of issues to be considered in evaluating equipment for approval if the equipment is not listed.

3. Verify that fountain equipment has GFCI protection unless supplied through a listed transformer at or below the low voltage contact limit.

Submersible equipment listed for operation at or below the low voltage contact limit is not required to have GFCI protection. Transformers and power supplies used to create such systems must be or include isolation-type transformers with grounded metal barriers (often referred to as *shields*) between primary and secondary windings. All submersible equipment that operates above the low voltage contact limit is required to be GFCI protected.

4. Check luminaires and equipment for compliance with voltage limitations.

Luminaires and other submersible equipment used in fountains must operate at 150 volts or less between conductors. Other submersible equipment, such as pumps, may operate at circuit voltages up to 300 volts but must be GFCI protected.

5. Verify that the top of luminaire lenses are below water level, unless listed for above-water locations.

Unless specifically approved for above-water use, fountain luminaires must be installed with the luminaire lenses submerged. Lenses facing upward, which are quite common in fountains, must be guarded from contact by persons or be listed for use without a guard. The lenses of luminaires operating underwater may have high surface temperatures even though they are covered by water.

6. Verify that equipment that depends on submersion for safe operation is protected by a low-water cutoff or other suitable means.

Some fountain luminaires are designed to be cooled by the water in which they are installed. If a luminaire depends on submersion to prevent overheating, a low-water cutoff or other approved means must be provided to de-energize the luminaire when it is not submerged. Float switch assemblies are available and are often used for this purpose.

7. Verify that equipment is equipped with threaded conduit entries, that cords are limited to 10 ft (3.0 m), and that equipment in contact with water is corrosion resistant.

Equipment that is not provided with a flexible cord must have threaded conduit entries. Metal parts in contact with water must be brass or other corrosion-resistant metal. Where cords are used, the cords must be suitable for the use. Not more than 10 ft (3.0 m) of a cord may be exposed in a fountain. The 10-ft (3.0-m) limit should be applied to each cord and not to the sum of the lengths of all cords in a fountain. Cords may be run from one or more junction boxes to a group of luminaires, each with its own cord. Cords that extend beyond the fountain must be enclosed.

8. Verify that equipment can be serviced without draining water from the fountain.

Luminaires and other equipment must be removable from the water for relamping or maintenance. Under this rule, permissible luminaire types roughly correspond to wet-niche and no-niche types used in swimming pools. The forming shell of a luminaire may be embedded in the fountain structure as long as the luminaire can be removed from the forming shell without draining the fountain water.

9. Check junction boxes and other enclosures for compliance with the requirements for swimming pools as well as requirements for underwater boxes.
Boxes used for junction boxes and other enclosures installed above the waterline must comply with the rules for boxes for swimming pools; that is, they must be listed as swimming pool junction boxes. See Item 6 in the checklist titled Initial Inspection: Prior to Pouring of Concrete or Burial. Submerged boxes are often used in fountains, and Part V of Article 680 provides specific rules for such boxes. Boxes used underwater must be:

- Submersible (for example, a Type 6P enclosure is intended for prolonged submersion)
- Furnished with provisions for threaded conduit entries or with compression glands or seals for flexible cord entry
- Copper, brass, or other approved corrosion-resistant material
- Filled with an approved potting compound to prevent moisture from entering the box
- Firmly attached to the fountain surface or box supports
- Bonded as required

Paraffin wax has been used as a potting compound but only if approved by the authority having jurisdiction (AHJ); there are listed potting compounds available. Corrosion-resistant metal conduit may be used to support an underwater junction box subject to the provisions of Article 314. Nonmetallic conduit cannot be used to support a box. Supplementary supports must be provided for boxes where PVC or RTRC is the wiring method.

10. Verify that all parts that are required to be grounded or bonded are grounded or bonded using appropriate methods.
The equipment grounding conductor(s) of the branch circuit(s) supplying a fountain must be used to bond equipment. The requirements of Part II for a bonding grid do not apply to fountains unless the fountains are part of a swimming pool. The equipment required to be bonded at a fountain includes all metal piping systems associated with the fountain. Equipment required to be grounded includes all equipment requiring grounding under Article 250. The following equipment is also specifically mentioned as requiring grounding at fountains:

- **A.** All electrical equipment in the fountain or within 5 ft (1.5 m) of the inside wall of the fountain except for low-voltage luminaires that do not require grounding in accordance with their listings
- **B.** All electrical equipment associated with the recirculating system
- **C.** All panelboards, other than service equipment, that supply fountain electrical equipment

Except for flexible cords, grounding methods and wiring methods are the same as for swimming pools, because specific sections of Part II are referenced in 680.55(A). See Item 17 in the Intermediate and Final Inspections checklist for swimming pools. These rules include the requirement for an insulated copper equipment grounding conductor, sized according to the circuit rating as provided in Article 250, except that equipment grounding conductors that are part of the branch circuit must not be smaller than 12 AWG. Flexible cords must be furnished with an insulated copper equipment grounding conductor that is part of the cord and is connected to a grounding terminal in the supply enclosure.

11. Check power-supply cords for proper type, sealing, and terminations.
Flexible cords must be identified for extra-hard usage and water resistance (indicated by a *W* suffix in the cord type). A potting compound must be used to seal the end of a cable

jacket to keep water from entering an enclosure through a cord or through the conductors in a cord. Potting compound must also be used to protect the grounding terminal connections in equipment from the effects of water. Terminations and cord connections must be permanent except where grounding-type attachment plugs are used. Attachment plugs are restricted to equipment that is not located in any part of a fountain containing water.

12. Verify that signs installed in fountains are GFCI protected, located more than 5 ft (1.5 m) from fountain walls, not portable, and otherwise comply with Articles 600 and 250.

See the definition of **Electric Sign** in the Key Terms section of this chapter. This rule applies only to electric signs in or within 10 ft (3.0 m) of the edge of a fountain. All signs are required to be protected by GFCI devices

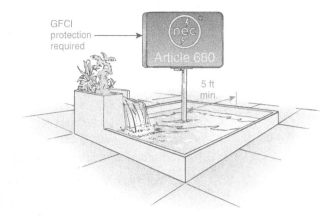

FIGURE 10-12   Location requirement for electric signs in fountains. (Source: *National Electrical Code® Handbook*, NFPA, Quincy, MA, 2014, Exhibit 680.12.)

(FIGURE 10-12). Signs within fountains must be at least 5 ft (1.5 m) from the outside edges of a fountain to keep them out of normal reach of people outside the fountain. Portable signs cannot be located in fountains and must be at least 5 ft (1.5 m) away from the inside walls of a fountain. Disconnects must be provided for signs in accordance with Article 600. According to Section 600.6, a sign disconnect may be remote from the sign if it can be locked in the off position. However, this rule also refers to 680.12, which requires a disconnect to be in sight from the equipment it controls. Grounding and bonding of signs must comply with Articles 600 and 250. (See the Commercial and Industrial Inspections chapter in this manual for checklists and requirements that apply to signs.)

Checklist 10-7: Therapeutic Pools and Tubs

1. Review definitions and determine the applicability of Part VI of Article 680.
See the definitions of **Packaged Therapeutic Tub or Hydrotherapeutic Tank Equipment Assembly**, **Permanently Installed Swimming, Wading, Immersion, and Therapeutic Pools**, and **Self-Contained Therapeutic Tubs or Hydrotherapeutic Tanks** in the Key Terms section of this chapter. Permanently installed therapeutic pools are treated as swimming pools and must comply with Parts I and II of Article 680. Although therapeutic pools and tubs in health care facilities are the focus of this checklist, Part VI covers the same types of equipment in other locations, such as gymnasiums and athletic training rooms. As used in Article 680, the term *therapeutic tub* includes hydrotherapeutic tanks. Portable therapeutic appliances such as those used in conjunction with ordinary bathtubs are covered as appliances by Article 422.

2. Check therapeutic pools and tubs for suitability for approval.
All equipment installed in the vicinity of, or as part of, therapeutic pools and tubs must comply with Article 680 and be suitable for the purpose. Generally, equipment listed or identified for the purpose should be approved. Because most inspectors are not equipped to test equipment for code compliance, often listing is the only reasonable way for an inspector to determine that equipment meets the requirements for approval under the *NEC*. Section 110.3(A) provides a list of issues to be considered in evaluating equipment for approval if the equipment is not listed.

3. Verify that permanently installed therapeutic pools comply with the applicable requirements for permanently installed pools (Parts I and II).
As is mentioned in Item 1 of this checklist, permanently installed (not readily disassembled) therapeutic pools are treated as swimming pools and must comply with Parts I and II of Article 680. Use the Therapeutic Pools and Tubs checklist for swimming pools.

4. Verify that outlets for therapeutic tubs are GFCI protected (separate or integral) unless supplying field-assembled tubs rated 3-phase or over 250 volts.

Outlets supplying self-contained, packaged, or field-assembled therapeutic tubs or hydrotherapeutic tanks are required to be GFCI protected. These may or may not be supplied through receptacle outlets. GFCI protection of the outlet is not required if integral GFCI protection is provided for all internal parts of self-contained or packaged equipment. GFCI protection of the supply is not required for field-assembled equipment that is 3-phase or rated over 250 volts, but the requirements for bonding and grounding the equipment still apply, as covered in Item 5.

5. Verify that all parts of therapeutic tubs that are required to be grounded or bonded are grounded or bonded using appropriate methods.

Bonding and grounding requirements for therapeutic tubs are essentially the same as those for indoor spas and hot tubs. The rules for swimming pools are modified and relaxed somewhat for therapeutic tubs. The distance within which items are required to be bonded to a therapeutic tub is reduced to 5 ft (1.5 m) in all directions from the 12-ft (3.7-m) vertical rule that applies to swimming pools. Like swimming pools, all metal parts of electric equipment and fittings associated with a tub must be bonded. However, the exemptions for small metal parts such as isolated drain fittings or towel bars that apply to swimming pools and spas or hot tubs are also applicable to therapeutic tubs. Small metal parts that are attached to metal framing, such as a mirror frame or a towel bar mounted to a metal frame wall do require bonding. Methods of bonding and grounding are modified and are similar to those for spas and hot tubs. For example, a number of items of electrical equipment may be bonded by mounting them to a common metal frame or base with direct metal-to-metal connections. Other bonding methods include interconnections of threaded metal piping, which is not required to be brass or equally corrosion-resistant, connections by metal clamps, and 8 AWG or larger solid copper bonding jumpers. Like swimming pools, separate insulated equipment grounding conductors are required for equipment associated with permanently installed therapeutic pools because these pools are required to follow the installation rules for permanently installed pools contained in Part II of Article 680. Grounding connections and methods meeting the requirements of Article 250 are acceptable.

6. Verify that all receptacles within 6 ft (1.8 m) of a therapeutic tub are GFCI protected.

Unlike hot tubs and swimming pools, receptacles are not required within any specific distance of a therapeutic tub. All receptacles that are installed within 6 ft (1.8 m) of a therapeutic tub must have GFCI protection.

7. Verify that all luminaires used in areas of therapeutic tubs are of the totally enclosed type.

The extent of the "area" is not defined in the *NEC*. Based on the other rules of Part VI of Article 680, the area above or within 5 ft (1.5 m) of a therapeutic tub would be a reasonable interpretation of the area where luminaires are required to be totally enclosed.

Checklist 10-8: Hydromassage Bathtubs

1. Review definitions and determine the applicability of Part VII of Article 680.

See the definition of **Hydromassage Bathtub** in the Key Terms section of this chapter. Hydromassage bathtubs are increasingly common in dwelling units and in some hotels and motels. Note that a primary difference between some hydromassage bathtubs and some spas is the discharge of water between uses. Spas and hot tubs usually retain the water for more than one use. Part VII is the only part of Article 680 that applies to hydromassage bathtubs. Part I does not apply, except for the definition of a hydromassage bathtub. The extra or special requirements for hydromassage bathtubs are primarily concerned with GFCI protection and with bonding of hydromassage bathtub equipment.

2. Check hydromassage bathtub equipment for suitability for approval.

All equipment associated with hydromassage bathtubs must comply with Article 680 and be suitable for the purpose. Generally, equipment listed or identified for the purpose should be approved. Because most inspectors are not equipped to test equipment for *Code* compliance, often listing is the only reasonable way for an inspector to determine that equipment meets the requirements for approval under the *NEC*. Article 110 provides a list of issues to be considered in evaluating equipment for approval if the equipment is not listed.

3. Check hydromassage bathtub equipment for readily accessible GFCI protection.

GFCI protection is required for hydromassage bathtub equipment. Properly rated GFCI receptacles that supply the equipment can provide GFCI protection. However, where a GFCI receptacle is used, it must be located where it is readily accessible. Since this means that it must be located where it can be reached quickly and without removing obstacles, it likely will not be acceptable to be located in the equipment space below the tub unless that space is readily accessible, perhaps because it is fitted with an access door that meets the definition of readily accessible in Article 100. Other devices are available that may be more convenient for users and that can be located where readily accessible. Receptacles that are rated 125 volts and up to 30 amperes must also be GFCI protected if installed within 6 ft (1.8 m) horizontally of the inside walls of the tub.

4. Verify that an individual branch circuit is used to supply a hydromassage bathtub.

Electrical equipment and wiring in the vicinity of a hydromassage tub must also meet the requirements for equipment and wiring installed in bathrooms. Equipment such as a hydromassage bathtub is not permitted on the circuit intended for bathroom receptacles, but must be supplied by a circuit that supplies no other loads.

5. Check other electrical equipment in the same room as a hydromassage bathtub for compliance with the ordinary rules of Chapters 1 through 4 of the *NEC*.

Electrical equipment and wiring in the vicinity of a hydromassage tub must also meet the requirements for equipment and wiring installed in bathrooms, whether or not the hydromassage bathtub is actually part of a bathroom area as defined in Article 100. Most hydromassage bathtubs are installed in bathroom areas, which would apply a GFCI requirement for all 125-volt, 15- or 20-ampere bathroom receptacles. However, in cases where such a tub is not actually part of a bathroom, the foregoing statements still apply. See Item 3 of this checklist.

6. Verify that hydromassage bathtub equipment is accessible without damaging the building structure or finish.

Access must be provided to hydromassage bathtub electrical equipment (**FIGURE 10-13**). The means of access cannot require damaging the building finish. For example, it should not be necessary to remove tiles or cut through wall finishes to gain access to hydromassage equipment. Access from outside the bathroom area may be preferred for aesthetic reasons. The receptacle for **cord-and-plug–connected** equipment may be accessible through a service access opening only if the receptacle face is in direct view and not more than 1 ft (300 mm) from the opening.

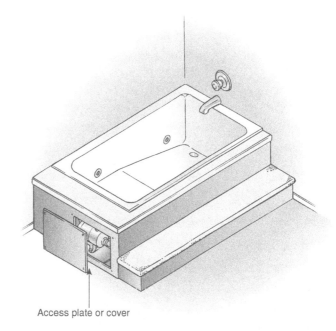

Access plate or cover

FIGURE 10-13   Access panel or plate for servicing of hydromassage bathtub electrical equipment. (Source: *National Electrical Code® Handbook*, NFPA, Quincy, MA, 2008, Exhibit 680.9.)

7. Verify that metal parts required to be bonded are connected together with a minimum 8 AWG solid copper bonding jumper.

Most of the electrical equipment and water piping associated with hydromassage bathtubs is located below some part of the tub. The various grounded metal parts of electrical equipment and piping systems and the bonding terminal on a circulating pump that is not double-insulated must be bonded together with 8 AWG solid or larger copper bonding jumpers. If a double-insulated motor is used, the 8 AWG solid bonding jumper must be connected to the equipment grounding conductor of the motor branch circuit and left long enough to connect to a possible replacement motor that is not double insulated. Where all of the piping is plastic, except for a few isolated fittings, those isolated fittings are not required to be bonded. Only those grounded electrical parts that are in contact with circulating water are required to be bonded. The grounded electrical parts in this context are those that are otherwise required to be connected to an equipment grounding conductor. Other electrical equipment may be required to be connected to an equipment grounding conductor by Article 250, but such equipment is not required to be bonded to the piping systems unless the electrical equipment is in contact with circulating water. This required bonding is to establish a limited equipotential system in the vicinity of the hydromassage bathtub similar to that required for swimming pools. The *Code* does not require the 8 AWG bonding conductor to be run back to the panelboard supplying the hydromassage bathtub branch circuit or to any electrode or service equipment.

KEY TERMS

The *NEC* defines the key terms pertaining to swimming pools and related installations as follows (all definitions are from Section 680.2 unless otherwise noted):

Bonded (Bonding) [*Definition from Article 100*] Connected to establish electrical continuity and conductivity. [*The bonding of metal parts required in the vicinity of a body of water is intended to eliminate any voltage differences between parts or within the body of water by establishing a single potential for all the conductive parts. Metal parts and equipment around swimming pools and similar areas are not bonded together to carry fault current or complete a path to the power source as they frequently are in Article 250; rather, they are bonded to maintain the same voltage potential of metal parts, or in the words of 680.26 ". . . to reduce voltage gradients in the pool area." Separate rules are provided for equipment that is required to be grounded by connection to an equipment grounding conductor, as the grounding and bonding paths to motors, heaters, and the like may well be expected to carry fault current and are intended to complete a path back to the grounding point of the power source.*]

Cord-and-Plug–Connected Lighting Assembly A lighting assembly consisting of a luminaire intended for installation in the wall of a spa, hot tub, or storable pool, and a cord-and-plug–connected transformer.

Dry-Niche Luminaire A luminaire intended for installation in the floor or wall of a pool, spa, or fountain in a niche that is sealed against the entry of water. [*These luminaires are often designed for service from the rear with access provided via a tunnel, passageway, or other method such that a lamp can be changed without entering or draining the pool. In some cases, water may have to be partially drained to service a dry-niche luminaire.*]

Electric Sign [*Definition from Article 100*] A fixed, stationary, or portable self-contained, electrically illuminated utilization equipment with words or symbols designed to convey information or attract attention.

Forming Shell A structure designed to support a wet-niche luminaire assembly and intended for mounting in a pool or fountain structure.

Fountain Fountains, ornamental pools, display pools, and reflection pools. The definition does not include drinking fountains.

Ground-Fault Circuit Interrupter (GFCI) [*Definition from Article 100*] A device intended for the protection of personnel that functions to de-energize a circuit or portion thereof within an established period of time when a current to ground exceeds the values established for a Class A device.

> *Informational Note:* Class A ground-fault circuit interrupters trip when the current to ground has a value in the range of 4 to 6 milliamperes. For further information, see UL943, *Standard for Ground-Fault Circuit Interrupters.*

[*This type of equipment comes in various forms, including circuit breakers, individually mounted devices, receptacles, and devices intended as part of a cord assembly or as internal components of equipment. All forms are commonly referred to as GFCI protection or GFCI devices. Many types of equipment associated with swimming pools, fountains, spas, hot tubs, and the like are required to be protected by GFCI devices. GFCI devices are distinguished from other devices such as ground-fault protection of equipment (GFPE) devices that trip at higher levels and are designed to protect conductors or equipment from damage rather than providing shock protection.*]

Hydromassage Bathtub A permanently installed bathtub equipped with a recirculating piping system, pump, and associated equipment. It is designed so it can accept, circulate, and discharge water upon each use. [*The discharge of water after each use is the most significant characteristic distinguishing hydromassage bathtubs from spas and hot tubs.*]

Low Voltage Contact Limit A voltage not exceeding the following values:
  (1) 15 volts (RMS) for sinusoidal ac
  (2) 21.2 volts peak for nonsinusoidal ac
  (3) 30 volts for continuous dc
  (4) 12.4 volts peak for dc that is interrupted at a rate of 10 to 200 Hz
  [*In earlier editions of the* NEC, *the value that was used to determine when a shock hazard existed at a pool was simply 15 volts. However, that level did not account for the actual shock hazard values from sources other than sinusoidal (RMS) alternating current (ac) voltages. This definition recognizes that the shock hazard level is different for different sources, where the lowest hazard voltage is actually associated with a pulsating direct current (dc) source.*]

Maximum Water Level The highest level that water can reach before it spills out.

No-Niche Luminaire A luminaire intended for installation above or below the water without a niche.

Packaged Spa or Hot Tub Equipment Assembly A factory-fabricated unit consisting of water-circulating, heating, and control equipment mounted on a common base, intended to operate a spa or hot tub. Equipment can include pumps, air blowers, heaters, lights, controls, sanitizer generators, and so forth.

Packaged Therapeutic Tub or Hydrotherapeutic Tank Equipment Assembly A factory-fabricated unit consisting of water-circulating, heating, and control equipment mounted on a common base, intended to operate a therapeutic tub or hydrotherapeutic tank. Equipment can include pumps, air blowers, heaters, lights, controls, sanitizer generators, and so forth.

Permanently Installed Decorative Fountains and Reflection Pools Those that are constructed in the ground, on the ground, or in a building in such a manner that the fountain cannot be readily disassembled for storage, whether or not served by electrical circuits of any nature. These units are primarily constructed for their aesthetic value and are not intended for swimming or wading.

Permanently Installed Swimming, Wading, Immersion, and Therapeutic Pools Those that are constructed in the ground or partially in the ground, and all others capable of holding water to a depth greater than 42 in. (1.0 m), and all pools installed inside of a building, regardless of water depth, whether or not served by electrical circuits of any nature. [*The addition of the word* Immersion *to this defined term and to the defined term* Storable Swimming, Wading, or Immersion Pool, *as well as in the definition of* Pool *was intended to clarify that Article 680 also covers bodies of water used for baptisteries and any similar purpose.*]

Pool Manufactured or field-constructed equipment designed to contain water on a permanent or semipermanent basis and used for swimming, wading, immersion, or therapeutic purposes. [*A hydromassage bathtub is not intended to be treated as a pool and is not required to meet the requirements for a pool.*]

Pool Cover, Electrically Operated Motor-driven equipment designed to cover and uncover the water surface of a pool by means of a flexible sheet or rigid frame.

Self-Contained Spa or Hot Tub Factory-fabricated unit consisting of a spa or hot tub vessel with all water-circulating, heating, and control equipment integral to the unit. Equipment can include pumps, air blowers, heaters, lights, controls, sanitizer generators, and so forth.

Self-Contained Therapeutic Tubs or Hydrotherapeutic Tanks A factory-fabricated unit consisting of a therapeutic tub or hydrotherapeutic tank with all water-circulating, heating, and control equipment integral to the unit. Equipment may include pumps, air blowers, heaters, light controls, sanitizer generators, and so forth.

Spa or Hot Tub A hydromassage pool, or tub for recreational or therapeutic use, not located in health care facilities, designed for immersion of users, and usually having a filter, heater, and motor-driven blower. It may be installed indoors or outdoors, on the ground or supporting structure, or in the ground or supporting structure. Generally, a spa or hot tub is not designed or intended to have its contents drained or discharged after each use. [*The discharge of water after each use is a defining difference and is a characteristic of a hydromassage bathtub.*]

Storable Swimming, Wading, or Immersion Pool or Storable/Portable Spas and Hot Tubs Those that are constructed on or above the ground and are capable of holding water to a maximum depth of 42 in. (1.0 m) or a pool, spa, or hot tub with nonmetallic, molded polymeric walls or inflatable fabric walls regardless of dimension.

Wet-Niche Luminaire A luminaire intended for installation in a forming shell mounted in a pool or fountain structure where the luminaire will be completely surrounded by water. [*Wet-niche luminaires are connected to cords, some portion of which is stored in the forming shell of the luminaire, so that the luminaire can be removed from its shell and extended outside of the water for relamping without draining any water from the pool.*]

KEY QUESTIONS

1. Which parts of Article 680 apply to the installation?
2. What wiring methods are used?
3. What electrical equipment is associated with the installation, and where is it located?
4. What type of underwater luminaires are used?
5. What equipment and structures are required to be bonded, and how will these items be connected together in order to establish equipotential bonding in the pool area?
6. What are the voltages of the equipment used in and around the body of water?

CHECKLISTS

	Item	Inspection Activity	*NEC* Reference	Comments
		Checklist 10-1: Initial Inspection: Prior to Pouring of Concrete or Burial		
	1.	Review definitions and determine the applicability of Article 680.	680.1, 680.2	
	2.	Check overhead conductor clearances for conformance with requirements.	680.8	
	3.	Check underground wiring for suitability, clearances from pool, and minimum cover requirements.	680.10	
	4.	Check underwater luminaires for locations, wiring methods, and connections to wiring methods.	680.23	
	5.	Check pool-related equipment for appropriate wiring methods.	680.21, 680.23(B) and (F), 680.25	
	6.	Check junction boxes and enclosures connecting to underwater luminaires for listing and labeling, appropriate size and location, and proper materials.	680.24	
	7.	Check forming shells and wiring methods for underwater audio equipment.	680.27	
	8.	Verify that metal parts of pools and other nearby electrical equipment and metal parts are connected together to create equipotential bonding in the pool area, using appropriate methods.	680.21(B), 680.26	

✔	Item	Inspection Activity	*NEC* Reference	Comments
		Checklist 10-2: Intermediate and Final Inspections		
	1.	Check pool equipment for suitability for approval.	110.3(A) and (B), 680.4	
	2.	Check transformers, power supplies, and GFCIs for identification and suitability for the purpose.	680.23(A)(2)	
	3.	Verify that conductors on the load side of transformers or GFCIs are separated from conductors not protected by GFCIs.	680.23(F)(3)	
	4.	Verify that general-use receptacles are not located within 6 ft (1.8 m) of pool walls and that all receptacles within 20 ft (6.1 m) of pool walls are GFCI protected.	680.21(C), 680.22(A)(1), (2), and (5)	
	5.	Verify that any receptacle installed between 6 and 10 ft (1.8 and 3.0 m) of pool walls that is used for pool equipment, is a single, grounding-type device, and is GFCI protected.	680.22(A)(1), (4), and (5), 680.22(B)	
	6.	Verify that at least one GFCI receptacle on a general-purpose branch circuit is installed between 6 and 20 ft (1.8 and 6.1 m) of pool walls at dwelling units.	680.22(A)(3)	
	7.	Verify GFCI protection of outlets for pool circulating pump motors that are supplied by single-phase, 15- or 20-ampere, 120- through 240-volt branch circuits.	680.21(C)	
	8.	Verify that luminaires and ceiling fans are located so that required clearances are maintained.	680.22(B)	
	9.	Verify that luminaires and ceiling fans are GFCI protected where GFCI protection is required.	680.22(B)(3) and (4)	
	10.	Verify that switches are located at least 5 ft (1.5 m) from pool walls or separated from the pool by a permanent barrier.	680.22(C)	
	11.	Verify that other outlets are at least 10 ft (3.0 m) from the pool.	680.22(D)	
	12.	Check flexible cords (where used) for compliance with equipment grounding conductor requirements and length limitations.	680.7	
	13.	Check underwater luminaires for locations, wiring methods, and connections to wiring methods.	680.23	
	14.	Check pool-related equipment for appropriate wiring methods.	680.21(A)	
	15.	Check junction boxes and enclosures connecting to underwater luminaires for listing and labeling, appropriate size and location, and proper materials.	680.24	
	16.	Check forming shells and wiring methods for underwater audio equipment.	680.27	

(continues)

Checklist 10-2: Intermediate and Final Inspections

✔	Item	Inspection Activity	*NEC* Reference	Comments
	17.	Verify that metal parts of pools and nearby equipment and metal parts are bonded to an appropriate equipotential bonding grid, using appropriate methods.	680.26	
	18.	Verify that all equipment required to be grounded is grounded by insulated copper equipment grounding conductors and/or bonding jumpers of the proper size.	680.6, 680.21(A)(1), 680.23(B), (C), (D), and (F)	
	19.	Verify that a disconnecting means is provided, accessible, located within sight of pool equipment, and not located within 5 ft (1.5 m) of pool walls.	430.102, 680.12, 680.22(C)	
	20.	Check electric pool heaters for subdivision of heating elements and sizing of branch-circuit conductors.	680.9	
	21.	Check equipment rooms or pits for adequate drainage.	680.11	
	22.	Check electrically operated pool covers for proper motor and controller location, motor enclosure, and GFCI protection.	680.27(B)	
	23.	Check deck-area heaters for suitability and proper clearances from pool.	680.27(C)	

Checklist 10-3: Storable Swimming Pools and Storable Spas and Hot Tubs

✔	Item	Inspection Activity	*NEC* Reference	Comments
	1.	Review definitions and determine the applicability of Part III of Article 680.	680.2	
	2.	Verify that cord-connected pool pumps are double-insulated and that internal metal parts are grounded through a grounding-type attachment plug.	680.31	
	3.	Verify that all electrical equipment associated with the storable pool is provided with GFCI protection.	680.31, 680.32	
	4.	Check luminaires for compliance with requirements based on the voltage of the luminaires.	680.33	
	5.	Verify that receptacles are not located within 6 ft (1.8 m) of the pool.	680.34	

Checklist 10-4: Spas and Hot Tubs: All Installations

✔	Item	Inspection Activity	*NEC* Reference	Comments
	1.	Review definitions and determine the applicability of Part IV of Article 680.	680.2	
	2.	Check spa and hot tub equipment for suitability for approval.	110.3(A) and (B), 680.4	
	3.	Review the checklist for permanent pools for compliance with the applicable provisions of Parts I and II of Article 680 (modified for indoor installations).	680.42	
	4.	Check outlets supplying a self-contained spa or hot tub or a packaged spa or hot tub assembly for integral or separately provided GFCI protection.	680.44	
	5.	Verify compliance with requirements for disconnecting means.	422 Part III, 680.12, 680.22(D)	
	6.	Verify that spas or hot tubs in other than single-family dwellings are provided with an emergency shutoff or control switch, as required.	680.41	
	7.	Check electric pool heaters for subdivision of heating elements, and sizing of branch-circuit conductors.	680.9	

Checklist 10-5: Spas and Hot Tubs: Indoor Installations Only

✔	Item	Inspection Activity	*NEC* Reference	Comments
	1.	Verify that a suitable wiring method of Chapter 3 of the *NEC* is used to supply and connect spas and hot tubs, unless cord-and-plug connections are permitted.	680.43	
	2.	Verify that at least one GFCI-protected receptacle on a general-purpose branch circuit is located between 6 and 10 ft (1.8 and 3.0 m) of the walls of the spa or hot tub.	680.43(A)	
	3.	Verify that any receptacles used to supply power to a spa or hot tub are GFCI protected.	680.43(A)(3)	
	4.	Verify that luminaires and paddle fans are spaced as required from spa or hot tub walls and above maximum water level and that GFCI protection is provided, as required.	680.43(B)	
	5.	Verify that switches are located at least 5 ft (1.5 m) from the inside walls of the spa or hot tub.	680.43(C)	
	6.	Verify that all parts that are required to be grounded or bonded are grounded or bonded using appropriate methods.	680.43(D), (E), and (F)	
	7.	Check underwater audio equipment for compliance with Part II.	680.43(G)	

✔	Item	Inspection Activity	*NEC* Reference	Comments
		Checklist 10-6: Fountains		
	1.	Review definitions and determine the applicability of Part V of Article 680.	680.2, 680.50	
	2.	Check fountain equipment for suitability for approval.	110.3(A) and (B), 680.4	
	3.	Verify that fountain equipment has GFCI protection unless supplied through a listed transformer at or below the low voltage contact limit.	680.51(A), 680.56(A)	
	4.	Check luminaires and equipment for compliance with voltage limitations.	680.51(B)	
	5.	Verify that the tops of luminaire lenses are below water level, unless listed for above-water locations.	680.51(C)	
	6.	Verify that equipment that depends on submersion for safe operation is protected by a low-water cutoff or other suitable means.	680.51(D)	
	7.	Verify that equipment is equipped with threaded conduit entries, that cords are limited to 10 ft (3.0 m), and that equipment in contact with water is corrosion resistant.	680.51(E)	
	8.	Verify that equipment can be serviced without draining water from the fountain.	680.51(F)	
	9.	Check junction boxes and other enclosures for compliance with the requirements for swimming pools as well as requirements for underwater boxes.	680.52	
	10.	Verify that all parts that are required to be grounded or bonded are grounded or bonded using appropriate methods.	680.53, 680.54, 680.55	
	11.	Check power-supply cords for proper type, sealing, and terminations.	680.56	
	12.	Verify that signs installed in fountains are GFCI protected, located more than 5 ft (1.5 m) from fountain walls, not portable, and otherwise comply with Articles 600 and 250.	680.57	

Checklist 10-7: Therapeutic Pools and Tubs

✔	Item	Inspection Activity	*NEC* Reference	Comments
	1.	Review definitions and determine the applicability of Part VI of Article 680.	680.2, 680.60 through 680.62	
	2.	Check therapeutic pools and tubs for suitability for approval.	110.3(A) and (B), 680.4	
	3.	Verify that permanently installed therapeutic pools comply with the applicable requirements for permanently installed pools (Parts I and II).	680.61	
	4.	Verify that outlets for therapeutic tubs are GFCI protected (separate or integral) unless supplying field-assembled tubs rated 3-phase or over 250 volts.	680.62(A)	
	5.	Verify that all parts of therapeutic tubs that are required to be grounded or bonded are grounded or bonded using appropriate methods.	680.62(B),(C), and (D)	
	6.	Verify that all receptacles within 6 ft (1.8 m) of a therapeutic tub are GFCI protected.	680.62(E)	
	7.	Verify that all luminaires used in areas of therapeutic tubs are of the totally enclosed type.	680.62(F)	

Checklist 10-8: Hydromassage Bathtubs

✔	Item	Inspection Activity	*NEC* Reference	Comments
	1.	Review definitions and determine the applicability of Part VII of Article 680.	680.2, 680.70	
	2.	Check hydromassage bathtub equipment for suitability for approval.	110.3(A) and (B)	
	3.	Check hydromassage bathtub equipment for readily accessible GFCI protection.	680.71	
	4.	Verify that an individual branch circuit is used to supply a hydromassage bathtub.	680.71	
	5.	Check other electrical equipment in the same room as a hydromassage bathtub for compliance with the ordinary rules of Chapters 1 through 4 of the *NEC*.	680.72	
	6.	Verify that hydromassage bathtub equipment is accessible without damaging the building structure or finish.	680.73	
	7.	Verify that metal parts required to be bonded are connected together with a minimum 8 AWG solid copper bonding jumper.	680.74	

Emergency and Standby Systems and Fire Pumps

OVERVIEW

This chapter covers the installation and wiring requirements for emergency systems, legally required standby systems, optional standby systems, and fire pumps. Critical operations power systems (COPS) are similar in many ways, and in a sense are a type of legally required standby system since the necessity for a COPS installation may be determined or approved by governmental authorities, but they are covered in a separate article. This manual does not address them specifically (see *National Electrical Code®* [*NEC®*] Article 708 for more information).

Emergency systems and legally required standby systems are similar, because they all require some permanently installed alternative power source or sources. Optional standby systems are permitted to be supplied by portable power sources. The alternative sources vary from large diesel generators to individual batteries for one or two lamps. Fire pumps do not necessarily require an alternative source, but alternative sources are often used to ensure the reliable power needed for fire pumps.

Fire pumps are, by their nature, an "emergency load" in common terminology, as they operate only when fire suppression water is being supplied to a building. However, the fire pump may or may not be classified as an *emergency system load*, as defined in the *NEC*, depending on the nature of the building and local requirements or interpretations. It is important to note that unless the authority having jurisdiction (AHJ) specifically classifies the electrical system supplying a fire pump as *emergency*, the requirements of Article 700 do not modify or add to the requirements of Article 695.

Emergency systems are primarily intended to provide for safe evacuation of buildings and as such, often only include exit and egress lighting. Other systems necessary for safe evacuation or to prevent a hazard from developing, such as some elevators or some ventilation systems, may be classified as emergency or as legally required standby systems by the AHJ. Optional standby systems are those not necessary for evacuation or for meeting any legal requirement, but those desired for convenience or continued operation during a power failure. Standby power systems for single-family dwellings are generally classed as optional standby systems.

SPECIFIC FACTORS

The *NEC* provides *installation* requirements for permanently installed emergency and standby systems. The requirements for *having* emergency and standby systems and equipment and the specific loads that must be connected to the systems are found in other codes and standards, such as the NFPA 101, *Life Safety Code*, or building codes. Similarly, fire pumps are not required by the *NEC*, but where they are required by other codes, the *NEC* provides requirements for wiring and supply systems to electrically driven fire pumps. Fire pumps driven directly by prime movers such as diesel fire pumps are not covered by the *NEC*, but some aspects of the engine control circuits are covered. In other than health care facilities, emergency loads are primarily those that are required for safe evacuation of a building. To help avoid some confusion, such systems in health care facilities are called "essential electrical systems." The *life safety branch* of an *essential electrical system* powers items that might otherwise be classed as emergency system loads. See the Special Occupancies chapter of this manual for more information on essential electrical systems. Legally required standby systems are those not necessary to the life safety of the building inhabitants, perhaps, but necessary for the safety of fire fighters or to avoid a hazardous condition from developing. Optional standby systems are intended for whatever loads the user selects.

The *NEC* establishes a graduated scale of requirements for standby power systems. Emergency systems are at the top of this scale. Requirements for emergency systems are more restrictive than for other standby systems. Emergency system wiring must be separated from other wiring, including wiring of other standby systems. This wiring separation requirement also applies where separate emergency systems are supplied from separate emergency sources. Although using two separate systems is not a common approach, the reliability of each separate emergency system is enhanced by maintaining a separation between the circuit conductors of each other emergency system. Emergency systems supplied by centralized power sources (rather than by individual, self-contained unit equipment) are required to have their own **transfer switches**, separate from legally required standby and optional standby transfer means, even though the same generator is permitted to be used to supply all three systems, and if necessary (generally because the generator is not large enough to simultaneously supply all of the connected loads) to ensure full power to all loads classed as emergency, automatic load shedding may be employed as a means to prioritize the power supplied from the generator. The transfer means must be automatic and identified and listed for emergency system use. Emergency systems are more restrictive because they are generally intended to supply only those loads required for the safe evacuation of a building or structure. The separation requirement is intended to minimize the effects of faults or failures in other circuits and systems on the critical functions of the emergency system and related emergency loads.

Legally required standby systems are the next level in the scale of requirements. Legally required standby systems are not required to be separated from other wiring. The transfer means must be automatic and is required to be identified and approved for legally required standby use, which, in the absence of a separate standard for legally required standby transfer switches, typically means the transfer switch will be listed for emergency use. Automatic transfer switches listed for use in emergency systems are also suitable for use in legally required standby systems and optional standby systems. Legally required and optional standby systems may use the same transfer equipment if the system capacity is adequate for both systems.

Optional standby systems are not required to have an automatic means of transferring load from one source to another; manual transfer switches may be used, but all transfer equipment must prevent the inadvertent interconnection of normal and standby sources. The user is permitted to select the entire load that will be connected. Portable or stationary generators that are the only source of power are generally treated as separately derived systems rather than as standby sources. Transfer equipment is not required unless an electric utility is either the normal or standby source. Isolation of a portable generator used for optional standby from utility sources may be provided by means other than transfer switches where only qualified persons service the installation.

The precise nature of the loads supplied by emergency and legally required standby systems is not always easily defined. Occasionally, an emergency system is expanded to include other loads. For example, as noted previously, in some cases designated elevators in high-rise buildings have been classified by the AHJ as emergency or legally required standby loads because they may be used for evacuation or firefighting. Fire pumps are a special type of load. Perhaps the most striking difference between fire pumps and most other motors is the limited overcurrent protection provided for fire pumps. Specifically, overload protection is not permitted for fire pump circuits. Fire pump feeders and branch circuits are required to be provided with short-circuit protection. The short-circuit protection will also provide some protection from high-level ground faults, at least in solidly grounded systems, but even where ground-fault protection of equipment (GFP or GFPE) would be required for feeders or services based on the voltage and current ratings of the equipment, it is not permitted for fire pumps. Fire pumps are intended to be run to failure rather than operate an overcurrent device in those instances when an overload occurs while the fire pump is in operation. Also, as with emergency systems, wiring for fire pumps is required to be separated and isolated from the wiring of other systems. This requirement is extended to the point that most fire pumps have their own services. Wiring methods that may be used with fire pumps are also restricted to methods that are highly resistant to physical damage and the effects of fire in an area of a building in which the fire pump circuit is installed.

KEY QUESTIONS

1. What loads are supplied by the emergency or standby system?

In some cases, standby systems are inaccurately named. Correct terminology and clearly defined systems are necessary to properly apply the *NEC* to emergency and standby systems. Emergency and standby systems are categorized by the loads they supply. Emergency loads are required to be served by an emergency system. Other loads may be served by legally required standby or optional standby systems. The user may select loads to be served by optional standby systems. Some jurisdictions, such as a local building department or fire marshal, will select nonemergency loads that require standby systems, putting those selected loads on legally required standby systems. Emergency systems require separate transfer equipment and isolated wiring. When both emergency and standby systems exist on the same project, the same source or different sources may be used for different systems. On the one hand, a single properly sized generator could serve emergency, legally required standby, and optional standby systems with more than one transfer switch. On the other hand, in some buildings, the emergency lights and exit lights may be served by **unit equipment**. Legally required and optional standby systems could then be served by a generator with a single transfer switch. In such a case, no special isolated wiring would be required except at or within the unit equipment.

2. What alternative power sources are being used for the emergency or standby system?

The suitability, capacity, location, and reliability of the alternative source are critical to the assessment of an emergency or standby system. Although the answer to this question is fairly simple, it is vital to the design and inspection of the system in question. More than one alternate source may be used. Unit equipment is sometimes used in conjunction with a generator because the unit equipment involves a much shorter transfer time, but the generator can produce power for a longer period, subject primarily to fuel capacity. A generator is a commonly used source for emergency and standby power (**FIGURE 11-1**). A single generator can be used to supply the loads of an emergency system, legally required standby system, and optional standby system, provided that it is adequately sized and will supply the loads in their order of importance.

3. Where is the transfer equipment located?

For emergency systems, the transfer equipment often marks the beginning of the independent wiring required for emergency systems. If emergency systems and other standby systems are supplied by the same power source, more than one transfer switch is usually required, and the point of separation may be at the supply source or may be at distribution equipment such as a switchboard from which individual feeders for emergency, legally required standby, and optional standby loads are supplied. (Health care facilities are a possible exception.) The transfer equipment is also a logical starting point for inspections of wiring methods, separation of wiring, and system capacity.

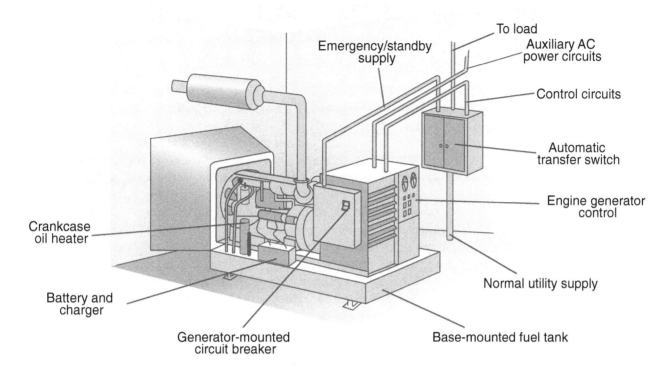

FIGURE 11-1   Electrical elements of a typical generator installation for supplying emergency or standby power.

4. How is the emergency or standby power distributed, and are the emergency system circuits separated from normal and other standby system circuits?

This question is most applicable to emergency systems when a central alternative source is used instead of individual unit equipment. Emergency system circuits are required to be installed using a completely independent wiring system and cannot share common raceways or cables with circuits not designated as emergency. Standby systems are not required to have their wiring separated from other systems. However, if many emergency or standby loads are supplied, a distribution system of feeders and branch circuits must be established, and this distribution scheme is central to the inspection of the system(s). This question must be answered to complete the checklist items that are concerned with the accessibility of switches and overcurrent protection. Where generators are used for emergency systems, extra loads, such as fuel pumps or the generator battery charger, are also added to the emergency system. In some assembly and high-rise occupancies, feeder and circuit wiring must be protected by a listed fire-rated assembly that has a 2-hour fire rating or equivalent. This requirement also applies to some generator control wiring.

5. What power source is being used for a fire pump?

This question is most applicable to a plan reviewer or to the person who must determine whether a proposed power source or combination of sources meets the requirement for a reliable source. Nevertheless, the answer to this question provides the starting point for inspection of the power supply or supplies to a fire pump.

6. Where is the fire pump located, and how is the power supply routed to the fire pump?

The reliability of the power source required for a fire pump is partly contingent on the independence of all aspects of the supply to the fire pump, including the supply wiring. Fire pump power supplies and wiring must be highly immune to disturbances or faults on other parts of the building wiring and power supply. The determination and acceptance of a reliable power supply for a fire pump will typically rest in the hands of fire officials and the electrical inspector or building official.

PLANNING FROM START TO FINISH

Emergency systems, standby systems, and fire pumps do not exist alone. They are always part of other installations, and they have many of the characteristics of other installations. Each may involve underground or embedded work, wiring methods that may be concealed, grounding requirements, and elements that cannot be checked until the final inspection. Some installations, especially high-rise projects, will include emergency systems, legally required standby systems, fire pumps, and perhaps optional standby systems. Therefore, the inspections of emergency and standby systems and fire pumps must be integrated with each other and with the other inspection phases on a particular job.

Early inspections, including rough and intermediate inspections, will include verifying the proper selection and use of wiring methods, checking for separation of wiring for emergency systems and fire pumps, and checking underground and embedded work. The proper use of wiring methods and many other aspects of emergency systems, standby systems, and fire pumps are covered in other chapters of this manual.

Because many standby systems involve the use of separately derived systems, grounding preparations can also be checked at an early stage and on subsequent inspections. Grounding must also be closely examined for proper system grounding connections in all systems, especially those systems that are not separately derived due to interconnected grounded conductors. See the Grounding and Bonding chapter of this manual for a discussion of grounding and, specifically, the grounding of separately derived systems.

Most of the details covered by the checklists in this chapter will be inspected at, or near, the end of a job. Acceptance testing is required of most systems covered in this chapter. Listing and identification of equipment are also important issues. The signs or signals required for power sources, grounding, and the presence of a separate service for a fire pump must also be checked in the late stages of a project. All of these issues depend on the equipment being in place and ready to operate. Whereas many checklist items can be covered prior to the application of power to the equipment, all acceptance testing requires energized equipment and completed installations, but generally does not require exposure to energized live parts, at least not for the inspector. Therefore, the testing phase is likely to be part of the last inspection prior to allowing users to occupy a building.

Because the requirements to provide an emergency system, a legally required standby system, or a fire pump are not dictated by the *NEC,* the electrical inspector must coordinate the use of the checklists in this chapter with the inspections made under the codes and standards that do require these systems. The electrical inspector must verify compliance with the *NEC* with regard to installation of the systems covered by this chapter, the equipment that must be installed, the locations of equipment, and the nature of the loads dictated by other codes. **TABLE 11-1** compares the requirements for systems covered by this chapter. As seen in this table, more stringent requirements apply to the systems that perform or preserve critical life safety functions.

WORKING THROUGH THE CHECKLISTS

Checklist 11-1: Emergency Systems

1. Determine the applicability of Article 700.
See the definition of **Emergency System** in the Key Terms section. Article 700 applies if the occupancy is determined to include emergency loads. However, because many occupancies have limited emergency loads that can be served entirely by unit equipment, generators are not always required.

TABLE 11-1	Emergency and Standby System Requirements

Requirements for Installation and Operation	Article 700 Emergency Systems	Article 701 Legally Required Standby Systems	Article 702 Optional Standby Systems
1. System required and classified by authority having jurisdiction (AHJ)	Yes	Yes	No
2. Equipment approved for the intended use	Yes	Yes	Yes
3. Acceptance and periodic testing	Yes	Yes	No
4. Capacity of source to supply all designated loads that will be operated simultaneously	Yes	Yes	Yes
5. Load-shedding equipment (where necessary to prioritize loads)	Automatic	Automatic	Automatic or manual
6. Identification of transfer equipment for class of service	Yes Listing required	Yes Listing required	Suitable for use
7. Transfer equipment electrically operated and mechanically held	Yes	Yes	No
8. Sign at electrical service indicating location of alternative source	Yes	Yes	Yes
9. Indication of ground fault for systems over 150 volts to ground, with protective devices rated 1000 amperes or greater	Yes	No	No
10. Transfer equipment limited to supplying only that specific class of load	Yes	No (can be combined with 702 loads)	No (can be combined with 701 loads)
11. Marking of boxes and enclosures to indicate use	Yes	No	No
12. Physical separation of wiring from other systems	Yes	No	No
13. Fire protection for equipment under specified conditions of occupancy	Yes, 2 hours	No	No
14. Power available to equipment time frame	10 seconds maximum	60 seconds maximum	No time frame specified
15. Capacity of storage batteries	1.5 hours at 87.5% voltage level	1.5 hours at 87.5% voltage level	Not specified
16. On-site fuel supply for prime mover	2 hours	2 hours	Not specified
17. Use of municipal systems for fuel or cooling water	No, unless approved by AHJ	No, unless approved by AHJ	Yes
18. Connection ahead of service disconnecting means as source	No	Yes	Yes
19. Access to branch-circuit overcurrent protective devices	Authorized personnel only	Authorized personnel only	Not specified
20. Power source permanently installed	Yes	Yes	Not required

2. Check equipment for suitability for approval.

All emergency system equipment and wiring must be approved. The *NEC* does not require listing of all emergency equipment, although certain items, such as transfer equipment and some forms of fire protection, are required to be listed for emergency use. Section 110.3(A) provides a list of considerations to be used in approving equipment when it is not available as a listed product. The inspector may require listed equipment where it is available, as inspectors generally are not able to conduct adequate field testing. A transfer switch is required to be listed for emergency system use where used in conjunction with an Article 700 emergency system. This labeling/listing must be provided by the switch manufacturer and must indicate compliance with applicable product standards. **FIGURE 11-2** shows an automatic transfer switch and associated control panel that are intended to be mounted in an enclosure.

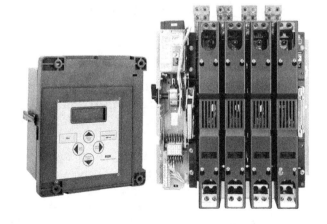

FIGURE 11-2   Electrical components of an emergency transfer switch and control panel. (Courtesy of ASCO Power Technologies.)

3. Review load calculations and verify that system capacity is adequate.

The system designer should furnish load calculations. Calculated capacity must be at least equal to the entire load to be connected at one time. Where motor loads are a significant portion of the emergency load, the power source may have to be oversized to accommodate the starting currents of the motors. Power sources, such as generators, may have to be derated for installation at high altitudes (typically above 3300 ft [1000 m]). Power sources are also required to be tested under maximum load conditions.

4. Verify that system capacity is adequate for any nonemergency loads it feeds or that automatic selective load pickup and load shedding are provided.

Emergency system power sources may feed nonemergency loads through separate transfer switches if they are capable of supplying the entire connected load or if they are equipped to shed the nonemergency loads when necessary. Loads are required to be served in order of priority. Emergency loads have the highest priority, followed by legally required standby loads, and then optional standby loads. Thus, if the connected load is greater than the system capacity, the optional standby loads must be shed first when the load becomes too great for the supply.

If a fire pump is part of the emergency system, the fire pump usually represents a significant portion of the emergency load. Most of the time, a transfer to emergency power is due to a power failure of one form or another. If the fire pump is not being used, the excess capacity may be used for other loads, such as optional standby loads. Generally, if a fire pump is connected to more than one source, it will require its own transfer switch. Such a system, which is set up to shed the optional loads if the fire pump starts, is an example of the load shedding contemplated by this rule. Similarly, a generator may be used for peak load shaving, where the generator is designed to operate in parallel with a utility source, but only when the load reaches some predetermined level. The same generator may be used as an emergency power source if equipped with appropriate controls.

5. Verify that power sources are suitable and capable of supplying the load within 10 seconds and maintaining the load for at least 1½ hours.

Power sources for emergency systems must be capable of providing current to the required loads within 10 seconds of the failure of the normal supply. For battery-supplied systems, including unit equipment, the power supply must be adequate for 1½ hours of operation. Systems supplied by storage batteries other than in unit equipment must maintain at least 87½ percent of normal voltage over the 90-minute period. Alternatively, unit equipment batteries must maintain at least 60 percent of the initial level of emergency illumination. Generators and fuel cells must have at least a 2-hour fuel supply. Uninterruptible power supply (UPS) systems used to supply emergency power must meet the appropriate requirement based on whether the UPS system is supplied by a generator or batteries or both. The transfer to battery systems may be instantaneous or nearly so, but prime-mover-driven systems require a few seconds to start and be ready for connection to the load.

6. Verify that generators or fuel cells, if used, have on-site fuel adequate for at least 2 hours of operation and that fuel pumps, if any, are supplied by emergency power.

As noted in Item 5, internal combustion engine generators must have an on-site fuel supply that will last for at least 2 hours under full-load conditions. The same requirement applies to the fuel supply for a fuel cell system. Base-mounted tanks or other tanks located near a generator that can store fuel adequate for at least 2 hours of operation may be the only tanks supplied. Many systems, especially those that are also designed to supply optional standby systems, have "day tanks" capable of storing a minimum amount of fuel combined with larger tanks that store enough fuel for an extended period. These systems usually have level controls and electric pumps to keep the day tank full (See Figure 11-1 for an illustration of a standby power source with a day tank). Such a pump system must be supplied by emergency power. Other standards, such as NFPA 110, *Standard for*

Emergency and Standby Power Systems, include requirements that specify longer time frames for the on-site fuel supply for specific emergency conditions.

To save space, fuel tanks and their associated pumps, sensors, and controls are often mounted on the same frame or base as the generator, frequently under the generator so that the tank forms part of the base for the generator as illustrated in Figure 11-1. Often the generator/tank assembly is mounted inside a building as shown in Figure 11-1. Since diesel is a common fuel source with a relatively high flash point (above typical temperatures in conditioned spaces of buildings) and the fuel is in contained systems, these installations usually do not create hazardous (classified) locations (see the Hazardous Locations chapter of this manual).

7. Verify that unit equipment, if used, is fixed in place and connected to the same circuit that supplies normal lighting to the area, ahead of any local switches.

Unit equipment is commonly used for emergency illumination, usually in the form of self-contained exit signs or emergency/egress lights. Unit equipment may also be fitted to individual luminaires (usually fluorescent) to power all or part of the luminaire during a power failure. Unit equipment must be fixed in place so that it is not portable. Cord-and-plug connections may be used if the cords are not longer than 3 ft (900 mm). In general, unit equipment must be connected to the same branch circuit that supplies lighting for the area served by the unit equipment. This rule is based on the fact that unit equipment that is fed by other circuits or from other panels might not operate on a localized failure of a panelboard supply or a local lighting circuit. However, if a separate uninterrupted area is served by three or more normal lighting circuits, none of which are part of a multiwire branch circuit, a separate circuit may be used for the unit equipment if it originates at the same panelboard as the normal circuits and the unit equipment is supplied by an over-current device that is equipped to be locked in the on position. Self-contained emergency lighting units are designed to automatically transfer to alternative power (battery) when the normal lighting branch circuit for an area loses power (**FIGURE 11-3**). Some units like these are designed so that "remote heads" can be connected to the alternative power source for lighting larger areas from a single source. Wiring to these remote heads is required to be installed using a wiring method from Chapter 3 of the *NEC*. Remote heads supplied by unit equipment that is connected to a lighting circuit for a space inside of a building may be used to provide lighting for the exterior of an exit door from that same space.

In some cases, unit equipment may be preferred or required even where an alternate source such as a generator provides emergency power. For example, in anesthetizing locations in health care facilities, unit equipment is required to provide immediate lighting (rather than waiting for 10 seconds for transfer to an alternate source) in case of a failure of the normal power source. In other cases, unit equipment may be desired for the same reason even where it is not required by the *NEC* (see Item 30 of the Health Care Facilities checklist in the Special Occupancies chapter of this manual).

8. Verify that transfer equipment is automatic, listed for emergency use, equipped with means for bypass, electrically operated, and mechanically held.

Transfer of emergency circuits to alternative power sources must be automatic and must be designed to ensure that interconnection of power sources will be avoided. Automatic transfer switches are commonly used for this

FIGURE 11-3 Self-contained, fully automatic unit equipment for operating emergency lighting. (Courtesy of Dual-Lite, Inc.)

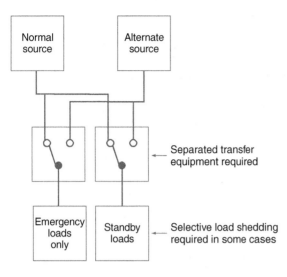

FIGURE 11-4   Supplying emergency and standby loads from a common power source.

purpose. The transfer equipment must be listed for emergency use. An automatic transfer switch must be electrically operated and mechanically held. Some means of bypassing and isolating the transfer equipment should be supplied in order to test the system without interruption of power. Bypass is permitted, not required; however, if the load is really critical, disconnecting the load and perhaps evacuating the building each time a test is performed may not be acceptable and bypass may be necessary.

Where the emergency source is also used for parallel power production, the requirements of Article 705 must also be met to ensure that the system can operate safely while connected to another source and that all sources can be isolated for servicing.

9. Verify that transfer equipment supplies only emergency loads.

Although an emergency power source may supply other loads, the transfer equipment used for emergency power may not supply any loads but emergency loads. Additional transfer equipment is required for legally required or optional standby loads. This rule is intended to increase the reliability of the emergency system by separating it from other systems, and it is closely related to the rules discussed in Items 13 and 15 of this checklist. (This rule is modified in some health care facilities, where all of the essential system, which includes emergency-type loads on the life safety branch, may be supplied by a single transfer switch when the total load is not over 150 kilovolt-amperes [kVA].) The required emergency loads must be supplied by a transfer switch or switches dedicated to the emergency system. A common power source is permitted to supply additional transfer switches for nonemergency loads. **FIGURE 11-4** shows the proper supply arrangement where emergency and other standby loads are supplied.

10. Verify that audible and visual signals are provided as required. (Automatic disconnection of emergency systems on ground faults is not required.)

Certain conditions of an emergency system are required to be reported by audible and visual signals. The signals are required "where practicable," which means where it is feasible or where it can be accomplished in practice, not where it is only theoretically possible or only where it is relatively easy or practical. The conditions that are to be reported are:

- *Derangement*—essentially any malfunction or other out-of-order condition of the power source
- *Carrying load*—applies to batteries when carrying load
- *Not functioning*—applies to battery chargers
- *Ground fault*—applies to emergency power supplies that, because of their ratings and characteristics, would otherwise require GFP for a service or feeder circuit

GFP is not prohibited on emergency systems, but rather than apply GFP to equipment that could disconnect emergency power, sensors that activate signaling devices on a ground fault of 1200 amperes or more must be supplied. This is a modification of the requirements for GFP for equipment found in 215.10 for feeders and 230.95 for services, but it does not apply to the normal power source, only to the emergency source. Instructions to personnel on the course of action to be followed in the event of a ground fault must be provided at or near the location of the sensor. Other generator status indications and controls are often provided and may be required by building or fire codes. (See NFPA 110 for more information on generator signals.)

11. Check for signs at service equipment indicating emergency system type and location and for signs at grounding locations indicating sources connected.

Where emergency power is provided by storage batteries, a UPS system, a generator, or other on-site centralized power source, a sign must be provided at the service equipment.

The sign must indicate the type of power system present and its location. This rule does not apply to facilities where emergency power is provided by unit equipment.

Signs are also required where the grounding electrode conductor connection to a grounded circuit conductor of an emergency system is located remote from the emergency system power source. For example, this often happens when the emergency system power source is not a separately derived system, such as where a 4-wire, generator-supplied system is connected through a three-pole transfer switch. Article 250 generally prohibits additional grounding connections to a grounded conductor once the system has been grounded at the service, so the grounding connection is usually not located at the generator in this example. This means the grounding reference for and fault current path to the alternate source are through the normal source equipment. So where the emergency power source is not separately derived and does not have its own electrode connection or system bonding jumper, disconnection of the grounding/bonding connection at the service also interrupts the connection to the emergency power source. Therefore, a warning sign must be posted at the normal source that reads as follows:

WARNING: SHOCK HAZARD EXISTS IF GROUNDING ELECTRODE CONDUCTOR OR BONDING JUMPER CONNECTION IN THIS EQUIPMENT IS REMOVED WHILE ALTERNATE SOURCE(S) IS ENERGIZED.

The sign must comply with the requirements of 110.21(B) with regard to adequacy, permanence, and durability.

12. Check boxes and enclosures for permanent identification as components of the emergency system.

The *NEC* does not specify how the enclosures and boxes are to be marked, other than requiring that the marking be permanent. Certainly, signs or stickers labeled "Emergency System" or with a similar legend could meet this requirement. Boxes and other enclosures are often painted red to identify them as emergency, but because this method is also sometimes used to identify fire alarm devices and systems, the inspector should verify that such identification is clear and not ambiguous. This rule helps to ensure separation of systems by identifying the enclosures that are required and intended to contain only emergency circuits (see Item 13).

Identification of raceways and cable assemblies is not required, because they do not fall under the definition of "enclosure" in Article 100. Nevertheless, raceway and cable identification may be desired and is permitted and can be required by specification or other contract documents even though it is not required by the *NEC*.

13. Verify that emergency wiring is entirely independent of other wiring, except as specifically permitted for common enclosures, luminaires, and boxes.

Emergency wiring must be separated from wiring of other systems. Emergency wiring is permitted to be in the same cable, raceway, or enclosure only where the wiring of the other system is essential for operation because both systems connect to the same equipment. Emergency wiring is permitted with other wiring in transfer equipment enclosures, in exit or emergency lighting luminaires that connect to two power sources, in the junction boxes attached to such exit or emergency luminaires, in listed load control relays supplying exit or emergency luminaires, or in boxes connected to unit equipment where the supply circuit and the emergency circuit supplied by the unit equipment are in the same enclosure. None of these exceptions permits mixing of emergency circuits with other circuits in raceways or cables—only in the luminaires, equipment, or junction boxes as necessary. The circuits supplied by the emergency system are required to be physically separated from the normal and standby circuit conductors (**FIGURE 11-5**).

Other alternatives are permitted for separation of wiring from an emergency source that also supplies legally required or optional standby systems or both. These alternatives

FIGURE 11-5   Separation of emergency system branch-circuit wiring from standby and normal power system wiring.

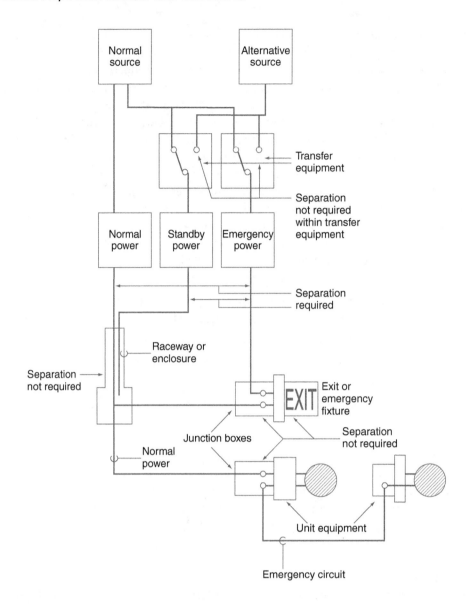

permit the separation to begin at the distribution equipment rather than the generator source, even, in some cases, where there is an overcurrent device at the source. These options are coordinated with the requirements for protection of generator conductors and generators found in 240.21, 445.12, 445.13, and 445.19.

A generator is permitted to supply a vertical switchboard with separate sections or individual enclosures with overcurrent protection in accordance with 445.19. In addition, Section 700.9(B)(5) permits an emergency source to supply "any combination of emergency, legally required, or optional loads" if one or more of four conditions are met (**FIGURE 11-6**):

- The different systems (types of loads) are supplied from separate vertical switchboard sections or from individual disconnects in separate enclosures.
- The common bus of the switchboard or the individual enclosures are supplied by single or multiple feeders from the power source without overcurrent protection at the source, or the overcurrent protection at the source is selectively coordinated with the downstream overcurrent protection in each system.
- Emergency loads are supplied by different vertical switchboard sections or by different individual disconnect enclosures from legally required and optional standby loads.

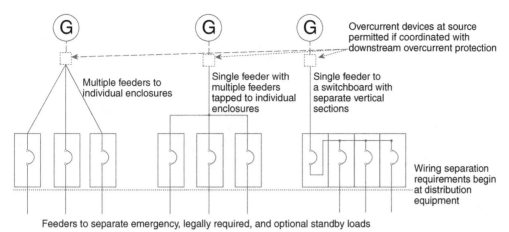

FIGURE 11-6   Separation of emergency system supply wiring from wiring of other standby systems can begin at distribution equipment.

- Single or multiple feeders are run between an emergency source and the supply distribution equipment where the systems are separated.

Figure 11-6 illustrates the first three of these arrangements. The fourth condition would allow multiples of one of the first three for large multibuilding campuses.

14. Verify that emergency feeder circuits and equipment in high-rise (over 75 ft) buildings and assembly occupancies of over 1000 persons have suitable fire protection.

In addition to the separation requirements listed in Item 13, emergency feeders and related feeder equipment must have fire protection to increase the reliability of such systems in fire conditions. The feeders must be protected by installing the feeders in a building that is fully protected by an approved automatic fire suppression system, or installing a listed electrical circuit protective system or listed thermal barrier system with a 2-hour fire rating to protect the circuit, or installing the circuit in a listed fire-rated assembly that contains only emergency circuits and has at least a 2-hour fire rating, or embedding the circuit in at least 2 in. (50 mm) of concrete. Type MI (mineral-insulated, metal-sheathed) cable is an example of a wiring method that has an inherent fire rating that meets the requirements for maintaining the circuit integrity requirement of this section without additional protection such as concrete encasement or installation in a fire-resistive envelope. Emergency feeder equipment such as transformers, panelboards, and transfer switches must be installed in spaces protected by a 2-hour fire-resistant rating or in spaces that are fully protected by automatic fire suppression systems. **TABLE 11-2** summarizes and compares the requirements for physical protection of emergency circuit conductors and distribution equipment and fire pump circuit conductors. These additional rules do not apply to buildings that do not fit the description in this checklist item.

15. Verify that emergency branch circuits supply only emergency loads.

Emergency lighting circuits may supply only emergency lighting loads. Nonemergency equipment may not be supplied by emergency circuits (see Item 9 of this checklist). The determination of what is an emergency load and what is not must be made by the AHJ.

16. Verify that power to emergency lighting in areas served by high-intensity discharge (HID) luminaires is maintained until normal illumination is restored.

HID lighting does not come back on (or restrike) immediately after power is restored and involves some cool-down and warm-up time to reach full intensity. Therefore, where HID lighting is used for normal illumination, another source is required to provide emergency illumination until the HID luminaires come back on. HID luminaires are often available with an optional "restrike" lighting feature that keeps an incandescent source, usually a quartz lamp, lit until the HID source restarts. Such luminaires, when supplied

TABLE 11-2	Protection Methods for Fire Pump Conductors, Emergency Feeders, and Emergency Distribution Equipment Where Routed or Located Inside Building

Method of Protection	Fire Pump Conductors per Article 695	Emergency Feeders for Assembly Occupancies Greater Than 1000 People or Above 75 ft (23 m) in Height per Section 700.10(D)	Emergency Distribution Equipment for Assembly Occupancies Greater Than 1000 People or Above 75 ft (23 m) in Height per Section 700.10(D)
1. Raceways embedded in 2-in. (50-mm) concrete	Permitted	Permitted	Not applicable
2. Located in buildings fully protected by automatic fire suppression system	Not permitted	Permitted	Permitted (sprinkler protection in equipment space)
3. Listed electrical circuit protective system with minimum 2-hour fire rating	Permitted (load side of service only)	Permitted	Not applicable
4. Listed thermal barrier system for electrical system components	Not permitted	Permitted	Not applicable
5. Listed fire-rated assembly with minimum 2-hour fire rating	Permitted (load side of service only)	Permitted (assembly restricted to emergency circuits only)	Permitted (fire-rated building space)

Source: *National Electrical Code® Changes*, NFPA, Quincy, MA, 1999, page 425, as modified.

with emergency lighting circuits, may function as both normal and emergency lighting sources. Other alternatives for providing maintained illumination are also permitted.

17. Verify that emergency lighting equipment is arranged so that an area will not be left in total darkness by the failure of a single lighting element.

Most unit equipment, such as emergency lights and exit lights, have more than one lamp so that the failure of a single lamp will not completely disable the unit equipment. Where other forms of emergency lighting are used, such as ballasts with batteries that power an individual lamp, the equipment must supply more than one lamp in an area or more than one arrangement of such equipment must be supplied. Equipment in adjacent areas may also be used to keep an area from being left "in total darkness." The "single elements" mentioned in this rule do not include the battery pack or ballast or the electrical or electronic circuitry that control the transfer to batteries. Lamps are specifically included in this rule because their eventual failure is a normal occurrence.

18. Verify that emergency lighting is supplied automatically on failure of the normal power supply.

Emergency lighting must be automatically connected to an emergency source on failure of the normal supply. More than one scheme is permitted, but automatic connection to the alternative power supply is required for emergency lighting and other equipment on the emergency system. This function should be confirmed and is easily verified by testing (see Item 21 of this checklist). Where automatic load control relays are used, the relay is not to be used as transfer equipment. See the definition of **Automatic Load Control Relay** in the Key Terms section. An automatic load control relay is connected to only one source and acts to energize emergency lighting loads that are normally off upon loss of normal supply.

19. Verify that any switches that may disconnect power to emergency lighting are conveniently accessible, but only to authorized persons.

Switches installed in emergency lighting circuits must be arranged and located so that only authorized persons can disconnect emergency lighting circuits. Switches that act to energize but not de-energize emergency lighting may be located where **accessible** to unauthorized persons. Three- and four-way switches or switches connected in series so that both switches must be turned on to energize the circuit may not be used in emergency lighting circuits. Generally, all switches that control emergency lighting must be conveniently accessible to those persons responsible for the emergency lights. In theaters and places of assembly, such switches must be located in or convenient to the lobby. Switches that can disconnect emergency lighting circuits may not be located in projection booths or on stages or similar platforms, but switches that can only energize emergency circuits are permitted in such locations.

20. Verify that branch-circuit overcurrent devices in emergency circuits are accessible only to authorized persons.
Emergency-circuit branch-circuit overcurrent devices must be located where they are accessible only to authorized persons so that the possibility of inadvertent or malicious disconnection is minimized.

21. Verify that all emergency system overcurrent devices are selectively coordinated.
Emergency overcurrent devices must be chosen and designed to be selectively coordinated with all supply side overcurrent devices to prevent a fault on the load side of any single overcurrent device from operating an upstream overcurrent device. This system coordination is intended to localize a faulted circuit without compromising the operation of the emergency system functions in other areas of the building. This will ensure that emergency systems remain operational for as long as possible when needed. The **selective coordination** must be selected by qualified persons, documented, and made available to the AHJ as well as others who design, install, inspect, maintain, or operate the system.

22. Verify that surge protection is provided on emergency system switchboards and panelboards.
A listed surge protection device (SPD) is required to be installed on all emergency system switchboards and panelboards. Like many other rules regarding emergency systems, this requirement is intended to enhance the reliability of the emergency system by reducing the likelihood that an electrical surge could result in an outage of the emergency system. See Article 285 for the type of SPD that should be used in a specific situation.

23. Verify and witness that testing is conducted as required and that a written schedule and record for periodic testing and maintenance are provided.
An emergency system must be tested when it is installed. The inspector must conduct or witness this test. Additional testing is required to be performed periodically on an approved schedule. A means of testing the system under full-load conditions must be provided. Full-load testing may actually energize the emergency system or may be simulated by connection of some phantom load that is at least equal to the anticipated emergency load. Periodic maintenance also may be required, especially where battery systems are involved. (Battery systems include batteries used for engine starting, control, or ignition.) Written records of testing and maintenance must be kept. Requirements and best practices on testing procedures and frequency of testing can be found in NFPA 110.

24. Check for compliance with NFPA 101, NFPA 110, and other applicable building codes.
NFPA 101, *Life Safety Code*, and NFPA 110, *Standard for Emergency and Standby Systems*, may or may not apply to a given installation, depending on which codes and standards are adopted and enforced by the AHJ. These documents do provide very useful information on the proper application of emergency systems and the requirements for emergency illumination. Whichever codes and standards are adopted must be observed.

Checklist 11-2: Legally Required Standby Systems

1. Determine the applicability of Article 701.
See the definition of **Legally Required Standby System** in the Key Terms section.

2. Check equipment for suitability for approval.

All legally required standby system equipment and wiring must be approved for the intended use. The *NEC* does not necessarily require listing of all standby equipment, although certain items—transfer equipment, for example—are required to be listed for emergency use (because there is no separate standard for evaluating legally required standby transfer switches). Section 110.3(A) provides a list of considerations to be used in approving equipment that is not available as a listed product. The inspector may require listed equipment where it is available, as inspectors generally are not able to conduct adequate field testing.

3. Review load calculations and verify that system capacity is adequate.

The system designer should furnish load calculations. Calculated capacity must be at least equal to the entire load to be connected at one time. Where motor loads are a significant portion of the standby load, the power source may have to be oversized to accommodate the starting currents of the largest motor. Power sources such as generators may also have to be derated for installation at altitudes above 3300 ft (1000 m) mean sea level (MSL). Power sources are required to be tested under maximum load conditions.

4. Verify that system capacity is adequate for any optional standby loads it feeds or that automatic selective load pickup and load shedding are provided.

Legally required standby system power sources may feed optional standby loads if they are capable of supplying the entire connected load or if they are equipped to shed the optional standby loads when necessary. Loads are required to be served in order of priority. Thus, if the connected load is greater than the system capacity, the optional standby loads must be shed when the load becomes too great for the supply. See Item 4 of the Emergency Systems checklist for a discussion of the requirements covering standby sources that supply emergency loads as well as standby loads.

5. Verify that power sources are suitable and capable of supplying the load within 60 seconds or less and maintaining the load for at least 1½ hours.

Power sources for legally required standby systems must be capable of providing current to the required loads within 60 seconds of the failure of the normal supply. Some applications may require standby power to be connected in less than 60 seconds. For battery-supplied systems, including unit equipment, the power supply must be adequate for 1½ hours of operation at not less than 87½ percent of the normal system voltage. Generators and fuel cells must have at least a 2-hour fuel supply. Uninterruptible power supply (UPS) systems used to supply standby power must meet the appropriate requirement based on whether the UPS system is supplied by a generator or batteries or both. The transfer to battery systems may be instantaneous or nearly so, but prime-mover-driven systems require a few seconds to start and be ready for connection to the load.

6. Verify that generators, if used, have on-site fuel adequate for at least 2 hours of operation.

As noted in Item 5, generators and fuel cell systems must have an on-site fuel supply that will last for at least 2 hours under full-load conditions. In most cases, on-site fuel supplies or a combination of on-site and utility supplies are required unless the AHJ judges a utility fuel system to be sufficiently reliable on its own.

7. Verify that unit equipment, if used, is fixed in place and connected to the same circuit that supplies normal lighting to the area, ahead of any local switches.

Unit equipment may be used for required standby lighting, usually in the form of self-contained luminaires. (If some equipment is required to be supplied by a legally required standby system, some lighting in the same area may be required to be similarly supplied. Such lighting would be classified as *legally required* rather than *emergency* if it is not for life safety or evacuation.) Unit equipment also may be fitted to individual luminaires (usually fluorescent) to power all or part of the luminaire during a power failure. Unit equipment must be fixed in place so that it is not portable. Cord-and-plug connections may be used if the cords are not longer than 3 ft (900 mm). In general, unit equipment must be connected to the same branch circuit that supplies lighting for the area served by the unit equipment. This rule is based on the fact that unit equipment that is fed by other circuits or from other panelboards might not operate on a localized failure of a panelboard supply or a local lighting circuit. However, if a separate uninterrupted area is served by three or more normal lighting circuits, a separate circuit may be used for the unit equipment if it originates at the same panelboard as the normal circuits and if it is supplied by an overcurrent device that is equipped to be locked in the on position.

8. Verify that transfer equipment is automatic, listed for emergency use, equipped with means for bypass, electrically operated, and mechanically held.
Transfer of standby circuits to alternative power sources must be automatic and must be designed to ensure that interconnection of power sources will be avoided. Automatic transfer switches are commonly used for this purpose. The transfer switches must be listed for standby use, but since there is not a separate listing standard for standby use, equipment listed for emergency use is generally required. Automatic transfer switches must be electrically operated and mechanically held. Some means of bypassing and isolating the transfer equipment should be supplied in order to test the system without interruption of power. Bypass is permitted, not required, but if the load is really critical, disconnecting the load each time a test is performed is likely not acceptable.

9. Verify that audible and visual signals are provided as required. (GFP of equipment on legally required standby systems is not required.)
Certain conditions of a legally required standby system are required to be reported by audible and visual signals. The signals are required "where practicable," which means where it is feasible or where it can be accomplished in practice, not where it is merely theoretically possible and not only where it is easily accomplished or "practical." The conditions that are to be reported are:
- *Derangement*—essentially any malfunction or other out-of-order condition of the power source
- *Carrying load*—applies to whatever power source is used to supply the load
- *Not functioning*—applies to battery chargers
- *Ground fault*—applies to standby power supplies that, because of their ratings and characteristics, would otherwise require GFP for a service or feeder circuit

The alternative source for a legally required standby system is not required to have GFP, even though such protection might be required for the normal source, based on the characteristics and rating of the feeder or service overcurrent device. GFP is not prohibited for legally required standby systems. However, rather than apply GFP to equipment that could disconnect legally required standby power, sensors that activate signaling devices on a ground fault at a maximum of 1200 amperes or more must be supplied. This is a modification of the requirements for GFP for equipment found in 215.10 for feeders and 230.95 for services, but it does not apply to the normal power source, only to the emergency and legally required standby sources. Instructions to personnel on the course of action to be followed in the event of a ground fault must be provided at or near the location of the sensor.

Other generator status indications are often provided and may be required by building or fire codes. See NFPA 110, *Standard for Emergency and Standby Systems*, for more information on generator signals.

10. Check for signs at service equipment indicating standby system type and location and for signs at grounding locations indicating sources connected.
Where legally required standby power is provided by a UPS system, generator, or other centralized power source, a sign must be provided at the service equipment. A similar

sign must also indicate the type of standby power system present and its location. This rule does not apply to facilities where standby lighting is supplied by individual unit equipment.

Signs are also required where the grounding electrode conductor connection to a grounded circuit conductor of a legally required standby system is located remote from the standby system power source. For example, this often happens when the standby system power source is not a separately derived system, such as where a 4-wire, generator-supplied system is connected through a three-pole transfer switch. Article 250 of the *NEC* generally prohibits additional grounding connections to a grounded conductor once the system has been grounded at the service, so the grounding connection is usually not located at the generator in this example. This means the grounding reference for and fault current path to the alternate source are through the normal source equipment. So where the legally required standby power source is not separately derived and it does not have its own electrode connection or system bonding jumper, disconnection of the grounding/bonding connection at the service also interrupts the connection to the alternate power source. Therefore, a warning sign must be posted at the normal source that reads as follows:

WARNING: SHOCK HAZARD EXISTS IF GROUNDING ELECTRODE CONDUCTOR OR BONDING JUMPER CONNECTION IN THIS EQUIPMENT IS REMOVED WHILE ALTERNATE SOURCE(S) IS ENERGIZED.

The sign must comply with the requirements of 110.21(B) with regard to adequacy, permanence, and durability.

11. Check wiring for compliance with the general requirements of Chapters 1 through 4 of the *NEC*. (Separation of standby wiring from other general wiring is not required.)
Wiring for legally required standby systems is required to meet all of the normal provisions of the *NEC* based on the voltage, type of use or supply, and wiring methods used. Separation from other power systems (other than emergency systems) is not required by Article 701.

12. Verify that overcurrent protection devices for legally required standby systems are accessible to authorized persons only.
Legally required standby branch-circuit overcurrent devices must be located where they are accessible to authorized persons only, so that the possibility of inadvertent or malicious disconnection is minimized.

13. Verify that all legally required standby system overcurrent devices are selectively coordinated.
Legally required standby system overcurrent devices must be chosen and designed to be selectively coordinated with all supply side overcurrent devices to prevent a fault on the load side of any single overcurrent device from operating an upstream overcurrent device. This system coordination is intended to localize a faulted circuit without compromising the operation of the standby system functions in other areas of the building. This will ensure that legally required standby systems remain operational for as long as possible when needed. The selective coordination must be selected by qualified persons, documented, and made available to the AHJ as well as others who design, install, inspect, maintain, or operate the system.

14. Verify and witness that testing is conducted as required and that a written schedule and record for periodic testing and maintenance are provided.
A legally required standby system must be tested when it is installed. The inspector must conduct or witness this test. Additional testing should be performed periodically on an approved schedule. A means of testing the system under full-load conditions must be

provided. Full-load testing may actually energize the standby system or may be simulated by connection of some phantom load that is at least equal to the anticipated standby load. Periodic maintenance also may be required, especially where battery systems are involved. (Battery systems include batteries used for engine starting, control, or ignition.) Written records of testing and maintenance must be kept. NFPA 110, *Standard for Emergency and Standby Systems,* provides additional information on testing and maintenance procedures for emergency and standby generators and other power sources.

Checklist 11-3: Optional Standby Systems

1. Determine the applicability of Article 702.
See the definition of **Optional Standby System** in the Key Terms section of this chapter. Optional standby systems supplied by portable or permanently installed power sources are covered by Article 702.

2. Check equipment for suitability for approval.
All optional standby system equipment and wiring must be approved for the intended use. The *NEC* does not necessarily require listing of optional standby equipment, but all equipment must be suitable for the use. Section 110.3(A) provides a list of considerations to be used in approving equipment when listed equipment is not available. Where listed equipment is available, the inspector may require listed equipment, as inspectors generally are not able to conduct adequate field testing.

3. Verify that transfer equipment is suitable for the intended use.
When the standby system is optional, the user can select the loads that will be supplied by the system. Load calculations are based on Article 220, the same as other load calculations, and equipment ratings should be based on that calculation. As noted in Item 2 of this checklist, all equipment, including transfer equipment, must be suitable for the use, but equipment for optional standby systems is not required to be specifically listed or identified for standby use. It must, however, be suitable for the maximum available short-circuit current. Some equipment, such as a manual transfer switch or a portable generator, may be listed, but not necessarily as standby equipment. Automatic transfer is permitted but not required for optional standby systems. Transfer equipment must be designed and installed so that inadvertent connections between normal and standby power sources cannot occur. Transfer equipment must be installed if a utility is either the standby or normal source so that utility workers will not be endangered by a backfeed from the standby system. However, separate transfer equipment is not required for the temporary connection of a portable generator where the normal source is isolated by disconnection of normal conductors or by a lockable disconnecting means and the installation and connection are done only by qualified persons to provide equivalent safety for utility workers.

4. Verify that audible and visual signals are provided as required.
Certain conditions of an optional standby system are required to be reported by audible and visual signals where the power source is permanently installed. The conditions that are to be reported are:
- *Derangement*—essentially any malfunction or other out-of-order condition of the optional standby power source
- *Carrying load*—applies to whatever power source is being used to provide standby power

The signals are required "where practicable," which means where it is feasible or where it can be accomplished in practice, not where it is merely theoretically possible and not only where it is easily accomplished or "practical." For example, if the source is portable and not normally connected or in place, the reporting of a "derangement" signal is not always possible or meaningful, so signals are not required for portable standby sources.

5. Check for signs at service equipment indicating standby system type and location and for signs at grounding locations indicating sources connected.

Where optional standby power is provided by a UPS system, generator, or other centralized on-site power source, a sign must be provided at the service equipment. The sign must indicate the type of standby power system and its location. This rule does not apply to facilities where standby power is provided by unit equipment. Since portable sources may not be "on-site," signs are not required to indicate their location. For example, in a single-family dwelling, as well as in many other occupancies, emergency illumination is not required, but luminaires designed to supply emergency lighting may be used to supply optional standby lighting if desired. For systems equipped with transfer means and intended for connection to portable equipment, a sign at the service indicating the intended connection point of the portable system would meet the intent of this section because the location of the portable equipment is not fixed.

Signs are also required where the grounding electrode conductor connection to a grounded circuit conductor of a standby system is located remote from the emergency system power source. For example, this often happens when the standby system power source is not a separately derived system, such as where a 4-wire, generator-supplied system is connected through a three-pole transfer switch. Article 250 of the *NEC* generally prohibits additional grounding connections to a grounded conductor once the system has been grounded at the service, so the grounding connection is usually not located at the generator in this example. This means the grounding reference for and fault current path to the alternate source are through the normal source equipment. So where the optional standby power source is not separately derived and does not have its own electrode connection or system bonding jumper, disconnection of the grounding/bonding connection at the service also interrupts the connection to the optional standby power source. Therefore, a warning sign must be posted at the normal source that reads as follows:

WARNING: SHOCK HAZARD EXISTS IF GROUNDING ELECTRODE CONDUCTOR OR BONDING JUMPER CONNECTION IN THIS EQUIPMENT IS REMOVED WHILE ALTERNATE SOURCE(S) IS ENERGIZED.

If the standby source is portable and connected to the premises wiring through a power inlet (a male attachment plug), a sign must be provided near the inlet to inform those making a connection of the type of system that may be connected based on the wiring of the transfer equipment. One of the following is required:

WARNING:
FOR CONNECTION OF A SEPARATELY DERIVED
(BONDED NEUTRAL) SYSTEM ONLY

or

WARNING:
FOR CONNECTION OF A NONSEPARATELY DERIVED
(FLOATING NEUTRAL) SYSTEM ONLY

The sign must comply with the requirements of 110.21(B) with regard to adequacy, permanence, and durability.

The 2014 *NEC* permits the use of a flanged inlet as the means of disconnecting a premises wiring system from a portable generator rated 15 kW or less. Prior to this change, a disconnecting means, in addition to the cord-and-plug connection, was required to be installed as the disconnecting means for the alternate source supply to a building or structure.

6. Check wiring for compliance with the general requirements of Chapters 1 through 4 of the *NEC*. (Separation of standby wiring from other general wiring is not required.)

Wiring for optional standby systems is required to meet all of the normal provisions of the *NEC* based on the voltage, type of use or supply, and wiring methods used. Separation from other power systems (other than emergency systems) is not required by Article 702.

7. Verify load calculation of the optional standby system and that source is fully rated for the load where automatic transfer equipment is used.

Sizing of equipment and circuits is based on the load selected by the user. Load calculations are made in accordance with Article 220, and the standby system is required to have sufficient capacity for all of the loads that will be in operation at any given time. The user of the optional standby system is permitted to manually connect the loads that are to be supplied by the standby power source.

Where automatic transfer equipment is used, it must be adequate for the full load that will be transferred unless an automatic load management system is provided to limit the load to no more than the rating of the transfer equipment and capacity of the standby source.

Checklist 11-4: Fire Pumps

1. Determine the applicability of Article 695.

Article 695 applies to electrically driven fire pumps (**FIGURE 11-7**), their power sources, interconnecting circuits, and switching and control equipment, including the control circuits and equipment for engine-driven fire pumps. Article 695 does not cover performance, maintenance, or acceptance testing of fire pumps, and it does not cover pressure maintenance pumps, also known as jockey or makeup pumps. Performance and testing requirements and additional information can be found in NFPA 20, *Standard for the Installation of Stationary Pumps for Fire Protection*, 2010 edition. The *NEC* does cover the performance of the wiring system and the supply source. For example, the source and wiring must be capable of reliably supplying the locked-rotor current (LRC) value of the motors indefinitely. Furthermore, the voltage at the controller is not allowed to drop more than 15 percent under starting conditions or more than 5 percent under 115 percent load conditions. This should be verified by written report or by observation and measurement.

Voltage drop under starting conditions is measured at the line terminals of the fire pump controller based on the rated voltage of the controller. Voltage drop measurements are made at the load terminals of the controller based on the voltage rating of the motor for running conditions.

2. Check equipment for listing.

Fire pump controllers, motors, and transfer switches must be listed for fire pump service. Only electric motors with specified locked-rotor characteristics can be used. **FIGURE 11-8** shows the markings of product testing organizations that enable the inspector to determine that equipment associated with a fire pump installation is listed for fire pump service. Although the *NEC* does not cover diesel engine fire

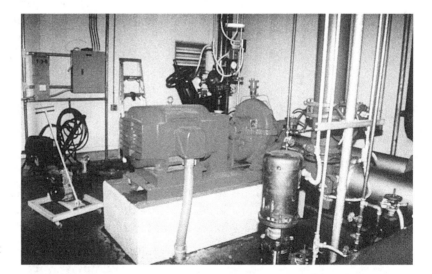

FIGURE 11-7   Electric motor-driven fire pump installation. (Courtesy of Firetrol Brand Products-Emerson Network Power.)

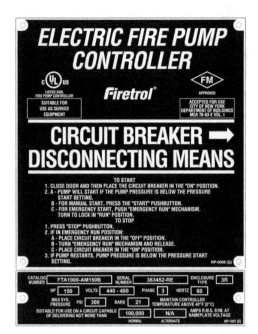

pumps, it does contain some requirements for wiring and installation of diesel engine fire pump controllers, including a requirement for listing of the controllers.

3. Verify that a reliable source of power is provided.

Power for fire pumps must be reliable and have adequate capacity. Reliable power may be supplied by a reliable utility source, an on-site power production facility, dedicated feeders, or a combination of two or more utility sources, on-site power production facilities, or feeders. An individual source that is deemed reliable must be capable of carrying the sum of the LRC of the fire pumps and pressure maintenance motors plus the full-load currents of associated loads indefinitely. A motor will not remain in a locked-rotor condition very long without damage that is likely to result in a short circuit or ground fault, but the motor must be allowed to run as long as possible or to failure. Reliable power can also be provided by a utility service, on-site power production facility, or feeder combined with an emergency or stand-by generator. Because any one source is subject to occasional failure, especially utility sources, more than one source is often required by the AHJ. In some areas, such as areas that are prone to earthquakes, operation of the fire pump may coincide with water, fuel, or power failures or interruptions in the same area, so on-site fuel, water, and power production is often required to ensure reliability. Note that the *NEC* does not specifically require an alternative source, but more than one source may be needed to provide the required reliability. **FIGURE 11-9** shows some acceptable arrangements of supplying reliable power to a fire pump. Section 695.3 provides details for various arrangements of individual or multiple sources, multibuilding campus-style complexes, and on-site standby generators.

The connections shown in Figure 11-9A could be supplemented with other alternative on-site sources if the listed fire pump controller were instead a listed combination fire pump controller and power transfer switch. Although Figure 11-9A shows the fire pumps being supplied by a separate set of service conductors run to separate locations, the connection (and the beginning of separation of conductors) may also be located ahead of and not within the same cabinet, enclosure, vertical switchgear section, or vertical switchboard section as the service disconnecting means. This would allow a single switchboard with a separate tap in the switchboard but in a separate vertical section and ahead of any other service overcurrent devices. In such a case the conductors tapped ahead of any other service overcurrent device or disconnect would still be considered to be service conductors and would still provide a separate service to the fire pump controller.

The generator in Figure 11-9B would have to be classed as an *on-site power production facility* to be the sole source. An *on-site standby generator* could be used only as an alternative source where multiple sources are used. (See the definitions of **On-Site Power Production Facility** and **On-Site Standby Generator**.)

The multiple feeders shown in Figure 11-9C must originate from independent sources in order to be considered reliable. For example, the illustration shows two utility sources, but one feeder is often supplied from a utility and another from an on-site source. The fire pump controller used for multiple sources would have to include a transfer switch.

4. Verify that continuity of power is ensured and supervised.

Continuity of power to fire pumps is critical. Therefore, power connections must be made directly to a listed fire pump controller, a listed combination controller and transfer switch, or a listed fire pump power transfer switch, or it may be made through no more than one disconnecting means that is supervised. Supervision may be provided by

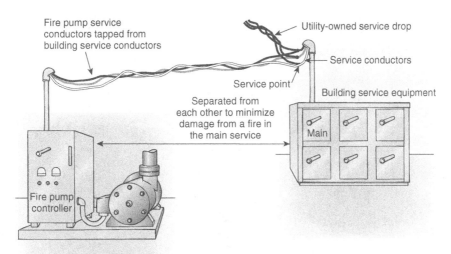

Fire pump service conductors tapped from building service conductors

Utility-owned service drop

Service conductors

Service point

Building service equipment

Separated from each other to minimize damage from a fire in the main service

Main

Fire pump controller

FIGURE 11-9A   Two different configurations permitted for connecting to electric utility-owned service drops.

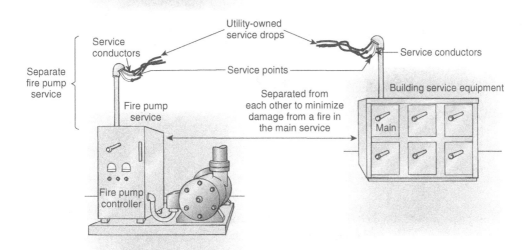

Utility-owned service drops

Service conductors

Service conductors

Separate fire pump service

Service points

Building service equipment

Fire pump service

Separated from each other to minimize damage from a fire in the main service

Main

Fire pump controller

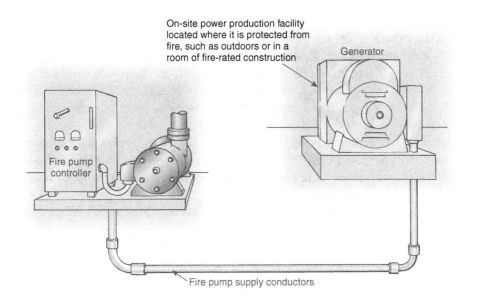

On-site power production facility located where it is protected from fire, such as outdoors or in a room of fire-rated construction

Generator

Fire pump controller

Fire pump supply conductors

FIGURE 11-9B   On-site power production facility as a source for an electric motor-driven fire pump.

FIGURE 11-9C Multiple feeder sources for electric motor-driven fire pumps located in buildings on a campus.

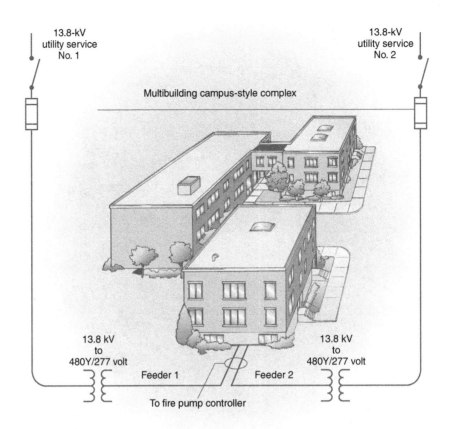

signaling services or devices that are constantly monitored either locally or remotely, by disconnects that can be locked in the on (closed) position, or by location in controlled enclosures or buildings combined with sealing of the disconnect and weekly recorded inspections.

Where there is a disconnect installed between an on-site generator and a controller/transfer switch, the disconnect must be in a separate enclosure from other generator disconnect switches. In other words, if a generator with disconnecting means is used to supply a fire pump and also emergency or standby loads, at least two generator disconnects must be provided.

Listed combination fire pump controllers with power transfer switches allow all power sources to be directly connected (**FIGURE 11-10**). The combination controller can also be used as service equipment. Because the controller must be located near and in sight of the pump motor, this arrangement requires only one piece of equipment to provide power and to control the operation of the fire pump. Although the *Code* permits other arrangements, the simplest arrangement is usually the most reliable and is therefore often preferred. The combination controller shown is also constructed to satisfy the requirements that the disconnecting means be suitable for use as service equipment, be lockable in the *closed* position, and be located remotely from and not in the same equipment as other disconnecting means.

To ensure continuity of power, overcurrent devices must be selected to carry full LRC of fire pump and pressure maintenance pumps. This means that overcurrent protective devices are intended to operate only under high-current fault conditions, not overload conditions. Conductors and other devices are not required to be sized for LRC and may be

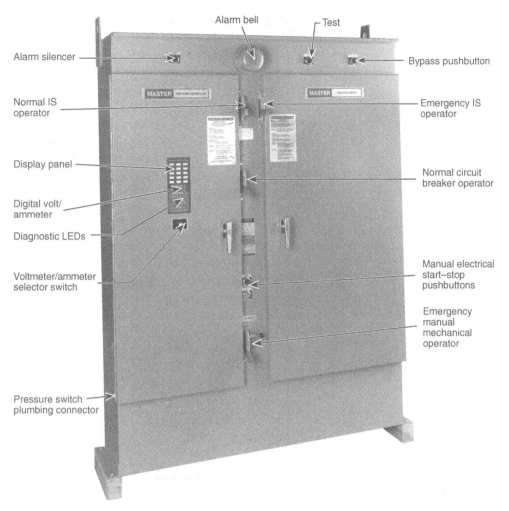

FIGURE 11-10   Combination fire pump controller and power transfer switch. (Courtesy of Master Control Systems, Inc.)

Alarm bell

Test

Alarm silencer

Bypass pushbutton

Normal IS operator

Emergency IS operator

Display panel

Normal circuit breaker operator

Digital volt/ ammeter

Diagnostic LEDs

Voltmeter/ammeter selector switch

Manual electrical start–stop pushbuttons

Emergency manual mechanical operator

Pressure switch plumbing connector

sized based on normal full-load current (FLC) [see 695.6(B)]. The 2014 edition of the *NEC* added other specific requirements for the overcurrent protective devices:

- They must be part of an assembly listed for fire pump service and must comply with four performance requirements or characteristics.
- The overcurrent device may not open within 2 minutes at 600 percent of the FLC, or open with a restart transient of 24 times the FLC, or open within 10 minutes at 300 percent of the FLC.
- For circuit breakers, the trip point may not be field adjustable.

Another change in the 2014 *NEC* clarified the acceptable use of instantaneous (magnetic trip only) circuit breakers in a fire pump circuit. Such circuit breakers are permitted to be used only in listed fire pump transfer or other listed fire pump control equipment and cannot be field installed as a separate circuit protective element. The ability to carry the higher LRC is a normal requirement for many elements of motor circuits and is considered in the testing and listing of controllers and disconnects for fire pumps.

5. Verify that transformers, other than utility or service transformers, are properly sized and protected.

Generally, fire pumps are supplied by separate services at the application voltage of the fire pump. Where the utility voltage is different from the voltage applied to a fire pump motor, transformers (and disconnects) may be installed between the power source and the

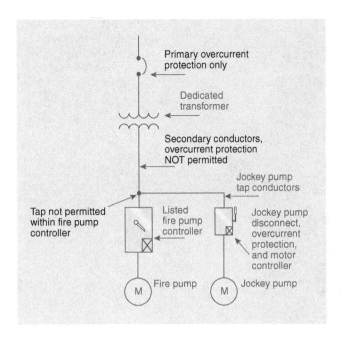

Primary overcurrent
protection only

Dedicated
transformer

Secondary conductors,
overcurrent protection
NOT permitted

Jockey pump
tap conductors

Tap not permitted
within fire pump
controller

Listed
fire pump
controller

Jockey pump
disconnect,
overcurrent
protection,
and motor
controller

M Fire pump

M Jockey pump

FIGURE 11-11   The overcurrent device in the primary of a dedicated transformer supplying a fire pump installation must be sized to carry the locked-rotor current motor(s) and associated fire pump accessory equipment indefinitely.

fire pump controller. This situation is most likely when the service voltage is greater than 600 volts, as in the example calculation in **FIGURE 11-11**. Transformers installed between the power source and a fire pump must be dedicated to the fire pump unless the power source is a feeder. (Utility transformers are not covered by this rule but must comply with the capacity requirements as discussed in Item 3 of this checklist.) Transformers must be rated at a minimum of 125 percent of the sum of the currents of the fire pump(s) and pressure maintenance pump(s), plus 100 percent of any fire pump accessory equipment, plus any other loads if the transformer supplies a feeder in a multibuilding arrangement. Transformer primary overcurrent protection must be capable of carrying the sum of the LRC of the fire pump and pressure maintenance motors plus the FLC of any related accessory equipment. No secondary overcurrent protection is permitted for the transformer. These rules modify Article 450 and are consistent with the requirements for ensuring continuity of power as covered in Item 4.

Figure 11-11 is an example illustration of a transformer and primary overcurrent device for an installation dedicated to a fire pump. The example shows the supply-circuit arrangement for the fire pump and jockey pump supplied by the dedicated transformer.

6. Verify that supply wiring is routed outside of buildings or is otherwise protected against damage and is independent of other wiring.

Supply wiring must be installed as service entrance conductors and kept outside of a building. This applies to conductors supplied from on-site power production facilities as well. Service entrance conductors are required to terminate in service equipment that is located outside or nearest the point of entrance of the conductors to a building. Conductors encased in at least 2 in. (50 mm) of concrete or masonry or located under at least 2 in. (50 mm) of concrete beneath a building are considered to be outside the building [230.6].

Supply conductors (feeders) connected on the load side of the permitted disconnecting means are permitted to be installed within a building or structure using one of the following methods:

A. Routed through the building with a minimum 2-in. (50-mm) concrete encasement

B. Routed within a building in a fire-rated assembly dedicated to the fire pump conductors and having not less than a 2-hour fire resistance rating

C. Installed within listed electrical circuit protective systems that have a minimum of 2-hour fire-resistive rating

Only the first item in this list is considered to make the feeder conductors outside the building, and therefore it is the only one of the three methods that also applies to service conductors. The three listed methods help protect the conductors from fire, but they must also be protected from potential damage due to structural failure and operational accident. Protecting the conductors from fire in the building does not necessarily make the conductors outside the building as required by Article 230 or 695.6(A)(1) for service conductors. Other fire pump circuits connected to the load side of the final disconnecting means may supply only loads directly related to the fire pump and must be kept entirely separate from all other wiring. The requirements for protection of conductors from fire are summarized and compared with similar requirements for emergency systems in Table 11-2.

7. Verify that appropriate wiring methods are used for power and control wiring.

Power and control wiring for fire pumps must be installed in rigid metal conduit (RMC), intermediate metal conduit (IMC), electrical metallic tubing (EMT), liquidtight flexible metal conduit (LFMC), liquidtight flexible nonmetallic conduit (LFNC-B), Type MC (metal-clad) cable with an impervious outer coating, or Type MI cable. These restrictions are intended to provide reasonable protection from mechanical injury for fire pump wiring. Like the wiring on the supply side of the controller/disconnect, load side power and control wiring must also be protected from physical damage. Details for the installation of listed electrical circuit protective systems up to a junction box ahead of the controller, for junction box type and installation, and for raceway terminations and maintenance of enclosure integrity are covered in 695.6(H), (I), and (J). Many of these requirements are extracted from NFPA 20, *Standard for the Installation of Stationary Pumps for Fire Protection*, 2010 edition.

8. Check equipment for appropriate locations and mounting.

Controllers and transfer switches for fire pumps must be as close as feasible to and within sight of the motors. The controller and transfer switch enclosures must be protected from water that might escape from pumps or pump connections. Energized equipment must be at least 12 in. (300 mm) above floor level. All control equipment must be securely mounted on noncombustible supporting structures. Separate structures are not required if substantial noncombustible walls are used. Similar requirements apply to controllers and batteries for diesel engine fire pumps. Generator control wiring that is routed through a building must be protected in the same manner as feeder conductors.

9. Verify that conductors and overcurrent protection are adequately sized.

Overcurrent protection must be sized to provide short-circuit protection only. Automatic overload protection is not permitted for fire pumps. GFPE as normally required by 230.95 for service conductors and 215.10 for feeders is not permitted in fire pump supplies. Overcurrent protection for the individual sources covered in 695.3(A)(1) through (3) must be sized to carry the full locked-rotor of the pump motor plus associated loads indefinitely. Conductors, however, are sized similarly to most other conductors that supply motors. That is, they must have an ampacity not less than 125 percent of the full-load current of the pump motor(s) and pressure maintenance motors plus 100 percent of the associated fire pump accessory equipment. The conductors are not required to be sized for the locked-rotor currents. Overcurrent protection for conductors supplied by an on-site standby generator does not have to be sized to carry locked-rotor current indefinitely. The overcurrent protection in such circuits is required to instantaneously pick up the full load of the fire pump and associated load supplied by the generator and cannot be larger than required by 430.62.

KEY TERMS

The *NEC* defines the key terms pertaining to emergency and standby systems and fire pumps as follows (all definitions are from Article 100 unless otherwise noted):

Accessible (as applied to equipment)   Admitting close approach; not guarded by locked doors, elevation, or other effective means.

Automatic Load Control Relay (Relay, Automatic Load Control)   [*Definition from 700.2*] A device used to set normally dimmed or normally off-switched emergency lighting equipment to full power illumination levels in the event of a loss of the normal supply by bypassing the dimming/switching controls, and to return the emergency lighting equipment to normal status when the device senses the normal supply has been restored.

> *Informational Note:* See ANSI/UL 924, *Emergency Lighting and Power Equipment*, for the requirements covering automatic load control relays.

Coordination (Selective)   Localization of an overcurrent condition to restrict outages to the circuit or equipment affected, accomplished by the selection and installation of overcurrent protective devices and their ratings or settings for the full range of available overcurrents, from overload to the maximum available fault current, and for the full range of overcurrent protective device opening times associated with those overcurrents. [*Selective coordination is very important and required for emergency and legally required standby systems and multibuilding feeder supplies to fire pumps to ensure that overcurrent response to a fault on the normal system will not cause automatic interruption of the alternative system or vice versa. However, selective coordination cannot be witnessed or tested, at least not completely, so compliance with this requirement must be based on a certification by the design engineer(s). The objective is to keep emergency and legally required electrical systems operational and limit any loss of power to the lowest level possible on the electrical distribution system.*]

Emergency System   [*Definition from 700.2*] Those systems legally required and classed as emergency by municipal, state, federal, or other codes or by any governmental agency having jurisdiction. These systems are intended to automatically supply illumination, power, or both, to designated areas and equipment in the event of failure of the normal supply or in the event of accident to elements of a system intended to supply, distribute, and control power and illumination essential for safety to human life.

> *Informational Note:* Emergency systems are generally installed in places of assembly where artificial illumination is required for safe exiting and for panic control in buildings subject to occupancy by large numbers of persons, such as hotels, theaters, sports arenas, health care facilities, and similar institutions. Emergency systems may also provide power for such functions as ventilation where essential to maintain life, fire detection and alarm systems, elevators, fire pumps, public safety communications systems, industrial processes where current interruption would produce serious life safety or health hazards, and similar functions.

[*For specification of locations where emergency lighting is considered essential to life safety, see NFPA 101*, Life Safety Code, *2012 edition. For further information regarding performance of emergency and standby power systems, see NFPA 110*, Standard for Emergency and Standby Power Systems, *2010 edition. For further information regarding wiring and installation of emergency systems in health care facilities, see Article 517 of the NEC. For further information regarding performance and maintenance of emergency systems in health care facilities, see NFPA 99*, Standard for Health Care Facilities, *2012 edition.*

The key to the definition of emergency systems is classification of the loads by the AHJ. Exit and egress lighting are generally recognized as emergency loads. Some other loads, such as certain ventilation fans, may be emergency, legally required standby, optional standby, or normal loads, depending on how the ventilation system is viewed by the AHJ.

Stairwell pressurization, ventilation for areas of refuge, and atrium exhaust may be seen as necessary to life safety or as necessary only to efficient firefighting. Thus, certain loads may be classified differently by different authorities, and the classification may depend on the codes or ordinances that are adopted and enforced.]

Legally Required Standby System   [*Definition from 701.2*] Those systems required and so classed as legally required standby by municipal, state, federal, or other codes or by any governmental agency having jurisdiction. These systems are intended to automatically supply power to selected loads (other than those classed as emergency systems) in the event of failure of the normal source.

> *Informational Note:* Legally required standby systems are typically installed to serve loads, such as heating and refrigeration systems, communications systems, ventilation and smoke-removal systems, sewage disposal, lighting systems, and industrial processes that, when stopped during any interruption of the normal electrical supply, could create hazards or hamper rescue or fire-fighting operations.

[*See the commentary under the definition of* Emergency System. *Some of the examples given as legally required standby loads may be considered emergency system loads by some authorities or in some situations. Fire pumps are a good example. The* NEC *does not require an alternative power source for fire pumps. However, some jurisdictions may treat the fire pump as an emergency load or a legally required standby load if the normal power source is not considered sufficiently reliable.*]

On-Site Power Production Facility   [*Definition from 695.2*] The normal supply of electric power for the site that is expected to be constantly producing power. [*On-site power production facilities are permitted as the primary source of power for a fire pump because they are always on-line. These sources may be in addition to a utility source or may be the only normal source. In contrast, an on-site standby generator is permitted only as a secondary source of power for a fire pump, in the event that the primary source is not considered sufficiently reliable.*]

On-Site Standby Generator   [*Definition from 695.2*] A facility producing electric power on site as the alternate supply of electric power. It differs from an on-site power production facility in that it is not constantly producing power.

Optional Standby System   [*Definition from 702.2*] Those systems intended to supply power to public or private facilities or property where life safety does not depend on the performance of the system. These systems are intended to supply on-site generated power to selected loads either automatically or manually.

> *Informational Note:* Optional standby systems are typically installed to provide an alternative source of electric power for such facilities as industrial and commercial buildings, farms, and residences, and to serve loads such as heating and refrigeration systems, data processing, and communications systems, and industrial processes that, when stopped during any power outage, could cause discomfort, serious interruption of the process, damage to the product or process, or the like.

[*Optional standby systems are, as the name implies, strictly optional from the standpoint of the AHJ and may be connected by either automatic or manual means. Owners of the property may consider these systems vital and may call them emergency systems, and they may supply the optional loads from the same sources as an emergency system, but the portion of the system that supplies loads that are not critical to life safety are not emergency systems as defined and will have to be supplied by separate wiring systems and transfer means. Optional standby systems include those that are supplied by a permanently installed standby power source as well as those supplied by portable sources that are not permanently connected. A generator may be a separately derived system, so parts of Articles 250 may also apply in addition to parts of Article 445.*

Systems powered by nonpermanent sources were added to the Scope of Article 702 because of the proliferation of these types of systems, especially in dwellings, and because of the risks to utility workers from improperly connected portable power sources.]

Transfer Switch **(Switch, transfer)** An automatic or nonautomatic device for transferring one or more load conductor connections from one power source to another. [*Emergency and legally required standby systems must use listed automatic transfer means, but optional standby systems may use either automatic or manual transfer switches.*]

Unit Equipment [*Definition derived from 700.12(F)*] Individual unit equipment consists of (1) a rechargeable battery; (2) a battery charging means; (3) provisions for one or more lamps mounted on the equipment, or terminals for remote lamps, or both; and (4) a relaying device arranged to energize the lamps automatically upon failure of the supply to the unit equipment. [*Unit equipment is commonly used where the only emergency power requirements are for exit and egress lighting. Unit equipment is relatively inexpensive, is easy to install, and, unless remote lamps are used, does not require dedicated or separate wiring systems. This definition is essentially the same as the definition of* Battery-Powered Lighting Units *from 517.2.*]

KEY QUESTIONS

1. What loads are supplied by the emergency or standby system?
2. What alternative power sources are being used for the emergency or standby system?
3. Where is the transfer equipment located?
4. How is the emergency or standby power distributed, and are the emergency system circuits separated from normal and other standby system circuits?
5. What power source is being used for a fire pump?
6. Where is the fire pump located, and how is the power supply routed to the fire pump?

CHECKLISTS

	Item	Inspection Activity	*NEC* Reference	Comments
		Checklist 11-1: Emergency Systems		
✔	Item	Inspection Activity	*NEC* Reference	Comments
	1.	Determine the applicability of Article 700.	700.1	
	2.	Check equipment for suitability for approval.	110.3(A) and (B), 700.5(A) and (C)	
	3.	Review load calculations and verify that system capacity is adequate.	700.4(A)	
	4.	Verify that system capacity is adequate for any nonemergency loads it feeds or that automatic selective load pickup and load shedding are provided.	700.4(B)	
	5.	Verify that power sources are suitable and capable of supplying the load within 10 seconds and maintaining the load for at least 1½ hours.	700.12	
	6.	Verify that generators or fuel cells, if used, have on-site fuel adequate for at least 2 hours of operation and that fuel pumps, if any, are supplied by emergency power.	700.12(B)(2) and (E)	
	7.	Verify that unit equipment, if used, is fixed in place and connected to the same circuit that supplies normal lighting to the area, ahead of any local switches.	700.12(F)	
	8.	Verify that transfer equipment is automatic, listed for emergency use, equipped with means for bypass, electrically operated, and mechanically held.	700.5	
	9.	Verify that transfer equipment supplies only emergency loads.	700.5(D)	
	10.	Verify that audible and visual signals are provided as required. (Automatic disconnection of emergency systems on ground faults is not required.)	700.6, 700.27	
	11.	Check for signs at service equipment indicating emergency system type and location and for signs at grounding locations indicating sources connected.	700.7	
	12.	Check boxes and enclosures for permanent identification as components of the emergency system.	700.10(A)	

(continues)

CHECKLISTS

✔	Item	Inspection Activity	*NEC* Reference	Comments
		Checklist 11-1: Emergency Systems		
	13.	Verify that emergency wiring is entirely independent of other wiring, except as specifically permitted for common enclosures, luminaires, and boxes.	700.10(B)	
	14.	Verify that emergency feeder circuits and equipment in high-rise (over 75 ft) buildings and assembly occupancies of over 1000 persons have suitable fire protection.	700.10(D)	
	15.	Verify that emergency branch circuits supply only emergency loads.	700.15	
	16.	Verify that power to emergency lighting in areas served by high-intensity discharge (HID) luminaires is maintained until normal illumination is restored.	700.16	
	17.	Verify that emergency lighting equipment is arranged so that an area will not be left in total darkness by the failure of a single lighting element.	700.16	
	18.	Verify that emergency lighting is supplied automatically on failure of the normal power supply.	700.17, 700.18	
	19.	Verify that any switches that may disconnect power to emergency lighting are conveniently accessible, but only to authorized persons.	700.20, 700.21	
	20.	Verify that branch-circuit overcurrent devices in emergency circuits are accessible only to authorized persons.	700.26	
	21.	Verify that all emergency system overcurrent devices are selectively coordinated.	700.28	
	22.	Verify that surge protection is provided on emergency system switchboards and panelboards.	700.8	
	23.	Verify and witness that testing is conducted as required and that a written schedule and record for periodic testing and maintenance are provided.	700.3	
	24.	Check for compliance with NFPA 101, NFPA 110, and other applicable building codes.	700.1, Informational Notes	

CHECKLISTS

✔	Item	Inspection Activity	*NEC* Reference	Comments
		Checklist 11-2: Legally Required Standby Systems		
	1.	Determine the applicability of Article 701.	701.1, 701.2	
	2.	Check equipment for suitability for approval.	110.3(A) and (B)	
	3.	Review load calculations and verify that system capacity is adequate.	701.4	
	4.	Verify that system capacity is adequate for any optional standby loads it feeds or that automatic selective load pickup and load shedding are provided.	701.4	
	5.	Verify that power sources are suitable and capable of supplying the load within 60 seconds or less and maintaining the load for at least 1½ hours.	701.12	
	6.	Verify that generators, if used, have on-site fuel adequate for at least 2 hours of operation.	701.12(B)(2)	
	7.	Verify that unit equipment, if used, is fixed in place and connected to the same circuit that supplies normal lighting to the area, ahead of any local switches.	701.12(G)	
	8.	Verify that transfer equipment is automatic, listed for emergency use, equipped with means for bypass, electrically operated, and mechanically held.	701.5	
	9.	Verify that audible and visual signals are provided as required. (GFP of equipment on legally required standby systems is not required.)	701.6, 701.26	
	10.	Check for signs at service equipment indicating standby system type and location and for signs at grounding locations indicating sources connected.	701.7	
	11.	Check wiring for compliance with the general requirements of Chapters 1 through 4 of the *NEC*. (Separation of standby wiring from other general wiring is not required.)	701.10	
	12.	Verify that overcurrent protection devices for legally required standby systems are accessible to authorized persons only.	701.25	
	13.	Verify that all legally required standby system overcurrent devices are selectively coordinated.	701.27	
	14.	Verify and witness that testing is conducted as required and that a written schedule and record for periodic testing and maintenance are provided.	701.3	

CHECKLISTS

✔	Item	Inspection Activity	NEC Reference	Comments
		Checklist 11-3: Optional Standby Systems		
	1.	Determine the applicability of Article 702.	702.1, 702.2	
	2.	Check equipment for suitability for approval.	110.3(A) and (B)	
	3.	Verify that transfer equipment is suitable for the intended use.	702.5	
	4.	Verify that audible and visual signals are provided as required.	702.6	
	5.	Check for signs at service equipment indicating standby system type and location and for signs at grounding locations indicating sources connected.	702.7	
	6.	Check wiring for compliance with the general requirements of Chapters 1 through 4 of the *NEC*. (Separation of standby wiring from other general wiring is not required.)	702.10	
	7.	Verify load calculation of optional standby system and that source is fully rated for the load where automatic transfer equipment is used.	702.4(B)	

✔	Item	Inspection Activity	*NEC* Reference	Comments
		Checklist 11-4: Fire Pumps		
	1.	Determine the applicability of Article **695**.	695.1, 695.2, 695.7	
	2.	Check equipment for listing.	695.10	
	3.	Verify that a reliable source of power is provided.	695.3	
	4.	Verify that continuity of power is ensured and supervised.	695.4	
	5.	Verify that transformers, other than utility or service transformers, are properly sized and protected.	695.5	
	6.	Verify that supply wiring is routed outside of buildings or is otherwise protected against damage and is independent of other wiring.	695.6(A)	
	7.	Verify that appropriate wiring methods are used for power and control wiring.	695.6(D), (F), (H), (I), and (J), 695.14(D), (E), and (F)	
	8.	Check equipment for appropriate locations and mounting.	695.12	
	9.	Verify that conductors and overcurrent protection are adequately sized.	695.4, 695.6(B), (C), and (G)	

CHECKLISTS

CHAPTER 12

Remote-Control, Signaling, and Fire Alarm Circuits and Optical Fiber Cables

OVERVIEW

This chapter covers installations of Class 1, Class 2, and Class 3 remote-control, signaling, and power-limited circuits, power-limited and nonpower-limited fire alarm circuits, and **optical fiber cables**. All of these subjects are covered in Chapter 7 of the *National Electrical Code®* (*NEC®*), and because they represent special conditions, the general rules of the first four chapters of the *NEC* are modified in significant ways. The special conditions for remote-control, signaling, fire alarm, and optical fibers are based on two primary differences from the general conditions of the *Code*: their uses and their energy limitations. Understanding the special uses, rules, and energy limitations associated with these limited-energy systems is the focus of this chapter.

Remote control and signaling are applications that do not usually require very much energy because their primary purpose is not to convert significant amounts of electrical energy to light, heat, or motion, but simply to transmit information to other **devices**. Certainly, these circuits and systems may include lights or sounds to convey information, but the amount of energy used and the functions are very different from light and power circuits and systems. A 120-volt **control circuit** for a motor, as an example, meets the definition of a **Class 1 circuit** although some people treat motor control circuits as their own separate type of circuit. If installed in a **raceway**, even with motor circuit conductors from the motor that it controls (3-phase motor plus three-wire), the conductors of the control circuit are usually not subject to the **ampacity** adjustment required in Section 310.15(B)(3)(a). (See 725.51.)

In a sense, other power-limited circuits and fire alarm circuits and systems are simply specialized applications of remote control and signaling. These special uses justify some modifications of the usual *Code* requirements, but the fact that energy levels are relatively low justifies other modifications as well. Most remote-control, signaling, and fire alarm circuits use small amounts of energy; however, the special rules are based not on energy used, but on energy available. So although two **remote-control circuits** may use similar amounts of energy and operate at similar **voltage** levels, one such circuit may be "energy limited" and another may not.

Energy (power) limitations are significant because when the available energy is limited, the shock and fire hazards that are a primary concern of the *NEC* are significantly reduced or in some cases eliminated. In the case of optical fibers, there is no electrical energy and therefore no shock hazard related to the fibers themselves. Light energy levels are usually limited as well, so there is no fire hazard and, for the most part, no other physical hazards presented by installed optical fiber cables. The light energy in optical fiber could be a hazard in some cases, but a person would not likely be exposed to that in an installed system. Optical fiber has the advantage of having tremendous bandwidth; that is, it is capable of handling extremely high volumes of information at very high speeds. The *NEC* covers optical fibers because they may be installed with or near electrical conductors and usually terminate in devices that are electrical in nature. In addition, the methods used for distributing optical fibers within and between **buildings** are mostly "wiring methods" covered by the *Code*. A major concern of Article 770, *Optical Fiber Cables*, is separation from electrical sources that could create a hazard, especially when the optical fiber cables also include conductors or conductive elements. Article 770 is also concerned with the proper application of both ordinary wiring methods and the special cables that contain the optical fibers. Proper application includes issues such as not interfering with other systems, not spreading fire or products of combustion, and not introducing fire or shock hazards because of the use of conductors or conductive elements in the optical fiber cables.

Remote-control, signaling, and fire alarm circuits bring with them all the issues mentioned for optical fiber cables, along with varying risks of shock or fire. The varying risks are based on the low energy requirements of the circuits and systems and the possible energy limitations. Most power-limited circuits (**Class 2 circuits**) have voltage and current limits that prevent them from being shock or fire hazards, and therefore are permitted to be installed using cables and conductors that are manufactured and **listed** for these specific limited-energy applications. **Class 3 circuits** may have energy levels high enough to be shock hazards, but with proper use and under normal conditions, they are not fire hazards. The nonpower-limited circuits covered here usually are significantly limited in available energy as compared to circuits used for power, light, and heat, but they remain shock hazards if they are not properly **enclosed** and have enough available energy to start a fire under certain conditions. Therefore, nonpower-limited circuits (Class 1 and **nonpower-limited fire alarm [NPLFA] circuits**) generally use ordinary wiring methods described in *NEC* Chapter 3, but rules about minimum wire sizes, the use of derating factors (correction and adjustments to ampacity), and requirements for **overcurrent** protection can be safely modified for these types of circuits. In addition to these modifications, power-limited circuits may also use special wiring methods and cable types that are not suitable for general use, may use smaller wires, and may not need overcurrent protection where the currents from the energy sources are inherently limited.

Because of the special uses and rules, and to maintain and ensure the energy limits of the circuits, the circuits covered in this chapter are also restricted from being in contact or close proximity to higher-powered circuits. In effect, these separation requirements allow the wiring system associated with limited-energy circuits to have lower levels of insulation and resistance to physical damage than power wiring methods. While wiring methods and some other requirements are somewhat relaxed for energy-limited circuits, restrictions on mixing of circuits are significantly greater, and these required separations become an important issue in inspections.

In this chapter, especially for purposes of applying separation requirements, power-limited circuits include Class 2 and Class 3 circuits, **power-limited fire alarm (PLFA) circuits**, communications circuits, community antenna television (CATV) circuits, low-power network-powered broadband communications, and, if only in the way separation requirements are applied, optical fiber cables. Nonpower-limited circuits include circuits for power and lighting as well as Class 1 circuits, NPLFA circuits, and medium-power network-powered broadband communications circuits. Class 1 circuits include both Class 1 remote-control circuits and Class 1 power-limited circuits, which, although they have separate names, are treated the same and often referred to collectively as just Class 1 circuits.

A significant change in the 2008 and 2011 editions of the *NEC* for **grounding** of communications systems is the intersystem bonding termination, defined in Article 100, and required by Sections 250.94, 800.100(B), 810.21, 820.100(B), and 830.100(B). This requirement aids in providing required grounding termination locations for bonding communications systems to the building or structure electrical system. The Grounding and Bonding chapter of this manual covers the requirement for providing an intersystem bonding termination. One final but important note is that abandoned cables are considered a fuel source by fire and building safety professionals and can contribute significantly to the fire load in buildings. Of particular concern is the increased fire load in spaces such as those above dropped or suspended ceilings. This, along with fire ratings and fire penetration concerns, is why abandoned cables are required to be removed unless needed and identified for future use.

SPECIFIC FACTORS

Articles 725, 760, and 770 of the *NEC*, the three articles covered by this chapter, all modify the general requirements of the code in significant ways. As an example, each of these articles provides modifications to the general wiring method requirements of Article 300. This is within the authority of the *NEC* Chapters 5, 6, and 7. This hierarchical arrangement is established in Section 90.3. The rules covered in this chapter are different from the general rules; specifically, Article 300 covers issues such as keeping circuit conductors together, burial depths, protection of conductors, requirements for boxes and **enclosures**, and raceway fill, among other issues.

Although Articles 725, 760, and 770 are much alike, they are also different in important ways. For example, the general requirement from 300.15 calling for boxes and other suitable enclosures at all splice locations is not applicable to Class 2 and Class 3 circuits because there is no specific reference in Article 725 that applies the 300.15 requirement to these circuits. Section 725.3 specifies that the only requirements from Article 300 that apply to remote-control, signaling, and power-limited installations are those specifically referenced in Article 725. However, Class 1 power-limited, remote-control, and **signaling circuits** are not exempt from the requirements of Article 300, because 725.46 specifically references Part I of Article 300. Because of the power limitations inherent to Class 2 and Class 3 circuits, the use of boxes or other enclosures is generally not necessary to prevent fire and shock hazards. This premise does not hold true in hazardous (classified) locations, and the installation of all wiring in these areas must comply with the applicable requirements in *NEC* Chapter 5. Other requirements from Article 300 that do apply are specified in 725.3(A), (B), (C), (H), (I), (J), (K), and (L) and elsewhere in Article 725. It should also be noted that Article 760 requires splices in PLFA circuits wired with PLFA cables to be made in boxes, in enclosures, or on fire alarm devices per the requirement in 760.130(B)(1). This requirement is not based on a concern with fire initiation or electric shock associated with PLFA circuits because neither is a significant hazard with such circuits; rather it is the concern regarding the vital life safety and property protection function of these circuits that is the basis for maintaining a higher degree of mechanical integrity by enclosing splices and terminations.

The obvious issue for inspectors of remote-control, signaling, and fire alarm circuits is the proper use of alternative wiring methods. Proper use requires that cables be selected not only for the class of circuit, but also for the intended location. Different locations, such as **plenum** ceilings or risers, require different types of cable or protective wiring methods. The same can be said of optical fiber cables. For nonpower-limited circuits such as Class 1, ordinary wiring methods are used, but different wire sizes and types (typically fixture wires covered by Article 402) may be permitted. Fire alarm circuits that are nonpower-limited may use either ordinary wiring methods with smaller or different wires like Class 1 circuits, or they may use special NPLFA cables.

To identify the circuits that need inspection under this chapter, markings on **equipment** must be observed. Because of the specialized nature of the circuits involved, remote-control, signaling, and fire alarm circuits, especially power-limited circuits, are frequently installed by people whose knowledge of the *NEC* is focused only on the requirements covering the limited-energy systems. These individuals may assume that as long as they are using the special cable types, pretty much anything goes, and it is all safe because it is "low voltage." The *NEC* covers these circuits because there still are hazards, and low voltage alone does not ensure safety. Limited-energy circuits are quite safe, but only if the specified energy limits are confirmed and maintained. Compliance with required circuit separations is critical to ensuring and maintaining the energy limits, and this is something else that is not well understood and therefore often misapplied.

The myth that low voltage is always safe is often a source of problems for more experienced installers as well. Many 24-volt control systems exceed the power limitations of Class 2 circuits, but installers sometimes think that because it is low voltage, they can use "bell wire" or "thermostat wire," both of which are common names for Type CL2 cables. A 24-volt transformer that is used to supply control or signal circuits, but is not listed as a Class 2 or Class 3 power supply, is a Class 1 source by default, and only Class 1 wiring (ordinary wiring methods of *NEC* Chapter 3) may be used. The voltage that is safe or considered to be "low" is also variable. Article 680 includes a definition of *Low Voltage Contact Limit* that is based on **wet locations** and varies according to the wave shape of the voltage. Similar information can also be found in the notes to Tables 11(A) and 11(B), which show that somewhat higher voltages are safe in **dry locations**.

Some issues that are critical to the proper installation and functioning of fire alarm circuits are not covered by the *NEC*. For one thing, the *Code* does not say where fire alarm systems are to be installed or how or when to test them. The *NEC* contains installation requirements and does not cover the design or functioning of fire alarm systems or panels. "Monitoring for integrity" or what used to be called "circuit supervision" is a function of the fire alarm panel. Nevertheless, the wiring must be laid out according to the manufacturer's and designer's instructions, or the circuits cannot be expected to function properly. For example, to maintain the integrity of a **grounded conductor** in a multiwire receptacle circuit, the grounded conductor is "pigtailed" rather than spliced on a wiring device. This method, called T-tapping in NFPA 72®, *National Fire Alarm and Signaling Code*®, is generally prohibited in fire alarm circuits because it interferes with the ability of a panel to be sure all devices are connected and operating properly. (T-tapping is permitted in addressable systems because the panel can interrogate each device individually and a proper response from each device verifies the integrity of the circuit.) The point is that requirements outside the *NEC* also play a part in the proper installation of fire alarm circuits.

Although this manual does not include checklists that specifically cover the communications systems and circuits covered by Article 800, many of the issues covered here also apply to communications. For example, communications systems are also permitted to use alternative cable types for wiring, and they must be kept separated from higher voltage systems and wiring. In fact, *NEC* Chapter 8 wiring methods are often permitted as substitutes for the wiring methods of Articles 725 and 760. However, *NEC* Chapter 8 requirements are also very different from the circuits discussed here. For example, *NEC* Chapter 8 does not modify the requirements of the rest of the *NEC* the way Articles 725, 760, and 770 do. Instead, the requirements of the rest of the *Code* apply only when Article 800 specifically references them. Also, grounding of communications systems is very different because communications circuits frequently originate outside of the buildings where they are installed. Therefore, grounding is required for protectors or cable shields and some entering raceways to provide a way to equalize or arrest voltages and currents that could otherwise be introduced to the building by the communications wiring, and this is the purpose of the intersystem grounding termination required by Article 250. In some cases, fire alarm or signaling circuits may be treated like communications circuits with regard to grounding and protectors if they are run outside of buildings where they are exposed to higher voltages from contact with power lines or from lightning strikes.

KEY QUESTIONS

1. Are remote-control or signaling circuits installed?

Most buildings and certainly most houses include remote-control and signaling circuits. Signaling circuits are commonly found in doorbell circuits in dwellings and in similar bell and annunciation systems in other buildings. Common applications for remote controls include motor controls, thermostats, door openers, and the like. Computer interconnections (network or data wiring), intercoms, and security systems are other common applications of Class 2 circuits, although this designation has more to do with the power supplies than with use as remote control or signaling.

2. Does the installation include a fire alarm system?

The answer to this question obviously determines whether there will be any inspection under Article 760. Dwelling units usually do not have fire alarm systems (though they may), and the interconnections between residential-type multiple station smoke alarms are not fire alarm circuits by definition because they are not powered by a fire alarm panel. Fire alarm systems are required in high-rise buildings and in many other occupancies. Other codes and standards dictate where fire alarm systems must be installed. Usually, where fire alarm systems are required or installed, the system will be shown on plans or indicated by a circuit for the fire alarm system in one of the panels.

3. Are optical fiber cables used?

Optical fiber cables are seeing increased use in buildings, including dwellings, because of their very high bandwidth. Occasionally, they are used as a backbone for data wiring within a building or may even be extended to individual computers. Generally, if such systems are installed, they will be shown on special systems plans, but they may be installed by specialty designers and contractors who produce their own plans, which are not part of the original layout of the building. Optical fiber cables, where used, may not appear on a job until near the end of construction. Perhaps the best way of answering this question is to examine the types of specialty cables that are actually installed.

4. Where are the power sources for control, signal, or fire alarm circuits?

Fire alarm circuits are powered by fire alarm systems, usually fire alarm panels, so for fire alarm systems, this question is answered by locating the fire alarm panel. Often, it is located in a maintenance office or similar area, but in high-rise or other large buildings, it will be located in a fire command center. In any case, required fire alarm systems will usually be shown on plans. Power sources for other control or signal systems may be, and usually are, located throughout a building, most commonly at the controlled equipment.

5. What types of special-purpose cables are being used?

This question must be answered by a site inspection. The critical questions that follow this one are in the checklists: Are the cables that are used appropriate for the circuits and locations?

6. What locations include special cables?

Since certain types of cables or wiring methods are required for locations such as plenums and risers, the answer to this question will determine what specific cable markings are required or what types of raceways may be used.

7. Do the special cables share raceways or enclosures with any other systems?

This question is especially important where power-limited circuits are used because of the need to maintain separations between the power-limited circuits and higher-powered circuits. Even where nonpower-limited circuits, such as Class 1 circuits, are used, restrictions are placed on the circuits that may be mixed in a raceway or enclosure. For convenience, control wiring for rooftop air-conditioning or heating units is often run through the same raceway and roof penetration as the power wiring, but unless the control circuits are Class 1 and are wired throughout the entire circuit by using ordinary wiring methods, such installations are not *NEC* compliant. A Class 2 circuit in a Class 2 cable may not be run in the same raceway with power conductors. If it is reclassified and installed as a Class 1 circuit and the Class 2 cable is replaced by ordinary wiring methods, mixing of the Class 1 and power circuits may be permitted. Even Class 1 circuits are permitted to be run only with power conductors with which they are functionally associated.

PLANNING FROM START TO FINISH

Many of the checklist items in this chapter can be inspected only at the end stages of a project. Much of the wiring and equipment for control, signaling, and fire alarm systems is not installed or available for inspection until the finish inspection is made. Inspections made prior to the installation of tiles in a suspended ceiling should include a check of the wiring installed in the ceiling space, especially if the space is used for handling air. These inspections could well involve all of the types of wiring discussed in this chapter. Often, some control and signaling wiring is installed as an afterthought, maybe as part of tenant improvements, so spot checking the ceiling spaces after installation of tiles is also good practice.

Inspections of raceways and boxes can be made at underground and rough stages, where ordinary wiring methods are used for Class 1 or NPLFA wiring or for enclosing all or part of other cabling systems. However, at these stages, the wiring methods for control, signaling, and fire alarm may not be readily distinguishable from the wiring for other systems, so inspections will be the same as for other systems. However, in dwelling units, wiring for doorbells, thermostats, and the like is often installed in **concealed** spaces and can be inspected only at the rough stage. The inspector should examine routing of the cables, not only for appropriate methods, support, and protection, but also for likely locations of terminations. For example, one indication that separation requirements are not being followed is Class 2 wiring for a doorbell being installed in the same junction box with 120-volt wiring.

Fire alarm systems are usually indicated on drawings, but an inspector is likely to discover wiring for some limited-energy circuits in a site inspection, whether or not such systems are shown on the plans. The first question following such a discovery should be, "What type of circuit am I looking at?" Since the circuit type is determined by the power supply and not by the cable type, the power source provides the answer to this question. Once the power source is identified, the type of circuit and appropriate cable types are easily determined. So the next step is to verify that the cable types used are appropriate for the circuit type and the locations where the cables are installed. Generally, there are no special cable types for Class 1 systems. Instrumentation tray cable (Type ITC) is used for certain circuits in industrial occupancies that are similar to Class 1 circuits, but the circuits covered by Article 727 are treated as a separate type of circuit, not a special type of Class 1 circuit. The inspector can also investigate issues other than cable types, once the power source has been located. These issues include verification of appropriate wiring methods for nonpower-limited circuits and for the wiring on the supply side of power supplies, as well as verification of proper overcurrent protection, correct installation of the power supplies, compliance with raceway fill where applicable, and conformity with separation requirements.

Regardless of whether control, signaling, or fire alarm systems are encountered, the power supply is the key to classification and represents the interface between the ordinary power system and the special-purpose wiring. Therefore, the power supplies are also the key to most inspections under this chapter. Optical fibers are somewhat different because they do not have "power sources" as such, but they do have some interface with other communications or control systems. The point of entrance is a good place to start inspections of optical fiber cables that originate outside a building. The point of entrance is the location of connections to grounding electrodes where such connections are required, and extensions of optical fiber cables within the buildings will usually originate near the point of entrance. The computer network or communications server or **controller** is the logical starting point for inspections of optical fiber systems that are used for communications or data backbones or for installations where fibers are run to the desk tops.

WORKING THROUGH THE CHECKLISTS

Checklist 12-1: Class 1, Class 2, Class 3, Remote-Control, Signaling, and Power-Limited Circuits

1. Verify that tubing, piping, or other mechanical system is not installed in raceways and cable trays containing conductors for Class 1, Class 2, and Class 3 circuits.

Steam, water, pneumatic, gas, and other nonelectrical systems are not permitted to be installed in the same raceway or cable tray with conductors of Class 1, Class 2, and Class 3 circuits. Installation of such other systems can impact raceway or cable tray fill, and failure of the nonelectrical systems creates the risk of compromising the integrity of the control and signaling circuits. Cable trays are particularly attractive to the installers of other systems as a convenient way of running the piping or tubing. Failure of flammable gas piping or tubing could cause an electrical failure, and a short circuit or ground fault in a power-limited system that would not normally be an ignition source could result in enough thermal energy to cause ignition of many flammable gases.

2. Verify that installation of cables does not prevent access through panels designed to provide access.

Cables that run above ceilings must not interfere with the removal of panels that would otherwise afford access to equipment above the ceiling. **FIGURE 12-1** illustrates acceptable and unacceptable methods for supporting cables above suspended ceilings.

3. Verify that cables have been supported to prevent damage under normal conditions.

Article 725 does not specify the maximum spacing between supports, so this is a subjective judgment for inspectors. Cables and conductors **exposed** on outer surfaces of ceilings and walls must be supported by hangers, straps, staples, cable ties, or other **approved** means. This support hardware must be attached to the structure of a building. Cables must be installed only in "a neat and workmanlike manner." Suspended ceilings are not structural components of a building, so a grid ceiling cannot be used to support cables above or exposed on the ceiling except as permitted in 300.11 for ceilings that are listed for supporting other wiring. Although the provisions of 300.11(A) do not apply to Class 2 and Class 3 circuits, Article 725 does not modify listing requirements of ceiling systems. The authority having jurisdiction (AHJ) must determine what constitutes a neat and workmanlike installation and may choose to refer to other consensus standards for information on standard or acceptable installation practices in making this determination.

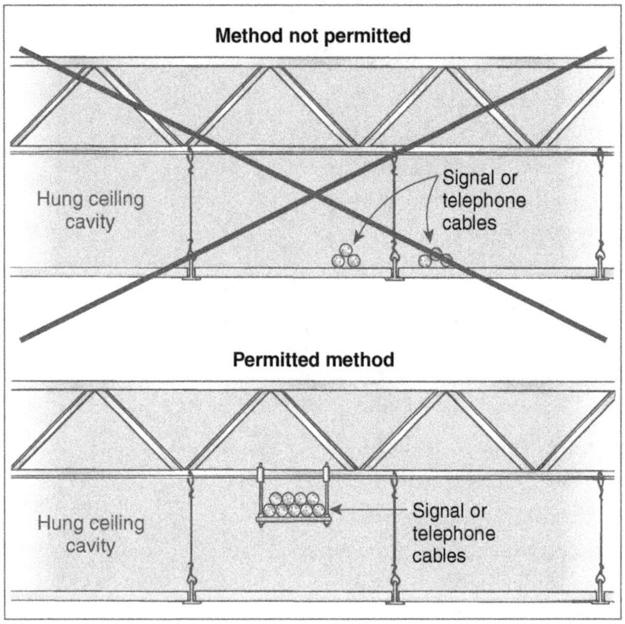

4. Identify safety-control circuits for Class 1 designation.

Some types of equipment rely on remote-control circuits for the safe operation of the equipment. Where the

FIGURE 12-1   Improper and proper installation of Class 1, Class 2, and Class 3 wires and cables above a dropped or suspended ceiling.

failure of a circuit in a safety control system would cause a failure in the equipment and the equipment failure would directly produce a fire or life safety hazard, the control circuits must be classified and installed as Class 1 circuits and be protected by being installed in electrical metallic tubing (EMT), intermediate metal conduit (IMC), rigid metal conduit (RMC), rigid nonmetallic conduit (polyvinyl chloride conduit [PVC] or reinforced thermo-setting resin conduit [RTRC]), Type MC (metal-clad) cable, or Type MI (mineral-insulated, metal-sheathed) cable or be otherwise protected from physical damage. This rule requires reclassification of the circuit, which means that all Class 1 circuit rules apply, so it is not just a requirement to change wiring methods. Short lengths of flexible metal conduit (FMC) or liquidtight flexible conduit (either metallic [LFMC] or nonmetallic [LFNC]) could be used if deemed by the AHJ to be suitably protected, but such wiring methods may be used only where not subject to physical damage.

FIGURE 12-2 shows a common example of a system that includes critical safety controls. In this case, the 120-volt circuit, where the safety controls are connected, is also used as a power circuit for the blower motor and, based on the voltage and current levels, could not be a Class 2 or Class 3 circuit. However, failure of overpressure or overtemperature limits would create a direct fire and life safety hazard in the form of a boiler failure or explosion. Conversely, a failure in the 24-volt Class 2 thermostat circuit might cause the room temperature to become uncomfortably hot or cold but would not itself create a fire or life safety hazard. A judgment of hazards in cases like this must also consider the failure modes of the circuit(s). For example, many overhead door operators use pressure sensors on the bottom of a door or infrared sensors to determine whether there is an obstruction that would interfere with the door closing. One assumption, especially in dwelling unit overhead doors, is that the obstruction might be a small child. The failure of such a safety system could cause the door to be a direct life safety hazard to the child. However, many such door control systems are designed to "fail safe" so that any failure in the sensor circuit will disable the door or allow it only to open and not to close. Circuits of this type do not need to be reclassified.

5. Verify Class 1, 2, or 3 circuit designation.

Class 1, Class 2, and Class 3 circuits are distinguished from other circuits by their use as remote-control or signaling circuits. The three classes are then distinguished from each other by their power limitations. The power limitations are established by the power sup-plies. Although some Class 1 circuits are "power lim-ited," the power limitations are not low enough to elimi-nate both fire and shock hazards, so all Class 1 circuits are treated the same as far as wiring methods are con-cerned. Usually, the obvious sign that Class 2 or Class 3 circuits are being used is the presence of the special Class 2 or Class 3 cables. These cables should be traced back to their power supplies to confirm the ratings of the power supplies. Generally, the power supplies for Class 2 or Class 3 circuits must be listed, and the labeling of the power supply usually answers the question of how the circuits should be classified (see Item 12 of this check-list). Some power supplies are Class 2 under some con-ditions (dry locations, for example) and Class 3 under other conditions (perhaps because supplied circuits extend into a wet location). **FIGURE 12-3** shows markings on power supplies that indicate that connected circuits should be classified as Class 2. Some power supplies are listed other ways but are still considered to supply Class 2 circuits. A common example is a limited power circuit

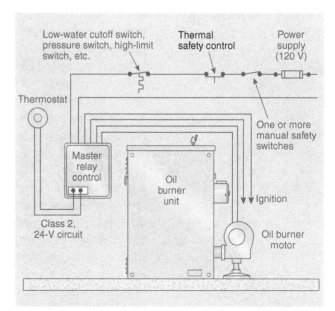

FIGURE 12-2   Elements of Class 1 critical safety circuits for an oil-fired boiler.

FIGURE 12-3   Class 2 control circuit marking on garage door opener.

from listed information technology equipment that supplies network data circuits. Many **nominal** 24-volt control systems are Class 1 circuits, because they are supplied from power sources that are not listed as Class 2 or Class 3.

6. Verify that Class 1, Class 2, and Class 3 circuits are adequately identified at junctions and terminations. The method of identification is not specified. However, the purpose is to "prevent unintentional interference with other circuits during testing and servicing." In other words, remote-control and signaling circuits are likely to require testing at startup and perhaps some servicing or troubleshooting over their installed life. So they must be **identified** in such a way that they will not be mistaken for power circuits and power circuits will not be mistaken for control or signaling circuits. For one thing, control and signal circuits typically have lower power levels than power circuits and therefore often are less hazardous than power circuits, and thus the need for a clear distinction is evident. The type of circuit will be obvious if special cable types are used throughout the circuit, and junction boxes are not required for Class 2 and Class 3 circuits. Class 1 circuit conductors are required to be installed using the same types of methods as power circuits, so markings to indicate the type of circuit at Class 1 circuit junction boxes and equipment terminations are necessary.

7. Verify overcurrent protection for 18 and 16 AWG Class 1 circuit conductors.
Class 1 circuit conductors are exempted from the general rule that requires conductors to be protected at their ampacities. The exemption is found in 240.4(G). Nevertheless, overcurrent protection is required for Class 1 circuit conductors, and for conductors 14 American Wire Gauge (AWG) and larger, the requirement for protecting the conductors at their ampacities is reinstated with modifications:

- For the purposes of selecting the overcurrent protective device, the conductors are not subject to the derating factors of 310.15. These derating factors include correction factors for temperature and adjustment factors for more than three conductors in a raceway. Adjustment factors do apply for the purposes of determining the load to be carried by the Class 1 circuit conductors if the Class 1 conductors are continuously loaded to over 10 percent of their ampacities. Although correction factors may not apply, temperature limits do, according to 310.10 and 402.5, so temperatures must be considered in the selection of insulation types, and correction factors may be applicable in some cases.
- The small conductor rule of 240.4(D) does not apply, so a 14 AWG conductor may be protected at its ampacity as determined by 310.15 and is not always restricted to a 15-ampere overcurrent device.
- Overcurrent protection rules are further modified for motor control circuit conductors that are tapped from the load side of a motor **branch-circuit** and short-circuit device in accordance with 430.72.
- In some cases, the overcurrent protection may be provided on the primary side of a Class 1 transformer or power supply (**FIGURE 12-4**). In this example, the ampacity of the secondary conductors is 8 amperes, and the ratio between the output and input is 1:4, so a 2-ampere fuse in the input circuit will limit the output circuit to 8 amperes.

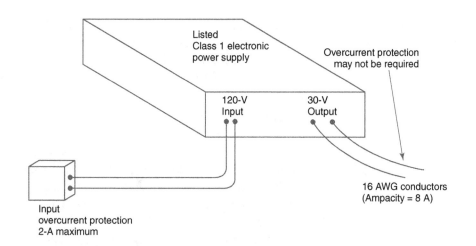

FIGURE 12-4   Example of a Class 1 secondary circuit that is considered to be protected by the primary supply overcurrent device.

- In power and light circuits controlled by Class 1 circuits, overcurrent protection may be provided by a branch-circuit device of the power or light circuit that is rated no more than 300 percent of the ampacity of the Class 1 conductors (**FIGURE 12-5**).

These modifications for 14 AWG and larger also apply to 16 and 18 AWG conductors, except that overcurrent protection for 18 and 16 AWG conductors is specified by 725.43: no

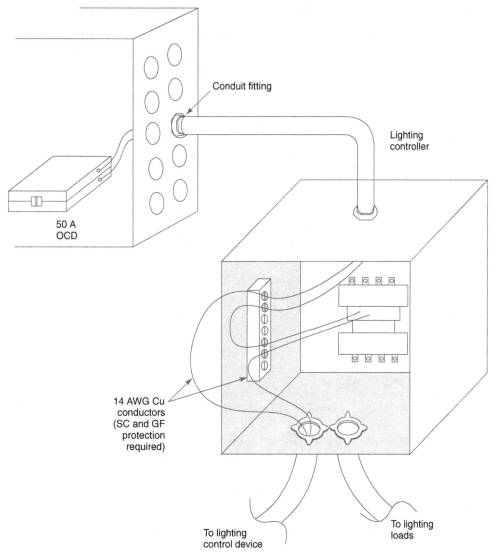

FIGURE 12-5   Example of a Class 1 circuit that is tapped from a branch circuit. The branch-circuit overcurrent device cannot exceed 300 percent of tap conductor ampacity. (20 amperes for a 14 AWG 75 degree C conductor).

greater than 7 amperes for 18 AWG and no more than 10 amperes for 16 AWG. These values are slightly higher than the ampacities for these sizes, which are 6 and 8 amperes, respectively, from Table 402.5 of the *NEC*. Notice that although 240.4(D) does not apply in this case, the overcurrent protection specified in 725.43 is the same as that required by 240.4(D).

8. Check for proper Class 1 wiring methods.

Part I of Article 300 is specifically referenced by Article 725 for Class 1 circuit conductors, and the ordinary wiring methods from Chapter 3 of the *NEC* must be used. Unlike Class 2 and Class 3 circuits, special wiring methods are not available. (Actually, Article 727 provides an alternative wiring method, Type ITC, for circuits that might otherwise be Class 1 circuits, but this method is restricted to industrial occupancies, the power supply is limited to 150 volts and 5 amperes, and the circuits are not considered to be Class 1 even though they may be used for similar purposes.) The ordinary rules of Chapter 3 of the *NEC* are modified with respect to separations from other circuits, sizes and types of wires, and the application of derating factors. For Class 1 circuits, 300.15, which covers the required locations for boxes, is reinstated. (These rules from Article 300 do not apply to Class 2 and Class 3 circuits.)

9. Check the insulation type of 18 and 16 AWG Class 1 circuit conductors.

Insulation types for conductors 14 AWG and larger are selected from Article 310. Conductors sized 18 or 16 AWG are selected from Article 402, but only certain types of fixture wires are included in the list of permissible types. All of the insulation types in the list are 600-volt rated insulations. Other types may be used if listed for Class 1 use.

10. Verify proper raceway fill and derating for Class 1 circuits.

Ampacity adjustment does not apply to Class 1 circuit conductors for the purposes of determining the maximum load that a conductor can supply, but temperature limits must be observed. The adjustment factors of 310.15(B)(3)(a) are specifically referenced. Most Class 1 circuit conductors are not heavily loaded, so the load current does not add significant heat to the conductors in a cable or raceway. This is the reason that adjustment factors do not apply when such conductors are lightly loaded. If anything, the lightly loaded Class 1 conductors may even act as a heat sink for other conductors in the assembly, and where the Class 1 conductors are loaded to more than 10 percent of their ampacities, the application of adjustment factors is more likely to require an increase in the size of the power conductors than an increase in the size of the Class 1 conductors.

Conduit fill limits of 300.17 and Table 1 in Chapter 9 do apply to Class 1 conductors whether they are in raceways with only other Class 1 conductors or share raceways with functionally associated power conductors. These rules restate, but do not modify, the raceway fill or "derating" factors for power or lighting conductors. In effect, Class 1 circuit conductors that are loaded to 10 percent or less of their ampacities are not considered current-carrying conductors for the purposes of applying adjustment factors. Therefore, where such Class 1 conductors are installed in a wireway or auxiliary gutter, they are not counted in the maximum of 30 current-carrying conductors that may occupy any cross-section of these raceways without derating, but they are counted for purposes of wireway or auxiliary gutter fill limitations.

11. Check on separation of Class 1 circuits from power circuits except where functionally associated.

Article 300 permits conductors of various circuits to occupy the same raceway or enclosure as long as all circuits are 600 volts or less. The insulation on all conductors must also be at least equal to the greatest circuit voltage of any circuit in the group. This rule is modified by Article 725 as noted in Item 8 of this checklist. Two or more Class 1 circuits may occupy the same raceway, cable, or enclosure, but all such conductors must be insulated for the maximum voltage of any circuit in the group. Since Class 1 circuits are limited to

600 volts and, as noted in Item 9 of this checklist, since all Class 1 conductors must be insulated for 600 volts, Class 1 conductors can easily and usually be mixed with other Class 1 circuit conductors, and the rule is essentially the same as specified in Article 300.

Mixing Class 1 conductors with power or lighting conductors is another matter. Class 1 and power or lighting conductors can only share a raceway, cable, or enclosure if they are functionally associated. Class 1 and power-supply conductors can also occupy the same control center. Although this usually implies functional association, a control center may include many circuits that are only vaguely associated. Functional association means that the Class 1 circuit directly affects the operation of the power circuit in some manner, such as a control circuit that is used to energize and de-energize the power conductors with which it is installed. In manholes, Class 1 circuits must be separated from other circuits by wiring method, by fixed nonconductors, or by being securely fastened in place. Class 1 circuits may be run in cable trays with circuits that are not functionally associated if they are separated by a barrier. The barrier may be a barrier installed in the cable tray, or the armor of a metal-enclosed cable such as Type MC cable. This rule does not modify the prohibitions on mixing conductors of circuits of less than 600 volts with conductors in circuits that are more than 600 volts.

12. Verify listing of Class 2 and Class 3 power sources.

With the exception of some dry cell batteries and thermocouples, Class 2 and Class 3 power sources must be listed. Listed Class 2 or Class 3 transformers are most common, but other listed and properly marked equipment may also be used. Other Class 2 sources may not be specifically listed or marked as Class 2 power sources, but instead may be limited power circuits of listed information technology equipment (ITE) or may be designated as limited power circuits of equipment that is listed as industrial control equipment.

FIGURE 12-6 is an example of equipment markings that imply Class 2 power limitations. Figure 12-6 shows a listed ITE power supply, in this case, the power supply for a laptop computer. In accordance with the Underwriters Laboratories (UL) listing standard, interconnecting wiring for equipment listed as ITE is assumed to be power limited unless the output connectors are marked otherwise. (Conductors for dial-up modems receive their power from the telecommunications network, so these would be classified as communications circuits.)

FIGURE 12-6   Example of marking on listed ITE.

The power limits for ITE power-limited circuits are actually somewhat lower than permitted for Class 2 circuits. (Although Article 725 considers most computer interconnections to be Class 2, much of the cabling that is installed is classified as communications cable [Type CM], which is an acceptable substitute for Class 2 cable.) In addition to checking for listings of power sources, the location of the power source must also be suitable. On one hand, because of the low shock hazard, Class 2 sources are often located where the load terminals are exposed. Many Class 2 transformers are designed this way, and such exposure is not a code violation. On the other hand, many Class 2 transformers become Class 3 if installed in a wet location or where circuits are subject to contact in a wet location. (See Figure 12-16 in the Key Terms section.)

13. Check wiring methods on the load side and supply side of Class 2 and Class 3 power sources.

Article 725 permits special methods to be used for wiring Class 2 and Class 3 circuits. It also permits the use of Class 1 methods. Three basic choices of wiring methods are given:

- Use a special cable type listed and marked for Class 2 or Class 3 use. This includes the use of suitable listed substitute cables and is the most common choice.
- Use Class 1 methods. This permits the use of ordinary wiring (*NEC* Chapter 3 methods) or perhaps a mix of Class 1 and Class 2 or 3 methods, although the circuit remains Class 2 or Class 3. The use of Class 1 methods is optional in this case.
- Reclassify the circuit as Class 1 and use Class 1 methods. This requires Class 1 methods throughout the circuit installation, but also permits the less restrictive separation rules for Class 1 methods to be used. Reclassification means the circuit *becomes* a Class 1 circuit; it is no longer Class 2 or 3, even though it may remain connected to a Class 2 or 3 power supply, and Class 2 or Class 3 cables may not be used for any part of the circuit. The markings supplying and identifying the circuits as Class 2 or Class 3 must be removed when the circuits are reclassified. Better still, though not required, would be to replace the markings with an indication that the circuits have been reclassified as Class 1.

Wiring methods on the supply side of a Class 2 or Class 3 power source must be the ordinary methods of Chapter 3 of the *NEC*. One significant restriction on the supply side is that overcurrent protection is limited to 20 amperes. The 20-ampere overcurrent device is not required to be a branch-circuit overcurrent device as defined in Article 100, so a supplementary fuse or circuit breaker could be used. Supplementary fuses are often used for this purpose in packaged equipment such as air-conditioning rooftop units.

14. Verify separation of Class 2 and Class 3 circuits from other circuits.

The requirements for separation of Class 2 and Class 3 circuits from each other and from other circuits are much more restrictive than the separation requirements for Class 1 circuits. The circuit separations are required to ensure that the energy limitations and connected components are not compromised by contact or close association with higher-energy circuits. Class 1, power, light, NPLFA, and medium-power network-powered broadband communications are all considered to be higher-power circuits for the purposes of separation. The rules are complex because they address many possibilities. They can be summarized as follows:

- Class 2 or Class 3 circuits may not share a *raceway* with higher-power circuits for general applications.
- Class 2 or Class 3 circuits may share an *enclosure* with higher-power circuits only where they must connect to the same equipment, in which case they may be separated by barriers; by raceways within enclosures; by maintaining a minimum 0.25-in. (6-mm) physical separation; or, where the higher-power circuits do not exceed 150 volts to ground, by cable jackets or by installing the Class 2 or Class 3 circuits using Class 1 wiring methods.

- Where enclosures have a single opening for both Class 2 or Class 3 conductors and higher-power conductors, a fixed nonconductor may be used to provide separations.
- In manholes, separation between Class 2 or Class 3 circuits and higher-power circuits must be maintained by specific wiring methods, by fixed nonconductors, or by racking or otherwise fastening in place.
- In cable trays, separation can be provided by barriers or by installing the Class 2 or Class 3 conductors in Type MC cable, in which case the metallic covering on the MC cable provides separation.
- In other applications, separations may be provided by the use of specific wiring methods, by fixed nonconductors, or by maintaining a minimum 2-in. (50-mm) physical separation.
- Class 2 circuits can be intermixed with other Class 2 circuits, and Class 3 circuits can be intermixed with other Class 3 circuits in the same raceway, cable, or enclosure.
- Class 2 and Class 3 circuits can be mixed in an enclosure, cable, or raceway if conductors and cables with Class 3 insulation are used for all circuits.
- Class 2 or Class 3 circuits can be in the same *cable* with communications circuits, if all circuits are reclassified as communications circuits and installed using communications cables, or they may share listed composite cables that consist of separate jacketed cables under an overall outer jacket.
- Class 2 or Class 3 jacketed cables can be installed in the same raceway or enclosure with the jacketed cables of PLFA, communications, CATV, or low-power network-powered broadband communications circuits or optical fiber cables.
- Class 2 and Class 3 circuits must be separated from audio system output circuits, even though the audio output circuits may be installed using Class 2 or Class 3 wiring methods.

A common violation of the separation requirements for Class 2 power supplies is created by installing a Class 2 transformer in a panelboard, box, or other enclosure where the supply-side and load-side wiring are mixed in the same enclosure. **FIGURE 12-7** illustrates

FIGURE 12-7   Class 2 circuit separated from power conductors by panelboard enclosure.

proper separations achieved by installing a Class 2 transformer on the outside of a panelboard with only the power leads installed with other power conductors in the panelboard. (See 312.8 for permitted use of panelboard cabinet wiring space to contain splices.) Special boxes are also available that maintain separations by including separate compartments for the Class 2 and power circuits. Some of these are made to be installed behind a doorbell assembly or "chime kit" so that all wiring is concealed but still separated.

15. Verify that Class 2 and Class 3 circuits are not supported from raceways or cables.
Article 300 prohibits the use of raceways and cables for support of any other systems including other electrical systems. This rule is repeated in Article 725, but Article 725 recognizes the special rule in Article 300 that permits Class 2 conductors to be attached to a raceway that contains power-supply conductors for the same equipment controlled by the Class 2 conductors. This rule is commonly applied where a raceway is used to supply power to heating, ventilation, and air-conditioning (HVAC) equipment and the Class 2 control circuit cable is attached to the power raceway. There is no such exception for attaching a Class 2 cable to a power cable, however; the special rule applies only to raceways.

16. Verify proper application of Class 2 and Class 3 cables.
Cables for Class 2 and Class 3 circuits are made for specific applications. All such cables must be listed. Cables installed in ducts, plenums, and other spaces used for environmental air must be listed for such use or be installed in metal raceways. Cable types CL2P and CL3P are intended for so-called plenum use. (*Plenum ceilings,* as the term is commonly used, are also called "other space for environmental air" in the *NEC.*) According to 725.3(C), 300.22 applies to these applications, and 300.22 prohibits any wiring in ducts used for loose stock or vapor. Wiring ("equipment") in other ducts and plenums is permitted only where necessary for connecting to equipment that acts on or senses the air in the duct or plenum. **FIGURE 12-8** illustrates the three types of spaces discussed here along with permitted wiring methods. **Accessible** portions of abandoned cables must be removed from all spaces including from plenums.

Cables installed in risers must have an *R* suffix, as in CL2R or CL3R, or be installed in metal raceways or fireproof shafts with firestops. Riser applications in one- and two-family dwellings may use ordinary Class 2 or Class 3 cables. Accessible portions of abandoned cables must be removed. In riser installations, cables are installed in vertical runs that penetrate more than one floor.

Any Class 2 or Class 3 cables may be used in cable trays indoors except those with an *X* suffix. Cables with *X* suffix are limited to dwelling units or installations in raceways.

FIGURE 12-8 Cable installation requirements within ducts, plenums, and other spaces used for environmental air.

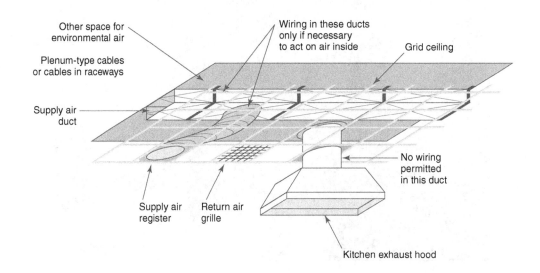

TABLE 12-1	Permitted Cable Substitutions for Class 2 and Class 3 Circuits		

Cable Type	Use	References	Permitted Substitutions
CL3P	Class 3 plenum cable	725.154(A)	CMP
CL2P	Class 2 plenum cable	725.154(A)	CMP, CL3P
CL3R	Class 3 riser cable	725.154(B)	CMP, CL3P, CMR
CL2R	Class 2 riser cable	725.154(B)	CMP, CL3P, CL2P, CMR, CL3R
PLTC	Power-limited tray cable	725.154(C) and (D)	
CL3	Class 3 cable	725.154(B), (E), and (F)	CMP, CL3P, CMR, CL3R, CMG, CM, PLTC
CL2	Class 2 cable	725.154(B), (E), and (F)	CMP, CL3P, CL2P, CMR, CL3R, CL2R, CMG, CM, PLTC, CL3
CL3X	Class 3 cable, limited use	725.154(B) and (E)	CMP, CL3P, CMR, CL3R, CMG, CM, PLTC, CL3, CMX
CL2X	Class 2 cable, limited use	725.154(B) and (E)	CMP, CL3P, CL2P, CMR, CL3R, CL2R, CMG, CM, PLTC, CL3, CL2, CMX, CL3X

Only Type PLTC (**power-limited tray cable**) may be used in outdoor cable trays. Again, the accessible portion of abandoned cables must be removed.

Other special uses and special cable types are permitted in specific situations and occupancies, and substitutions are permitted in accordance with the substitution hierarchy shown in **TABLE 12-1**.

Article 725 provides more than one way to determine the suitability of a specific cable type for a specific application and circuit type. Section 725.135 provides a list of applications based on occupancy type and location with the cable types that are suitable for that application. A similar list is provided in Table 725.154 that provides permitted cable types based on application. Both refer to 300.22 for the restrictions on cabling in various types of air-handling spaces.

17. Check listing and marking of Class 2 and Class 3 circuit conductors.
Section 725.179 describes the markings and characteristics of various Class 2 and Class 3 cable types. The cables are required to be marked CL2 for Class 2 circuits and CL3 for Class 3 circuits. In addition, a *P* suffix for *plenum,* an *R* suffix for *riser,* or an *X* suffix for *limited use* may be used. Type PLTC cable may be used for Class 2 or Class 3 circuits. The required markings are specifically intended to make inspection and identification easy, once the inspector or installer has determined the permissible types. As noted in Item 16, permissible types are listed by application, although generally, any type may be used, even in plenum or riser applications, if installed in a metallic raceway that is suitable for the occupancy and environment.

Checklist 12-2: Fire Alarm Systems

1. Verify that tubing, piping or other mechanical system is not installed in raceways and cable trays containing conductors of Class 1, Class 2, and Class 3 circuits.
Steam, water, pneumatic, gas and other nonelectrical systems are not permitted to be installed in the same raceway or cable tray with conductors of NPLFA and PLFA circuits. Installation of such a system can impact raceway or cable tray fill, and failure of the nonelectrical systems creates the risk of compromising the integrity of the fire alarm circuits. Cable trays are particularly attractive to the installers of other systems as a convenient way of running the piping or tubing. Failure of flammable gas piping or tubing could cause an electrical failure, and a short circuit or ground fault in a power-limited system that would not normally be an ignition source could result in enough thermal energy to cause ignition of many flammable gases.

2. Verify that installation of cables does not prevent access through panels designed to provide access.

Cables that run above ceilings must not interfere with the removal of panels that would otherwise afford access to equipment above the ceiling. (See Figure 12-1; although it is illustrating Class 1, Class 2, and Class 3 circuits, it also can be used as an illustration of acceptable and unacceptable methods for supporting fire alarm cables above suspended ceilings.)

3. Verify that cables have been supported to prevent damage under normal conditions.

Except for circuit integrity (CI) cable, Article 760 does not specify the maximum spacing between supports, so this is usually a subjective judgment for an inspector. The spacing for supports of CI cable may not exceed 24 in. (610 mm) generally. If located below 7 ft (2.1 m) above the floor, the supports may not exceed 18 in. (450 mm). The cable support and fasteners for CI cable must be steel. Other cables and conductors exposed on outer surfaces of ceilings and walls must be supported by hangers, straps, staples, or other approved means. This support hardware must be attached to the structure of a building. Cables must be installed in "a neat and workmanlike manner." (Suspended ceilings are not structural components of a building, so a grid ceiling cannot be used to support cables above or exposed on the ceiling. Section 300.11 does not apply to Article 760, but the listing and use requirements for grid ceilings are not modified by Article 760.) The AHJ must determine what constitutes a neat and workmanlike installation and may choose to refer to other consensus standards for information on standard or acceptable installation practices in making this determination. Raceways used for fire alarm cables or circuit conductors must be installed in accordance with the applicable article from Chapter 3 of the *NEC* because these rules are not modified by Article 760.

4. Check on proper identification of fire alarm circuits at junction and terminal locations.

To minimize unintentional interference with fire alarm systems, all junctions and terminal locations in fire alarm systems must be identified. The means of identification is not specified, but labels or paint are often used. Other equipment such as raceways and cables are not required by the *Code* to be identified as part of fire alarm systems. The purpose of this identification is to prevent unintentional signals on the fire alarm system from being caused by work on other systems. Such signals could be produced by shorting or opening conductors of fire alarm circuits, which would be more likely if the fire alarm circuits were mistaken for power circuits or conductors of other systems.

5. Distinguish nonpower-limited from power-limited fire alarm circuits and verify power supply is supplied by a branch circuit that supplies only loads associated with the fire alarm system and is not protected by GFCI or AFCI devices.

Most modern fire alarm circuits are supplied from fire alarm panels that include power supplies. In fact, according to the definition of **Fire Alarm Circuit** in the Key Terms section of this chapter, fire alarm circuits are fire alarm circuits because they are powered and controlled by fire alarm systems. The power supplies and the labeling of the fire alarm panel provide information about the types of circuits supplied. Systems often use power-limited circuits for initiating circuits (detection circuits) because very little power is used in such circuits. However, to supply sufficient power to many **notification appliances** such as strobes, NPLFA circuits are often needed for the notification circuits in large systems.

Classification of circuits is usually found in fire alarm panel labeling. **FIGURE 12-9** illustrates the various types of circuits (e.g., fire alarm circuits and power or control circuits) that may be connected to a fire alarm panel. In this case, the panel is a household fire warning and burglar alarm panel, and both the initiating and notification circuits are power-limited. Only the battery circuit is nonpower-limited in this example, and the battery wiring does not leave the panel.

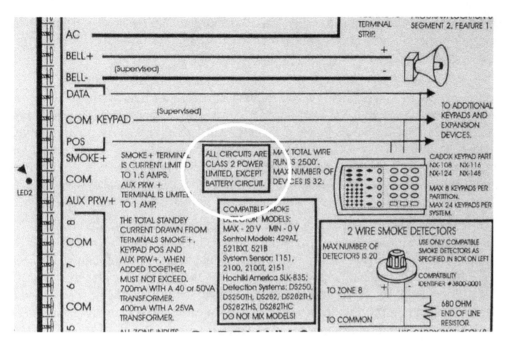

FIGURE 12-9 Example of power-limited marking on listed combination fire/burglar alarm panel.

The fire alarm power sources may not be supplied by **ground-fault circuit interrupter (GFCI)**– or **arc-fault circuit interrupter (AFCI)**–protected circuits, and the branch circuits must be "dedicated" circuits—that is, circuits used to supply no other loads. However, in dwelling units, the omission of AFCI protection on these circuits will require the use of wiring methods that provide some physical protection for the branch circuit to the fire alarm panel if the fire alarm panel is installed in one of the rooms where AFCI protection is otherwise mandated by 210.12. The required methods include RMC, IMC, EMT, steel-sheathed Type AC or type MC cables, and metal **outlet** and junction boxes. Fire alarm panels installed in rooms where AFCI protection is not required do not have to be supplied with the specified wiring methods. Most of the spaces where AFCI protection is not required are covered by 210.8(A) for dwellings, where GFCI protection is normally required. The omission of GFCI protection is recognized only in unfinished basements in 210.8(A). The disconnecting means for the circuit—likely a circuit breaker in a panelboard—must be identified with a red color and **labeled** "FIRE ALARM CIRCUIT."

6. Verify overcurrent protection for 18 and 16 AWG nonpower-limited fire alarm circuit conductors.

Overcurrent protection requirements and the determination of ampacity are identical for NPLFA and Class 1 circuits. Overcurrent protection is based on the ampacity of the conductors without applying adjustment or correction factors. See the discussion in Item 7 of the preceding checklist. However, as noted in Item 5 in this checklist, power sources for the fire alarm circuit are part of the fire alarm panel, which also includes overcurrent protection for the supplied fire alarm circuits. Fire alarm panels will also include instructions on maximum loading (maximum number of **initiating devices** or indicating appliances) on the supplied circuits. Conductor sizing must also consider voltage drop to comply with NFPA 72.

7. Check for proper nonpower-limited wiring methods.

NPLFA circuits are usually installed using ordinary wiring methods of Chapter 3 of the *NEC*. The rules requiring seals in conduit exposed to different temperatures to avoid condensation, requiring boxes, limiting raceway fill, requiring secured and supported wiring methods, prohibiting the use of suspended ceiling grids for support, providing adequate support for conductors in vertical runs, and permitting the use of separate wires for support, are all reinstated from Article 300. The articles that apply to specific wiring methods also apply. All of this is similar to Class 1 circuit requirements. However, Article 760 also permits some special wiring methods for nonpower-limited circuits—specifically, listed multiconductor NPLFA cables.

8. Check insulation type and rating of nonpower-limited circuit conductors.
Insulation types for conductors 14 AWG and larger are selected from Article 310. Conductors sized 18 or 16 AWG are selected from Article 402, but only certain types are included in the list of permissible types. All of the insulation types in the list are 600-volt rated insulations. Other types may be used if listed for NPLFA use. Special insulation types such as PAF and PTF are available for high-temperature applications. Fire alarm circuits must be installed using solid or stranded copper conductors. Restrictions on the use of or number of strands in stranded wire have been eliminated in the *NEC*.

9. Verify proper raceway fill and derating for nonpower-limited circuits.
Ampacity adjustment factors apply to NPLFA circuit conductors for the purposes of determining the maximum load that a conductor can supply, and temperature limits must also be observed. Only the adjustment factors of 310.15(B)(3)(a) are specifically referenced. Most NPLFA circuit conductors are not heavily loaded, so the load current does not add significant heat to the conductors in a cable or raceway. (Maximum numbers of devices and circuit lengths for given supplied circuits and wire sizes are usually furnished as part of the installation instructions of fire alarm panels.) Ampacity adjustment factors do not apply to conductors for the purposes of selecting the overcurrent protective device because the conductors are lightly loaded. Conduit fill limits do apply to NPLFA conductors whether they are in raceways with other NPLFA circuits or Class 1 conductors or, where permitted, share raceways with power conductors. These rules restate, but do not modify, the raceway fill or ampacity adjustment factors for power or lighting conductors. In effect, NPLFA circuit conductors that are loaded to 10 percent or less of their ampacities are not considered current-carrying conductors for the purposes of applying adjustment factors, but they are counted in raceway fill calculations. (Power conductors and NPLFA conductors are permitted in the same raceway, cable, or enclosure only where they connect to the same equipment. This is a somewhat more restrictive rule than the functional association rule that applies when Class 1 and power circuits are mixed together.)

10. Verify proper application of listed nonpower-limited fire alarm cables.
Listed cables for NPLFA circuits are available in three basic types:
  A. Type NPLFP for plenum applications
  B. Type NPLFR for riser applications
  C. Type NPLF for general-purpose use and in raceways
Type NPLFP can be substituted for Type NPLFR, and Types NPLFP or NPLFR can be substituted for Type NPLF. Any of these types may include the additional suffix *CI* for *circuit integrity* cable. (See **Fire Alarm Circuit Integrity [CI] Cable** in the Key Terms of this chapter.) Because fire alarm systems are critical to life safety and are required to perform certain functions during a fire, the use of the listed cable types is restricted in some locations. For example, where exposed or fished into concealed spaces, all splices must be in boxes or other suitable enclosures or on devices, and the cables must be protected from physical damage. NPFLA cable must be installed in a metal raceway or rigid nonmetallic conduit up to a height of 7 ft (2.1 m) above the floor unless concealed (protection is afforded by building construction) or otherwise protected. Where installed in a hoistway, the cable must be installed in RMC, rigid nonmetallic conduit, IMC, LFNC, or EMT unless installed in accordance with 620.21 for elevators, where some cords and cables are permitted.

NPLFP cable is prohibited from being installed in ducts and plenums specifically fabricated for air-handling purposes unless they are installed in a raceway and comply with 300.22. Section 300.22 applies to these installations by reference in 760.3(B). Section 760.3(B) lists exceptions to recognize the uses in 760.53, but exposed NPLFP cables are permitted only in other spaces for environmental air, the so-called plenum ceilings. Since 300.22 applies, NPLF cables may be installed in ducts or plenums where the cable is run in a metal raceway, but only where necessary to connect to equipment in the duct or plenum that acts on or senses the contained air. Most such equipment, such as a duct smoke detector, is more likely to be outside the duct with only a sampling tube extending into the duct.

11. Verify listing of power-limited source and that the power supply is supplied by a branch circuit that supplies only loads associated with the fire alarm system and is not protected by GFCI or AFCI devices.

As noted in Item 5 of this checklist, fire alarm circuits are those powered and controlled by a fire alarm system, so the labeling on the fire alarm panel is the place to find the class and listing information about the power supply. PLFA sources are not always listed as PLFA sources; they may be listed as Class 3 power supplies or transformers. Most large panels include an integral power supply, but many small ones use separate listed plug-in power supplies. The fire alarm power sources may not be installed on GFCI- or AFCI-protected circuits and must be supplied by circuits that supply no other loads. The circuit supplying a fire alarm panel must be identified with a red color and labeled "FIRE ALARM CIRCUIT." See the discussion in Item 5 of this checklist. The rules for power supply to the fire alarm panel are the same for both NPFLA and PLFA circuits, both of which may be supplied by the same fire alarm panel.

12. Check equipment marking to indicate power-limited circuit.

This rule says that the equipment supplying PLFA circuits must be marked to indicate that it is a PLFA circuit. The intent of this rule and the listing requirements is to mark only the panel terminals. This makes sense in light of the fact that the PLFA designation is determined by the power supply, which is usually part of the panel. The locations where PLFA circuits originate are the reasonable place to identify the type of circuit. If the circuits are reclassified as NPLFA circuits, these markings must be obliterated or otherwise eliminated. This requirement should not be confused with the requirement to mark boxes and other terminal locations as covered in Item 4 of this checklist. Terminal and junction locations must be marked to indicate only that they are fire alarm circuits, not necessarily that they are NPLFA or PLFA circuits.

13. Verify proper wiring methods on the load side and supply side of a power-limited source.

Article 760 permits special methods to be used for wiring PLFA circuits. It also permits the use of NPLFA methods. Three basic choices of wiring methods are given:

- Use a special cable type listed and marked for PLFA circuit use. This includes the use of suitable listed substitute cables.
- Use NPLFA methods. This permits the use of ordinary (*NEC* Chapter 3) wiring methods as permitted by 760.25 for NPLFA circuits or NPLFA cable types while the circuit remains PLFA. The use of NPLFA methods is optional in this case. This option is often used when the owner or designer wants to specify separate wires in raceways or to comply with other local codes that govern fire alarm wiring methods. However, if a PLFA circuit runs outside between buildings, the use of NPLFA methods may be desirable for the outside portion in order to use a wiring method that is suitable for the conditions (**FIGURE 12-10**). Communications wiring methods substituted for PLFA cable types are another option.
- Use a combination of nonpower-limited and power-limited wiring methods for the same circuit, if the circuit is not reclassified as NPLFA. Another option is to reclassify the circuit as NPLFA. This requires NPLFA methods throughout the circuit installation but also permits the more lenient separation rules for NPLFA methods to be used. Reclassification requires the elimination of the PLFA markings discussed in Item 12 because the circuit becomes an NPLFA circuit throughout.

FIGURE 12-10 Outside PLFA circuits wired with NPLFA methods.

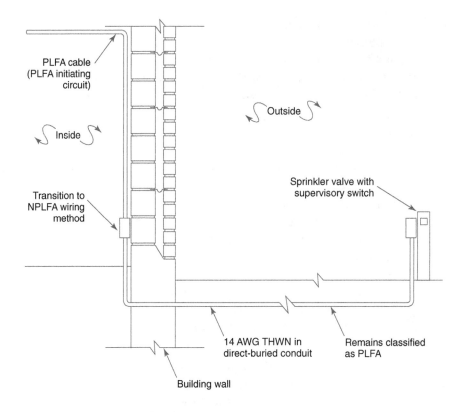

PLFA cable
(PLFA initiating
circuit)

Outside

Inside

Sprinkler valve with
supervisory switch

Transition to
NPLFA wiring
method

14 AWG THWN in
direct-buried conduit

Remains classified
as PLFA

Building wall

Wiring methods on the supply side of a PLFA power source must be the ordinary methods of Chapter 3 of the *NEC*. One significant restriction on the supply side is that overcurrent protection is limited to 20 amperes. NFPA 72, *National Fire Alarm and Signaling Code*, 2013, Section 10.6.5.1, requires the primary power to a fire alarm panel to be supplied by a branch circuit that supplies no other loads. A dedicated circuit is not required for household fire alarm systems according to Section 29.6.3(4), but dedicated branch circuits are required for all occupancies under the *NEC*. NFPA 72 also requires a secondary source, but this source is often supplied with the fire alarm panel. As is mentioned in Items 5 and 11 of this checklist, fire alarm power sources may not be installed on GFCI- or AFCI-protected circuits. These restrictions do not apply to the single- and multiple-station smoke alarms most commonly used in dwellings, but Section 29.6.3(5) in NFPA 72 requires a backup (secondary) power source where such devices are installed on GFCI- or AFCI-protected circuits. This secondary source usually takes the form of a battery installed in each smoke alarm. "Dedicated circuit" in this context means a circuit that supplies no load other than the fire alarm system, but it could supply more than one component of a fire alarm system such as a panel and a separate secondary power source. In that case it would not be an "individual branch circuit" as defined in Article 100 of the *NEC*.

Because fire alarm systems are critical to life safety and are required to perform certain functions during a fire, the use of the listed cable types is modified for some locations. Where exposed or fished into concealed spaces, all splices must be in boxes or other suitable enclosures or on devices, and the cables must be protected from physical damage. PFLA cable must be installed in a metal raceway or rigid nonmetallic conduit up to a height of 7 ft (2.1 m) above the floor unless concealed or otherwise protected. Where installed in a hoistway, the cable must be installed in RMC, rigid nonmetallic conduit, IMC, or EMT, unless installed in accordance with 620.21 for elevators, where some cords and cables are permitted.

14. Verify separation of power-limited fire alarm circuits from other circuits.

The requirements for separation of PLFA circuits from other circuits are much more restrictive than the separation requirements for NPLFA circuits. The circuit separations are needed to be sure the energy limitations are not compromised by contact or close association with higher-energy circuits. NPLFA, Class 1, power, light, NPLFA, and medium-power network-powered broadband communications are all considered to be higher-power circuits for the purposes of separation. The rules are complex because they address many possibilities. They can be summarized as follows:

- PLFA circuits may not share a *raceway* with higher-power circuits for general applications.
- PLFA circuits may share an *enclosure* with higher-power circuits only where they must connect to the same equipment, in which case they may be separated by barriers; by raceways within enclosures; by maintaining a minimum 0.25-in. (6.0-mm) physical separation; or, where the higher-power circuits do not exceed 150 volts to ground, by cable jackets (**FIGURE 12-11**) or by installing the PLFA circuits using NPLFA wiring methods.
- Where enclosures have a single opening for both PLFA conductors and higher-power conductors, a fixed nonconductor may be used to provide separations.
- In other applications, separations can be provided by the use of specific wiring methods, by fixed nonconductors, or by maintaining a minimum 2-in. (51-mm) physical separation.
- PLFA circuits can be intermixed in the same raceway, cable, cable tray, or enclosure with other PLFA circuits or with Class 3, communications, or low-power network-powered broadband communications circuits.
- PLFA and Class 2 circuits can be mixed in an enclosure, cable, cable tray, or raceway if conductors and cables with insulation equal to that required for PLFA circuits are used for all circuits.
- PLFA circuits must be separated from audio system output circuits, even though the audio circuits may be installed using Class 2 or Class 3 wiring methods.

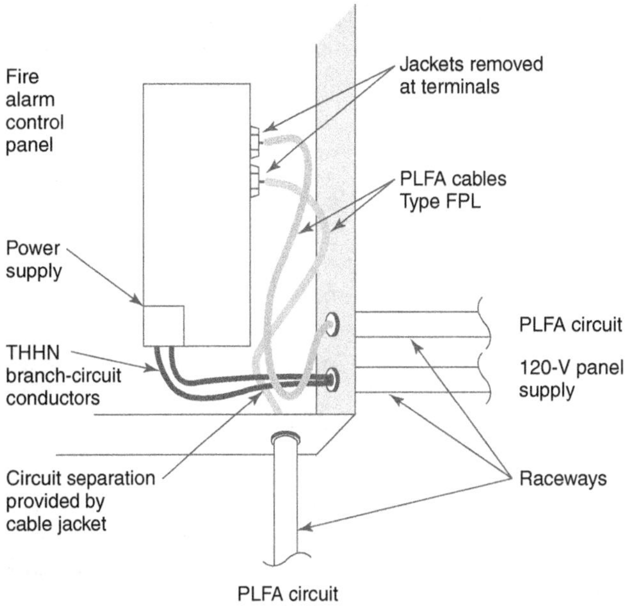

These rules are very similar to the separation requirements for Class 2 and Class 3 circuits but are a little different to accommodate only those types of installations and intermixing that are likely to be needed with PLFA circuits.

FIGURE 12-11   Separation of power-limited and nonpower-limited cables and conductors within an equipment enclosure.

15. Verify that power-limited fire alarm circuits are not supported from raceways.

Article 300 prohibits the use of raceways and cables for support of any other systems including other electrical systems. This rule is repeated in Article 760 for PLFA circuits, and 300.11 is referenced in 760.25 for NPLFA circuits. Unlike Class 2 circuits, there is no exception for PLFA circuits where associated power conductors are contained in the raceway or cable.

16. Verify proper application of listed power-limited fire alarm cables.

Cables for PLFA circuits are made for specific applications. Cables installed in ducts, plenums, and other spaces used for environmental air must be listed for such use or be installed in metal raceways. Type FPLP (fire power-limited plenum) cable is intended for so-called plenum use. According to 760.3(B), 300.22 applies to these applications, and 300.22 prohibits any wiring in ducts used for loose stock or vapor. Wiring in other ducts and plenums specifically fabricated for air-handling purposes is permitted only where necessary for connecting to equipment that acts on or senses the air in the duct or plenum.

However, 760.3(B), Exception, permits plenum applications of FPLP cable by reference to 760.154(A), but the 300.22 restrictions must be followed according to 760.154(A). Therefore, FPLP cables used for PLFA circuits may not be installed in the spaces covered by 300.22(A) (ducts that handle loose stock or vapor), they may be installed in the spaces covered by 300.22(B) (fabricated ducts and plenums) only to connect to equipment in a duct that acts on the contained air, and they may be installed in the other spaces for environmental air (also called plenums or more typically plenum ceilings) covered by 300.22(C) without further restriction. According to 760.25, accessible portions of **abandoned fire alarm cables** must be removed from ducts and other spaces for environmental air. Most equipment to which a fire alarm circuit would be connected and that acts on contained air is usually mounted outside a duct. However, line-type fire detectors may be a type of heat-sensitive cable that is sometimes installed in a duct and does actually act on or detect conditions in the duct, although such detectors are more commonly used in open areas.

Cables installed in risers (vertical runs that penetrate more than one floor) must be FPLR (the R is for "riser") or must be installed in metal raceways or fireproof shafts with firestops. Riser applications in one- and two-family dwellings may use ordinary FPL cables. Abandoned cables must be removed.

Generally, Type FPL cable (or a permitted FPL substitute) must be used in other applications, but substitutions of communications cables may be used, including substitutions for *P* or *R* suffix cables. Other methods, including installations in raceways, and special methods for portable fire alarms for motion picture stages and sets, are permitted. Plenum or riser cables are permitted to be run exposed, but exposed lengths of the cables are limited to 10 ft (3.0 m) in length.

Article 760 provides more than one way to determine the suitability of a specific cable type for a specific application and circuit type. Section 760.135 provides a list of applications based on occupancy type and location with the cable types that are suitable for that application. A similar list is provided in Table 760.154 that provides permitted cable types based on application. Both refer to 300.22 for the restrictions on cabling in various types of air-handling spaces.

17. Check listing and marking of listed power-limited fire alarm cables.
Section 760.179 describes the markings and characteristics of various PLFA cable types. The cables are required to be marked FPLP for plenum applications, FPLR for riser applications, or FPL for general-purpose use. Coaxial cables, usually listed as Type CATV or Type CM, may also be listed as FPLP, FPLR, or FPL. In addition, any of the mentioned cables may also have a *CI* suffix for circuit integrity cable. Insulated continuous line-type fire detectors must be listed and marked the same as other cables, except for the CI marking. See the definition of *Fire Alarm Circuit Integrity (CI) Cable* in the Key Terms section. The markings are specifically intended to make selection, inspection, and identification easy, once the determination of permissible types has been made, based on where the cable is to be installed and whether the performance requirements for circuit integrity have to be met. Listing and marking of NPLFR cables are covered by 760.176. The *NEC* does not require circuit integrity cable. The locations that require such wiring are dictated by NFPA 72. Where survivability is required by NFPA 72, the additional requirements of 760.176(F) apply. The survivability obtained by the use of CI cable may also obtained through the use of an electrical circuit protective system. One type of this system uses MI cable to provide a 2-hour fire rating.

Checklist 12-3: Optical Fiber Cables and Raceways

1. Verify that installation of cables does not prevent access through panels designed to provide access.
Cables that run above ceilings must not interfere with the removal of panels that would otherwise afford access to equipment above the ceiling. (See Figure 12-1; although it is

illustrating Class 1, Class 2, and Class 3 circuits, it also can be used as an illustration of acceptable and unacceptable methods for supporting optical fiber cables above suspended ceilings.)

2. Verify that cables have been supported to prevent damage under normal conditions.
Article 770 does not specify the maximum spacing between supports, so this is a subjective judgment for an inspector. Cables and conductors exposed on outer surfaces of ceilings and walls must be supported by hangers, straps, cable ties, staples, or other means. This support hardware must be attached to the structure of a building. Nonmetallic cable ties and other nonmetallic cable accessories used to secure and support cables in other spaces used for environmental air (plenums) must be listed as suitable for plenum-type installations. They must have low smoke and heat release properties like other "plenum-rated" wiring materials.

Cables must be installed only in "a neat and workmanlike manner." However, Section 300.11 also applies, and this section requires all wiring to be secured in place. (Suspended ceilings are not structural components of a building, and a grid ceiling cannot be used to support cables above or exposed on the ceiling unless the ceiling is listed for that purpose.) The AHJ must determine what constitutes a neat and workmanlike installation. Because of the fragile nature of individual optical fibers, the manufacturer may require or recommend specific maximum support spacings. Accepted industry installation practices can also be found in other ANSI consensus standards that are referenced in informational notes. Raceways used for optical fiber cables must be installed and supported in accordance with the applicable article from Chapter 3 of the *NEC,* except that only certain sections of Article 362, which covers electrical nonmetallic tubing (ENT), are applicable to listed nonmetallic optical fiber raceways. The rules about maximum and minimum sizes and raceway fill are not generally applicable to listed nonmetallic optical fiber raceways.

3. Check for proper raceway fill where optical fiber cables are installed with electric conductors.
The general raceway fill requirements found in 300.17 do not apply to Article 770 according to 770.3. However, raceway fill is applied where **nonconductive optical fiber cables** are installed in the same raceway with electric conductors. Section 770.110 refers directly to Chapter 3 and Chapter 9 in the *NEC* for these requirements. Raceway fill does not apply to optical fiber cables without electric conductors. Any rules about raceway fill in individual raceway articles are modified by this rule because it appears in *NEC* Chapter 7.

Only nonconductive optical fiber cables (Type OFN) are mentioned in 770.110, because **conductive optical fiber cables** are not permitted in the same raceway with non-power-limited circuits [770.133(A)]. Other optical fiber cables are permitted in raceways with nonpower-limited circuits only if functionally associated. However, *either* conductive optical fiber cables (Type OFC) *or* Type OFN may be installed in the same raceway with power-limited circuits such as Class 2, PLFA, communications, and CATV. **FIGURE 12-12** illustrates and summarizes this discussion of the application of raceway fill limits.

Listed optical fiber raceways may also be installed as innerduct within raceways, per Chapter 3 of the *NEC,* but these raceways within raceways may only be used for optical fiber cables, so raceway fill limits do not apply. The raceway must be selected and used in accordance with the same rules for the cables themselves.

In all cases, the uses of optical fiber cables are also subject to the manufacturer's instructions. Most manufacturers will specify maximum pulling tensions, and for long runs or multiple bends, reduced raceway fill may be required to avoid excessive pulling tension and damage to the cables. Note that these are not necessarily safety issues and are therefore not necessarily *NEC* issues. They are performance issues, which may become safety issues in some cases, depending on the intended use of the optical fibers.

FIGURE 12-12 Permitted installations of optical fiber cable with cables of other systems.

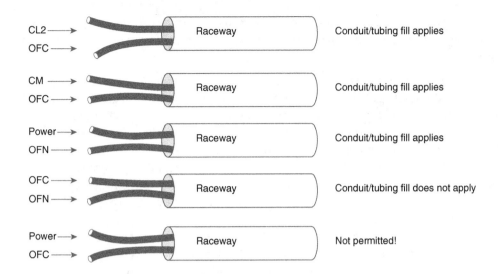

CL2 → OFC →	Raceway	Conduit/tubing fill applies
CM → OFC →	Raceway	Conduit/tubing fill applies
Power → OFN →	Raceway	Conduit/tubing fill applies
OFC → OFN →	Raceway	Conduit/tubing fill does not apply
Power → OFC →	Raceway	Not permitted!

4. Check for grounding and bonding of non-current-carrying metallic members of optical fiber cables.

Non-current-carrying metallic members of optical fiber cables must be grounded or interrupted by an insulating joint at the point of entrance to a building. (See **Point of Entrance** in the Key Terms section.) This requirement also applies to similar cables that terminate on the outside of buildings. In either case, it applies only where the cable is exposed to contact with electric light and power conductors and is meant to reduce or eliminate the introduction of foreign voltages to the building. Non-current-carrying metal members of Type OFC cables inside buildings must be grounded and/or bonded to keep those conductive parts from becoming and remaining **energized**. Grounding and bonding of conductive members is generally required to be done by connection to the intersystem bonding termination required by 250.94 in new installations, but where there is no such termination, as in existing buildings, other electrodes and connection points are specified in 770.100(B)(2) or (B)(3). Requirements for the conductor and electrode to be used and the point of connection to the building system for grounding and bonding are detailed in 770.100 (770.106 for mobile homes) and are almost identical to the requirements for the grounding/bonding required for communications systems in Chapter 8 of the *NEC*.

5. Check for proper application of listed optical fiber cables.

Cables for optical fibers are made for specific applications. Cables installed in ducts, plenums, and other spaces used for environmental air must be listed for such use or be installed in metal raceways. Cable Types OFCP (plenum-rated conductive optical fiber cable) and OFNP (plenum-rated nonconductive optical fiber cable) are intended for plenum use. In effect, 770.113 repeats the requirements of 300.22 without referencing 300.22 except for descriptions of the different types of ducts, plenums, and other spaces for environmental air. Wiring in fabricated ducts and plenums is permitted only if the cables are directly associated with the air distribution system, and then the plenum-type cables are either limited to 4 ft (1.22 m) in length or must be installed in raceways that comply with 300.22(B)—that is, metal raceways. Plenum cables, metal conduits, and listed plenum (nonmetallic) optical fiber raceways may also be used in plenum ceilings, but the listed plenum raceways may contain only plenum cables, whereas other types of cables may be used in metal raceways or totally enclosed metal cable trays (trays with solid bottoms and covers). According to 770.25, accessible portions of abandoned cables must be removed in all locations. (See the discussion under the definitions of **Abandoned Class 2, Class 3, and PLTC Cable** and **Abandoned Optical Fiber Cable** in the Key Terms section).

Cables installed in risers (vertical runs that penetrate more than one floor) must be OFNR or OFCR or must be installed in metal raceways or fireproof shafts with firestops. Listed plenum or riser optical fiber raceways or riser **cable routing assemblies** may also be used but must contain only plenum- or riser-rated cables. If the risers are in fireproof shafts with fire stops, cables, optical fiber raceways, and cable routing assemblies may be general-purpose, plenum, or riser types.

Generally, Type OFC, OFN, OFCG, or OFNG (conductive or nonconductive optical fiber cables, where the suffix *G* indicates general purpose) cable must be used in other applications, but substitutions of plenum or riser optical fiber cables may be used. Any type of OFC or OFN may be used in cable trays, subject to location restrictions. For example, a ventilated or ladder-type cable tray in a plenum ceiling would have to contain plenum-rated cables or wiring methods. But general-purpose or riser types of cables could be used in totally enclosed metal cable trays.

See **TABLE 12-2** for a summary of the applications of optical fiber cables. This table also appears as Table 770.154(a) in the *NEC*.

Since optical fiber cables do not provide an ignition source, any type may be used in hazardous (classified) locations subject to the wiring method and sealing requirements of Chapter 5 of the *NEC*. For example, the cables or the penetrations or raceways through which the cables pass must not provide a means for gas or vapor to be transmitted from a classified area to a nonclassified area. All of the applicable requirements for wiring in classified areas apply to **composite optical fiber cables** because they do contain conductors that are capable of carrying current.

Composite cables are listed according to the type of electrical conductors or circuits in the cables, including any suffix for plenum, riser, or general-purpose use, with the additional suffix *OF* for optical fiber. As such, they must be installed in conformance with both Article 770 and the article(s) applicable to the conductors with regard to separations and in accordance with the applicable article(s) for the conductors with regard to location and cable listing.

6. Verify appropriate separations between optical fiber cables and electrical conductors.

Optical fibers themselves cannot be energized by electrical circuits because they are nonconductive. Separation requirements are therefore based on whether or not conductors or conductive elements are also present and on the power levels of the conductors. Conductors are present in composite cables, and metallic armor or strength members not intended as conductors may be present in OFC cables. Composite cables containing non-power-limited circuits are permitted only where the functions of the circuit conductors and the optical fibers are associated. The composite cables themselves are permitted to occupy a cable tray, raceway, or termination enclosure with other nonpower-limited conductors. Except in limited industrial applications, the voltage limit for composite cables is 1000 volts.

Type OFN (nonconductive) cables may be mixed with nonpower-limited circuit conductors and cables in raceways or cable trays. Type OFN cables may not occupy enclosures where the nonpower-limited circuits are terminated unless the optical fibers and circuit conductors are functionally associated.

Type OFC (conductive) cables may not occupy the same raceway or cable tray with nonpower-limited circuits. Optical fibers are permitted in the same cable, and both OFC and OFN types of cables are permitted to occupy the same raceway, cable tray, or enclosure with power-limited circuits. Power-limited circuits include Class 2 and Class 3, PLFA, communications, CATV, and low-power network-powered broadband communications.

| TABLE 12-2 | Applications of Listed Optical Fiber Cables in Buildings |

Applications		Cable Type		
		OFNP, OFCP	OFNR, OFCR	OFNG, OFCG, OFN, OFC
In specifically fabricated ducts as described in 300.22(B)	In fabricated ducts	Y*	N	N
	In metal raceway that complies with 300.22(B)	Y*	Y*	Y*
In other spaces used for evironmental air as described in 300.22(C)	In other spaces used for evironmental air	Y*	N	N
	In metal raceway that complies with 300.22(C)	Y*	Y*	Y*
	In plenum communications raceways	Y*	N	N
	In plenum cable routing assemblies	NOT PERMITTED		
	Supported by open metal cable trays	Y*	N	N
	Supported by solid bottom metal cable trays with solid metal covers	Y*	Y*	Y*
In risers	In vertical runs	Y*	Y*	N
	In metal raceways	Y*	Y*	Y*
	In fireproof shafts	Y*	Y*	Y*
	In plenum communications raceways	Y*	Y*	N
	In plenum cable routing assemblies	Y*	Y*	N
	In riser communications raceways	Y*	Y*	N
	In riser cable routing assemblies	Y*	Y*	N
	In one- and two-family dwellings	Y*	Y*	Y*
Within buildings in other than air-handling spaces and risers	General	Y*	Y*	Y*
	Supported by cable trays	Y*	Y*	Y*
	In distributing frames and cross-connect arrays	Y*	Y*	Y*
	In any raceway recognized in Chapter 3	Y*	Y*	Y*
	In plenum communications raceways	Y*	Y*	Y*
	In plenum cable routing assemblies	Y*	Y*	Y*
	In riser communications raceways	Y*	Y*	Y*
	In riser cable routing assemblies	Y*	Y*	Y*
	In general-purpose communications raceways	Y*	Y*	Y*
	In general-purpose cable routing assemblies	Y*	Y*	Y*

Used with permission from NFPA 70®-2014, *National Electrical Code®*, Copyright © 2013, National Fire Protection Association, Quincy, MA. Table 770.154(a). This reprinted material is not the complete and official position of the NFPA on the referenced subject, which is represented only by the standard in its entirety.

Note: An "N" in the table indicates that the cable type shall not be permitted to be installed in the application. A "Y*" indicates that the cable shall be permitted to be installed in the application subject to the limitations described in 770.110 and 770.113.

Informational Note No. 1: Part V of Article 770 covers installation methods within buildings. This table covers the applications of listed optical fiber cables and raceways and cable routing assemblies in buildings. The definition of *Point of Entrance* is in 770.2. Optical fiber entrance cables that have not emerged from the rigid metal conduit (RMC) or intermediate metal conduit (IMC) are not considered to be in the building.

Informational Note No. 2: For information on the restrictions to the instsllation of optical fiber cables in fabricated ducts, see 770.113(B).

Informational Note No. 3: Cable routing assemblies are not addressed in NFPA 90A-2012, *Standard for the Installation of Air Conditioning and Ventilations Systems*.

KEY TERMS

The *NEC* defines the key terms pertaining to remote-control, signaling, and fire alarm circuits and optical fiber cables as follows (all definitions are from Article 100 unless otherwise noted):

Abandoned Class 2, Class 3, and PLTC Cable   [*Definition from 725.2*] Installed Class 2, Class 3, and power-limited tray cable (PLTC) that is not terminated at equipment and not identified for future use with a tag.

Abandoned Fire Alarm Cable   [*Definition from 760.2*] Installed fire alarm cable that is not terminated at equipment other than a connector and not identified for future use with a tag.

Abandoned Optical Fiber Cable   [*Definition from 770.2*] Installed optical fiber cable that is not terminated at equipment other than a connector and not identified for future use with a tag.

> *Informational Note:* See Article 100 for a definition of *Equipment*.

[*These three definitions of abandoned cables differ primarily in the class of cable. Otherwise, the definitions are essentially the same. These definitions are used in rules that require the accessible portion of abandoned cables to be removed. Similar definitions and requirements also apply to audio cables, cables in information technology rooms, communications, CATV, and network-powered broadband communications. However, in some cases, such as in Article 645 for Information Technology Rooms, the tagging requirements for abandoned cables of similar circuits are much more detailed and include the intended future use and dates of that intended use as well as dates the tags were placed.*]

Accessible (as applied to wiring methods)   Capable of being removed or exposed without damaging the building structure or finish or not permanently closed in by the structure or finish of the building. [*To fully understand this definition, it must be used in conjunction with the definitions of Concealed and Exposed (as applied to Wiring Methods). Wiring methods located behind removable panels designed to allow access are not considered permanently enclosed or concealed and are considered exposed. Although a raceway may be exposed and accessible, the conductors inside are considered to be concealed; thus, abandoned cables in raceways are not required to be removed because they are not considered accessible even though they could be withdrawn from the raceway.*]

Ampacity   The maximum current, in amperes, that a conductor can carry continuously under the conditions of use without exceeding its temperature rating.

Approved   Acceptable to the authority having jurisdiction.

Arc-Fault Circuit Interrupter (AFCI)   A device intended to provide protection from the effects of arc faults by recognizing characteristics unique to arcing and by functioning to de-energize the circuit when an arc fault is detected. [*Fire alarm system power sources are not permitted to be supplied through AFCIs. This covers only the supply to the fire alarm system control panel itself, and not to initiating devices, notification appliances, annunciators, or the like, that are powered and controlled by the fire alarm panel. Article 760 does not cover the power supply to single- and multiple-station smoke alarms. These types of smoke alarms commonly used in dwelling units are supplied by branch circuits that are covered by the requirements of Article 210. Since most branch circuits or outlets typically used to supply fire alarm panels are required to be protected by either AFCIs or GFCIs in dwelling units, individual or dedicated branch circuits are needed to supply fire alarm panels. A circuit that supplies power to a fire alarm panel is not permitted to be AFCI- or GFCI-protected. In fact, branch circuits that supply no other loads are required by Article 760 for both power-limited and nonpower-limited fire alarm power sources in all occupancies.*]

Branch Circuit The circuit conductors between the final overcurrent device protecting the circuit and the outlet(s). [*Branch circuits are used to supply power to fire alarm systems and remote-control and signaling systems, but the fire alarm circuits and remote-control or signaling circuits themselves are not usually considered branch circuits. A branch circuit that supplies a smoke alarm(s) like those commonly used in dwellings and guest rooms is a branch circuit but not a fire alarm circuit.*]

Building A structure that stands alone or that is cut off from adjoining structures by fire walls with all openings therein protected by approved fire doors.

Cable Routing Assembly [*Definition from 770.2*] A single channel or connected multiple channels, as well as associated fittings, forming a structural system that is used to support, route and protect high densities of wires and cables, typically communications wires and cables, optical fiber and data (Class 2 and Class 3) cables associated with information technology and communications equipment. [*Optical fiber/communications cable routing assemblies are evaluated under UL2024a,* Outline of Investigation for Optical Fiber Cable Routing Assemblies, *which states, "These assemblies are nonmetallic U-shaped wiring troughs that may or may not have covers." They are permitted in Article 770 as general purpose or riser types. The lack or possible lack of a cover distinguishes them from optical fiber or communications raceways.*]

Class 1 Circuit [*Definition from 725.2*] The portion of the wiring system between the load side of the overcurrent device or power-limited supply and the connected equipment.

> *Informational Note:* See 725.41 for voltage and power limitations of Class 1 circuits.

Class 2 Circuit [*Definition from 725.2*] The portion of the wiring system between the load side of a Class 2 power source and the connected equipment. Due to its power limitations, a Class 2 circuit considers safety from a fire initiation standpoint and provides acceptable protection from electric shock.

Class 3 Circuit [*Definition from 725.2*] The portion of the wiring system between the load side of a Class 3 power source and the connected equipment. Due to its power limitations, a Class 3 circuit considers safety from a fire initiation standpoint. Since higher levels of voltage and current than Class 2 levels are permitted, additional safeguards are specified to provide protection from an electric shock hazard that could be encountered. [*The three classes of circuits defined here are together different from other circuits primarily in the way they are used. The difference between the three types of circuits is in the power limitations or, in the case of Class 1 circuits, in the fact that there may be no power limitation. In spoken language, these sound identical to Class I, Class II, and Class III hazardous locations, and may be confused with Class I flammable liquids, or Class II and Class III combustible liquids, unless the context is clear.*]

Composite Optical Fiber Cable [*Definition from 770.2*] A cable containing optical fibers and current-carrying electrical conductors. [*These cables may also contain non-current-carrying conductive members such as metallic strength members and metallic vapor barriers. Composite optical fiber cables are classified as electrical cables in accordance with the types of electrical conductors they contain.*]

Concealed Rendered inaccessible by the structure or finish of the building.

> *Informational Note:* Wires in concealed raceways are considered concealed, even though they may become accessible by withdrawing them.

[*See the discussion under the definition of* Accessible (as applied to wiring methods).]

Conductive Optical Fiber Cable [*Definition from 770.2*] A factory assembly of one or more optical fibers having an overall covering and containing non-current-carrying conductive member(s) such as metallic strength member(s), metallic vapor barrier(s), metallic armor or metallic sheath.

Control Circuit The circuit of a control apparatus or system that carries the electric signals directing the performance of the controller but does not carry the main power current.

Controller A device or group of devices that serves to govern, in some predetermined manner, the electric power delivered to the apparatus to which it is connected. [*The circuits that direct the operation of a controller are usually* remote-control circuits *as defined in this section. Controllers may also be the source of power for Class 1, Class 2, or Class 3 circuits. Controllers represent a location where control circuits and power circuits are likely to have to occupy the same enclosure, and where separation requirements are likely to be an issue.*]

Device A unit of an electrical system, other than a conductor, that carries or controls electric energy as its principal function. [*Components such as switches, circuit breakers, fuseholders, receptacles, attachment plugs, and lampholders that distribute or control but do not consume electricity are considered devices, as are many components of remote-control, signaling, or fire alarm systems. For the purposes of Article 760, typical devices include smoke detectors, heat detectors, and pull stations. These types of equipment are referred to as* initiating devices *in NFPA 72,* National Fire Alarm and Signaling System Code, *2010 edition.* **FIGURES 12-13A** and **13B** *illustrate typical initiating devices.*]

FIGURE 12-13A   Fire alarm system initiating device (heat detector). (Courtesy of GE Security.)

Enclosed Surrounded by a case, housing, fence, or wall(s) that prevents persons from accidentally contacting energized parts.

Enclosure The case or housing of apparatus or the fence or walls surrounding an installation to prevent personnel from accidentally contacting energized parts or to protect the equipment from physical damage.

   Informational Note: See Table 110.28 in the *NEC* for examples of enclosure types.
   [*Raceways are generally not considered to be enclosures although they are sometimes used to protect conductors or cables from physical damage.*]

Energized Electrically connected to, or is, a source of voltage.

Equipment A general term including fittings, devices, appliances, luminaires, apparatus, machinery, and the like used as a part of, or in connection with, an electrical installation.

Exposed (as applied to live parts) Capable of being inadvertently touched or approached by a person nearer than a safe distance.

   Informational Note: This term applies to parts that are not suitably guarded, isolated, or insulated.
   [*The live parts of many power-limited circuits are energized, but are permitted to be exposed to contact because in most situations the voltage levels are low enough to eliminate shock hazards.*]

Fire Alarm Circuit [*Definition from 760.2*] The portion of the wiring system between the load side of the overcurrent device or the power-limited supply and the connected equipment of all circuits powered and controlled by the fire alarm system. Fire alarm circuits are classified as either nonpower-limited or power-limited. [*Fire alarm circuits are defined and differentiated under NFPA 72, primarily by function, but are further classified by the way they are monitored for integrity*

FIGURE 12-13B   Fire alarm system initiating device (smoke detector). (Courtesy of Mammoth Fire Alarms, Inc.)

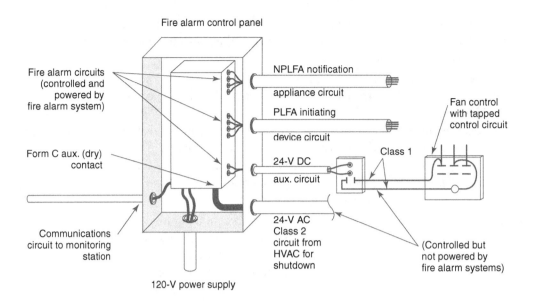

FIGURE 12-14 Fire alarm system and associated system circuits.

(formerly called supervision). In the NEC, *fire alarm circuits are defined and classified by their power supplies, and the power supplies are usually part of or located at the fire alarm control panel.* **FIGURE 12-14** *shows an illustration of the different types of circuits commonly associated with a fire alarm system. Note that wiring to smoke detectors powered by a fire alarm system is connected to a fire alarm circuit, but the wiring to residential-type smoke alarms that are not connected to a fire alarm control panel is not.*]

Fire Alarm Circuit Integrity (CI) Cable [*Definition from 760.2*] Cable used in fire alarm systems to ensure continued operation of critical circuits during a specified time under fire conditions. [**FIGURE 12-15** *illustrates the construction of circuit integrity cable. The* NEC *does not require the use of this cable type; it simply permits its use where needed to satisfy the performance requirements of NFPA 72. Some CI-Type cable gets its rating only when installed in specific types of raceway and other types may be used without raceways. NFPA 72 requires this type of cable or equivalent protection primarily for systems used for partial evacuation or relocation of occupants where "evacuation signaling zones" are needed such as in some very large facilities, some high-rise buildings, and some hospitals. NFPA 72 specifies that circuits used for such purposes have to meet survivability requirements. Survivability requirements may also apply to mass notification systems. Somewhat similar cables used for other types of circuits are covered by Article 728, Fire Resistive Cable Systems.*]

Grounded (Grounding) Connected (connecting) to ground or to a conductive body that extends the ground connection.

Grounded Conductor A system or circuit conductor that is intentionally grounded. [*A grounded system is a system that includes a grounded conductor. Most fire alarm circuits are ungrounded, as are the most commonly encountered remote-control and signaling circuits, such as those used for doorbells and many nominal 24-volt thermostats. Conductors used to ground equipment are not grounded conductors in most cases; they are equipment grounding conductors or bonding jumpers.*]

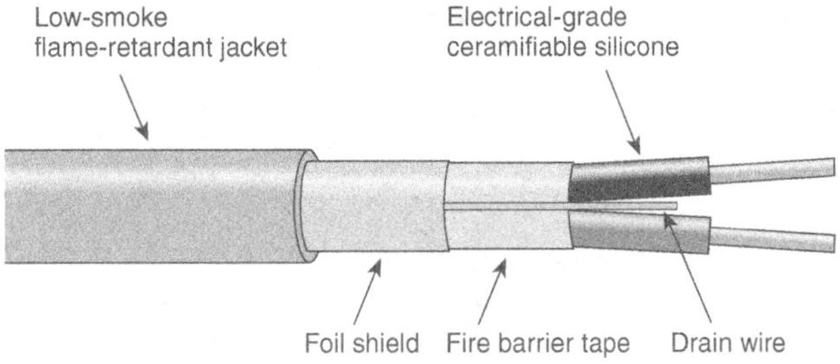

Ground-Fault Circuit Interrupter (GFCI) A device intended for the protection of personnel that functions to de-energize a circuit or portion thereof within an established period of time when a current to ground exceeds the values established for a Class A device.

Informational Note: Class A ground-fault circuit interrupters

Low-smoke flame-retardant jacket

Electrical-grade ceramifiable silicone

Foil shield   Fire barrier tape   Drain wire

FIGURE 12-15 Components of CI cable.

trip when the current to ground is 6 milliamperes or higher and do not trip when the current to ground is less than 4 milliamperes. For further information, see UL 943, *Standard for Ground-Fault Circuit Interrupters.*

[*Fire alarm system power sources are not permitted and household burglar alarm panels are not required to be supplied through GFCIs, even though the supply may be from a receptacle that is in a location (such as an unfinished basement) where GFCI protection is otherwise required.*]

Identified (as applied to equipment) Recognizable as suitable for the specific purpose, function, use, environment, application, and so forth, where described in a particular *Code* requirement.

> *Informational Note:* Some examples of ways to determine suitability of equipment for a specific purpose, environment, or application include investigations by a qualified testing laboratory (listing and labeling), an inspection agency, or other organizations concerned with product evaluation.

Initiating Device [*Definition from NFPA 72, 2013 3.3.132*] A system component that originates transmission of a change-of-state condition, such as in a smoke detector, manual fire alarm box, or supervisory switch. [*See the definition of* Device. *This definition includes Analog Initiating Devices, such as a heat detector that provides a heat level rather than just responding at a fixed set point; Automatic Extinguishing System Supervisory Devices, such as a device that communicates that status of a fire pump; Nonrestorable Initiating Devices, such as a melting link-type heat detector; Restorable Initiating Devices, such as common smoke detectors; and Supervisory Signal-Initiating Devices, such as a sprinkler valve "tamper switch."*]

Labeled Equipment or materials to which has been attached a label, symbol, or other identifying mark of an organization that is acceptable to the authority having jurisdiction and concerned with product evaluation, that maintains periodic inspection of production of labeled equipment or materials, and by whose labeling the manufacturer indicates compliance with appropriate standards or performance in a specified manner.

Listed Equipment, materials, or services included in a list published by an organization that is acceptable to the AHJ and concerned with evaluation of products or services, that maintains periodic inspection of production of listed equipment or materials or periodic evaluation of services, and whose listing states that the equipment, material, or services either meets appropriate designated standards or has been tested and found suitable for a specified purpose.

> *Informational Note:* The means for identifying listed equipment may vary for each organization concerned with product evaluation, some of which do not recognize equipment as listed unless it is also labeled. Use of the system employed by the listing organization allows the authority having jurisdiction to identify a listed product.

Location, Dry A location not normally subject to dampness or wetness. A location classified as dry may be temporarily subject to dampness or wetness, as in the case of a building under construction.

Location, Wet Installations under ground or in concrete slabs or masonry in direct contact with the earth; in locations subject to saturation with water or other liquids, such as vehicle washing areas; and in unprotected locations exposed to weather. [*Some limited-energy circuits that are not normally a shock hazard may present a shock hazard in a wet location. Therefore, some power supplies that are normally Class 2 are Class 3 if used in a wet location.* **FIGURE 12-16** *shows an example of a common type of plug-in power supply with this type of labeling. The label in this case does not mean the transformer is necessarily suitable for a wet location; it means that the power supply is Class 3 if persons are subject to contact with the circuits in wet locations. Class 2 alternating current circuits suitable for wet location contact are limited to 15 volts root-mean-square (rms).*]

FIGURE 12-16   Power supply listed for the supply of Class 2 circuits in locations that are not wet and for the supply of Class 3 circuits in wet locations.

Nonconductive Optical Fiber Cable [*Definition from 770.2*] A factory assembly of one or more optical fibers having an overall covering and containing no electrically conductive materials.

Nonpower-Limited Fire Alarm (NPLFA) Circuit [*Definition from 760.2*] A fire alarm circuit powered by a source that complies with 760.41 and 760.43.

Notification Appliance [*Definition from NFPA 72, 2010 3.3.160*] A fire alarm system component, such as a bell, horn, speaker, light, or text display, that provides audible, tactile, or visible outputs, or any combination thereof. [*An appliance as defined in the NEC is utilization equipment that is built in a standardized size and that performs a function such as clothes washing, air conditioning, or food mixing. For the purpose of Article 760 and NFPA 72, the term is used in a more specific application. Notification appliances utilize electric energy to operate lights, bells, horns, text display, or other similar equipment for alerting building occupants of a fire emergency.* **FIGURES 12-17A** *and* **17B** *show examples of notification appliances. This definition includes Audible Notification Appliances, Exit Marking Audible Notification Appliances, Textual Audible Notification Appliances, Tactile Notification Appliances, Visible Notification Appliances, and Textual Visible Notification Appliances, all of which are defined in more detail in NFPA 72.*]

Optical Fiber Cable [*Definition from 770.2*] A factory assembly or field assembly of one or more optical fibers having an overall covering.

> *Informational Note:* A field-assembled optical fiber cable is an assembly of one or more optical fibers within a jacket. The jacket, without optical fibers, is installed in a manner similar to conduit or raceway. Once the jacket is installed, the optical fibers are inserted into the jacket, completing the cable assembly.

Outlet A point on the wiring system at which current is taken to supply utilization equipment. [*In general, appliances such as fire alarm horns and strobes are utilization equipment, and devices such as smoke detectors are not. However, in the context of the definition of a branch circuit, the power supply or fire alarm panel is the utilization equipment and the fire alarm circuits or signaling circuits are not treated as branch circuits, and thus the connections to initiating devices and notification appliances are not outlets.*]

FIGURE 12-17A   Audible and visible fire alarm system notification appliance. (Courtesy of Mammoth Fire Alarms, Inc.)

FIGURE 12-17B   Visible fire alarm system notification. (Courtesy of Gentex Corporation.)

Overcurrent Any current in excess of the rated current of equipment or the ampacity of a conductor. It may result from overload, short circuit, or ground fault.

> *Informational Note:* A current in excess of rating may be accommodated by certain equipment and conductors for a given set of conditions. Therefore the rules for overcurrent protection are specific for particular situations.

Plenum A compartment or chamber to which one or more air ducts are connected and that forms part of the air distribution system. [*In the context of the* NEC, *a plenum is generally thought of as something specifically fabricated for handling air and a so-called plenum ceiling is called either an "other space for environmental air" or a plenum. Also, so-called plenum-rated cable is generally restricted to the other spaces for environmental air and is generally not allowed in fabricated plenums or ducts.*]

Point of Entrance (as applied to optical fiber cables) [*Definition from 770.2*] The point within a building at which the optical fiber cable emerges from an external wall, from a concrete floor slab, from a rigid metal conduit (Type RMC), or from an intermediate metal conduit (Type IMC). [*Point of entrance is defined in the same manner for communications circuits. Although the term is also used with respect to service conductors, outside feeders, transformer secondary conductors, and taps, it is not defined specifically for those purposes. Article 230 considers conductors to be outside a building if they are encased in concrete or remain under a building slab. Using this definition, 770.46 permits the point of entrance to be extended into a building by enclosing the cable in rigid metal conduit or intermediate metal conduit that is grounded by connection to a grounding electrode, or in new installations, to the intersystem bonding termination required by 250.94.* See **FIGURE 12-18**.]

Power-Limited Fire Alarm (PLFA) Circuit [*Definition from 760.2*] A fire alarm circuit powered by a source that complies with 760.121. [*The* NEC *defines classes or types of power-limited circuits by their power supplies. PLFA supplies are essentially the same as Class 3 power supplies from the standpoint of energy limitations. In many cases, voltages are low enough not to present a shock hazard, but the current limitations are not as low as for Class 2 circuits, so they could provide an ignition source in some circumstances. Since Class 2 sources are more limited than Class 3 sources, they may also be used, and sometimes are used for fire alarm power sources, especially small fire alarms systems for dwellings.*]

Power-Limited Tray Cable (PLTC) A factory assembly of two or more insulated conductors rated at 300 V, with or without associated bare or insulated equipment grounding conductors, under a nonmetallic jacket.

Raceway An enclosed channel of metallic or nonmetallic materials designed expressly for holding wires, cables, or busbars, with additional functions as permitted in this *Code*.

> *Informational Note:* A raceway is identified within specific article definitions.

[*Raceways include, but are not limited to, rigid metal conduit, rigid nonmetallic conduit, intermediate metal conduit, liquidtight flexible conduit, flexible metallic tubing, flexible metal conduit, electrical nonmetallic tubing, electrical metallic tubing, underfloor raceways, cellular concrete floor raceways, cellular metal floor raceways, surface raceways, wireways, and busways.*]

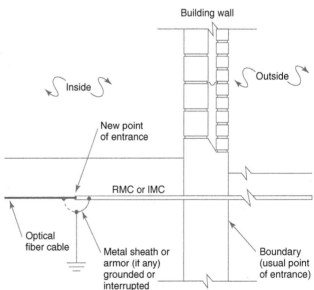

FIGURE 12-18   Point of entrance for optical fiber cables.

Remote-Control Circuit   Any electrical circuit that controls any other circuit through a relay or an equivalent device. [*Remote control and signaling are the primary uses for Class 1, Class 2, and Class 3 circuits. See the definition of* Signaling Circuit. *Temperature controls and controls for garage door openers are common examples of remote-control circuits in dwelling units.*]

Signaling Circuit   Any electric circuit that energizes signaling equipment. [*Doorbells are probably the most common example of a signaling circuit.*]

Voltage (of a circuit)   The greatest root-mean-square (rms) (effective) difference of potential between any two conductors of the circuit concerned.

> *Informational Note:* Some systems, such as 3-phase four-wire, single-phase 3-wire, and 3-wire dc, may have various circuits of various voltages.

[*In much of the* NEC, *the voltage that is considered the threshold for a shock hazard is 50-volt alternating current (ac). In NFPA 70E,* Standard for Electrical Safety in the Workplace, *2012 edition, equipment operating at 50 volts or less is not treated as energized equipment from the standpoint of establishing a safe working condition unless the available current is sufficient to create an arc-flash hazard. This "traditional" voltage threshold for shock hazards has been shown to be somewhat high under test conditions. Therefore, for some purposes, such as the voltage limits on Class 2 transformers or the voltage level of low-voltage lighting, the maximum voltage may be 42.4 volts peak, 30 volts rms, or 60 volts dc or sometimes less. These values are based on actual testing and coincide with the UL extra-low-voltage (ELV), safety extra-low voltage (SELV), and telecommunications network voltage (TNV) levels. Higher voltages are considered hazardous from a shock standpoint. As noted in the definition, the term* rms *refers to the effective voltage of a sinusoidal ac voltage. The peak voltage in a sinusoidal system is the rms voltage times the square root of 2. Thus, 30 × 1.414 = 42.4 volts. In situations where contact is likely to be made in a wet location, the voltages are halved, so the maximum voltages for Class 2 sources in wet locations are 15 volts rms and 21.2 volts peak. These levels coincide with the permissible levels for lighting in swimming pools where GFCI protection is not provided. The relationships between rms and peak voltages are shown in* **FIGURE 12-19**. *See also the discussion under* Location, Wet, *and the definition of* Low Voltage Contact Limit *in Article 100 in the* NEC *and in the Swimming Pools and Related Installations chapter of this manual.*]

Voltage, Nominal   A nominal value assigned to a circuit or system for the purpose of conveniently designating its voltage class (e.g., 120/240 volts, 480Y/277 volts, 600 volts).

> *Informational Note No. 1:* The actual voltage at which a circuit operates can vary from the nominal within a range that permits satisfactory operation of equipment.

> *Informational Note No. 2:* See ANSI C84.1-2006, *Voltage Ratings for Electric Power Systems and Equipment (60 Hz).*

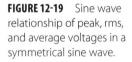

FIGURE 12-19   Sine wave relationship of peak, rms, and average voltages in a symmetrical sine wave.

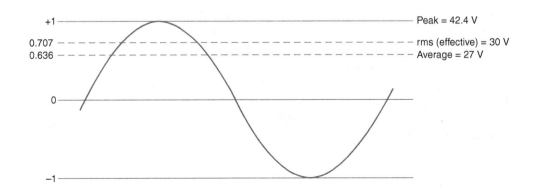

KEY QUESTIONS

1. Are remote-control or signaling circuits installed?
2. Does the installation include a fire alarm system?
3. Are optical fiber cables used?
4. Where are the power sources for control, signal, or fire alarm circuits?
5. What types of special-purpose cables are being used?
6. What locations include special cables?
7. Do the special cables share raceways or enclosures with any other systems?

CHECKLISTS

✔	Item	Inspection Activity	NEC Reference	Comments
		Checklist 12-1: Class 1, Class 2, Class 3, Remote-Control, Signaling, and Power-Limited Circuits		
	1.	Verify that tubing, piping or other mechanical system is not installed in raceways and cable trays containing conductors for Class 1, Class 2, and Class 3 circuits.	700.3(K) 300.8	
	2.	Verify that installation of cables does not prevent access through panels designed to provide access.	725.21	
	3.	Verify that cables have been supported to prevent damage under normal conditions.	725.24	
	4.	Identify safety-control circuits for Class 1 designation.	725.31	
	5.	Verify Class 1, 2, or 3 circuit designation.	725.41, 725.121	
	6.	Verify that Class 1, Class 2, and Class 3 circuits are adequately identified at junctions and terminations.	725.30	
	7.	Verify overcurrent protection for 18 and 16 AWG Class 1 circuit conductors.	725.43, 725.45, 725.49	
	8.	Check for proper Class 1 wiring methods.	725.46	
	9.	Check the insulation type of 18 and 16 AWG Class 1 circuit conductors.	725.49	
	10.	Verify proper raceway fill and derating for Class 1 circuits.	725.51	
	11.	Check on separation of Class 1 circuits from power circuits except where functionally associated.	725.48	
	12.	Verify listing of Class 2 and Class 3 power sources.	725.121	

(continues)

CHECKLISTS

Checklist 12-1: Class 1, Class 2, Class 3, Remote-Control, Signaling, and Power-Limited Circuits

✔	Item	Inspection Activity	*NEC* Reference	Comments
	13.	Check wiring methods on the load side and supply side of Class 2 and Class 3 power sources.	725.127, 725.130, 725.154	
	14.	Verify separation of Class 2 and Class 3 circuits from other circuits.	725.133, 725.136, 725.139	
	15.	Verify that Class 2 and Class 3 circuits are not supported from raceways or cables.	300.11(B)(2), 725.143	
	16.	Verify proper application of Class 2 and Class 3 cables.	725.3, 725.154, 725.135	
	17.	Check listing and marking of Class 2 and Class 3 circuit conductors.	725.179	

Checklist 12-2: Fire Alarm Systems

✔	Item	Inspection Activity	*NEC* Reference	Comments
	1.	Verify that tubing, piping or other mechanical system is not installed in raceways and cable trays containing conductors for Class 1, Class 2, and Class 3 circuits.	760.3(G) 300.8	
	2.	Verify that installation of cables does not prevent access through panels designed to provide access.	760.21	
	3.	Verify that cables have been supported to prevent damage under normal conditions.	760.24	
	4.	Check on proper identification of fire alarm circuits at junction and terminal locations.	760.30	
	5.	Distinguish nonpower-limited from power-limited fire alarm circuits and verify power supply is supplied by a branch circuit that supplies only loads associated with the fire alarm system and is not protected by GFCI or AFCI devices.	760.41, 760.121, 210.12(A)	
	6.	Verify overcurrent protection for 18 and 16 AWG nonpower-limited fire alarm circuit conductors.	760.43, 760.45	
	7.	Check for proper nonpower-limited wiring methods.	760.46	
	8.	Check insulation type and rating of nonpower-limited circuit conductors.	760.49	
	9.	Verify proper raceway fill and derating for nonpower-limited circuits.	760.48, 760.51	

✔	Item	Inspection Activity	NEC Reference	Comments
Checklist 12-2: Fire Alarm Systems				
	10.	Verify proper application of listed nonpower-limited fire alarm cables.	760.53	
	11.	Verify listing of power-limited source and that the power supply is supplied by a branch circuit that supplies only loads associated with the fire alarm system and is not protected by GFCI or AFCI devices.	760.121	
	12.	Check equipment marking to indicate power-limited circuit.	760.124	
	13.	Verify proper wiring methods on the load side and supply side of a power-limited source.	760.127, 760.130, 760.154	
	14.	Verify separation of power-limited fire alarm circuits from other circuits.	760.136, 760.139	
	15.	Verify that power-limited fire alarm circuits are not supported from raceways.	760.143	
	16.	Verify proper application of listed power-limited fire alarm cables.	760.154, 760.176	
	17.	Check listing and marking of listed power-limited fire alarm cables.	760.179	

✔	Item	Inspection Activity	NEC Reference	Comments
Checklist 12-3: Optical Fiber Cables and Raceways				
	1.	Verify that installation of cables does not prevent access through panels designed to provide access.	770.21	
	2.	Verify that cables have been supported to prevent damage under normal conditions.	770.24, 770.110	
	3.	Check for proper raceway fill where optical fiber cables are installed with electric conductors.	770.12, 770.110, 770.133	
	4.	Check for grounding and bonding of non-current-carrying metallic members of optical fiber cables.	770.93, 770.100, 770.114	
	5.	Check for proper application of listed optical fiber cables.	770.113, 770.154, 770.179	
	6.	Verify appropriate separations between optical fiber cables and electrical conductors.	770.133	

Index

Note: Page numbers followed by *b*, *f* and *t* indicate material in boxes, figures and tables respectively.

CPSIA information can be obtained
at www.ICGtesting.com
Printed in the USA
BVHW011952020819
554780BV00013B/315/P